The Biology of the
Naked Mole-Rat

MONOGRAPHS IN BEHAVIOR AND ECOLOGY

Edited by John R. Krebs and
Tim Clutton-Brock

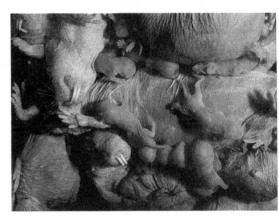

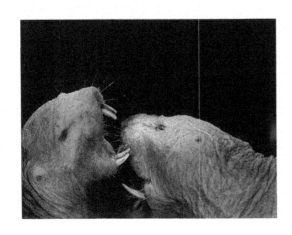

The
Biology of
the Naked
Mole-Rat

Edited by
PAUL W. SHERMAN
JENNIFER U. M. JARVIS
RICHARD D. ALEXANDER

Princeton University Press
Princeton, New Jersey

Published by Princeton University Press,
41 William Street, Princeton, New Jersey

In the United Kingdom Princeton University Press,
Oxford

Library of Congress Cataloging-in-Publication Data

The Biology of the naked mole-rate / edited by Paul W
Sherman, Jennifer U M. Jarvis, Richard D. Alexander
p   cm. — (Monographs in behavior and ecology)
Includes bibliographical references.
Includes index.
ISBN 0-691-08585-4 (cloth : alk. paper) — ISBN (invalid)
0-691-02498-0 (paper · alk paper)
1. Naked mole rat—Africa, Northeast   2 Naked mole
rat—Africa, Northeast—Behavior. I. Sherman, Paul W.,
1949– . II. Jarvis, J. U. M. (Jennifer U M.), 1939–
III Alexander, Richard D.   IV. Series.
[DNLM: 1 Behavior, Animal   2. Rodentia.
QL 737.R628 B615]
QL 737 R628B56   1991
599 32'34—dc20
DNLM/DLC   90-8555

Color plate photos: R A. Mendez and J.U M Jarvis

This book has been composed in Adobe Times Roman

Princeton University Press books are printed on acid-free
paper, and meet the guidelines for permanence and
durability of the Committee on Production Guidelines for
Book Longevity of the Council on Library Resources

Printed in the United States of America by
Princeton University Press, Princeton, New Jersey

10 9 8 7 6 5 4 3 2 1
(Pbk.)
10 9 8 7 6 5 4 3 2 1

Contents

Preface

This book summarizes our current knowledge of the biology of the naked mole-rat (*Heterocephalus glaber*). The genesis of the book occurred at a meeting in August 1985, at Ann Arbor, Michigan, among R. D. Alexander, R. A. Brett, J.U.M. Jarvis, E. A. Lacey, and P. W. Sherman (plate P-1). At that gathering an outline was tentatively agreed upon and writing assignments were made. Since then the book has grown and changed, partly because of the evolving interests of the participants and also because of our having solicited contributions from everyone active in naked mole-rat research. The resulting volume covers a broad range of topics, including systematics, evolution, ecology, behavior, genetics, reproductive biology, and husbandry; continuity is provided by the common focus on a single species, the use of common terminology, and by frequent cross-referencing among chapters.

This is truly a collaborative book: most chapters have more than one author, and even those that do not have benefited from the input of ideas and often data from several individuals. However, authors have not attempted to reach consensus, and indeed some important disagreements remain. This is to be expected, since the biology of *H. glaber* has been studied quantitatively for only about 20 years. We believe that the differing opinions presented will ultimately enhance rather than detract from the book's value. By highlighting controversies as well as agreements, we hope to point the way for future research on these unusual and fascinating little rodents.

History of Recent Naked Mole-Rat Research

In 1967, J.U.M. Jarvis, who was a graduate student at the University of Nairobi in Kenya, began studying *Heterocephalus glaber* as a part of her dissertation on the mole-rats of eastern Africa. The animals were known to be colonial, and she observed that they apparently cooperated in excavating extensive systems of subterranean burrows. She found no pregnant females in monthly field samples, suggesting that breeding was opportunistic, perhaps depending on rainfall.

By 1974, Jarvis had joined the faculty at the University of Cape Town in South Africa. She returned to Kenya to collect naked mole-rats for physiological studies and attempted to establish breeding groups in the laboratory at Cape Town. Because she was unable to collect entire colonies in the field, Jarvis combined individuals from several different wild-caught colonies. Initially there was considerable aggression, especially among females, in these

Plate P-1 The group that gathered in August 1985 at Ann Arbor, Michigan to discuss naked mole-rat biology, the genesis of this volume occurred at the meeting. From left: R A Brett, P W Sherman, J U M. Jarvis, E. A Lacey, and R. D. Alexander

mixed colonies. The conflicts gradually tapered off once a female began breeding; thereafter, breeding was usually restricted to one female in each colony. Whether or not the results from these mixed laboratory colonies reflected the situation in nature was at that point unclear.

In the mid-1970s, R. D. Alexander, a professor at the University of Michigan, lectured on the evolution of eusociality at several universities in the western United States. In an effort to explain why vertebrates had apparently not evolved eusociality, he hypothesized a fictitious mammal that, if it existed, would be eusocial. This hypothetical creature had certain features that patterned its social evolution after that of termites (e.g., the potential for heroic acts that assisted collateral relatives, the existence of an ultrasafe but expansible nest, and an ample supply of food requiring minimal risk to obtain it). Alexander hypothesized that this mythical beast would probably be a completely subterranean rodent that fed on large tubers and lived in burrows inaccessible to most but not all predators, in a xeric tropical region with heavy clay soil.

When Alexander presented his lecture at Northern Arizona University in May 1976, T. L. Vaughan was in the audience. After the lecture, Vaughan astonished Alexander by saying "Your hypothetical eusocial mammal is a perfect description of the naked mole-rat of Africa." Vaughan showed Alexander some preserved specimens of *H. glaber*, and provided him with Jarvis's address.

Alexander wrote to Jarvis, posing a series of questions about naked mole-rats designed to test whether or not they were indeed eusocial. Jarvis responded to Alexander's letter in detail, and all her answers tantalizingly seemed to be in the appropriate direction to suggest eusociality. Jarvis later commented to L. Gamlin of *New Scientist* magazine (July 30, 1987, no. 1571, p. 41) "I realized eventually that there was always only one breeding female per colony, and I knew there was division of labor — the penny was beginning to drop. But it was only when I received a letter from Dick Alexander in 1976 that the significance of it really became clear."

P. W. Sherman had been a doctoral student with Alexander and had studied ground squirrel social behavior for his dissertation; in 1976 he began a post-doctoral fellowship at the University of California (Berkeley). In June 1977, he attended the annual meeting of the American Society of Mammalogists at Michigan State University and heard a lecture by M. W. Hildebrand on adaptations for fossorial life. The behavior of the naked mole-rat featured prominently in this lecture, and a freeze-dried specimen was exhibited. The telephone lines between Berkeley and Ann Arbor buzzed. Sherman and Alexander decided to look at naked mole-rats in the field, and together with their wives, C. Kagarise Sherman and L. K. Alexander, they organized an expedition to Kenya. Jarvis invited them to visit Cape Town on the way.

Alexander and Sherman met Jarvis for the first time in late November 1979. Upon their arrival, Jarvis presented the two Americans with a rough draft of her (now famous) paper on the question of eusociality in *H. glaber* (eventually published in *Science* magazine in 1981). Alexander and Sherman commented extensively on the manuscript and invited Jarvis to join them in Kenya. The five expedition members got together in Nairobi a week later and drove 200 km southeast to the town of Mtito Andei, the site where much of the field research and collecting described in this volume have taken place. In three weeks at Mtito Andei, the researchers gathered basic ecological information (including colony distributions and food plants) and captured individuals from six *H. glaber* colonies, which were returned alive to the United States. These animals and their descendants have been studied at Michigan and Cornell University (where Sherman moved in 1980) ever since.

Over the years, Alexander has been particularly interested in what naked mole-rats can tell us about the evolution of eusociality (chap. 1), hairlessness, and altriciality (chap. 15). From 1980 to 1989 he sponsored four postgraduate students who studied the Michigan colonies. S. Finger and S. Isil investigated division of labor within colonies, with Isil receiving a master's degree for this work. S. H. Braude completed an undergraduate honors thesis on burrowing behavior in the lab (see Chapter 8) and subsequently began a study of behavioral ecology in the field (chap. 6) for his doctoral dissertation. More recently, J. W. Pepper has studied mole-rat vocalizations in the lab at Michigan (see Chapter 9).

Sherman has worked with numerous undergraduates and three postgraduate students in studying the social behavior, colony structure, and genetics of the Cornell colonies. In particular, E. A. Lacey completed an undergraduate honors thesis on the animals (chap. 10), and S. F. Payne received a Master's degree for her similar behavioral studies. Recently, H. K. Reeve has studied the genetic structure of colonies and the role of the breeding female in colony organization (chap. 11). Given the common focus of Alexander's and Sherman's groups on behavior, they joined forces to develop ethograms of vocal (chap. 9) and nonvocal (chap. 8) behaviors.

In 1980, Jarvis spent 6 weeks in the field, accompanied by R. Buffenstein, K. C. Davies, and M. Griffin. This expedition gathered information on the distribution of naked mole-rats in Kenya, mapped burrows, and explored food abundance and distribution. Three complete colonies and individuals from three others were collected and returned alive to Cape Town. Techniques for maintenance and husbandry were developed (Appendix), and these colonies have thrived and reproduced. Their reproductive biology has been studied by Jarvis (chap. 13) and by four postgraduate students: B. Broll, who is interested in chemical control of reproduction and age polyethism, and R. Buffenstein, E. McDaid, and M. J. O'Riain, who together with P. Ross (University of Witwatersrand) are collaborating to study the factors affecting growth and body size (chap. 12). Jarvis has also sponsored three other postgraduate students, K. C. Davies, N. C. Bennett, and B. G. Lovegrove, who have investigated the ecology, behavior, and physiology of other less social genera and species in the family Bathyergidae (chap. 3).

R. A. Brett came on the mole-rat scene in 1982. Until then, most behavioral data had been obtained from captive animals. For his dissertation, Brett studied the population structure, food distribution, burrowing patterns, and demography of H. glaber near Mtito Andei, and radio-tracked individuals to investigate activity patterns and to map a complete burrow system. Key portions of his Ph.D. dissertation are presented as chapters 4 and 5 in this volume. In the course of his studies, Brett took four incomplete colonies to the Institute of Zoology in London. The reproductive physiology and the social and hormonal correlates of reproductive suppression in these animals and their descendants have been investigated by D. H. Abbott, L. M. George, and two postgraduate students, C. H. Liddell and C. G. Faulkes (chap. 14).

In 1983, R. L. Honeycutt (then at Harvard) began studying the systematics and phylogeny of the rodent family Bathyergidae in collaboration with D. A.

Schlitter and several postgraduate students. During four expeditions to Kenya, they collected specimens and tissue samples from several hundred naked mole-rats in six localities for genetic studies. These have been analyzed using DNA sequencing technology, mitochondrial DNA (chap. 2), and, in collaboration with K. Nelson and Sherman, using starch-gel electrophoresis (chap. 7).

Taken together, the results presented in this volume thus summarize the first two decades of concentrated research on naked mole-rats. This book is intended to provide a working base of empirical information and a common terminology for future studies of these remarkable creatures and for discussions of the intriguing theoretical issues raised by various facets of their social and reproductive biology.

Organization of the Book

The chapters in this volume fall, roughly, into four sections. The first section (chap. 1–3) introduces readers to the study animals and the many reasons for our interest in them. Chapter 1 is a theoretical treatise on the evolution of eusociality and a discussion of the pivotal position occupied by the naked mole-rat in evolutionary hypotheses for its existence. Chapters 2 and 3 are more empirical, deal specifically with mole-rats, and introduce readers to the systematics and evolution of the rodent family Bathyergidae and to the behavior and ecology of the 5 genera and 12 species in the family. These three opening chapters set the stage for a closer look at *Heterocephalus glaber*.

The second section (chap. 4–7) comprises field studies. These are intended to give readers an understanding of what naked mole-rats are like in nature and to serve as a basis of comparison with the laboratory studies that make up the latter half of this volume. Chapters 4 and 5 present information about colony sizes and compositions, spacing between colonies, the extent and layout of burrow systems, the timing of breeding, colony expansion and fissioning, the types and distributions of food sources, and predation. The following chapter (chap. 6) is a detailed look at participation in "volcanoing" (kicking out excavated soil from within the burrow system onto the surface), which is one of the most important and (apparently) dangerous of mole-rat behaviors. The second section of the book ends with chapter 7, a discussion of the population genetics of *H. glaber*, focusing on inter- and intracolonial comparisons based on data from nuclear and mitochondrial genomes.

Chapters 8–11, the book's third section, describe naked mole-rat behavior and the social organization of colonies. Qualitative descriptions of the animals' social and nonsocial behaviors (chap. 8) and vocalizations (chap. 9) are presented, along with hypotheses about the functional significance of each. Using these descriptions as a starting point, chapter 10 investigates whether *H. glaber* colonies are collections of individuals that do approximately the same things or instead are organized into subgroups with specialized functions. In Chapter 11, the behavior of the breeding female, the central figure in the colony, is investigated and her role in maintaining reproductive suppression and in colony dynamics is addressed.

In chapters 12–14, the book's final section, the reproductive biology of captive naked mole-rats is considered. In chapter 12, interindividual variability in growth and behavior is explored, and in chapter 13, various reproductive parameters are described, including the process by which breeding females and males are "chosen," breeding behavior, litter sizes, sex ratios, inter-litter intervals, the development of pups, and reproductive maturity and longevity. Finally, some of the hormonal correlates underlying divisions of labor in reproduction and colony maintenance tasks are explored in chapter 14.

Obviously, questions about the functioning of naked mole-rat societies would best be answered with field data. However, this is impossible because the animals rarely come aboveground, and they abscond if their burrows are tampered with. Thus, chapters in the third and fourth sections are based on laboratory studies. In all cases, the colonies were housed in transparent artificial burrow systems and maintained in quiet, warm, humid, and dimly illuminated rooms. Despite efforts to mimic natural conditions, however, our laboratory set-ups have at least three important differences: (1) artificial tunnel systems are 0.1%–3.0% as long as those dug by wild *H. glaber* colonies; (2) lab tunnels are not earthen, and the mole-rats cannot disperse, expand, or otherwise modify their structure (all attempts at "ant farm" arrangements to house naked mole-rats have failed); and (3) no predators harass lab colonies, no rainstorms afflict them, and they are constantly supplied with large quantities of fresh, succulent foods at predictable locations. Mindful of these differences between the laboratory and nature, authors of chapters in the third and fourth sections of this volume have attempted to relate their findings to field data whenever possible. Nonetheless, their results should be viewed as hypotheses to be tested, eventually, in wild colonies.

The volume concludes with a coda (chap. 15) discussing some of the most intriguing evolutionary questions about naked mole-rats that have not been adequately addressed. This chapter points the way to exciting future research possibilities. To facilitate such endeavors, we have included an Appendix explaining how to capture and transport naked mole-rats and how to maintain them in captivity. The husbandry of this mammal is a bit tricky, and we felt that a summary of our experiences would encourage readers and help them get started on their own field and laboratory studies.

Acknowledgments

The Office of the President of Kenya and the Wildlife Conservation and Management Department kindly granted various authors in this volume permission to conduct research in Kenya and to export live naked mole-rats. Many citizens of Kenya helped us capture mole-rats and excavate burrow systems; in particular, we thank A. S. Ndalinga Chondo, S. Komu Nzioki, M. Jackson, A. Lerewan, and N. Mathuku. Numerous other people have assisted us in various ways; in addition to those acknowledged in individual chapters, we wish to thank I. Aggundey, H. Bally, P. Bally, R. Bally, P. Chaponnière, S. E. Glickman, G. Kinoti, A. Meyers, R. Obanda, F. A. Pitelka, P. Woodley, and the staff at the East African Herbarium in Narobi. All of the chapters in this volume were read and commented on by one or more other scientists. In that context, the editors and the authors thank the following external referees: F. H. Bronson, R. Buffenstein, C. S. Carter, T. H. Clutton-Brock, S. T. Emlen, G. Epple, R. G. Harrison, D. W. Leger, E. Nevo, J. L. Patton, O. J. Reichman, E. F. Rissman, T. D. Seeley, A. van Tienhoven, J. G. Vandenbergh, and two anonymous reviewers selected by the Princeton University Press. We also thank E. Wilkinson of Princeton for advising and encouraging us and for shepherding the book through the press.

More than sixty photographs illustrate the volume. The majority of these were taken at Cornell by Raymond A. Mendez, and the authors and editors sincerely thank him for donating his work. Additionally, the authors thank the following individuals for supplying the photographs that illustrate their chapters: R. Borland, R. A. Brett, C. G. Faulkes, M. Griffin, J.U.M. Jarvis, P. W. Sherman, and C. Springmann.

Field studies by the authors of this volume have been supported by the National Geographic Society, the U.S. National Science Foundation, the Zoological Society of London, and the University of Michigan. Financial support for laboratory research has been obtained from the U.S. National Science Foundation, the South African Council for Scientific and Industrial Research, MRC/AFRC at the Institute of Zoology (London) and the following universities: California (Berkeley), Cape Town, Cornell, Harvard and Michigan.

The volume was assembled at Cornell University. P. W. Sherman thanks T. M. Natoli and S. L. Mancil for typing and retyping manuscripts, T. Carson for expert copyediting, and all three for their patience and cheerfulness in the face of apparently insurmountable odds. Sherman also thanks T. D. Seeley, R. R. Hoy, and especially C. Kagarise Sherman for friendship and support, and his children (L. S. and P. G. Sherman), his parents (P. S. and K. P. Sherman), and his parents-in-law (R. E. and B. S. Kagarise) for encouragement and for accommodating naked mole-rats in their bathtubs at one time or another.

Contributors

DAVID H. ABBOTT, MRC/AFRC Comparative Physiology Research Group, Institute of Zoology, Regents Park, London, NW1 4RY, United Kingdom

RICHARD D. ALEXANDER, Museum of Zoology, University of Michigan, Ann Arbor, Michigan 48109

MARC W. ALLARD, Department of Wildlife and Fisheries Sciences, Texas A&M University, College Station, Texas 77843

NIGEL C. BENNETT, Zoology Department, University of Cape Town, Rondebosch 7700, South Africa

ROBERT A. BRETT, Institute of Zoology, Zoological Society of London, Regents Park, London NW1 4RY, United Kingdom

STANTON H. BRAUDE, Museum of Zoology, University of Michigan, Ann Arbor, Michigan 48109

BERNARD J. CRESPI, Zoology Department, University of Oxford, South Parks Road, Oxford OX1 3PS, United Kingdom

SCOTT V. EDWARDS, Zoology Department, University of California, Berkeley, California 94720

CHRISTOPHER G. FAULKES, MRC/AFRC Comparative Physiology Research Group, Institute of Zoology, Regents Park, London NW1 4RY, United Kingdom

LYNNE M. GEORGE, MRC/AFRC Comparative Physiology Research Group, Institute of Zoology, Regents Park, London NW1 4RY, United Kingdom

RODNEY L. HONEYCUTT, Departments of Wildlife and Fisheries Sciences and Genetics, Texas A&M University, College Station, Texas 77843

JENNIFER U. M. JARVIS, Zoology Department, University of Cape Town, Rondebosch 7700, South Africa

EILEEN A. LACEY, Museum of Zoology, University of Michigan, Ann Arbor, Michigan 48109

CAROLINE E. LIDDELL, MRC/AFRC Comparative Physiology Research Group, Institute of Zoology, Regents Park, London NW1 4RY, United Kingdom

ELIZABETH McDAID, Zoology Department, University of Cape Town, Rondebosch 7700, South Africa

KIMBERLYN NELSON, Department of Biological Chemistry and Molecular Pharmacology, Harvard University Medical School, Boston, Massachusetts 02115

KATHARINE M. NOONAN, 609 Kains Avenue, Albany, California 94706

M. JUSTIN O'RIAIN, Zoology Department, University of Cape Town, Rondebosch 7700, South Africa

JOHN W. PEPPER, Division of Biological Sciences, University of Michigan, Ann Arbor, Michigan 48109

HUDSON K. REEVE, Section of Neurobiology and Behavior, Cornell University, Ithaca, New York 14853

DUANE A. SCHLITTER, Section of Mammals, The Carnegie Museum of Natural History, 4400 Forbes Avenue, Pittsburgh, Pennsylvania 15213

PAUL W. SHERMAN, Section of Neurobiology and Behavior, Cornell University, Ithaca, New York 14853

The Biology of the
Naked Mole-Rat

1 The Evolution of Eusociality

Richard D. Alexander, Katharine M. Noonan, and Bernard J. Crespi

Eusociality is a remarkable topic in evolutionary biology. The term, introduced by Michener (1969), refers to species that live in colonies of overlapping generations in which one or a few individuals produce all the offspring and the rest serve as functionally sterile helpers (workers, soldiers) in rearing juveniles and protecting the colony. The wasps, bees, ants, and termites known to live this way had previously been called the "social" insects.

The recent discovery of eusociality in aphids (Aoki 1977, 1979, 1982) and naked mole-rats (Jarvis 1981, this volume) has provided biologists with new impetus to understand more fully the origins and selective background of this phenomenon, which has already played a central role in the analyses of sociality in all animals (Hamilton 1964) and, indeed, of evolution itself (Darwin 1859). These two new instances both broaden the search for correlates of eusociality in the widely different groups in which it has evolved independently and stimulate comparative study of related species of insects and vertebrates with homologous behaviors verging on eusociality (Eickwort 1981; J. L. Brown 1987; Lacey and Sherman, chap. 10).

An unusual and complicated form of sociality has thus evolved independently in four different groups, and in one, the Hymenoptera, has persisted from perhaps a dozen independent origins (F. M. Carpenter 1953; Evans 1958; Michener 1958; Wilson 1971). Explaining this phenomenon requires attention to a number of different questions. Darwin (1859) answered the basic one, How can natural selection produce forms that would give up the opportunity to reproduce, instead using their lives to contribute to the success of the offspring of another individual?

Darwin's Question: How Can Sterility Evolve?

Darwin used the origin of sterile castes as a potential falsifying proposition for his theory of evolution by natural selection. He referred to "the neuters or sterile castes in insect-communities . . . [which] from being sterile . . . cannot propagate their kind" as "the one special difficulty, which at first appeared to me insuperable, and actually fatal to my whole theory" (1859, p. 236). To solve the problem of how the sterile castes could evolve, he generated the magnificent hypothesis, which still stands, that if sterility (or any trait of a sterile form) can be carried without being expressed, then if those who express it contribute enough to the reproduction of others who carry the trait but do not express it, the trait itself can be "advanced by natural selection" (p. 236). In other words, if functionally sterile individuals help relatives produce offspring and thereby cause enough copies of the helping tendency to be created, then the tendency (ability, potential) can spread. Darwin was particularly concerned with how the sterile castes could evolve their own sets of attributes; his statements indicate that when he spoke of selection at the level of the "family" and "community" in eusocial insects, he was referring to the spread and preservation of traits that exist among the members of groups of related individuals. Thus, in the same context, he noted that "A breed of cattle always yielding oxen [castrates] with extraordinarily long horns could be slowly formed by carefully watching which individual bulls and cows, when matched, produced oxen with the longest horns; and yet no one ox could ever have propagated its kind" (p. 238). Similarly, he remarked that tasty vegetables could be produced by saving seeds from relatives of the vegetables already tasted or eaten and therefore unable to produce seeds. He also noted that cattle with "the flesh and fat . . . well marbled together" could be bred although "the animal has been slaughtered" if "the breeder goes . . . to the same family" (p. 238).

Darwin's hypothesis could scarcely be improved on today, even though, not knowing about genes, he had to rely on the concept of trait survival, and he had no way of being quantitative. His various remarks taken together are quite close to what modern investigators such as Hamilton (1964) and D. S. Wilson (1980) mean when they refer, respectively, to "inclusive-fitness maximizing" and "trait-group selection." Darwin's "family" method of selection to preserve traits is one of those long advocated by agricultural scientists (e.g., Lush 1947). His remarks cited here demonstrate the error of assertions either that Darwin invoked (a simplistic and unsupportable kind of) group selection to explain eusociality or that he did not discuss selection above or below the level of the individual. Darwin also showed in these statements that he understood how organisms can carry the potential (which we now know to be genetic) for varying their phenotypes between profoundly different states, depending on environmental circumstances.

Fisher (1930, p. 177) began the quantification of Darwin's idea of reproduction via collateral relatives (although he gave no evidence of being aware of Darwin's discussion when he did so) by developing a hypothesis to explain how bright coloration that attracted (and taught) predators could evolve in distasteful or poisonous caterpillars. He noted that if bright coloration were to spread among distasteful or poisonous caterpillars traveling in sibling groups, then a caterpillar with a new allele making it slightly more noticeable and thus more likely to give its life being tested could thereby teach a predator to avoid the entire sibling group. But, that caterpillar would have to save more than two full siblings, since each would have only a 50% chance of carrying the same allele for brighter color. (Using phylogenetic inference, Sillén-Tullberg [1988] argued that distastefulness and bright coloration often preceded gregariousness in lepidopterous larvae, but this argument does not negate the possibility of continued exaggeration of these traits among gregarious forms.) Fisher (1930, p. 181) also remarked that tendencies of humans to risk their lives in heroic acts are most likely to have spread and become exaggerated because of the beneficial effects on copies of the genes responsible located in the collection of the hero's relatives.

Haldane (1932) carried the arguments about reproduction via collateral relatives further and also related them to the eusocial insects. (Haldane is reported to have commented [Maynard Smith 1975; pers. comm.] that we should expect individuals in species like our own to have evolved to give their lives only for more than two brothers or more than eight cousins, since brothers have a one in two chance of carrying alleles for such bravery and cousins a one in eight chance. This comment is said by Maynard Smith to have been made sometime in the early 1950s in a pub with only Maynard Smith and Helen Spurway Haldane present [see also Haldane 1955]. The close resemblance of this reported statement to Hamilton's [1964] statement has aroused some attention [see also Hamilton 1976]. In any case, the original idea of reproduction via collateral relatives was Darwin's, its initial quantification was by Fisher, and, as discussed later, Hamilton [1964] first developed it extensively.) Williams and Williams (1957) discussed the evolution of eusocial insects, approximately in Darwin's terms, but they were unaware of Fisher's discussions (G. C. Williams, pers. comm.) and added no new arguments.

Hamilton (1964) not only developed the ideas of Darwin, Fisher, and Haldane extensively, but he also showed that maximization of what he called *inclusive fitness* (a process some others have called *kin selection*, following Maynard Smith 1964) really applies to all social species. The general principle, familiar now to nearly all biologists, is that one can reproduce not only by creating and assisting descendants but also by assisting nondescendant or collateral relatives, and, other things being equal, it pays more to help closer relatives than to help more distant ones.

The Taxonomic Distribution of Eusociality

While these discussions of the process or the mechanics of the evolution of eusociality were going on, another virtually independent discussion of the patterns (phylogeny) of evolutionary change in eusocial forms and their relatives was taking place (Wheeler 1923, 1928; F. M. Carpenter 1930, 1953; Evans, 1958; Michener 1958; Wilson 1971; West-Eberhard 1978a,b; J. M. Carpenter, in press). This series of studies proceeded primarily by description and comparison of eusocial forms with their closest noneusocial relatives and by the techniques of phylogenetic reconstruction. Thus, Evans (1958) and Michener (1958) provided excellent reviews of the probable phylogenies of social behavior in bees and wasps, respectively. These various comparative and phylogenetic studies revealed that eusociality has persisted from at least 12 or 13 independent evolutionary origins in the Hymenoptera and from only 1 (or possibly 2 or 3; see Noirot and Pasteels 1987) origin in all other arthropods. (An exception is the clonal forms in aphids [Aoki 1977, 1979, 1982], which, according to Hamilton [1987] may have originated many times but suffered frequent extinctions because of diseases and parasites.) This finding shows that Darwin's theoretical answer to the general question of how sterility can evolve is only a beginning. It does not tell us why eusociality appeared or succeeded in the particular taxonomic groups in which it occurs today, and why it either did not evolve or did not persist in any other organisms.

Hymenoptera: The Haplodiploidy Argument

As Hamilton (1964) pointed out, special genetic systems can increase the reproductive benefit from tending collateral rather than descendant relatives. Hamilton showed that the Hymenoptera have a peculiar genetic asymmetry because of their haplodiploid method of sex determination. Because males are haploid, all of a male's sperm are genetically identical. Thus, when a female is monogamous, her daughters share all the genes from their father and half the genes from their mother. In respect to genes identical by immediate descent, daughters share an average of three-fourths rather than the usual half, even though they still share only half the genes of their own daughters. Hamilton offered the reasonable suggestion that this genetic asymmetry may have contributed to the tendency of the Hymenoptera to become eusocial. He added that it might also help explain why hymenopteran workers are essentially all females, since, on the average, one-fourth of a male's genes are identical to those of his sisters.

Hamilton's (1964) papers caused a surge of attention to the question of how and why eusociality evolved, and especially why it evolved more times in the Hymenoptera than in all other animals combined. His cautious and conservative suggestions about the effects of haplodiploidy on relatedness between helpers and their siblings and offspring, and about the prevalence of females among workers in the Hymenoptera, were widely accepted, turned into dogma, and "came to dominate many textbook and popular accounts" (Andersson 1984, p. 166). Indeed, for a while it seemed that most biologists believed that to explain sterile castes one had to locate a genetic asymmetry like that in the Hymenoptera, in which siblings are genetically more similar to each other than to parents and offspring. Bartz for example, writing on the evolution of termites, stated that unless parents are related, their (inbreeding) offsprings' offspring will not be more closely related to their sibs than to their offspring, so that "the selection pressure to remain and raise siblings disappears in a single generation" (1979, p. 5765).

This argument has been doubted on several different grounds (Hamilton 1964; Lin and Michener 1972; Alexander 1974; Ghiselin 1974; West-Eberhard 1975, 1978a; Trivers and Hare 1976; Evans 1977; Craig 1979, 1980, 1982; Eickwort 1981; Andersson 1984). Thus, (1) multiple matings by females reduce the closeness of relationship between sisters, (2) males, which also must be tended as juveniles, are not as closely related to females as their sisters, and (3) early in the development of each colony, queen control of sex ratios causes her interests in this regard, rather than those of daughters, to be served. All of these effects (and others that may or may not be relevant to the origins of eusociality, such as multiple reproductive females or short-lived reproductive females, either of which may produce workers that assist reproductives other than their mothers; see, e.g., West-Eberhard 1978a) tend to erode the advantage to potential helpers from haplodiploidy. Moreover, social interactions cannot be predicted from genetic relationships (Hamilton's r) alone (including those caused by sex-ratio biases within broods in haplodiploid forms); to suppose that they can is to ignore variables of age, life stage, and environment that also adjust reproductive costs and benefits ($b + c$ in Hamilton's expression $rb - c > 0$). If nepotism toward collateral relatives required that individuals be more closely related to those relatives than to their own offspring, then nepotism would not be expected to extend beyond the nuclear family except for sisters in haplodiploid, monogamous species.

Following Hamilton's (1964) development of the concept of inclusive fitness, models to help account for the restricted distribution of eusociality were almost invariably developed explicitly either to help explain eusociality in the Hymenoptera or to account for its existence in the Isoptera by incorporating some mechanism that gives an effect paralleling that of haplodiploidy. But haplodiploidy does not occur in two of the three major groups that evolved eusociality (Isoptera and Rodentia). Moreover, haplodiploidy occurs in all Hymenoptera and several other groups of arthropods (Hamilton 1964, 1967; Borgia 1980; Andersson 1984), not merely in the Hymenoptera that became eusocial. Finally, some individuals among the progeny of a eusocial colony do not help rear siblings but become reproductive adults that found new colonies. Unless such individuals are, for some reason, less closely related to their siblings than to their offspring, we need to know about something other than genetic relatedness to explain even haplodiploid eusocial systems.

Haplodiploidy, then, is neither necessary nor sufficient to account for the appearance and maintenance of eusociality. As many authors have recently suggested (as did Hamilton 1964), we are required to search for additional contributing factors. Did the ancestors of termites and naked mole-rats possess traits that have the same genetic effect as haplodiploidy (see, e.g., Hamilton

1972; Bartz 1979; Lacy 1980)? Do some members of these groups possess other distinctive traits, or live in some special circumstances, that contributed to eusociality, as a result of effects different from those of haplodiploidy? Are there relevant features, or combinations of features, common to all eusocial forms or their ancestors and not exclusive to the Hymenoptera? Why was it profitable for the ancestors of eusocial forms to begin to live in groups, and in what kinds of groups did they live? What causes some offspring to remain in the parents' nest? Why do they begin helping? What happened within the social groups in which the ancestors of eusocial forms lived to cause them to continue to evolve along the route to eusociality, and what particular steps occurred along the evolutionary routes leading to the current diversity of eusocial forms? We consider these various questions in order.

Do Termites and Naked Mole-Rats Mimic Haplodiploidy?

As Hamilton (1964) realized, the termites, which have diplodiploid sex determination, represented an embarrassment to the haplodiploid aspect of his argument. Hamilton (1964, 1972), Taylor (1978), Bartz (1979), and Lacy (1980) tried to solve this problem by postulating situations in termites that would mimic the consequences of haplodiploidy. Bartz (1979), for example, argued that if male and female mates in termites are each highly homozygous but unrelated, then their offspring may be more closely related to each other than to their parents because the different gametes of each sex will be very similar genetically. The offspring will also be extremely heterozygous. To re-create high levels of homozygosity, Bartz postulated that within each colony the original parents typically die and are replaced by secondary reproductives from their brood, which inbreed as brother and sister. He argued that, through successive inbreeding, genetic drift (the result of reproduction by only a few individuals) would re-create homozygous genotypes in the eventual reproductives that would found new colonies through outbreeding.

Even if Bartz's (1979) theoretical argument accurately describes life in modern termites, the requirement that nests last several generations to re-create homozygosity suggests that it has little bearing on how helpership and eusociality originated in the orthopteroid (or rodent) line, because we might expect nests to have lasted but a single generation in their noneusocial ancestors, as, for example, in their distant relative, the subsocial wood cockroach *Cryptocercus* (Nalepa 1988). Before dismissing the argument too quickly, however, we must consider the possibility that termites (and naked mole-rats) did indeed live in long-lasting, multigenerational inbreeding groups before they were eusocial and that they later evolved to form new groupings or new colonies at intervals through outbreeding. This could have occurred because of the kinds of places in which they lived. In other words, underground tunnels (naked mole-rats) or the interior of logs (termites) could provide abundant local food supplies and unusual safety and be long-lasting and expansible, thereby meeting the needs of an enlarging social group (see later arguments). Thus, they could represent niches that would lead to just the situation postulated by Bartz (1979). Naked mole-rats, at least, are apparently extremely inbred (Reeve et al. 1990; see also Honeycutt et al., chap. 7). If Bartz's hypothesis were correct, however, we would predict that naked mole-rats establish colonies by extreme outbreeding. If they do not (there is no evidence of it yet, and Reeve et al. believe that their data indicate continuous rather than cyclic inbreeding), their ability to achieve and maintain eusociality diminishes the significance of Bartz's hypothesis.

Lacy (1980) proposed that if a large part of the primitive termite genome were sex-linked, a significant asymmetry in the coefficients of relationship would have resulted, causing early termites, like haplodiploid forms, to be more closely related to same-sex siblings than to their own offspring. But termite workers are both male and female, and there is no indication as yet that workers of either sex favor siblings of their own sex (see discussion and references in Andersson 1984; Crozier and Luykx 1985). It appears that the evolution of termite eusociality is unlikely to have been based on a male haploid analogy.

Are There Other Traits Relevant to Eusociality?

Although it may seem doubtful that the repeated evolution (or persistence) of eusociality in the Hymenoptera occurred solely because of their haplodiploidy, we still must ask why it happened there so many times and only once in the other 90% of the insects. In the arguments that follow, we are in no way doubting that genetic relatedness (kin selection, maximizing inclusive fitness) is central in explaining cooperation, helping behavior, and the evolution of eusociality. The genetic question addressed is the narrower one of whether the closer relatedness of full sisters, as compared with parents and offspring, in monogamous, haplodiploid forms is sufficient to account for 12 or more separate origins of hymenopteran eusociality, as compared to 2 origins representing all other animals.

Testing the connection between haplodiploidy and the prevalence of eusociality in the Hymenoptera involves determining the relative chances that the Hymenoptera and the rest of the insects, or the entire animal kingdom, would become eusocial, independent of the genetic asymmetry of haplodiploidy. Hamilton's (1964) arguments implied that without the effect of haplodiploidy the hymenopteran and orthopteroid lines would have been equally likely to produce eusocial forms, or at least that the Hymenoptera would not have been 12 times as likely to do so. If we doubt that haplodiploidy accounts for the greater number of origins (or retentions) of eusociality in Hymenoptera, then we must ask if there are other correlates of eusociality that would have given the Hymenoptera an advantage. Or, is there a correlate of eusociality in the Hymenoptera that is more important than haplodiploidy? The answer to both questions, we believe, is yes.

Subsociality as a Universal Precursor of Eusociality

An old argument in the eusocial insect literature about whether eusociality evolved through a semisocial or a subsocial precursor (see, e.g., Michener 1958) has recently been revived in a slightly different form (Lin and Michener 1972; West-Eberhard 1975, 1978a). In Wheeler's (1923) usage, *subsocial* meant parental, referring to social groups made up of parents and offspring. Wheeler (1928, p. 12), however, restricted the term to forms in which the parent "continuously feeds the . . . [offspring] with prepared food (progressive provisioning)." The offspring of subsocial forms are thus tended or provisioned, though not necessarily all the way to adulthood, and they do not become sterile helpers. *Semisocial*, also an old term in entomology, meant that individuals of the same stage and age aggregate or herd together with (in the usage of Michener 1958) "division of labor (often weak or temporal) or cooperative activity" and (also Michener's usage) "without parent-offspring relationship" (p. 441). Michener (1969) introduced the term *parasocial* to include *semisocial, communal,* and *quasi-social,* all of which refer to particular kinds of social activity in bees, involving individuals of the same general age and stage, sometimes sisters. He used *semisocial* to refer to small colonies showing "cooperative activity and division of labor among adult bees as in true social groups" and *subsocial* as "family groups each consisting of one adult female and a number of her immature offspring which are protected and progressively fed by the adult" (p. 304). Here we argue for slightly less specific meanings of subsocial (any species with parental care), thus including Wheeler's (1928, p. 13) "infrasocial stages" 4, 5, and 6, rather than 6 alone, which requires continuous feeding with prepared food (progressive provisioning) and semisocial (aggregations of individuals of approximately the same age and stage) (note that parasociality can be substituted for semisociality in the statements that follow with little change of meaning).

In terms of the origins of eusociality, the contrasting of subsociality and semisociality may have been misleading, because, as Michener (1958) pointed out, all of the so-called semisocial bees that can be used as examples are also already subsocial (so, it appears, are the semisocial wasps). Female Hymenoptera that group or share nests are thus already parental. Michener believed that species that preprovision and seal the cells of their offspring could not have been subsocial before they were semisocial because they never associate with any but their adult offspring, but this point only bears on the question of whether social interactions between juveniles and their parents preceded social interactions between adult siblings or vice versa. The question that Wheeler (1923) first raised must be rephrased to ask, not whether semisociality or sub-sociality leads to eusociality, but rather whether subsociality (parental care, whether it is progressive provisioning involving social interactions or not) leads to eusociality directly or through semisociality. In other words, to what extent did interactions among adults, taking place in species with adults that were already parental, affect the likelihood that eusociality would evolve? Has cooperative group-nesting among adult bees and wasps facilitated the evolution of eusociality? Did helpers initially aid younger siblings in growing up in nests founded by their mother (or both parents), or did helping first occur among sisters after the mother was dead, so that helpers in fact aided primarily nieces and nephews? Or did both patterns exist during the evolution of eusociality in different forms?

In some modern social wasps, inseminated females found nests together, with only one producing the eggs and the others serving as workers (West-Eberhard 1969; Noonan 1981); in others, nests are founded by multiple queens, which are at least sometimes sisters (West-Eberhard 1978a, pers. comm.), and swarms of workers. In honey bees (*Apis*) and stingless bees (Meliponinae), nests are founded by single queens and swarms of workers. In army ants, which do not have subterranean nests, colonies form by fission. Fission also occurs in "polydomous" ants, and sometimes in termites when colony tunnels become very long, so that the first workers are siblings of the (new) reproductives. In ground-nesting bees, females (sometimes sisters) may cooperate in digging tunnels and guarding communally used nest entrances; some associations seem to involve reproductive division of labor. In some bees, subterranean nests are founded by lone females, which die, leaving their adult daughters functioning in a group in much the same way as multiple foundresses (Michener 1969, 1974, 1985).

These various examples of cooperation among (sometimes) sister reproductives and helpers—without the mother present and sometimes without evidence of age or size differences—raise the question of whether worker castes may not have evolved from helping other individuals (sometimes, at least, sisters) of about the same age, as in nest founding (Lin and Michener 1972; West-Eberhard 1978a). At first this may seem particularly likely, given that, to specialize as workers, helpers require juveniles that need helping *throughout their reproductive lives*. (Specialized workers differ from workers that help briefly but retain a strong capacity to become an independent reproductive, therefore retaining a phenotype virtually indistinguishable from that of individuals that never help.) This condition is facilitated by queens evolving to live longer than their helper daughters. To take the most favorable case, if a female and her worker sisters mature at the same time, she does not need to outlive them to provide sufficient eggs to use all of their reproductive effort, as would their mother, who would be much older than they.

However, evolution of phenotypic divergence into worker and queen castes (i.e., evolutionary inception of eusociality) in these circumstances would seem unlikely for four reasons. First, as cooperative foundresses, incipient workers would not realize the savings in time possible to matrifilial workers because, upon their emergence as adults, they would not be given the headstart of being provided with eggs, ready to be helped (see also Queller 1989). Second, taking up workership after joint nest founding (as opposed to laying one's own eggs) usually would not circumvent the added time and risks of mating and colony founding (benefits to helpers that simply stayed in the mother's nest), though it may reduce them. Some of these problems are circumvented in groups of cooperative sisters using their deceased mother's nest, but in such cases one has to ask if helping began in the context of aiding the mother's younger offspring (as in the usual lone-foundress matrifilial model) and was later transferred to the sisters' offspring or to even more distant relatives. Third, various unavoidable uncertainties, such as mortality during dispersal and nest founding or in overwintering, would presumably cause facultative helping to be favored over obligate workership. That is, it would appear more difficult, or more indirect, to evolve profound caste differences and to drive the initiation of caste divergence back into the early stages of development, as is the case in highly eusocial modern species (see below). Finally, incipient workers helping nieces and nephews or cousins would have to be able to give much help inexpensively, or else have little chance of reproducing by independent nesting, to make helping pay genetically.

In summary, in group-nesting species with parental females, such as halictine bees (Michener 1969), sisters (or even nonrelatives) may cooperate and show extensive parental care, but most modern eusocial forms tend to have single queens. Group founding of nests in ants and social wasps is often followed by severe aggression, eliminating all but one queen (West-Eberhard 1978b; Rissing and Pollock 1986). Because founding by swarms is derived, helping of nonsiblings by tropical polistine workers (West-Eberhard, pers. comm.) is probably not relevant to the origins of helping. In such forms, presumably, all effort is devoted to developing a colony large enough to resist predation before any reorganization that could result in workers' tending only relatives can take place.

Helpers in eusocial forms typically contribute to the success of younger siblings, not same-age or older siblings and not nieces and nephews. Generally speaking, parental care (subsociality) preadapts species for the evolution of forms of eusociality in which older individuals help younger ones. West-Eberhard (1978a) believed that the first workers in polygynous wasp societies evolved in forms in which the lone founding mother died, leaving groups of sisters (or sisters and granddaughters), some of which lay eggs while others work. Such forms would seem likely to have preceded others in which the offspring in such a nest formed a group comprising workers and one breeding female. However, they may also be regarded as derived forms (derived from a matrifilial society in which adult females reared or protected juvenile sisters), exhibiting a particular form of group nesting in response to predation or other difficulties of independent nest founding.

Group founding of nests in eusocial forms, it would seem, occurs for one or more of three reasons: (1) new single-queen nests are sometimes vulnerable to predators and parasites, especially when foraging away from the nest is mandatory; (2) subordinate females may be able to replace a queen either by helping or simply by lurking on or near her nest during its early stages (Noonan 1981); and (3) subterranean cavities suitable for nests may be expensive to locate or excavate. All three situations could contribute to cooperation among founding sisters. The question is, did helping among potential queens (sisters or not) contribute to eusociality, or did eusociality actually stem from parental care, where group and swarm nesting involving multiple reproductives are secondary, and cooperation among parental sisters rarely if ever leads to worker specialization.

The implication of these combined considerations is that some small-colony multiple-foundress social wasps and ground-nesting bees have remained in an apparently primitive social (subsocial or small-colony eusocial) condition partly because the selective situations that would lead them to phenotypic divergences paralleling other (small- and large-colony) eusocial insects are less likely to occur among sister foundresses or same-age sisters in their dead mother's nest. In other cases (e.g., honey bees, stingless bees, some tropical wasps that use swarms to found nests), nests initiated by single foundresses have become too vulnerable to predation. These types of nests are especially vulnerable, probably because of the food value of a large colony of juveniles and stored food. Obviously, nest founding by swarms cannot be a primitive trait in the evolution of eusociality. There seems to be no particular evidence that group founding by sisters has simply persisted as a primitive trait, gradually evolving into swarm founding. Neither are there cases of univoltine eusocial species with multiple foundresses. Such cases might be expected from the "semisocial" hypothesis, but they could not occur in species with matrifilial origins of eusociality. It seems to us, therefore, that the paltry evidence available tends to return us to the matrifilial family, founded by lone females or monogamous pairs, as the likely primitive condition preceding the evolution of eusociality in both Hymenoptera and Isoptera (primitively, the founding female need live only long enough to provide her first generation of offspring with eggs or dependent offspring).

In other words, excluding aphids (Aoki 1977, 1979, 1982) and other clone-forming species (e.g., polyembryonic forms), sterile castes may always have evolved in forms that were already extensively parental, whether or not they have always been preceded in evolution by helpers at the nest that were tending younger siblings. The original social groupings from which eusociality evolved in the Hymenoptera, Isoptera, and Rodentia, according to this hypothesis, would have been composed of parents and their offspring, whether or not groupings of nesting females also occurred.

Parent-Offspring Groups as Ancestral to Eusocial Forms

Parental care can be viewed as a kind of social grouping between parents and offspring. Reasons for group living have been discussed by several authors (Alexander 1974, 1977, 1979, 1987, 1989; Wilson 1975; Hoogland and Sherman 1976; Gamboa 1978; Rubenstein and Wrangham 1987). Alexander and Hoogland and Sherman argued that there are few primary reasons for group living (i.e., selective situations that could account for the origins of group living, as opposed to secondary effects deriving from it or involved only in maintaining or furthering it): (1) clumping on clumped resources (initially involving competitive effects, rather than cooperation, unless cooperative group living had already evolved for other reasons); (2) "selfish" herds (Hamilton 1971) in which individuals use others to facilitate their own safety from predators (also not initially cooperative); or (3) cooperative efforts to secure elusive or powerful prey or to combat some other extrinsic threat, such as predators (or a cooperative effort such as huddling together during winter by flying squirrels; Alexander 1977, 1989). There seem to be no other likely reasons for expecting parent-offspring groups to form. Because parents and offspring are closely related, however, and because such groups presumably form as a part of parental care, some kind of cooperative or helpful effect in respect to either predators or food seems likely always to be the primary reason for the grouping, as is assumed in the above arguments.

Presumably, parents of any species evolve temporary groupings with their offspring because the offspring are thereby protected from predators or can be fed, or both, since feeding offspring is itself likely to be a direct or indirect protection against predators, and protection from predators may facilitate feeding. Thus, a parent that protects its offspring by placing it in a safe place, such as a nest (e.g., monotremes, reptiles, and birds), is likely to create a situation in which feeding the offspring is beneficial because food is probably not maximally available at safe nest sites. Similarly, any parent that simply keeps its offspring nearby (e.g., mammals, many parental insects) also may benefit from providing food, since food suitable for the offspring is often not optimal at locations where it is optimal for the parent. Finally, a parent that places its offspring where food is optimal for the offspring may be constrained to protect the offspring as it feeds (including providing protective nests or other structures, as in some wasps), since it is unlikely that food resources and predator protection for the offspring are optimal in the same places and times; when they are, parental care presumably does not evolve. Once juveniles are concentrated in locations with abundant food, however, predators are likely to concentrate on the locality, and adding parental care may sufficiently alleviate predator effects so as to enhance the parents' reproduction.

Expanding these considerations may help explain the evolution of parental care in diverse groups such as nesting birds for example, in comparing nesting birds with altricial and precocial young, or mammals that hide their offspring with those that take them along from birth. It may also bear directly on the evolution of eusociality. We argue below that some forms evolved eusociality partly because parent-offspring groups happened to begin living in those rare microhabitats where both food and protection from predators were enhanced by parental care for multigenerational periods.

Taxonomic Distribution and Antiquity of Subsociality

As already suggested, subsociality (parental care involving direct interaction between parent and offspring) may be a universal (and perhaps obligate) precursor of eusociality in sexually reproducing forms. To consider the significance of this argument for the taxonomic distribution of eusociality, we must address some additional, difficult questions. What is the distribution of subsociality in the Hymenoptera compared with the rest of the insects, arthropods, or animals in general? What is the relationship between the distribution of haplodiploidy and the distribution of subsociality outside the Hymenoptera? How many species, in other words, possess each of these two apparent preadaptations for the evolution of eusociality, and how many possess both? We assume that the more widespread a supposed evolutionary precursor of any derived condition in a taxonomic group, the more chances for the appearance of the derived condition.

Far more subsociality is known in the Hymenoptera than in all the rest of the insects (or arthropods) combined, indeed, probably more than in all other animal species. Spradbery (1973) indicated that there are around 35,000–40,000 species of aculeate Hymenoptera exclusive of the 10,000–15,000 eusocial ants. Fewer than 5,000 species of wasps and bees are eusocial (Wilson 1971), and most of the remaining forms are parental; about 10,000–20,000 species carry enough food to their young to take them all the way to adulthood. Not all parents in these subsocial groups interact with their offspring, but at least the stage is set for that possibility (Wilson 1971).

In contrast, fewer than 300 orthopteroids and a handful of other diplodiploid insects (Wilson 1971; Eickwort 1981) are known or thought to be extensively parental. The relatives of termites (cockroaches and webspinners) include fewer than 8,000 estimated species, subsocial or not, of which 3,700 have been described (Roth and Willis 1960; Borror and DeLong 1964).

On the basis of numbers of extant species, eusociality evolved once for perhaps every 2,500 modern species of subsocial Hymenoptera and once for every 300 modern species of subsocial orthopteroids. Even without the presumed advantage of haplodiploidy, then, on the basis of frequency of subsociality we might have expected the Hymenoptera to produce eusocial forms almost 100 times as often as the orthopteroids. Roughly speaking, the Hymenoptera include up to 99% (all but 300 of 30,000–40,000) of the modern subsocial species and account for 92%–93% of the origins of eusociality. These figures are approximately what would be expected if subsociality were an essential prerequisite of eusociality, and haplodiploidy (or something else correlated with it, which we argue below is, for the Hymenoptera, complete metamorphosis) had a somewhat negative effect on its likelihood of appearance.

These comparisons, however, involve only the relative numbers of extant species and the supposed numbers of independent origins of eusociality necessary to account for extant forms. It would be more accurate, but obviously impossible, to take into accurate account the relative numbers of subsocial species in hymenopteran and orthopteroid lines throughout geological history, their relative antiquities, and the total number of origins of eusociality. We can state, however, that orthopteroids are considerably older than hymenopterans, the fossil record of cockroaches extending to the Carboniferous (ca. 300 million years before the present [M.Y.B.P.]; F. M. Carpenter 1930) and that of the wholly subsocial order Embioptera (not thought, however, to be ancestral to termites) to the Permian (ca. 260 M.Y.B.P.; Reik 1970). The oldest Hymenoptera are from the Triassic (ca. 220 M.Y.B.P.; Burnham 1978).

Orthopteroids were probably also relatively much more abundant in earlier geological periods, the situation reversing itself at some unknown time (F. M. Carpenter 1930). According to Carpenter, cockroaches made up 80% of the Upper Carboniferous insect fauna, and Burnham (1978) regarded ants as the most abundant insects in Tertiary deposits. The earliest ant and termite fossils are of similar age (ca. 135 M.Y.B.P.; Reik 1970; Burnham 1978). The oldest hymenopteran fossil of the suborder Apocrita (parasitic and parental forms) is from the Jurassic (ca. 180 M.Y.B.P.; Reik 1970; Rasnitzyn 1975, 1977), whereas the oldest bee fossils appear now to date not from the Oligocene (ca. 34 M.Y.B.P.; Burnham 1978), as long believed, but from 100 M.Y.B.P. (a worker of the genus *Trigona*; Michener and Grimaldi 1988). The oldest evidence of eusociality in wasps is from the Oligocene (ca. 34 M.Y.B.P.; Burnham 1978). The antiquity of subsociality in orthopteroids is unknown. There are no fossilized hymenopterans or orthopteroids suggesting origins of eusociality additional to those suggested by extant species (Burnham 1978). As Evans (1977) pointed out, many eusocial lines could have been lost without a trace, but there is no reason to expect such losses to have been biased by taxonomic group.

Finally, one must also take into account that once eusociality has evolved in a particular form, additional origins may be less likely. The abundance and diversity of ants, for example, surely affects the likelihood that eusocial forms resembling them, either in taxonomy or in life-style, will evolve today; moreover, ants represent a fearsome source of predation for any incipiently eusocial arthropods that begin accumulating food and vulnerable juvenile stages in stationary locations.

Thus, in something less than 180 million yr, subsocial Hymenoptera gave rise to at least 12 different eusocial lines, and in something less than 280 million yr, subsocial orthopteroids gave rise to at least 1 eusocial line. Subsociality may be one and a half to two times older in orthopteroids than in hymenopterans, whereas eusociality may be of equal age in the two groups, though probably younger in bees and wasps than in ants and termites. These figures do not tell us how many subsocial species actually existed in each group across geologic history. For subsociality to account for a 12:1 ratio in appearance of eusocial forms, assuming a 2:1 advantage in time for orthopteroids and no advantage from haplodiploidy for hymenopterans, the Hymenoptera would be expected to have at least 24 times as many subsocial species as orthopteroids. This figure may be accurate for all of geologic time, or even low, but today the Hymenoptera probably have about 100 times as many subsocial species.

Except for not requiring interactions between parents and offspring in defining subsociality, we have not biased the figures against Hamilton's (1964) suggestion; in fact, the opposite is more likely. If, for example, we followed Hamilton (1978; 1980 lecture delivered to the Animal Behaviour Society in Seattle, Washington) and included the parasitic Hymenoptera as possible direct precursors of eusocial forms, we might have expected the Hymenoptera to have evolved eusociality several hundred times as often as the orthopteroids did. Moreover, if we limit our search for subsociality outside the Hymenoptera to the groups that are likely ancestral to termites, we find not 300 cases of subsociality but fewer than 50 actual reported cases.

These calculations are obviously too crude and approximate to be very useful, and no one would have thought to attempt them if the dogma had not been generated that haplodiploidy is sufficient to explain the apparently disproportionate number of origins of eusociality in Hymenoptera. The comparisons just made merely show that there is no empirical evidence that haplodiploidy gave a net advantage to the Hymenoptera in the likelihood of evolving sterile castes and that simply comparing numbers of independent origins of sterile castes does not constitute such evidence. Indeed, the figures just reviewed imply that, to whatever extent haplodiploidy favored the evolution of eusociality in the Hymenoptera, some as yet unknown preadaptations favored the evolution of eusociality in the ancestors of termites. We believe that such preadaptations did exist in the ancestors of termites, and we develop the argument below. First, however, we comment further on haplodiploidy.

Haplodiploidy and Subsociality Outside the Hymenoptera

Because of widespread association of haplodiploidy and subsociality (Borgia 1980), if all haplodiploid and diplodiploid arthropods were considered, haplodiploidy would probably appear to have promoted eusociality even less readily than is implied above. As Hamilton (1967) first pointed out, haplodiploidy occurs in many subsocial mites, beetles, thrips, and other arthropods outside the Hymenoptera (see Andersson 1984, table 1). Indeed, in arthropods, haplodiploidy seems more closely correlated with subsociality than with eusociality. There is a likely reason for this correlation. If siblings live in groups by themselves, as occurs in many parental organisms (one correlate being that otherwise parents are required to evolve ways of avoiding tending someone else's young), they sometimes may have no one to mate with but one another. Again, as Hamilton (1967) showed, when brother-sister matings are the rule (and males are not parental), it pays a female to make only enough males to inseminate her daughters. The haplodiploid female can accomplish this because she controls the sex of each offspring by controlling the fertilization of each egg as it is laid.

As Borgia (1980) noted, the first time a haploid male was produced, it would have been a macromutation, and we might wonder how such a novelty competed initially. In a sibling group (e.g., of a subsocial form), however, such a male would not have to compete with unrelated, normal, diploid males in the population at large, and, as concerns sexual competition, it would tend to have its sisters all to itself.

Therefore, in all animals, subsociality may frequently have led to local mate competition, and vice versa; and local mate competition, whether preceded by subsociality or not, may have facilitated the preserving of haploid males (e.g., in ancestral Hymenoptera). We hypothesize that while such transitions were occurring, subsociality was here and there giving way to eusociality. Haplodiploidy, when present, almost surely contributed to this situation, especially in species with monogamous females.

Why Are Hymenopteran Workers Female, Those of Termites and Naked Mole-Rats of Both Sexes?

By denying that an advantage from closer relatedness among sisters was the principal reason for the evolution of helpers and workers in Hymenoptera, the above arguments leave unanswered why hymenopteran workers are female, whereas those of other eusocial forms are approximately equally divided between the sexes. Hamilton (1964, 1972) suggested an answer for the Hymenoptera (see also Lin and Michener 1972; Alexander 1974; West-Eberhard 1975; Andersson 1984). Throughout the Hymenoptera, with rare exceptions (e.g., Cowan 1978; Eickwort 1981), only females show parental behavior. Only the females, in other words, are subsocial. The first helping at the nest in Hymenoptera was probably done by recently emerged adult females. It would seem that natural selection would have favored females that increased the proportion of females in their broods when such early helping was useful or likely (e.g., in first broods). If, in this manner, males were eliminated by their manipulative mothers from the situation in which helping was reproductive, then they would have had little or no opportunity to evolve the ability to become workers.

Female helping in Hymenoptera must have been promoted by the female hymenopteran's powerful flight and her sting (Alexander 1974; West-Eberhard 1975; Andersson 1984; Starr 1985). Evolutionarily, stings were initially ovipositors, then prey paralyzers, then defensive (and less often prey-carrying) devices (Snodgrass 1935; Evans and West-Eberhard 1970). As special aspects of parental care, they are possessed only by females. Because of the widespread divergence in life spans (senescence patterns) between reproductives and workers in many eusocial forms (see below), we believe that nest defense was a central aspect of early helping behavior. Females of the suborder Apocrita possessed the sting and powerful flight abilities — both presumably evolved in the context of parental behavior, primarily as means of finding, subduing, and transporting food to offspring in safe locations — as well as other parental tendencies and abilities. From the start, females of Apocrita were uniquely equipped to be helpers at the nest, and their mothers were preadapted to perpetuate the sex difference in helping by adjusting the sex ratios of their offspring appropriately.

Kukuk et al. (1989) denied the significance of the sting in the evolution of eusociality, but they accomplished this largely by denying it a function except repulsion of vertebrate predation on eusocial nests, which they argue would have been restricted to large-colony derived forms. If stings were used against arthropods, or against small vertebrates such as mice and shrews, however, then even small-colony forms may have benefited. In any case, one must find an adaptive reason for the maintenance of the female sting as a weapon through whatever stages and times were necessary for the evolution of large-colony eusocial forms, assuming that the initial eusocial forms lived in small colonies (see also Starr 1985, 1989).

Why are the workers of both termites and naked mole-rats composed of both sexes? In ancestral termites and naked mole-rats, the female may have been somewhat more parental than the male or even the sole tending parent. Unlike the larvae and pupae of Hymenoptera, which are tended directly or indirectly until adulthood, the juveniles of termites and naked mole-rats are not helpless for long. They quickly become active and relatively independent (naked mole-rats begin working as small juveniles, about 30 days of age; Jarvis 1981; Lacey and Sherman, chap. 10; Jarvis et al., chap. 12). Moreover, parental care in modern eusocial termites and naked mole-rats, except for nursing and grooming in the latter, is carried out mainly by small (young) animals (see Lacey and Sherman, chap. 10; Jarvis, chap. 13); this surely was not the case in the subsocial ancestors of termites and naked mole-rats. Unlike adult hymenopteran workers, which must have evolved sterility through redirection of already evolved parental abilities, the parental abilities of juvenile termites and naked mole-rats must have evolved concomitantly with the evolution of eusociality or as a part of it. Even if one sex of juvenile termites or naked mole-rats was initially more amenable to the evolution of quasi-parental care, the ancestral termite and naked mole-rat females were not preadapted to adjust the sexes of their offspring easily and quickly to meet changes in the immediate situation, as do hymenopteran females. All of these facts would tend to favor the evolution of more or less equal helper abilities in the two sexes of termites and naked mole-rats.

Why Are Helper Sex Ratios Male-Biased Outside the Hymenoptera?

In diplodiploid species, at least three factors are important in considering likely patterns of altruism between same-sex siblings and between different-sex siblings: (1) sexual (mate) competition, (2) avoidance of deleterious inbreeding, and (3) degree of relationship between the altruist and the assisted offspring relative to the degree of relationship between the altruist and its own offspring (or those of its mate).

Two helper situations are possible in family groups (with one mother): assistance to offspring of siblings or assistance directly to siblings. Sexual competition is greater between siblings (or between parent and adult offspring) when they are of like sex, but deleterious inbreeding can occur only between individuals of different sexes. Thus sexual competition reduces the likelihood of cooperative breeding involving individuals of the same sex, and the risk of inbreeding reduces the likelihood of cooperative breeding involving relatives of different sexes. Because sexual competition is more intense among males, a greater tendency to disperse may be characteristic of females in situations involving a high risk of inbreeding, whereas lowered success in breeding may characterize young adult males. Both factors will tend to produce a male bias among helpers at the nest.

Female vertebrates and insects alike are generally more confident of their parenthood than are males, because a female can usually be more certain that an offspring or an egg came from her body than a male can be that it came from his sperm. (The exceptions are certain externally fertilizing fish and amphibians in which the male is involved more directly than the female in the act of fertilization, and in which, as expected, the male is also more parental than the female [Williams 1966; Alexander 1974].) Therefore, both males and females are, on the average, more closely related to their sister's offspring than to their brother's. Helpers of siblings are most likely to be brothers or sisters of the mother. This bias is most trivial in the case of the ensconced termite king and queen, where the male's confidence of parenthood very likely approaches that of the female; this supposition is reinforced by the presence of nonmotile sperm (Sivinski 1980) and simplified genitalia (Eberhard 1985) in at least some termites. It is difficult, on this basis, to find any reason from kin selection for expecting a bias in the sex ratio of sterile termites.

Let us apply these considerations to the data available for vertebrates, chiefly birds and pack-living canids. The probability of constant association of bird or canine siblings in family groups from hatching or birth to adulthood implies that mechanisms reducing deleterious inbreeding can easily evolve (i.e., individuals ought to be able to recognize siblings as such, if it is important, and to behave appropriately). If so, then sexual competition between sisters might become more important than inbreeding between brothers and sisters in inhibiting helping and close interaction among adults. Moreover, a female's brother should be more willing to invest in her offspring than her sister will be, since, on the average, the brother is less closely related to his mate's offspring than the sister is to her own offspring. The effect is increased whenever a male's ability to sequester a mate and prevent other males' access to her is reduced. This situation is in turn likely whenever the male involved is not a clear dominant or must mate within a group (e.g., a canine pack) in which sexual monopolization of females is difficult or impossible. These facts play a role in the quasi-parental attention shown in some human societies by the mother's brother (Alexander 1974, 1977, 1979; see also Greene 1978; Kurland 1979; Flinn 1981).

Female offspring are also less satisfactory than male offspring as auxiliaries to the reproduction of the original parents. The average relationship of females to their own offspring is greater than their relationship to their mother's offspring because of the possibility of multiple mating and different fathers. This possibility might lead to selection that favors or reinforces monogamy in parents that are evolving to secure an increasing amount of auxiliary parental care from their broods.

All arguments appear to support the notion that in vertebrate families increased parental investment involving nonbreeding adults behaving parentally most often involves a male's rearing of the offspring of his sister or mother. This is true only if one does not include situations in which one of the conditions of a male's becoming a helper is access (even if secondary) to the reproducing female, as in some human families (Berreman 1962) and in Tasmanian native hens (Maynard Smith and Ridpath 1972).

Once a reproductive pattern involving auxiliary parents has been established, plasticity in reproductive rates matching fluctuations in environmental resources may be accomplished in part through variations in clutch and litter sizes without restricting parental care to the actual parents, especially if only groups are able to capture an abundance of game (as in canines) or to defend a territory (as in birds). Apparently, these conditions could lead to gain from the frequent production of broods containing single females (or a small number, depending on the likelihood of mortality and of beneficial pack fission) and several (more) males. This situation has been recorded rather frequently in wild pack-living canines (Estes and Goodard 1967; Lawick-Goodall and Lawick-Goodall 1970; Lawick-Goodall 1971; Mech 1970; Schaller 1972). Males in such circumstances may more often move between packs singly, though females also do so, apparently as a result of being ostracized by other females; and the presence of two or more females in the pack may often be responsible for large packs splitting into two or more smaller packs. Furthermore, the above situation may account for reports that males other than dominants are sometimes the sole breeders in canine packs containing one female and several males (in this hypothesis her brothers) (Murie 1944). Such a male may be an unrelated joiner of the pack, and the other males may benefit by allowing him to father the offspring of their sister as an alternative to inbreeding.

Occasionally sex ratios favoring females might occur if environmental resources fluctuate such that, after a period favoring auxiliary parents and male-biased sex ratios, monogamous breeding is favored. In a male-biased population in which two parents are sufficient, parents should gain by producing female-biased broods. Maturation would have to occur within a season, or predictability of the quality of seasons would have to extend beyond a year.

The model proposed here to explain sex ratios in temporary helpers at the nest among vertebrates does not incorporate the possibility of sex ratios' being affected by local mate competition or direct differential parental investment in the two sexes. Neither does it deal with the difficult question of the effects of parental investment extending beyond the onset of the offspring's reproduction, a virtual certainty in many mammals. Nevertheless, it appears to account for several observations on vertebrates: (1) a preponderance of males serving as auxiliary parents to the offspring of relatives (several birds and canines); (2) male-biased sex ratios (e.g., several birds and canines; also naked mole-rats; see Brett, chap. 4; Jarvis, chap. 13); (3) a high frequency of litters containing one or two females and several males (African hunting dogs; Estes and Goddard 1967; Lawick-Goodall and Lawick-Goodall 1970; Lawick-Goodall 1971; Schaller 1972); (4) increases in male biases in the sex ratio during poor seasons or in dense populations (wolves; Mech 1970); (5) significant female biases in the sex ratio during good times (wolves; Mech 1970); (6) movement of lone females as well as males between packs, even though all males are not breeding (wolves; Mech 1970); and (7) occasional nondominant males siring the offspring of single females in packs containing other more dominant males (wolves; Mech 1970).

If all juveniles passed through a period during which they acted as helpers to their parents, dimorphism between helpers and independent breeders would not necessarily be expected, and such dimorphism may be absent in most or even all cooperatively breeding vertebrates. Dimorphism may yet be discovered among some facultatively cofounding or lone-founding *Polistes* queens, in which smaller individuals might be likely to serve as workers to larger ones except during unusually good years or following unusually high winter mortality, when the number of superior nest sites exceeds the number of surviving queens (West-Eberhard 1969; Gibo 1974; Noonan, unpubl. data). To our knowledge, no one has examined the possibility that vertebrates may have consistently different phenotypes correlating with tendencies to produce their own offspring or to assist other relatives in breeding.

Do Orthopteroids and Vertebrates Have Special Advantages?

Two new questions arise out of our arguments concerning the importance of subsociality to the evolution of eusociality. First, what still undiscovered traits or situations enabled or caused the ancestors of termites to evolve eusociality, given that their prospects appear so poor on the basis of their diplodiploid sex determination, the absence of powerful defensive devices, and the relative rarity of subsociality (compared with Hymenoptera) in their ancestors? Except for the efforts to invoke some parallel to the effects of haplodiploidy (above), this topic has been little discussed. The higher vertebrates are nearly all subsocial; birds and mammals are all parental. The second question, then, is: If special genetic asymmetries are not required, why haven't birds and mammals evolved eusociality repeatedly?

We believe that termites and naked mole-rats had two remarkable advantages over the Hymenoptera in evolving eusociality: (1) their gradual metamorphosis; and (2) the distinctively safe, long-lasting, expansible, and food-rich locations that they began to inhabit. To explain this, we must use still another theory to which Hamilton has been a major contributor (Hamilton 1966), Williams's (1957) pleiotropic theory for the evolution of senescence (for a general review, see Alexander 1987).

Gradual Metamorphosis

It may be supposed that the evolution of eusociality requires merely an overlap between the reproductive life of the mother and the helping ability of the oldest offspring. For extensive or irreversible worker specialization to be advantageous, however, the parent must live long enough to provide opportunities for the helper to use all of its reproductive effort, its whole lifetime, in helping its siblings. If helpers cannot use their whole lifetimes in helping, they should not evolve to be extensively or irreversibly specialized as helpers, but they should retain the ability to become reproductive (adults) quickly and elaborate the tendency to test continually the existing reproductives and their potential replacements. The alternative is that opportunities for some kind of truly remarkable heroism permit the saving of large numbers of more distant relatives; such a situation may have been involved in wasps that have large numbers of outbreeding queens and highly specialized and irreversible worker-soldiers (West-Eberhard, pers. comm.). However, these wasps may merely illustrate the importance of predators in shaping founding by swarms and avoidance of small-sized funnels in colony formation. Predators also affected the specialization of worker-soldiers, which originally evolved as a result of the care of closer relatives.

One possible solution to the dilemma posed above is for a female to produce a large single brood of offspring that could benefit from assistance across a period approximately equivalent to the helping lifetime of the older sibling. This is roughly what happens each season with the north-temperate-zone paper wasp, *Polistes fuscatus* (West-Eberhard 1969; Noonan 1981). Another solution is that the mother could produce successive, smaller broods of offspring that could be helped, as occurs in most modern large-colony eusocial forms (e.g., termites, ants, honey bees); obviously, this possibility has been enhanced by the evolution of relatively longer lifetimes in reproductive individuals. When mothers do not consistently provide siblings throughout the lifetimes of helping offspring, and helpers retain their ability and tendency to reproduce on their own at some point, situations like those existing in cooperatively breeding birds and mammals prevail.

These considerations lead us to hypothesize that organisms with gradual metamorphosis, such as termites, birds, and mammals, have an inherent advantage in evolving eusociality over organisms with complete metamorphosis, such as the Hymenoptera. Gradual metamorphosis means that juveniles more or less resemble adults and change more gradually into the adult form and function. For example, juvenile termites and naked mole-rats, unlike juveniles of the subsocial Hymenoptera, become self-sufficient at early ages. Because they are more nearly active miniatures of the adults, they could start helping younger siblings while they themselves were still immature and improve steadily in helping ability as they matured. They might also need less help than juvenile hymenopterans, although this assumption depends on the kind of help needed (e.g., defending the nest as compared to supplying food) and on the manner and extent of change during the juvenile life (i.e., as sociality advanced, very young termites and naked mole-rats could have evolved to use more assistance, and older juveniles could have evolved to become more independent). The overlap of lifetimes required to favor evolution of functional sterility in helpers is more likely if helping begins in juveniles.

Williams (1957), Medawar (1957), and Hamilton (1966) all argued that senescence in all organisms, including ourselves, occurs because of the accumulation of deleterious gene effects late in life and that this accumulation occurs because selection is less potent later in life. Genes acting later in life affect less of each living individual's reproduction and do not affect at all the reproduction of individuals that have died as a result of accidents, predators, or parasites. Either genes with good early effects and bad later ones, alleles with good early effects and no later ones, or genes with the same phenotypic effects but different reproductive effects across adult life, then, would lead to senescence. Despite such deleterious effects, these sorts of genes would persist unless there were alternative alleles whose effects were sufficiently beneficial throughout adult life for them to outcompete genes beneficial early in life and deleterious later. This is an unlikely possibility, especially in long-lived organisms with complex and sequentially patterned adult lives (for a review of the topic of senescence, including much of the recent literature, see Alexander 1987).

Reproductive effort in the form of helping by juveniles would lower the residual reproductive value of helpers and tend to raise mortality, causing the onset of senescence in the juveniles themselves. The result would be a ballooning of the importance of modified juvenile attributes and an even earlier onset of senescence. This process could continue until the juvenile termite or naked mole-rat had evolved never to reach adulthood under ordinary circumstances. It is significant for this argument that termite workers have frequently been described as permanent juveniles (Kennedy 1947; Wilson 1971) and that juvenile hormone promotes worker differentiation in termites (Luscher 1972, 1977; Wanyonyi 1974).

In contrast, hymenopterans, with complete metamorphosis involving a maggotlike larva followed by an inactive pupal stage, cannot begin helping on a large scale until they have emerged in the adult form. Moreover, even if the larva evolves some helping ability (such as silk production in some ants; Wilson and Hölldobler 1980), it cannot gradually improve such workership during development toward the adult stage as can the nymphal termite juvenile. This means that, compared with termites or naked mole-rats and barring differences in opportunities for heroic nest defense, young hymenopterans gain primarily a slight timing advantage from the early onset of reproductive effort by helping younger siblings rather than by reproducing themselves; even this effect can be significant, emphasizing the importance of the ecological correlates of eusociality (Queller 1989). Although hymenopteran siblings may be needier than termite or naked mole-rat siblings with respect to worker help, a longer period of sustained "parental" effort on the part of helpers would be required for the help to pay off. Thus, in Hymenoptera, selection for the early exertion of reproductive effort (directed toward siblings rather than offspring) would be much less effective than in ancestors of termites in accomplishing intraspecific divergence of life lengths. The divergence is necessary to provide adult offspring with alternatives to independent reproduction that would consistently use all the offspring's reproductive effort.

Sterility is not an all-or-nothing phenomenon. Differing proportions of helpers without offspring may die because they helped; and most individuals in eusocial castes actually have some ability to make their own offspring in special circumstances. For a reproductive to live long enough to enable a helper to use all of its reproductive effort in helping, reproductive phenotypes must evolve to senesce more slowly than worker and soldier phenotypes, leading to an overlap of the reproductives' lives with the helper stages of at least the first individuals to undertake workerlike activities (Wheeler 1928; Evans 1958; Alexander 1974; Breed 1975, 1976). In modern eusocial insects this overlap is often extensive. Short-lived helpers and long-lived reproductives characterize all Isoptera and most modern Hymenoptera. This generality links processes of senescence fundamentally to the evolution of eusociality and helps explain why helpers become more resigned to workership in some social species than in others (see below). It also explains the longstanding observation that when mothers and their offspring occur together in the same nest they do not both produce offspring; instead, the situation evidently always involves matrifilial eusociality (Wheeler 1928; Evans 1958, 1977; Alexander 1974).

In some eusocial forms, queens do not live much (or any) longer than the workers. In some temperate forms, such as *Polistes fuscatus*, queens evidently have not evolved to live through a second winter, and they can make all the eggs for new reproductives by middle or late summer without living much (or any) longer than their first-generation offspring (workers). The workers are left with no option but to assist their mother's reproductive offspring, because they emerge too late to produce adult offspring of their own in time to mate and overwinter (West-Eberhard 1969; Noonan 1981). Why founding females have not evolved to live longer in the bees in which groups of sisters compete (and cooperate) in connection with reproduction in the same nest (Michener 1969, 1974, 1985; Lin and Michener 1972; West-Eberhard 1978a) appears moot.

According to the present model, disruptive selection in effort patterns occurs when parents are able to provide certain of their offspring (at least the firstborn) with opportunities to spend some of their reproductive effort on siblings before they would be able to reproduce on their own without incurring the risks of mating and establishing a new nest. Inclusive-fitness savings in time and a reduced risk of death before reproducing could even compensate for drops in relatedness to juveniles tended by offspring (such as the necessity of tending half siblings or even nieces or nephews). Both patterns of exerting reproductive effort (on offspring and collateral relatives) would persist in offspring (as facultative developmental alternatives), however, because they could be reproductively equivalent at any time, and the relative advantages of the two patterns usually fluctuate with some predictability during the life of the colony.

SAFE OR DEFENSIBLE, LONG-LASTING, INITIALLY SMALL, EXPANSIBLE, FOOD-RICH NEST SITES

In addition to gradual metamorphosis, termites and naked mole-rats have the advantage of a safe niche (microhabitat, nest) from which there is no necessity to exit because food is abundant within the site and because the niche is both long-lasting and expansible to accommodate a growing social group. Thus, many termites live within log fortresses, which are also their food. The nest or niche expands as the termites excavate the log, and they may also locate additional logs by burrowing underground and enhance defensibility by thickening or reinforcing walls with mud. Many species have evolved the ability to construct mud tunnels to additional food sources; some also live underground and forage outside on grasses (evidently secondarily; Wilson 1971). Naked mole-rats live underground, feeding primarily on large tubers, which must be approached and located by digging but which provide continuing food sources that do not require exit from the relative safety of underground tunnels (see Brett, chap. 5). At least in termites, nests typically begin small and, in some cases, can be expanded to accommodate thousands or millions of individuals, with abundant food still available locally.

These conditions are unlike those of virtually all social and solitary (nest-building) Hymenoptera, which must locate and transport food back to the nest, often by flying. We suggest that the peculiar combination of nest-site attributes shared by termites and naked mole-rats represents an important contribution to the likelihood of their evolving eusociality, compared with the Hymenoptera and with cooperatively breeding birds and mammals. For the most part, subterranean mammals either do not have abundant food supplies that can be located and used without emerging from the safety of the underground tunnels, or their food is distributed such that, even if they forage underground, the formation and maintenance of groups larger than a parent and its offspring are inhibited (e.g., moles that feed on insects, earthworms, or small subterranean parts of dispersed plants). Similarly, most birds and nonsubterranean mammals live or nest in locations that either are not defensible across generations or cannot be expanded to accommodate large social groups and still be defensible. A few species, such as hunting dogs, beavers, dwarf mongooses, and hole-nesting birds, produce offspring in relative safety and have evolved ways of moving significant amounts of food back to the den (transport, regurgitation, helper lactation). These are the vertebrate forms that most closely approach eusociality (see also Lacey and Sherman, chap. 10). Presumably, if their niches were expansible and their food supplies sufficiently abundant and localized around the nest site, some of them would have continued to evolve toward large-colony eusociality.

Four conditions can therefore be postulated that might lead to incipient eusociality. All depend on a safe, maintainable, or improvable (and costly or unlikely) nest site. (The third condition assumes monogamy and haplodiploidy; the others assume monogamy but do not require closer relatedness between siblings than between parent and offspring.)

1. Young are produced faster in the incipient eusocial colony even though all or virtually all emigrating nonsocial parents find suitable nest sites and produce viable young. In other words, expanding and improving a particular kind of nest site after it has been located and started is better (for the mother, as manipulator, or for the mother and all participating individual offspring) than distributing descendants among an adequate number of nest sites suitable for the raising of a single brood.

2. Young are produced faster in the incipient eusocial colony, but only because most emigrating nonsocial founders fail to reproduce. In other words, nest sites (or suitable nest sites) are severely limiting (Emlen 1981, 1984; Koenig and Pitelka 1981).

3. Young are not produced faster or saved in higher proportions in the incipiently eusocial colonies, but they are more closely related to helpers than are offspring. Thus, staying home and helping is genetically more profitable than starting a new family if the two alternatives produce the same number of descendants.

4. Young are not produced faster in the incipiently eusocial colony, but they are saved and helped enough to cause their producers to outreproduce noncolonial competitors. In other words, one must imagine that per capita reproduction becomes increasingly effective with three, four, or even up to hundreds of thousands of caretakers (parents and alloparents) as compared with one or two parents.

Nest sites meeting one or more of the above requirements must continue to be safe for multigenerational periods. If new colonies are initiated by individuals or pairs, as in most eusocial forms, nest sites may initially be hidden or inconspicuous or simply not valuable enough as food sources to attract certain kinds of predators. If eusocial colonies continue to increase in size, however, the nest must become physically or behaviorally more defensible because larger colonies of organisms with many juveniles are more attractive and detectable to parasites and predators. Structural defensibility can be enhanced by extending tunnels and making them more complex (enabling flight or delaying predators), minimizing sizes and numbers of openings into the nest, and enhancing the strength of walls. Behavioral defensibility can be enhanced by evolving tendencies and abilities of helpers to ward off attackers and by increasing the numbers of such defenders. Structural and behavioral defensibility can evolve together as access to a nest is restricted to passages defensible by individuals or small numbers of individuals (e.g., the enclosed paper nests of bald-faced hornets) and as individuals evolve increasingly effective defenses (Wilson 1971) for the particular kinds of structures they defend (e.g., enlarged heads and jaws; expellers of toxic substances as in squirt-gun termites, *Nasutitermes*). There is a sense here in which eusociality is indeed a continuation of parental care of offspring hidden or otherwise made safe in a nest.

Most eusocial forms live in the soil. Underground nests can be relatively invulnerable and also difficult to locate. Aside from army ant colonies, (up to 700,000 individuals), the largest eusocial colonies (ants, termites; up to 10 million) either live primarily in the soil or extend their nests into it (Wilson 1971). Moreover, most eusocial forms that maintain nests in the open (primarily wasps) live in the smallest and least permanent colonies. Their relatives with large colonies (e.g., tropical wasps, honey bees, and stingless bees) invariably enclose the nest, either in a cavity or an enveloping structure (West-Eberhard, pers. comm.). In addition, they have evolved the ability to eliminate the small-colony vulnerable stages from their nesting cycle by swarming to found new colonies, and they are particularly aggressive and feared by humans (and probably other vetebrates). Army ants, which are nomadic and fearsome even to large vertebrates, also fission to start new colonies. Fallen tree trunks appear to rank next to soil as nesting sites meeting the above requirements.

Nesting sites that promote eusociality must also be places where a single female can monopolize the production of offspring and the use of helpers during the early stages in the evolution of eusociality. If our scenario emphasizing such origins is appropriate, these requirements appear to rule out locations, such as caves, where multiple safe and proximal sites for single-female or pair nesting prevent such monopolization.

It seems to follow from the argument thus far that small animals are more likely than large ones to evolve eusociality. We speculate that large animals, such as birds and mammals, may not be able to increase the value of logs and tree trunks sufficiently to allow them to evolve eusociality in such places and that nest-site limitations were thus crucial in such forms. Several predictions about vertebrate sociality follow. First, the most nearly eusocial vertebrates should be expected to live in the soil, in large hollow trees or logs, or in constructed dens with similar characteristics (as do beavers). Second, if, for example, giant hollow trees and, say, hole-nesting social woodpeckers or kingfishers coexisted long enough, our argument would predict the evolution of eusociality. Third, if caves typically had structures in them, such as hollow spheres with small openings (spheres that could be expanded), then either birds or bats might have become eusocial.

Many small organisms live in apparently suitable sites yet have not evolved eusociality. Some may have failed to do so because parental care is of little or no value to them. Others, such as subsocial Embioptera, Gryllidae, Dermaptera, Hemiptera, Coleoptera, Scorpionida, and Arachnida that live subsocially in seemingly appropriate sites (but which, for one reason or another, may be too short-lived), may lack the ability to initiate evolution of adequate defense of a nest site or may not have been subsocial long enough. Many of these small forms are semelparous, and it seems obvious that the ancestors of all eusocial forms were iteroparous. Semelparous adults are not likely to improve nesting sites significantly or to create conditions leading their offspring to tarry at the nest. Moreover, even if some offspring did tarry, there would be no younger siblings to help unless the parents were iteroparous.

It may seem that eusociality should evolve much more easily in the tropics, because it is easier to establish there the kind of more or less continuous breeding that accompanies increasing colony size and continued nest defense. The life cycle of temperate insects may usually be so set by the seasons as to make it quite difficult to initiate continuous breeding as an aspect of the initiation of eusociality. This speculation seems to predict that persistent subsociality in the soil and in wood may be more prevalent in temperate regions than in the tropics (when it occurs in the tropics it is more likely to change to eusociality) and that eusocial insects evolved in the tropics. However, the possibility of seasonality yielding the selective situation that would lead to obligate workership in first broods without altering life spans in workers or queens, as described above for *Polistes fuscatus*, represents a counterargument.

Further Comments on Vertebrate Eusociality

It may be an oversimplification to assume that there are no eusocial vertebrates except naked mole-rats (see also Lacey and Sherman, chap. 10). African hunting dogs and wolves live in packs that hunt cooperatively. In some cases, one female and one male have pups, and their offspring from the last season or two help them rear the young, carrying back meat that they regurgitate for the pups and probably protecting them and their parents from some kinds of danger (Lawick-Goodall and Lawick-Goodall 1970; Mech 1970, 1988). Surely, helping in some of these species regularly causes helpers to produce no offspring. But the social groups are smaller than those of the eusocial insects, and there is no evidence yet of morphological divergence of parental and helper phenotypes.

Some cooperatively breeding birds behave like the social canines (Emlen 1984; J. L. Brown 1987) and, possibly, beavers (Wilson 1975), dwarf mongooses (Rood 1978), and naked mole-rats (Jarvis 1981; Lacey and Sherman, chap. 10; Jarvis et al., chap. 12; Faulkes et al., chap. 14). Some of these mammals and birds are similar to some wasps and bees, in which groups are small, phenotypes have diverged little or not at all among castes, obvious competition occurs among potential breeders, and high proportions of helpers seem to be waiting and watching in case they get the chance to breed.

In contrast to mammals, birds would appear to be significantly hampered because they cannot simultaneously expand nest sites to accommodate large numbers of individuals and defend them in stationary locations on a multigenerational basis. They do not possess sting equivalents to deal with the kinds of predators that wasps and bees are able to deter, and, as a consequence, they are not able to construct and use expansible nests equivalent to the exposed paper and mud nests of Hymenoptera.

Helper and parental phenotypes may also have failed to diverge in vertebrates because the jobs of parents and helpers do are very similar. Vertebrate workers may not have the same opportunities as eusocial insects for magnificently reproductive (family-saving) suicidal acts (probably in defense against vertebrates) and the specializations improving the ability to do them (West-Eberhard 1975). Canines probably lack the kinds of predators that could guide such evolution. Birds may have the predators but nothing paralleling the venomous sting of female Hymenoptera. One hymenopteran worker can deter either a huge predator (like a human or a bear) that can destroy its whole family (of hundreds or thousands) in one swipe, or a bumbler that could do it only by accident. By plugging a break in the nest fortress, one termite can also deter a predator. It is more difficult for most vertebrates to be such heroes, though such opportunities may exist for naked mole-rats when predatory snakes enter their burrows (see Jarvis and Bennett chap. 3; Brett, chap. 4; Braude, chap. 6).

Mammalian and avian social groups (other than "selfish herds") never get as big as those of the eusocial insects, and this also restricts the opportunities for superreproductive heroism. The ultimate heroes among eusocial forms are the polistine wasp and honey bee soldier-workers whose barbed stings cannot be extracted, making their attacks on predators irreversibly suicidal. One predicts that barbed stings will be used for defense only in species that form new colonies in swarms, such as honey bees and some tropical wasps. In very small colonies, workers are too valuable for suicidal attacks to be beneficial. The only other barbed stings are those of some ants, which evidently use them to kill prey (A. Mintzer, pers. comm.), and those of the wasp genus *Oxybelus*, which uses them to carry prey (Evans and West-Eberhard 1970); the prediction thus seems to be met.

Another reason why the vertebrate reproductive and worker failed to diverge sufficiently could be the relatively great behavioral plasticity of vertebrates, which reduces the likelihood of the evolution of alternative phenotypes (separate and discontinuous; behavioral, physiological, and/or morphological).

(Environmentally determined alternative phenotypes have evolved thousands of times in insects, not merely in connection with social life, but much more frequently in regard to dispersal in species in short-lived habitats, e.g., the phases of migratory locusts, alary morphs in Orthoptera and Hemiptera, alternative phenotypes in successive generations or on different hosts in aphids.) Assuming that vertebrate helpers at the nest improve the reproduction of their parents or siblings, their failure to evolve sterile castes may result from the absence of long-term predictable fluctuations in the reproductive value of helping versus reproducing directly. Again, the reversible flexibility of the individual vertebrate phenotype may be partly responsible for damping the effective severity of such fluctuations, and the relatively long lives and the iteroparity of vertebrates may have reduced the number of such fluctuations.

Causes and Effects of Queenship: Tracing Probable Changes as Eusociality Evolves

WHY DO SOME OFFSPRING TARRY IN THE PARENTS' NEST?

The point at which offspring leave the parent's care is a dangerous one. It would not be surprising to find offspring sometimes remaining in a parent's proximity after parental care had diminished to virtually nothing, particularly if the parent locates or builds a nest that is somewhat safer than the rest of the world. In other words, if the parent owns a relatively safe nest or home site, then an offspring can prolong parental care merely by remaining there. Even if the parent no longer gives benefits directly to the offspring, merely tolerating its presence increases the offspring's safety from predators. As a result, adults temporarily unable to locate suitable nest sites or mates may profit by spending time at the natal nest.

An adult offspring tarrying in the parent's nest would thereby be in a position to aid the parent in tending younger siblings. Thus, one might expect that helpers at the nest would appear in species with relatively safe nest sites (or the ability to protect offspring that stay nearby), species for which it is often temporarily difficult or dangerous to begin new nests, and especially, species for which both conditions exist.

As many authors have suggested (e.g., Emlen 1981; Koenig and Pitelka 1981; Woolfenden and Fitzpatrick 1984), starting new nests may be difficult or expensive because the habitat is already "filled" with nesting pairs or families. This would be especially likely if safe new nest sites, such as hollow trees, decaying logs, or particular kinds of underground niches, were a scarce resource. For some species, new nests are always expensive because older nests become safer through the efforts of their owners. This alone could create conditions in which helping might pay off genetically for some offspring, specifically those that mature at the opportune times. As argued earlier, improvements in parents' reproductive situations (nest safety, food supply) could make it profitable for older offspring to stay in the natal nest to feed or protect siblings, whether or not they were as closely related as their own offspring, rather than to attempt starting a new nest (Alexander 1974; Andersson 1984). Part of this advantage could come simply from the parents' being able to provide juveniles that can profit from assistance either more quickly or in greater numbers than the newly adult offspring can provide for itself.

How Queenship Begins: Asymmetry in Relatedness of Mother and Daughter to Helped Individuals

Let us try to reconstruct the sequence of steps by which queenship, and therefore eusociality, is initiated in matrifilial societies. When offspring initially start to help at the nest, they may be presumed to be unspecialized for helping and thus to have phenotypes similar to those of their parents. Females can either be inseminated or not. Assuming that at least sometimes they are inseminated and therefore can lay eggs (in Hymenoptera they could lay male-producing eggs even if unmated), what will happen if, say, a mother and daughter are both producing eggs? As Charnov (1978) has suggested, in matrifilial colonies, mothers that suppress their daughters' reproduction—for example, by eating their daughters' eggs (egg eating is a phenomenon commonly observed on wasp nests)—gain over those that do not, because daughters' eggs produce grandchildren that share only one-fourth of the mother's genes, whereas the mother's eggs produce daughters that share half the mother's genes. Therefore, eating of daughters' eggs by mothers is expected to spread.

Daughters that eat their mother's (female-producing) eggs do not gain genetically if their mothers are monogamous, because their mother's eggs produce sisters to the daughters, which, on the average, share half (termites) or three-fourths (Hymenoptera) of the daughters' genes and the daughter's own offspring also share only half of her genes. Sisters that lay eggs, however, produce nieces that share only one-fourth (or three-eighths) of the genes of a potential egg-eating daughter. Therefore, if they can make the distinction, daughters should be expected to eat their sisters' eggs but not their mother's (in haplodiploid forms, this argument applies only to the female-producing eggs of mothers; see Ratnieks 1988).

We can extend the egg-eating example or generalize from it: Mothers should evolve to prevent their daughters from attaining reproductive maturity or reproductive condition. Prevention could include a variety of activities, such as suppressing hormone production or interfering with the daughters' likelihood of being inseminated. Daughters, on the other hand, are expected to be passive about becoming reproductive, so long as their mothers have throughout history been monogamous, their sisters are not likely to become reproductive, and their mother is providing them with all the siblings they are able to tend and is likely to do so for the daughters' entire adult lives (i.e., the mother gives evidence of being healthy and vigorous).

One expects, then, an asymmetry in the behavior of mother and daughter from the beginning. The mother is expected to prevent her daughter from producing offspring, and the daughter is not expected to resist. The same mother-daughter asymmetry prevails under both haplodiploidy and diplodiploidy, and with respect to fathers and sons under diplodiploidy.

The mother is also expected to resist taking on risky tasks that might cause her own death and leave her daughter in charge of the nest, because the daughter can only produce offspring half as much like the mother as her own offspring. Mothers thus gain from avoiding dangerous tasks that can be assumed by daughters, tasks like foraging and defending the nest. Before any specialization of mothers as offspring-producers (that do not defend the nest or forage) and of daughters as workers or soldiers (i.e., when daughters are potentially just as reproductive as their mothers), daughters (under diplodiploidy) presumably have the same interest in avoiding dangerous tasks as the mother. Once the slightest difference between mothers and daughters has appeared — even if it is only a matter of individual experience that makes the daughter slightly better at foraging or defense or the mother slightly better at egg production — the daughters are expected to be immediately more willing to undertake riskier tasks than the mother. They should explicitly be more willing to undertake such tasks when doing so decreases the risk to the mother. The first reason for risk taking is that the mother has now become a better producer of juveniles that share half (or more) of the daughter's genes than the daughter herself. She can even continue this activity after the daughter is dead, should the daughter lose her life protecting the mother. Second, the daughter does not gain from protecting some of her sisters if she and her mother both lose their lives as a result (as opposed to the daughter's losing her life protecting her mother while her sisters also lose theirs). This is true because if the daughter and mother both die and the colony lives on, from a nonreproductive daughter's viewpoint, its offspring will be nieces produced by sisters. The survival of a worker's mother is more important than that of her sisters. Although workers in a eusocial colony appear mainly to be tending siblings, their primary duty, other than defending the nest, is evidently to protect their mother, the queen.

The above asymmetry presumably begins because mother-daughter teams that assume the above relationships to one another reproduce more effectively than those assuming symmetrical or other relationships. Presumably, mother-daughter teams do not form except when the pair can outreproduce other mother-daughter pairs that breed independently of one another. The second general part of this model (below) attempts to identify situations in which this condition prevails.

REASONS FOR DIVERGENCE OF PHENOTYPES OF
MOTHER AND HELPER

Offspring taking up defensive or foraging activities on behalf of sibs in their parental nests before leaving to reproduce independently experience an earlier onset of reproduction (by helping sibs). They may also experience higher mortality from extrinsic causes immediately after the onset of reproduction than do their contemporary siblings that leave to reproduce independently as soon as they mature. These two important parameters help shape senescence by natural selection because mortality rates affect the potency of selection across lifetimes (Williams 1957; Hamilton 1966; Alexander 1987). The differences in the two parameters for helping and nonhelping offspring would tend to accelerate the senescence of helper phenotypes compared to reproductive ones and thereby diminish the importance of any direct reproduction that helpers achieve later after they leave their parental nests. The self-aggravating nature of senescence would cause the lifetimes of workers to continue to diverge through shortening of worker lifetimes. At the same time, the longevity of reproductives increases if workers consistently assume the riskier parental duties. The reproductives are then freed to use more completely durable, defensible nest sites, and the durability of these sites may even be extended as a result of worker labor and defense. In turn, the benefits to helpers of staying at the parental nest and exerting even more reproductive effort on behalf of siblings would be enhanced. The positive feedback just described potentially can cause divergence of helper and reproductive phenotypes to the point at which the reproductive lifetimes of the helper and its parent overlap completely, and direct reproduction later in life becomes so negligible that helpers gain reproductively by becoming effectively or even obligately sterile. This divergence can occur even if lifetimes of both helpers and reproductives are lengthened as a result of a shifting of colonies into safer locations (burrows, logs) as sociality evolves.

These arguments, and those given earlier regarding the evolution of senescence, mean that, in a cooperatively breeding species, even slight divergence between mother and daughter with respect to ability to lay eggs and help at the nest, respectively, will in many situations set into motion a continuing selection for divergence in their phenotypes. The higher mortalities of helpers and the correlative lower mortalities of mothers will lead to differences in their senescence patterns, and the greater the divergence between the two kinds of life patterns, the more effective will be the selection for divergence. For example, when the mother has evolved to be somewhat less than twice as good as her daughters at reproduction, leaving aside the daughters' ability to forage and defend as compared to the mother's, the daughters would be expected to be indifferent about replacing even a promiscuous mother (although they are not indifferent about whether or not she is promiscuous). More precisely, this situation would occur when the mother and daughter team of baby-producer and worker is twice as good as the mother and daughter operating separately or with their roles reversed. The limits on divergence between queen or worker phenotypes will be set by the point at which optimal helping and optimal queenship are achieved, given particular extrinsic conditions and the relationship of queenship and helping to one another in the particular kind of social life involved.

Under complete metamorphosis, the hymenopteran mother, in contrast to termite and naked mole-rat mothers, appears to be able initially to give to her helping daughter only a small timing advantage, since the daughter cannot help until she has emerged in adult form. Even if reproductive female hymenopterans were always monogamous so that their incipient worker offspring were more closely related to sisters than to their own offspring, how could the hymenopteran mother evolve a lifetime long enough to use all of any of her offsprings' reproductive effort? All social Hymenoptera forage for their food and the food of their offspring outside the safety of the nest. Many fly during this foraging, and many have stings and aggressively defend the (often exposed) nest. Hymenopteran helpers thus tend not to be soldiers and workers but foraging soldier-workers. Soldiering is presumably the most dangerous task of all.

Naked mole-rats and termites, in contrast, protect themselves by living underground or inside their food (wood), which itself presents a barrier to predators, and all have relatively nonaggressive workers (and, in termites, separate soldier castes). Some termites that dwell and forage underground lack a soldier caste (Sands 1972), although young termite colonies tend to invest earlier and more heavily in soldiers than do ant colonies (Krishna and Weesner 1969; Haverty 1977).

Both long-distance foraging and aggressive nest defense are high-risk tasks, and specializations for their efficient execution should lead to rapid senescence in worker phenotypes because of the effect of high mortality rates on reproductive value (see also Oster and Wilson 1978). At the same time, the relief of the hymenopteran parent from these same high-risk duties would lower mortality and select for long life in the reproductive phenotype. The same divergence has obviously occurred in termites, but it involved replacing the adult stage in evolving workers by extending juvenile stages, including stationary molts in lower termites; reducing the numbers of juvenile instars in higher termites; and "workerizing" effects of juvenile hormone, in contrast to effects of juvenile hormone that bias morphology and physiology toward the reproductive stage in hymenopterans (Luscher 1977).

The above discussion still leaves bothersome questions. If the exposed nests of Hymenoptera tended to cause shortening of worker-soldier lifetimes, and the protected nests of termites and naked mole-rats tended to inhibit divergences of lifetimes in reproductives and workers because of lowered risk to workers, then how have the astonishing differences in life spans of termite workers and queens come about, and why is there as yet no evidence of divergences of life spans in naked mole-rats?

It seems necessary to postulate that, even though termite nests are relatively safe from many or most kinds of predators, certain threats have allowed highly reproductive heroic acts by helpers. Perhaps such acts involve primarily arthropod predators and include soldiers placing their heads in openings in the nest (as with nasutes) or workers repairing a nest break. Single individuals or small numbers might thereby save the entire colony from arthropod invaders. Perhaps naked mole-rats have indeed evolved divergent life spans between breeding females and workers. We still do not have enough data to show this effect, although the continual monitoring behavior and high levels of activity and aggression of the breeding female (Reeve and Sherman, chap. 11) suggest otherwise. More likely, perhaps, naked mole-rats have consistently lacked the kinds of predators that caused situations in which dramatic heroism could be repeated, yielding dramatically shortened life spans in workers. It is important that heroism here be understood as involving a high likelihood of mortality in members of one caste as a result of acts that have a high likelihood of protecting members of another caste.

If the above arguments concerning the importance of early reproductive effort and senescence in the evolution of workers are correct, then certain general predictions about queen and worker life spans are possible. Hymenopteran females undertake considerable risks in hunting and subduing prey. The evolution of eusociality in such forms should involve increased risk of extrinsic mortality to helpers and decreased risk to reproductives, leading to a reduction in worker life span and an increase in the life span of the queen compared with those of solitary relatives. Maximum life spans of eusocial hymenopteran queens (determined in the laboratory) are much greater than those of solitary relatives (Wilson 1971), whereas those of workers are generally similar (in ants) or shorter. Termites invading sound wood from habitats under bark filled with competitors and predators (Hamilton 1978) probably experienced a reduction in extrinsic mortality rates of both helpers and reproductives. As would be predicted, the life spans of termite kings and queens are much longer than the adult life spans of cockroach species, the longest known of which is that of the semelparous subsocial American woodroach, *Cryptocercus punctulatus*, which seems to tend its brood somewhat longer than 3 yr (see Nalepa 1988). Life spans of termite workers and pseudergates (so-called false workers) are probably slightly longer than those of most cockroaches.

The longest-lived workers among termites should be found in the wood-eaters inhabiting large logs, the shortest in grass-eating forms with foraging workers (cf. wood-dwelling *Kalotermes*, *Neotermes*, and *Reticulitermes* with *Mastotermes* and other species that have open-foraging workers; see Oster and Wilson 1978). A problem with this comparison is that the species with open-foraging workers generally inhabit large, complex colonies that are notoriously difficult to maintain in the laboratory. Therefore, the recorded maximum life spans for these species are more likely to be underestimates than are those for the wood-dwelling species. Among eusocial hymenopterans, the workers of flying wasps and bees should have the shortest life spans, especially when nests are exposed and the workers are highly specialized for defense (e.g., honey bees, which have barbed stings). Comparisons among different ant species should parallel those predicted for termites. It might be expected that the least aggressive, most helpless, most protected reproductives (e.g., those that found new colonies by fission or are otherwise "claustral" as opposed to foraging ant queens) should be the longest-lived reproductives. In fact, claustral lone foundresses are not longer-lived than foraging lone foundresses. Foraging by ant queens, usually confined to the period when the first workers are raised, may have little effect on their senescence rates because the initial workers have come to represent the queen's somatic rather than reproductive effort (Lin and Michener 1972; West-Eberhard 1975) and senescence begins only after the onset of exerting reproductive effort (Williams 1957; Hamilton 1966). The same problem applies to a comparison of claustral queens founding colonies with swarms founding colonies.

SENESCENCE AND DEVELOPMENTAL PATHWAYS IN TERMITES

Schedules of extrinsic mortality associated with different tasks, through their effects on life patterns of reproductive effort, should influence developmental pathways in termites in which flexibility is possible through larval, stationary, and nymphal molts (castes often vary with instars in termites; Wheeler 1928; Oster and Wilson 1978). Thus, among termites, workers in wood-dwelling species should be most capable of becoming soldiers because they are more apt than open-foraging workers to have long, somewhat indefinite life spans (Miller 1969). Soldiers in all termite species should be less apt to become workers than vice versa (Hewitt et al. 1969; Miller 1969; Noirot 1969; Wilson 1971; Watson and Abbey 1977). According to Gay (1968), soldier-nymph intercastes are more common than worker-nymph intercastes, but both are aberrations and exceedingly rare. Similarly, the ability of workers (or pseudergates) to molt into reproductives (neotenics or imagos) after functioning as workers should be reduced or suppressed in species whose foragers experience high extrinsic mortality. In fact, workers (pseudergates) have the capability of molting into reproductives only in lower termites, which do not forage openly, and soldiers in these species apparently lack the capability of becoming reproductives (Wilson 1971). A parallel in ants occurs in some species in which workers have rudimentary spermathecae and soldiers do not (Wheeler 1910). Recent views suggest that determinate workers may be associated with exposed foraging in both lower and higher termites (Hewitt et al. 1969; Noirot

1969; Watson et al. 1977), and indeterminate neuter castes may be restricted to the wood-dwelling Kalotermitidae (Watson et al. 1977).

REASON FOR AGGRESSION BETWEEN MOTHER AND HELPERS IN SMALL-COLONY EUSOCIAL FORMS

Eusocial species may be divided into small-colony forms (up to ca. 1,000 individuals) and large-colony forms (from ca. 1,000 individuals to several million). In small-colony forms (e.g., sweat bees, allodapine carpenter bees, bumblebees, some paper wasps, and naked mole-rats), the queen is the most active and aggressive individual in the colony; we believe she is continually monitoring the reproductive status of others and defending her right to produce offspring. (Reeve and Sherman [chap. 11] believe that queen aggression in small-colony forms may also incite workers to greater activity.) In large-colony forms (e.g., termites, honey bees, stingless bees, most ants), the queen may be passive, or even helpless, with workers not only tending her but coming to her for the pheromones that determine their nature as sterile helpers. In small-colony forms, castes are typically undifferentiated in morphology, physiology, or life span; in large-colony forms, there is typically dramatic differentiation in all these regards. In small-colony forms, the queen appears to control production of offspring, including sex ratios; in large-colony forms, workers sometimes control the proportion of males produced and may even produce the males. Because this division into two main categories does not appear to have been made previously, we do not know how many colonies lie between these extremes, nor have we yet attempted to describe the combinations of the above conditions expected in colonies of intermediate sizes. Wilson (1971) divided eusocial colonies into four sizes and discussed primarily degrees of caste differentiation and queen dominance.

From the outset, it is important to a daughter that the mother remain reproductive and healthy. If the mother is failing in these regards, then the daughter gains from assuming the role of reproductive female herself, without a delay that reduces her overall reproduction. Moreover, when more than one daughter is present, it is important to each daughter that she, rather than one of her sisters, be the replacement for her mother when and if the mother's reproductive powers wane. Accordingly, one expects daughters to monitor their mother more or less continually and retain the ability to shift rapidly into a reproductive mode, because the juveniles produced and reared by the new reproductive are twice as much like her as the offspring she would have reared if her sister had assumed the reproductive role.

So long as the colony remains small and the life spans of reproductive individuals and helpers are not very different, then helpers may be expected to resist evolutionary specialization as workers that significantly reduces or eliminates their ability to replace their mother rapidly. In small colonies, each helper has a relatively high chance of being the replacing reproductive if the mother dies, and if life spans are more or less the same between reproductive and helpers, the likelihood of a replacement of the reproductive during each helper's lifetime remains high.

As a corollary of the above, in small-colony eusocial forms, reproductive individuals (mothers) may be expected both to monitor their daughter-helpers' reproductive condition more or less continually and to demonstrate repeatedly their likelihood of remaining healthy and reproductive. In other words, one expects evidence of aggression, but not all-out or damaging aggression, between mothers and their helper or worker-soldier daughters in an incipiently eusocial colony from the start. This appears to be a good description of interactions between the breeding female and other individuals in a naked mole-rat colony (e.g., Isil 1983; Reeve and Sherman, chap. 11; Jarvis, chap. 13). Sisters, however, may be expected to fight, even until one is killed, for control of a nest in which the mother has died (e.g., Lacey and Sherman, chap. 10).

REASON FOR ABSENCE OF AGGRESSION BETWEEN MOTHER AND HELPER, AND CONTINUED DIVERGENCE OF PHENOTYPES, IN LARGE-COLONY EUSOCIAL FORMS

As eusocial colonies become large and long-lasting, mothers become increasingly specialized as producers of young, and offspring as workers and soldiers. Mothers become increasingly capable of providing larger numbers of helpers with sufficient numbers of close relatives to tend and of doing so for longer periods. As this happens, daughters have fewer opportunities to become replacement reproductives, both because their mother lives longer and because so many other individuals are available to replace her; the identity of a queen's replacement thus becomes a sweepstakes, with each individual having a chance of, say, one in a million, or even less.

The changes with colony size can reduce or eventually eliminate the monitoring of helpers by reproductives and cause the reproductive female to specialize so thoroughly in offspring production that she becomes increasingly dependent upon her helper offspring. One result is that the nonreproductive castes in large-colony forms, far from making any effort to escape the pheromones and other influences that cause them to remain nonreproductive, now exert effort to obtain the pheromone that causes them to take the worker or soldier route of development. This is true in part because an offspring with little likelihood of becoming a reproductive when changeovers occur gains from causing or enabling its mother to continue as the reproductive rather than being replaced by a sister of the offspring. Eusocial castes specializing extensively or irreversibly as workers or soldiers are therefore expected to occur only in large-colony eusocial forms.

REASON FOR WORKER CONTROL WHEN INTERESTS OF
QUEEN AND WORKER CONFLICT IN
LARGE-COLONY EUSOCIAL FORMS

In large-colony forms following the specialization of workers and soldiers and the corresponding relaxation of aggressive control over the reproductive states of workers and soldiers, workers can gain control of sex ratios and other features of colony life in which their interests differ from those of the queen. Some reproductive abilities will be retained by workers during the evolution to large-colony status because they are valuable to the queen. Thus, in Hymenoptera, the ability of workers to make males parthenogenetically is valuable to the queen if she dies and cannot be replaced. In termites, the ability of workers to become supplementary reproductives in parts of the tunnel system distant from the royal pair can be useful to the royal pair itself. Reproductives maintain relaxed control over the activities of workers because, in general, workers in large colonies do not gain by competing with reproductives. This lack of control can, however, also lead to easy and simple ways for nonreproductive castes to act in their own interests contrary to the interests of the reproductives. This argument predicts that worker-queen conflict will be restricted to certain activities, such as producing males in Hymenoptera (Bourke 1988) and establishing peripheral colonies by fission in termites.

Because first helper castes may, in evolutionary terms, have become the somatic effort of the queen, the queen has evolved to grow and increase in reproductive value, sometimes dramatically (in termites and ants), after adulthood and after colony foundation. Such changes in the queen can be considered one of the ways in which a parent colony can be improved as a reproductive resource from the viewpoint of offspring that are deciding whether to emigrate or to become helpers.

Summary

Darwin solved the general problem of worker sterility in eusocial forms by noting that if the trait of sterility can be carried without being expressed, and if those who express it sufficiently help those who carry it but do not express it, then the trait can spread by natural selection. W. D. Hamilton developed the idea of reproduction via collateral relatives and noted a relationship among haplodiploidy, closer relationships between full-sisters than between mothers and daughters, workers being restricted to females, and multiple origins of eusociality in the Hymenoptera.

We believe that the favorable combination of traits and circumstances that enabled (or caused) termites and naked mole-rats to evolve eusociality included gradual metamorphosis, subsociality (extensive parental care), and life in long-lasting, expansible niches (nests or microhabitats) safe from predation and rich with food that does not require exiting the safety of the niche to obtain it. In Hymenoptera haplodiploidy complements the preadaptations of widespread subsociality, powerful flight, and the presence of stings in females. It may have helped to overcome the disadvantages of complete metamorphosis for the evolution of eusociality: juveniles (larvae) are poor worker prospects, and the intervening pupal stage and the extensiveness of morphological and physiological transformation prevent gradual improvement in juvenile ability to help during development toward adulthood. Except for eusocial clones (e.g., aphids), eusociality is apparently always preceded evolutionarily by subsociality (extensive parental care); subsociality is at least as concentrated in the Hymenoptera as are independent origins of eusociality.

The sting is also regarded as instrumental in enabling Hymenoptera to evolve eusociality repeatedly in exposed locations (vulnerable to predators). Hymenoptera nesting in the open were able to create expanding and more or less permanent (usually paper) nests. Termites and naked mole-rats depended on locating abundant food in the vicinity of hidden or fortresslike, long-lasting, and expansible nest sites (fallen trees, patches of underground tubers).

Also contributing to hymenopteran eusociality were the powerful flight capabilities of the order Apocrita, derived from a history of carrying food to ensconced juveniles and useful for bringing food to young and for building and defending nests. Eusocial Hymenoptera that moved into niches resembling those of termites and naked-mole rats (except that Hymenoptera usually forage externally) have in some cases (ants and a few bees) lost both flight and the sting.

High risks associated with external foraging and defending exposed nest sites (probably from vertebrate predators) may have contributed to the evolution of eusociality in some Hymenoptera by causing disruptive selection on life-effort patterns in reproductives and workers (actually worker-soldiers), leading to life-span differences that permitted irreversible worker specialization. Incipient queens thus became long-lived enough to supply dependent juvenile relatives throughout the lives of their incipient worker offspring (seasonality can permit such ability without life-span differences). Other dangers (probably arthropodan), reducible through parental care and helping, must also have occurred in the microhabitats of termites (and probably ants) to bring about divergence in life lengths and the evolution of soldiers. Consistent with expectations from senescence theory, life spans of workers and soldiers in most eusocial Hymenoptera, under relatively high rates of mortality, have probably shortened, compared with noneusocial members of related groups; conversely, life spans of all castes of termites, naked mole-rats, and some ants have probably lengthened.

Birds and mammals are like termites in metamorphosis and parental care but less like them in ability to locate safe or defensible, expansible, permanent nest sites with ample food for large colonies. Apparently as a result, cooperatively breeding vertebrates have been halted evolutionarily in relatively small groups, usually with temporary helpers, and evidently always before divergence of phenotypes between reproductives and helpers.

Cooperatively breeding vertebrates resemble small-colony eusocial forms (those with up to several hundred individuals, e.g., *Polistes* wasps, halictine bees, naked mole-rats), in which the breeding female and male (female only, in the Hymenoptera) are the most active and aggressive members of the colony. Reproductives in such colonies seem required to defend their positions by monitoring the reproductive states of offspring and demonstrating their own reproductive health (through vigor and dominance), presumably because in small colonies it is profitable for each worker to retain the ability to replace swiftly an inadequate reproductive individual. In colonies with thousands or millions of individuals, replacement of reproductives becomes a sweepstakes, with the evident correlate that worker castes tend to specialize completely as helpers and reproductives become passive and dependent. Phenotypic divergence of reproductives and other castes, especially in life spans, is virtually restricted to large-colony forms. Divergence of life spans, in particular, would appear to exaggerate and perpetuate the differences in selective regimes between small-colony and large-colony eusocial forms because long-lived queens are unlikely to be replaced within the life span of any individual short-lived worker or soldier.

Acknowledgments

This paper was begun about 1975; the arguments on subsociality and gradual metamorphosis were developed by early 1976, when R.D.A.'s construction of a hypothetical eusocial mammal as a subterranean tropical rodent feeding on underground tubers caused T. A. Vaughan of Northern Arizona University to tell him that this hypothetical species was "a perfect description" of naked mole-rats. That conversation led to subsequent correspondence between J.U.M. Jarvis and R.D.A. and initiated the cooperation leading to the current volume (see the Preface).

Arguments about nest sites and their significance in relation to senescence and heroism were initiated in 1975 and developed considerably during a graduate seminar on eusociality held at Michigan in 1983. Materials on naked mole-rats, and the comparisons of small- and large-colony eusociality, were added in 1984–1985. Some of the arguments in the early part of the paper appear in the excellent review article of Andersson (1984). We have attempted to avoid repetition yet retain enough of the general arguments to provide over-all coherence.

For criticisms of the manuscript, we are grateful to S. H. Braude, G. C. Eickwort, J. Herbers, P. Pamilo, D. Queller, T. D. Seeley, P. W. Sherman, J. E. Strassmann, and, especially, M. J. West-Eberhard.

2 Systematics and Evolution of the Family Bathyergidae

Rodney L. Honeycutt, Marc W. Allard, Scott V. Edwards, and Duane A. Schlitter

The mole-rat family Bathyergidae has an exclusively African history dating to the early Miocene. Taxa in this family are endemic to sub-Saharan Africa with at least three species in two genera (*Georychus* and *Bathyergus*) restricted to South Africa, two monotypic genera (*Heterocephalus* and *Heliophobius*) restricted to eastern Africa, and one genus (*Cryptomys*) comprising seven species having a broader distribution in western, eastern, and southern Africa. The Bathyergidae is not to be confused with the Spalacidae; this latter family of mole-rats, comprising a single genus (*Spalax*), occurs in the eastern Mediterranean region, eastern Europe, and southern Russia (see Jarvis and Bennett, chap. 3). The systematics and evolutionary biology of the Bathyergidae relative to other rodent families has been debated for more than 80 years, resulting in many different classification schemes. Even within the family, information pertaining to the relationships among genera and the overall taxonomy is incomplete.

From an evolutionary standpoint, the Bathyergidae is an interesting group in several respects. First, the family represents an early rodent radiation in Africa, and its distribution and endemism suggest that this radiation was complex. Second, as discussed in detail in the following chapter, population structure within the family ranges from solitary to the highly structured social system of the naked mole-rat (Jarvis and Bennett, chap. 3). Third, the family is unique among the Rodentia in exhibiting several combinations of traits that confuse their placement within the order. Systematic and evolutionary studies on this family should therefore prove enlightening with respect to the historical biogeography of Africa, the evolution of eusociality, and the classification and patterns of morphological evolution in rodents.

Here we provide an overview of bathyergid systematics and evolution by discussing (1) the relationship of the Bathyergidae to other rodent families, (2) bathyergid taxonomy, (3) the relationships of genera and species within the family, and (4) the zoogeography and paleontological history of the family. In the next chapter, Jarvis and Bennett provide an overview of the behavior and ecology of the Bathyergidae.

Interfamilial Relationships and Classification

The order Rodentia is divided into 32 families containing almost half of all living species of mammals. Rodents show considerable diversity in form, habitats, behaviors, and life histories. The primary characters uniting the Rodentia as a monophyletic group reflect specializations in the masticatory apparatus (incisors, cheek teeth, and musculoskeletal features of the jaw and skull). Naturally occurring variations in the masticatory apparatus have been used to establish key characters that define major groups within the order, and these are the basis for most existing classifications. The key characters, however, may have evolved several times within the higher taxa of rodents (Luckett 1985; Wood 1985). Therefore, rodent taxa may share similarities as a result of convergent or parallel evolution rather than shared common ancestry. For this reason, there is confusion about the taxonomic placement of some groups.

It is particularly difficult to place the Bathyergidae relative to other families within the order Rodentia because their masticatory system cannot be unequivocally related to any other single taxonomic group. A historical overview of classifications (De Graaff 1979) indicates that the bathyergids have been variously placed in each of the three suborders of rodents, as well as in their own suborder, the Bathyergomorpha.

Traditional approaches to the classification of the Rodentia recognize three suborders: Hystricomorpha, Sciuromorpha, and Myomorpha (Brandt 1855; Simpson 1945). The primary feature that defines these three groups is where the masseter muscles originate and insert relative to the infraorbital foramen, zygomatic arch, and rostrum. The three suborders are (see fig. 2-1; Wood 1985): (1) Hystricomorpha: hystricomorphous with the anterior masseter medialis muscle originating largely on the rostrum, passing through an enlarged infraorbital foramen, and descending to insert on the lower jaw; (2) Sciuromorpha: sciuromorphous with the masseter lateralis muscle arising on the rostrum in front of the orbit or zygomatic arch and descending to insert on the lower jaw; the infraorbital foramen is small, with no muscle passing through it; (3) Myomorpha: the masseter medialis originates on the rostrum and passes through a narrow infraorbital foramen; anterior masseter superficialis originates on rostrum and masseter lateralis on zygomatic arch.

Unfortunately, bathyergids do not fit well into any of these three subdivisions. The reason for uncertainty is that in most bathyergids the infraorbital foramen is small with little or no evidence of penetration by the anterior masseter medialis muscle. Nevertheless, studies of the fossil genus *Bathyergoides* and the extant genus *Cryptomys*, as well as recent ontogenetic studies of the infraoribital muscles, indicate that penetration of the foramen by the masseter muscle does occur in certain taxa (Wood 1985; Maier and Schrenk 1987). This suggests a somewhat closer affinity of bathyergids to the Hystricomorpha. If

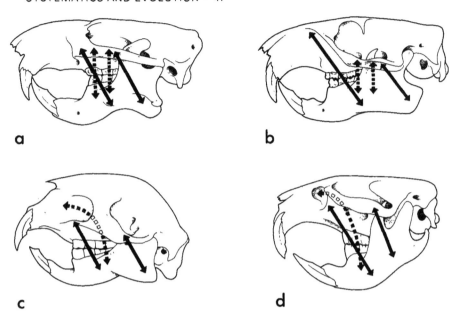

Fig 2-1 Diagrammatic representation of the origin and insertion of masseter muscles relative to the zygomatic arch, infraorbital foramen, and lower jaw in mammals a, Protrogomorphous (e g , *Aplodontia*), with masseter muscles originating on the zygomatic arch b, Sciuromorpha (*Marmota*); c, Hystricomorpha (*Hystrix*); d, Myomorpha (*Ondatra*). *Dashed arrows*, The internal and anterior masseter medialis muscles *Solid arrows*, The external masseter superficialis and masseter lateralis muscles

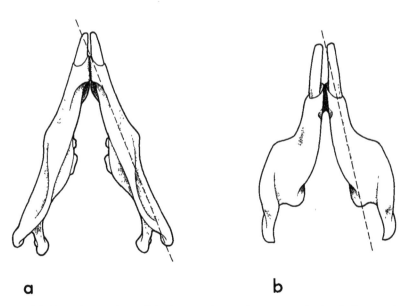

Fig 2-2. Ventral view of mandible showing the angular process, the key feature used by Tullberg (1899) to classify rodents into two major groups: a, Sciurognathi (e.g., *Marmota*); b, Hystricognathi (e.g., *Bathyergus*).

bathyergids are hystricomorphs, however, the protrogomorphous condition in most taxa (i.e., reduced infraorbital foramen and no penetration by the masseters) either reflects a reversal from a hystricomorphous state (Luckett and Hartenberger 1985) or hystricomorphy arising several times independently in different lineages (Wood 1985).

Another feature of the masticatory system that has been used in rodent classification is the angle of the jaw relative to the plane of the incisors. Two types of jaw angulation occur in rodents, hystricognathy and sciurognathy (see fig. 2-2; Tullberg 1899). In sciurognathy the angle arises from the ventral side of the horizontal ramus in the plane of the incisors, whereas in hystricognathy the angle arises from the lateral side of the mandible (Wood 1985). The Bathyergidae have a highly flared angle on the mandible, suggesting that they are more similar to the Hystricognathi than to the Sciurognathi (fig. 2-2).

Although bathyergids are not strictly hystricomorphic (the infraorbital foramen has been reduced), the consensus among most recent systematists is that the Bathyergidae belongs in the rodent suborder Hystricognathi, which also includes the Caviomorpha (cavylike rodents, e.g., New World porcupines, guinea pigs, and chinchillas), Hystricidae (Old World porcupines), and Thryonomyidae (cane rats). Monophyly of the Hystricognathi is supported by several shared, derived characters including the hystricognathous mandible, several features of the fetal membrane, the sacculus urethralis, a fused malleus and incus, an anterior opening in the pterygoid fossa, multiserial incisor enamel, the loss of the internal carotid artery, and albumin immunology (Luckett 1985; Luckett and Hartenberger 1985; Sarich 1985; Wood 1985). Despite hystricognathid monophyly, however, there is no clear evidence establishing a sister-group relationship between the Bathyergidae and any single lineage within the Hystricognathi.

Intrafamilial Relationships and Systematics

Beyond the establishment of the family Bathyergidae as a monophyletic group, there are no detailed systematic treatments of the family. Monophyly is substantiated morphologically, with all five genera sharing the following three synapomorphies (figs. 2-1, 2-2; Woods 1975; Wood 1985): (1) the highly flared angle of the lower jaw; (2) structures of the hyoid, laryngeal, and pharyngeal region; and (3) protrogomorphy (i.e., a secondarily reduced infraorbital foramen relative to the hystricomorphs, which have an enlarged infraorbital foramen).

The five genera of bathyergids have been grouped into two subfamilies (Roberts 1951; De Graaff 1981). The subfamily Bathyerginae includes a single genus, *Bathyergus*, with two species. Characteristics of the subfamily include grooved upper incisors and ungrooved lower incisors, upper incisors with roots originating above the molars, enlarged, flared, and posteriorly extending angular processes of the mandible; enlarged forefeet and claws, and large body size (e.g., old adults may reach 1,800 g). The second subfamily, the Georychinae, includes four genera: *Cryptomys, Georychus, Heliophobius,* and *Heterocephalus* (De Graaff 1981). This subfamily is characterized by having ungrooved incisors, upper incisors with roots originating behind the molars in the pterygoid region of the skull, less directly posterior orientation of the small angular process of the mandible, unenlarged forefeet and claws, and small body size (e.g., old adults are not more than 400 g).

Intrageneric Relationships

The most thorough systematic accounts of the Bathyergidae deal with individual genera (e.g., Roberts 1951; Ellerman et al. 1953; De Graaff 1981; Smithers 1983); surprisingly, even at this level, the taxonomy is unclear. Here we present a summary of the taxa, as they are now thought to be construed. To help clarify the current taxonomy, we have included synonymies for each mole-rat species. (Complete citations for these synonymies are found in Allen [1939] and De Graaff [1981] and are not repeated in this volume.)

CRYPTOMYS GRAY, 1864

The genus *Cryptomys* (common mole-rats) is distributed from Ghana to the extreme southwestern Cape Province of South Africa (fig. 2-3). This genus has long been a problematic group for systematists, primarily because extreme variation in external and cranial morphology makes it difficult to determine if the 49 named forms in Africa are actually species, subspecies, or local variants (Thomas 1917; Genelly 1965; Rosevear 1969). For instance, in South Africa alone the number of species recognized within this genus varies from 1 (Smithers 1983) to 18 (Ellerman 1940; Roberts 1951).

The genus *Cryptomys* is characterized by (1) ungrooved upper and lower incisors; (2) four simplified molars with only traces of reentrant folds on the labial surface in relatively unworn teeth, a core of dentine surrounded by a ring of enamel, and molars decreasing in size from anterior to posterior; (3) relatively well-developed bullae; and (4) a small infraorbital foramen, varying from circular to elliptical in shape. Species in the genus are variously colored silver gray, tan, charcoal, or black; several have a white forehead patch, and another white patch occurring either on the chest or the belly.

Previous disagreements notwithstanding, our review of characters examined by others (Hill and Carter 1941; De Graaff 1964a; Smithers 1971; Ansell 1978; Smithers and Wilson 1979; Williams et al. 1983; Nevo et al. 1987) and an examination of museum voucher specimens leads us to recognize seven

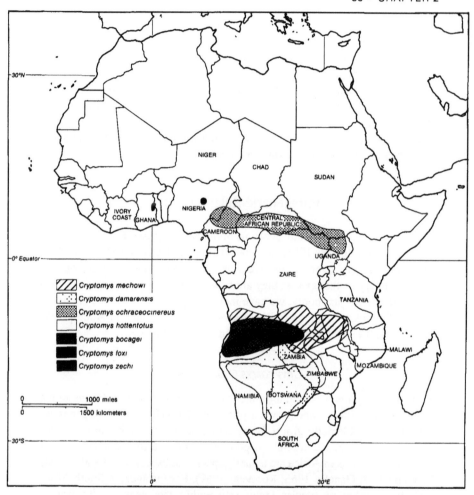

Fig 2-3 Geographic distribution of *Cryptomys* in Africa

species. These seven species can be divided into two groups based on the overall size and shape of the infraorbital foramen and the thickness of its outer wall. The three disjunct species occurring north of the equator (*C. foxi, C. ochraceocinereus*, and *C. zechi*; fig. 2-3) have small and thick-walled infraorbital foramina. *Cryptomus bocagei, C. damarensis*, and *C. mechowi* also exhibit the same condition. In contrast, the southern and eastern African species *C. hottentotus* has an infraorbital foramen with a thin external wall that is elliptically shaped.

Species of *Cryptomys* vary in size from *C. mechowi*, which measures up to 270 mm in head and body length (Hill and Carter 1941), to *C. hottentotus* which measures 90–135 mm in head and body length (De Graaff 1964a). The three northern species (fig. 2-3) vary from the small *C. foxi* of Nigeria (head and body length of 135–159 mm) to *C. ochraceocinereus* of Cameroon (head and body length of 160–190 mm) and Zaire (head and body length > 200 mm; Rosevear 1969; Williams et al. 1983). Synonymies for the seven species we recognize in the genus *Cryptomys* follow.

1. *Cryptomys bocagei* (De Winton, 1897). Includes *Georhychus* (sic) *kubangensis* Monard, 1933 as a synonym.

2. *Cryptomys damarensis* (Ogilby, 1838). Includes as synonyms *Georychus lugardi* De Winton, 1898; *Georychus micklemi* Chubb, 1909; and *C. ovamboensis* Roberts, 1946.

3. *Cryptomys foxi* (Thomas, 1911).

4. *Cryptomys hottentotus hottentotus* (Lesson, 1826). Includes as synonyms *Bathyergus caecutiens* Brants, 1827; *Bathyergus lugwigii* A. Smith, 1829; *Georychus holosericeus* Wagner, 1843; *Georychus exenticus* Trouessart, 1899; *Georychus hottentottus* (sic) *talpoides* Thomas and Schwann, 1906; *Georychus jorisseni* Jameson, 1909; *Georychus albus* Roberts, 1913; *Georychus vandami* Roberts, 1917; *Georychus vryburgensis* Roberts, 1917; *Georychus natalensis pallidus* Roberts, 1917; *C. transvaalensis* Roberts, 1924; *C. cradockensis* Roberts, 1924; *C. bigalkei* Roberts, 1924; *C. orangiae* Roberts, 1926; *C. vetensis* Roberts, 1926; *C. natalensis nemo* Allen, 1939; and *C. holosericeus valschensis* Roberts, 1946.

Cryptomys hottentotus amatus (Wroughton, 1907). Includes *Georychus molyneuxi* Chubb, 1908 as a synonym.

Cryptomys hottentotus darlingi (Thomas, 1895). Includes as synonyms *Georychus nimrodi* De Winton, 1897; *Georychus beirae* Thomas, 1908; and *C. zimbitiensis* Roberts, 1946.

Cryptomys hottentotus natalensis (Roberts, 1913). Includes as synonyms *Georychus jamesoni* Roberts, 1913; *Georychus arenarius* Roberts, 1913; *Georychus anomalus* Roberts, 1913; *Georychus aberrans* Roberts, 1913; *Georychus pretoriae* Roberts, 1913; *Georychus mahali* Roberts, 1913; *Georychus komatiensis* Roberts, 1917; *Georychus rufulus* Roberts, 1917; *Georychus stellatus* Roberts, 1917; *Georychus palki* Roberts, 1917; *C. junodi* Roberts, 1926; *C. melanoticus* Roberts, 1926; *C. montanus* Roberts, 1926; *C. langi* Roberts, 1929; *C. natalensis streeteri* Roberts, 1946; and *C. komatiensis zuluensis* Roberts, 1951.

Cryptomys hottentotus whytei (Thomas, 1897). Includes *C. h. occlusus* Allen and Loveridge, 1933 as a synonym.

5. *Cryptomys mechowi* (Peters, 1881). Includes as synonyms *Georychus ansorgei* Thomas and Wroughton, 1905; *Georychus mellandi* Thomas, 1906; and *C. blainei* Hinton, 1921.

6. *Cryptomys ochraceocinereus* (Heuglin, 1864). Includes as syn-

onyms *Georychus lechei* Thomas, 1895; *Georychus kummi* Thomas, 1911; and *C. ochraceocinereus oweni* Setzer, 1956.

    7. *Cryptomys zechi* (Matschie, 1900).

Recent studies of the relationships among *Cryptomys* taxa and the systematic status of named forms have focused on patterns of genetic variation (in chromosomes, Nevo et al. 1986; allozymes, Nevo et al. 1987; mitochondrial DNA, Honeycutt et al. 1987). These studies indicate that *C. damarensis* and *C. hottentotus* should be recognized as distinct species. In addition, the subspecies *C. hottentotus natalensis* is apparently divergent from the other two species and may represent a third distinct taxon.

    The genetic data provide some indication of the relationships among these three South African taxa. *Cryptomys damarensis* is the most divergent and differs from the other two taxa as follows: (1) the percentage of nucleotide divergence (δ) between mitochondrial DNA (mtDNA) haplotypes is 20%, which is as divergent as that seen among the genera *Georychus*, *Heliophobius*, and *Bathyergus*; (2) its standard karyotype has a 2N of 74 or 78, as opposed to a 2N of 54 for both *C. h. hottentotus* and *C. h. natalensis*; and (3) the Nei's *D* (genetic distance) value is 1.25 for allozyme divergence. Although *C. h. hottentotus* and *C. h. natalensis* have similar standard karyotypes, these two taxa are at least subspecies, since they have Nei's *D* value of 0.57 and a δ value of 15.7%.

GEORYCHUS ILLIGER, 1811

The genus *Georychus* (Cape mole-rats or blesmols) is monotypic with the single species, *G. capensis*. This species is found along the coastal zone of the southern and southwestern Cape Province of South Africa as far as coastal Transkei, in isolated populations in the northwestern Natal Province near the Lesotho border, and in the southern highveld of the Transvaal Province (fig. 2-4). This disjunct distribution does not seem to be an artifact of collecting or habitat availability, because *G. capensis* occurs in a variety of habitats; however, the animal seems to prefer sandy soils and alluvium in areas with 400–500 mm of rainfall per year.

    The characters for the genus and the single species are the same. *Georychus capensis* has a strikingly marked head and face; there are white patches on the muzzle, eyes, ears, and head that contrast sharply with the black to charcoal-colored head. The skull is arched in lateral profile. The upper and lower incisors are white and ungrooved; the upper incisors are rooted behind the molars. Four cheek teeth are present, and the upper ones have a narrow lingual and labial fold in the enamel; the lower cheek teeth have a persistent labial fold. The palate is narrow and extends behind the tooth row. Synonyms for the species follow.

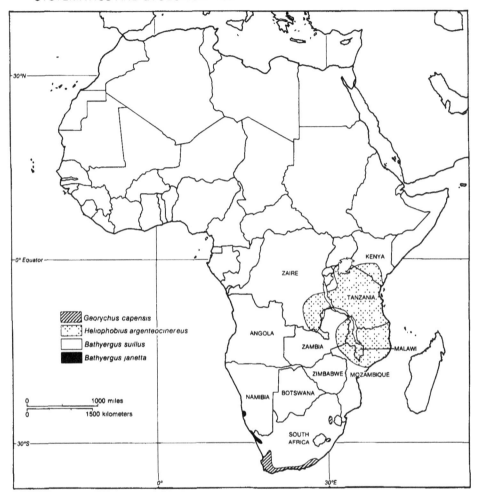

Fig 2-4. Geographic distribution of *Georychus, Bathyergus,* and *Heliophobius* in Africa.

Georychus capensis (Pallas, 1778). Includes as synonyms [*Mus*] *buffonii* F. Cuvier, 1834; *Fossor leucops* Lichtenstein, 1844; *G. c. canescens* Thomas and Schwann, 1906; and *G. yatesi* Roberts, 1913.

Georychus capensis also exhibits considerable geographic variation, with disjunct populations in South Africa (Nottingham Road vs. Cape Town and Du Toit's Pass) differing markedly in their mtDNA sequences ($\delta = 12.8\%$) and allozyme frequencies (Nei's $D = 0.43$; Honeycutt et al. 1987; Nevo et al. 1987). Although *G. capensis* is considered a monotypic species, the observed level of genetic variation seen between these two regions suggests the possibility of two species in South Africa.

BATHYERGUS ILLIGER, 1811

Two species differing in size and pelage color are recognized in the genus *Bathyergus* (Cape dune mole-rats); both occur in southern Africa (fig. 2-4). *Bathyergus janetta* lives in areas of sandy soils with less than 250 mm of rainfall per year, along the coast of southwestern Namibia around Luderitz and the mouth of the Orange River, and in the northwestern Cape Province from Orangemund southward to the Olifants River and inland as far as Ezelfontien. This species is smaller than the conspecific *B. suillus*, and it rarely reaches more than 230 mm in head and body length. The pelage is drab gray to buff dorsally, with a broad gray-brown band along the dorsal midline extending nearly to the rump; the head is black. *Bathyergus suillus* is the largest species in the family, with some individuals reaching 1,800 g. It occurs in soft sandy soils along the western Cape Province from south of the Olifants River to the Cape, and along the southern Cape coast to (at least) as far as Knysna in suitable sandy habitats with 350–500 mm of rainfall per year.

Both species of *Bathyergus* are characterized by light-colored areas around the muzzle, eyes, and ears (these are white and quite pronounced in *B. janetta*) and a white head patch (which is more pronounced in *B. suillus*). Both the upper and lower incisors are ungrooved and white. The palate is relatively narrow and the upper cheek teeth have a narrow lingual and labial fold forming reentrant angles in the enamel. In young animals a similar labial and even a lingual fold can be found in the lower cheek teeth. Synonymies for the two species in this genus follow.

1. *Bathyergus janetta* Thomas and Schwann, 1904. Includes as synonyms *B. j. inselbergensis* Shortridge and Carter, 1939; and *B. j. plowsi* Roberts, 1946.

2. *Bathyergus suillus* (Schreber, 1782). Includes as synonyms *Mus maritimus* Gmelin, 1788; *Arctomis* (sic) *africana* Lamarck, 1796; and *B. s. intermedius* Roberts, 1926.

On the basis of mtDNA studies, the two *Bathyergus* taxa are distinct. The δ value for mtDNA haplotype diversity is 7.6% (a value three times lower than that among species of *Cryptomys*), and the Nei's *D* value is 0.55.

HELIOPHOBIUS PETERS, 1846

The genus *Heliophobius* (silky or silvery mole-rats) occurs in well-drained sandy soils in savannas and woodlands at 750–1500 m elevation from southern Kenya to central Mozambique, in areas where rainfall varies from 250 to 600 mm (fig. 2-4; Kingdon 1974). Although two species have been recognized in this genus by De Graaff (1971), after examining the holotype of *H. spalax* and additional specimens from Taveta and the Taita Hills of Kenya, we conclude that the characters used by Thomas (1910) to separate the nominal taxon *H. spalax* from *H. argenteocinereus* are a result of age variation. Thus, our opinion is that this genus is monotypic.

Heliophobius argenteocinereus has silver to tan pelage reaching lengths of 25 mm; this is the longest pelage of any species in the family. Cranially, *Heliophobius* is characterized by six upper and lower cheek teeth, although not all are present at once because the anterior premolars are shed before the last molars erupt. The upper incisors are rooted behind the molars in the pterygoid region. The infraorbital foramen is relatively small. The angular process of the mandible does not flare far posteriorly. The palate is narrow and does not extend beyond the tooth row. Synonymies for the species and genus follow.

> *Heliophobius argenteocinereus* Peters, 1846. Includes as synonyms *Georychus albifrons* Gray, 1864; *Georychus pallidus* Gray, 1864; *H. marungensis* Noack, 1887; *H. emini* Noack, 1893; *H. robustus* Thomas, 1906; *G. kapiti* Heller, 1909; *H. spalax* Thomas, 1910; *Myoscalops mottoulei* Schouteden, 1913; and *H. angonicus* Thomas, 1917.

HETEROCEPHALUS RÜPPELL, 1842

Naked mole-rats occur in the hot dry regions of the Horn of Africa, from the Rift Valley of Ethiopia eastward into northern Somalia, from Lake Turkana in Kenya in a line to Mt. Kenya, around the eastern slopes to as far south as Tsavo Park, and then northeastward to the Somali coast (fig. 2-5; see also Jarvis and Bennett, chap. 3). This area is characterized by irregular rainfall, which generally averages 200–400 mm per year. Soil types throughout the range of *Heterocephalus glaber* vary from relatively soft sandy soils such as that found north of Mt. Kenya to extremely hard lateritic soils in southern and eastern Kenya (e.g., near Mtito Andei; see Brett, chap. 5).

This monotypic genus is characterized externally by skin that is not covered by pelage, although scattered hairs are present; from this lack of fur derives the animal's common name, the naked mole-rat. Indeed *H. glaber* cannot be mistaken externally for any other mammal (plate 2-1). Cranially, naked mole-rats are characterized by having three molars (although in some populations they have two). The molars are simplified enamel rings with no reentrant folds present. The upper and lower incisors are ungrooved, and the uppers are rooted behind the last upper molar; one of the most unusual features of the animal are its procumbent incisors, extending well outside the mouth (plate 2-1). The skull is relatively small for the family. The palate is narrow and extends beyond the tooth row.

At present, the genus *Heterocephalus* is considered monotypic, although it was marked by a proliferation of named taxa at the turn of the century. These names were based on differences in size and molar counts (three vs. two).

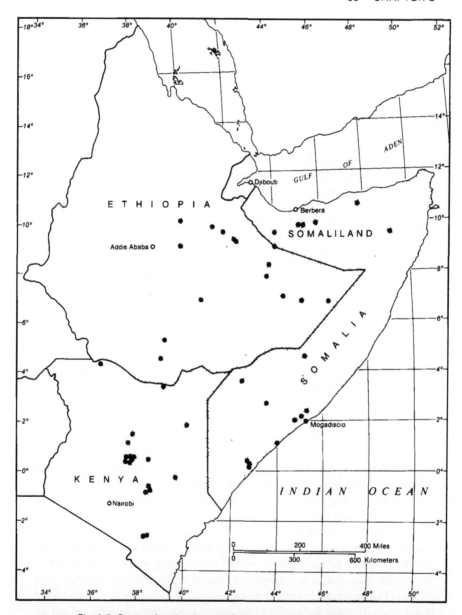

Fig 2-5. Geographic distribution of *Heterocephalus* in Africa.

Plate 2-1 *Top*, A naked mole-rat yawning (see Lacey et al , chap 8) Note the loose, wrinkled, virtually hairless skin, tiny eyes, reduced ear pinnae, small front and hind feet, and the procumbent incisors Also note that the hair-fringed lips close behind the incisors, this apparently keeps soil out of the esophagus and trachea during digging (see Jarvis and Bennett, chap 3) *Bottom*, Close-up of the head of a naked mole-rat. Photos. J U.M Jarvis, R A Mendez.

Plate 2-1.

Allen (1939) recognized a single species, *H. glaber* with two subspecies. However, a thorough revision is needed because size differences may relate to nutrition (Brett, chap. 4; Jarvis, chap. 13), age (Lacey and Sherman, chap. 10), or social circumstances (Jarvis and Bennett, chap. 3; Jarvis et al., chap. 12). Additional field studies and collections are needed from Somalia and Ethiopia, especially in the Ogaden region of eastern Ethiopia. Unfortunately, this geographic area has been politically unstable for decades. Although we list only a single species with all nominal taxa as synonyms, we acknowledge that this may need revision when new material becomes available.

> *Heterocephalus glaber* Rüppell, 1842. Includes as synonyms *H. phillipsi* Thomas, 1885; *H. ansorgei* Thomas, 1904; *H. dunni* Thomas, 1909; *H. stygius* Allen 1912; *H. g. progrediens* Lonnberg, 1912; and *H. g. scorteccii* de Beaux, 1934.

As shown in chapter 7, genetic data do suggest two distinct geographic groups of naked mole-rat populations in Kenya. In terms of mtDNA divergence, the δ value of 5.4% for animals collected about 300 km apart approaches that for the two species of *Bathyergus*; in contrast, allozyme divergence is very small. There is, however, some evidence of chromosomal differences between *H. glaber* populations. Individuals karyotyped from Kenya and Somalia differed in the length of their chromosomal short arms, but all possessed a 2N of 60 (George 1979; Capanna and Merani 1980). In addition, preliminary results from comparisons of hypervariable minisatellite DNA revealed some differences between the two collecting localities in Kenya (Reeve, pers. comm.).

Intergeneric Relationships

Two genetic studies have addressed phylogenetic relationships among bathyergid genera. The first study, by Nevo et al. (1987), involved an examination of allozyme variation among *Georychus capensis*, *Bathyergus suillus*, *B. janetta*, *Cryptomys damarensis*, *C. hottentotus hottentotus*, and *C. h. natalensis*. Twenty presumptive genetic loci were used to evaluate levels of allozyme variation, and phenograms depicting levels of genetic divergence were produced using Nei's genetic distances. We have reanalyzed these data cladistically using the criterion of parsimony (i.e., the shortest tree is the best tree). For this analysis, each allele at a locus was treated as a character and coded as present or absent. The resulting data matrix was used in a Phylogenetic Analysis Using Parsimony (PAUP, version 2.3, developed by D. Swofford). The "branch and bound" option was used, thus ensuring that the shortest tree would be obtained, with tree length defining the most parsimonious solution. The overall length of the phylogenetic tree (fig. 2-6a) was 111, and the consis-

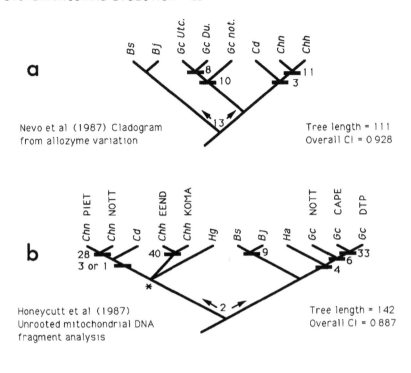

a

Nevo et al (1987) Cladogram
from allozyme variation

Tree length = 111
Overall CI = 0 928

b

Honeycutt et al (1987)
Unrooted mitochondrial DNA
fragment analysis

Tree length = 142
Overall CI = 0 887

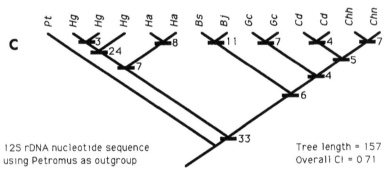

c

12S rDNA nucleotide sequence
using Petromus as outgroup

Tree length = 157
Overall CI = 0 71

Fig 2-6 Molecular phylogenetic hypotheses for the relationships of taxa within the family Bathyergidae. Overall tree length and consistency index (CI) for each tree is presented at the far right. Species abbreviations as in table 2-1 a, A cladogram of six bathyergid taxa based on allozyme variation (from Nevo et al 1987). b, A strict consensus tree, based on mtDNA restriction fragments, for the two bathyergid phylogenetic trees of equal length. The two values for the *Cryptomys hottentotus damarensis/C. h. natalensis* clade are the number of synapomorphies represented in the two trees Asterisk, The major point of disagreement between the two trees (after Honeycutt et al 1987). c, The single, most parsimonious tree based on preliminary 12S rRNA nucleotide-sequence data, using *Petromus* as the outgroup. The numbers at the internodes represent synapomorphies defining clades.

tency index was 0.928. The consistency index represents the amount of homo-plasy (reversals, parallelism, and convergence) found in the distribution of characters along branch lengths; the value of 0.928 was quite high (i.e., close to 1.00).

The results of the allozyme analysis suggested that *C. h. hottentotus* and *C. h. natalensis* are more closely related to each other than either is to *C. damarensis* and that the genus *Bathyergus* is the most divergent taxon, with *Georychus* being closer to *Cryptomys*. This analysis has two major limitations, however. First, *Heliophobius* and *Heterocephalus* were not included. Second, the phylogenetic tree was not rooted relative to an outgroup, thus making the establishment of character-state polarity (direction of character changes) difficult.

The second genetic study, by Honeycutt et al. (1987), involved a detailed analysis of mtDNA restriction-fragment variation within the Bathyergidae. A total of 493 distinct restriction fragments were produced using 15 restriction endonucleases to digest mtDNA from *Cryptomys h. hottentotus*, *C. h. natalensis*, *C. damarensis*, *Georychus capensis*, *Bathyergus janetta*, *B. suillus*, *Heliophobius argenteocinereus*, and *Heterocephalus glaber*. Of the restriction fragments generated, 126 were used in a cladistic analysis. Fragments were coded as present or absent and treated as unordered characters. An undirected parsimony analysis, using the branch-and-bound option in PAUP (version 2.3), was performed to find the most parsimonious tree that was rooted at the midpoint between the most divergent taxa (this procedure was used because the analysis did not include an outgroup).

Figure 2-6b represents a strict consensus tree for the two trees of equal length (142 steps) and consistency (0.887). Basically, a consensus tree depicts similar patterns of relationships shared by all trees of equal length. As can be seen in figure 2-6b, two major groups, *Cryptomys/ Heterocephalus* and *Georychus/Heliophobius/Bathyergus*, were defined. In addition there was a clear grouping of the two *Bathyergus* species; *Georychus* and *Heliophobius* grouped together as well. The pattern of relationships derived from the mtDNA fragment data is similar to that derived from the allozyme data in that both data sets recognize *Bathyergus* as being divergent from the other genera. The relationships among the *Cryptomys* taxa, however, are not resolved by the mtDNA data.

The phylogenetic relationships depicted in figure 2-6b allow us to hypothe-size the biogeography of speciation events that may have led to the present distributions of sister taxa. Genera in each of the two groups, *Cryptomys/ Heterocephalus* and *Georychus/Heliophobius*, have a similar pattern of disjunct distribution in eastern and southern Africa. As can be seen in figures 2-4 and 2-5, both *Heterocephalus* and *Heliophobius* are endemic to eastern Africa, whereas their closest relatives, *Cryptomys* and *Georychus*, occur predominantly in southern Africa. Similar disjunct distributional patterns between related groups occupying parts of eastern and southern Africa have been seen in many other taxa (e.g., fossil mammals, Van Couvering and Van Couvering 1976; passerine birds, Winterbottom 1967; xerophytic shrubs, Bakker 1967; other mammals, Roberts 1937). Collectively, these distributions suggest that at one time there was an ecological link between eastern and southern Africa. Such a link may have involved a drought corridor (i.e., an area where rainfall is less than 10 mm/mo) extending from southwest Africa, northeast to Somalia, the "Horn of Africa," and into Saudia Arabia (Balinsky 1962). The varying degrees of differentiation displayed by many organisms endemic to eastern and southern Africa imply that the corridor has been closed and reopened several times since the Miocene through fluctuating humid and arid periods (Wells 1967; Winterbottom 1967; Verdcourt 1969). Given the apparent age and past geographic distribution of certain fossil bathyergids in the lower Miocene (Lavocat 1978) and the current level of differentiation and disjunct distribution of related bathyergid taxa, it seems possible that the closing and reopening of this drought corridor during geologic history might explain the current distribution of *Georychus/Heliophobius* and *Cryptomys/Heterocephalus*.

Another interesting aspect of the phylogenetic relationships depicted in figure 2-6b is an implication about the evolution of complex sociality in *Heterocephalus glaber*. Although most bathyergid taxa are solitary, the genus *Cryptomys* is known to be colonial, with *C. damarensis* and *C. h. hottentotus* colonies composed of extended families (Bennett and Jarvis 1988; Jarvis and Bennett, chap. 3). The close relationship between *Heterocephalus* and *Cryptomys* implies that these taxa may have been derived from a common ancestor that was already on its way to becoming social.

Although the phylogenetic relationships revealed by the mtDNA fragment data (fig. 2-6b) provide some interesting insights, the results should be considered equivocal for two reasons. First, the level of divergence between genera is high, increasing the probability that some fragments that were scored as identical in mobility may, in fact, be convergent, thus increasing the likelihood of misidentifying shared characters. Second, because the phylogenetic tree shown in figure 2-6b was not rooted with an outgroup, character-state polarity could not be established.

Because of these limitations, we have begun a detailed study of mtDNA nucleotide-sequence variation within the Bathyergidae. This sequencing study involves the amplification of double- and single-stranded DNA using the polymerase chain reaction (PCR), and the nucleotide sequencing of the amplified DNA using the dideoxy chain termination method (Gyllensten and Erlich 1988; Saiki et al. 1988; Kocher et al. 1989). A 280-base-pair region of the mitochondrial 12S rRNA gene has been sequenced in both directions for all bathyergid taxa (table 2-1). Each nucleotide position was treated as an unordered discrete character with five possible states: A, C, G, T, or * (* indicates a gap). Again, a phylogenetic analysis using parsimony was performed; the

Table 2-1

Nucleotide-Sequence Data for Bathyergid 12S rRNA Gene

Species	Nucleotide Sequence

```
Species  Nucleotide Sequence

Md   cagagaactactagccatagcttaaaactcaaaggacttggcggtactttatatccatctagaggagcctgttctataatcgataaaccccgctctacct
Hg   ..............a.c............a..g.....a.c...............................................t.aa.........
Hg   ..............a.c............a..g.....a.c...............................................t.aa.........
Hg   ..............a.c............a..g.....a.c...............................................aa...........
Ha   ...........g...a............g...........c........................................a.aa...............
Ha   ...........g...*.a..........g...........c........................................a.aa...............
Bs   ..............c...a.........g...c.c.c.gc........................................a.aa................
Bj   ..............cg..a.........g...c.c.c.c.....................................tt...a.aa...............
Cd   ..............cg..a.........g...c.c.c.c.....................................t....a.aa...............
Cd   ..............cg..a......*...g...c.c.c.c........................................a.aa................
Chn  .......c..a.cg..a...........g...c.c.c.gc..............................g.........a.aa................
Chn  ....g...c..a.c..a...........g...c.c.c.c........................................a.aa.................
Gc   ....g...t.g..a......*........g...c.c.c.c..............................g.........a.aa................
Gc   ....g......a.................g...c.c.c.c..............................g...c.....a.aa................
Pt   .........a....c.............g...........c..........................................tc..a.aa.........


Md   caccatctcttgctaattcagcctatataccgccatcttcagcaaacctaaaaaggtattaa*agtaagcaaagaatcaaa*cataaaacgttaggt
Hg   ......t.g....c....t...........................g....tct.c.g..g.ca.*.......tc..t.*.*tgt...........
Hg   ......t.g....c....t...........................g....tct.c.g..g.ca.*.......tc..t.*.*tgt...........
Hg   ......t.g....c....t...........................g....tct.c.ga.g.ca.*.......tc..t.*.*tgt...........
Ha   ....ct.t.....c...t............................t...t.tt.a.a.ca.*........c..t.**t..t.............
Ha   ....ct.t.....c...t............................t...t.tt.a.a.ca.*........c..t.**t..t.............
```

```
Bs    t....ct.t.........a...t.................t..c..a...a.*.......gc.t.**t.t.g.........
Bj    t....ct.t.........a...t.................ttt.a.a..a.*.......gc.t.**t.t.g.........
Cd    .....ct.t.........a.....................g..t..t.a..aa.....c.tc.**c.a.......g....
Cd    .....ct.t.........a.....................g..t..c.a..aa.....c.tc.**c.a.......g....
Chn   .....ctat.........a...tt................t..c..a.aa.*.......c.tc.**c.g.......g....
Chh   .....ctat.........a...t.................g..ttt.a.aa.*......c.tc.**c.g.......g....
Gc    .....ct.t..c..a........................g..t..c.a.a..a.*.....c.tc.**t.t.g.........
Gc    .....ct.t..c...........................t..c..a.agaa.*.......c.tc.**c.t..........
Pt    t..gc.c..c.....c..t....................t..c..a...a.*.......g..c..g.t.tc.g...c....

Md    caaggtgtagccaatgaaatgggaagaaatgggctacattttc*ttataaagaaca**t**tactata*cccttatga
Hg    .........a.............................*.......c......*..aa..g...*gttacc.....
Hg    .........a.............................*.......c......*..aa..g...*gttacc.....
Hg    .........a.............................*.......cc.....*..aa..g...*gttacc.....
Ha    .....................g.................*......t..tc...*t.aaa..g.a*gt.ac......
Ha    .....................g.................*......t..tc...*t.aaa..g.a*gt.ac......
Bs    ....t..a.g...ag........................*......tac.....*.c.taa..gga.*a.ta.....
Bj    ....t..a.g...ag........................*......tat.....*.c.taa..gga.*a.ta.....
Cd    .........a.g...a..t....................*......gacg....*t.taa..g.a.g*.ta......
Cd    .........a.g...a..t....................*......gacg....*t.taa..g.a.g*.ta......
Chn   .........a.g..........................*.......acg....*.tctaa..g.g.*a.ta......
Chh   .........a.g..........................*.......ac.....*.tctaa..g.a.g...a..taa*
Gc    .........a.g..........................*......c*.t....*.*.aca..g.a.*a.ta......
Gc    .........a.g..........................*.......cgtt....*.*ctaa..g.a.*a.ta..g..
Pt    .....a..c...gggc...g.*.................*..g.*.t........gt.*c..g.*ta..a.......
```

NOTE: The fragment starts at *Mus* number 555 and ends at number 826 (Bibb et al 1981). Alignment was determined using Needleman and Wunsch (1970). Md, *Mus domesticus*; Hg, *Heterocephalus glaber*; Ha, *Heliophobius argenticinereus*; Bs, *Bathyergus suillus*; Bj, *B. janetta*; Cd, *Cryptomys damarensis*; Chh, *C. hottentotus hottentotus*; Chn, *C. h. natalensis*; Gc, *Georychus capensis*; Pt, *Petromus typicus*. Unless noted the sequence is identical to *Mus*. a, adenine; c, cytosine; g, guanine; t, thymine. An asterisk indicates that a space has been introduced to maintain alignment.

tree was rooted by comparison to an outgroup, the rodent genus *Petromus* (African rock rats), believed to be related to the Bathyergidae (Sarich 1985). The resulting phylogenetic tree (fig. 2-6c) is not congruent (in agreement) with the tree produced from the fragment data (fig. 2-6b). In the nucleotide-sequence tree, *Georychus*, *Bathyergus*, and *Cryptomys* are seen to be closely related, whereas *Heliophobius* and *Heterocephalus* group together. *Cryptomys h. hottentotus* and *C. h. natalensis* are also seen as more closely related to each other than either is to *C. damarensis*. If one uses the nucleotide-sequence data to construct the tree shown in figure 2-6c, the overall tree length is 172, therefore rendering this solution less parsimonious.

Thus, the nucleotide-sequence data lead to a somewhat different hypothesis for the biogeographic history of the Bathyergidae and the evolution of sociality. First, the genera cluster in a manner one would expect given their current distributions in Africa (see figs. 2-3–2-5), thus obviating the need for an explanation involving a drought corridor and multiple exchanges between eastern and southern Africa. Second, the independent origins of *Cryptomys* and *Heterocephalus* suggest the possibility that sociality is quite ancient in the family, with similar social systems evolving in parallel among bathyergid lineages.

We caution the reader not to overinterpret these preliminary sequence data. There are some problems with the outgroup in that when *Mus* was used in conjunction with *Petromus*, *Heterocephalus* became more divergent from the other genera of bathyergids than was *Petromus*. Therefore, *Petromus* may not be an appropriate outgroup. This can be tested by examining other related rodent lineages including the Hystricidae, Caviomorpha, and Thryonomyidae, and will be the subject of our continuing research efforts.

Summary

Although there has been considerable research and debate over the placement of the Bathyergidae relative to other rodent families, little information exists about relationships among genera or the taxonomy of any species. This chapter reviews the taxonomic history of the family and more recent genetic data (mtDNA, allozymes, and chromosomes) that are pertinent to the systematics, biogeography, and evolution of the Bathyergidae.

The genetic data have yielded three basic observations. (1) The genus *Cryptomys* in southern Africa consists of several divergent taxa probably representing species level differences, and *C. h. hottentotus* and *C. h. natalensis* are probably each other's closest relatives. (2) *Georychus capensis* and *Heterocephalus glaber* exhibit marked geographic variation; whether or not specific- or subspecific-level differences exist is uncertain. (3) The relationships among genera in eastern and southern Africa are somewhat uncertain. Morphologically, *Bathyergus* is the most divergent genus; genetically, the results are equivocal. On the one hand, mtDNA fragment data suggest that *Heterocephalus* and *Cryptomys* are each other's closest relatives and that *Georychus* and *Heliophobius* are closely related to each other; *Bathyergus* is the most divergent according to these mtDNA data. On the other hand, nucleotide-sequence data from the 12S rRNA gene suggest that the eastern and southern African genera form separate monophyletic groups. Different interpretations of the evolution of sociality in bathyergids and their biogeographic history can be derived depending on which phylogenetic hypothesis one accepts. These results point to the need for an expanded systematic study of the family Bathyergidae.

Acknowledgments

We thank P. Gathinji, K. M. Nelson, M. B. Qumsiyeh, and E. Nevo for assistance in collecting specimens. R. E. Leakey, I. Aggundey, and colleagues at the National Museum of Kenya made this research possible through support and field assistance. We also thank the Office of the President and National Parks and Wildlife, Nairobi, for issuing research clearance and collecting permits. Comments on the manuscript were given by J.U.M. Jarvis, P. W. Sherman, and two anonymous referees. Fieldwork was supported by the Barbour and Richmond Funds from the Museum of Comparative Zoology at Harvard University, the Netting Research Fund, Cordelia Scaife May Charitable Trust, and the O'Neil Fund of The Carnegie Museum of Natural History. Research was supported by a National Institutes of Health Predoctoral Traineeship in Genetics to M.W.A. and a National Science Foundation grant to R.L.H.

3

Ecology and Behavior of the Family Bathyergidae

Jennifer U. M. Jarvis and Nigel C. Bennett

As indicated in the previous chapter, the family Bathyergidae comprises 5 genera and 12 species. Bathyergid mole-rats exhibit the widest range in mean body size and social structure of any of the 7 families or subfamilies of subterranean rodents (table 3-1). The family is endemic to Africa and occurs from the southern tip of the continent to about $10°$ north of the equator (De Graaff 1981). Bathyergids are found in a wide variety of soils, altitudes, and climates. Their local distribution is often patchy, but where conditions are favorable, they may occur in high densities.

The bathyergids of the genera *Bathyergus*, *Georychus*, and *Heliophobius* are solitary, whereas *Cryptomys* and *Heterocephalus* are social. In general, the solitary mole-rats are larger in size and occur in regions with a higher annual rainfall (> 350 mm/yr) than the social bathyergids (De Graaff 1981; Honeycutt et al., chap. 2). In arid areas, some of the social species occur in habitats where many of the plants are widely spaced, patchy in distribution, and have large subterranean storage organs (Jarvis and Sale 1971; Jarvis 1978; Brett 1986, chap. 5; Lovegrove and Painting 1987; Bennett 1988). These social bathyergids have the largest number of individuals in their colonies and show division of labor.

Because of the diversity in social structure found within the family Bathyergidae, which displays a virtual continuum from a solitary to eusocial life-style, the extreme but by no means exclusive sociality exhibited by the naked mole-rat must be examined within the framework of the social systems found in the other genera. We present an overview of the behavior and ecology of the Bathyergidae, briefly examine the possible determinants of sociality in the family, and place *Heterocephalus glaber* into perspective with respect to the other genera.

A Brief Description of the Family Bathyergidae

The systematics and evolution of the Bathyergidae were dealt with by Honeycutt et al. in chapter 2 and will not be reiterated here. The family is divisible into two subfamilies, the Bathyerginae and the Georychinae. This subdivision is based on the way the animals dig, on features associated with their methods of digging, and on body size and skull and incisor morphology.

Table 3-1
The families, subfamilies and genera of subterranean rodents

Family/Subfamily	Genera	Distribution
Bathyergidae mole-rats	*Bathyergus* *Cryptomys* *Georychus* *Heliophobius* *Heterocephalus*	Africa
Cricetidae voles	*Myospalax* *Ellobius* *Prometheomys*	Asia
Ctenomyidae tuco-tucos	*Ctenomys*	South America
Geomyidae pocket gophers	*Cratogeomys* *Geomys* *Heterogeomys* *Macrogeomys* *Orthogeomys* *Pappogeomys* *Thomomys* *Zygogeomys*	North America
Spalacidae mole-rats	*Spalax*	Europe, Asia, North Africa
Rhizomyidae bamboo/root rats	*Cannomys* *Rhizomys* *Tachyoryctes*	Africa, Asia
Octodontidae octodonts	*Spalacopus*	South America

NOTE· Information modified from Nevo (1982).

The subfamily Bathyerginae contains one genus (*Bathyergus*) with two species, both of which are solitary. *Bathyergus suillus*, the Cape dune mole-rat, is the largest of the Bathyergidae, and adult males may attain a body mass of 1,800 g. *Bathyergus janetta*, the Namaqua dune mole-rat, rarely exceeds 750 g. Both species are sexually dimorphic with males larger than females. The distribution of the dune mole-rats is limited to soft arenosols in which they dig with strong large-clawed forefeet.

The mole-rats in the subfamily Georychinae are smaller (≤ 350 g) than those in the Bathyerginae. They are also distinctive in that they excavate the soil with their incisors, and the claws of their forefeet are small. Mole-rats in the genera *Georychus* and *Heliophobius* are solitary, whereas *Cryptomys* and *Heterocephalus* are social.

The taxonomy of the genus *Cryptomys* is problematic, and in this chapter we follow Honeycutt et al. (chap. 2) in recognizing seven species, two of which (*C. damarensis* and *C. hottentotus*) occur in southern Africa. After revision of the taxonomy of this genus is complete, at least one of the subspecies of *C. hottentotus* may be elevated to specific status (Honeycutt et al., chap. 2).

Early Descriptions of the Naked Mole-Rat

Brett (1986) summarized the early descriptions of *Heterocephalus glaber*. A precis of his findings is presented here (table 3-2). Naked mole-rats were first collected in Ethiopia and the type specimens were described by Rüppell (1842). Initially it was thought that these hairless rodents were the young of a haired adult, but subsequent expeditions to Italian Somaliland by Captain V. Bottego (Thomas 1885a,b) and Parona and Cattaneo (1893) showed conclusively that adult *H. glaber* were hairless except for a sparse covering of sinusoidal (sensory) hairs.

The early collections contained a wide range of sizes of naked mole-rats, some of which also varied in the number of cheek teeth present. Thomas (1885a,b, 1895, 1897, 1903, 1904, 1909) described a number of species, and even proposed a new genus, *Fornaria* (Thomas 1903), for one of them. Allen (1939) questioned the validity of all these taxa. It is now clear that even within a single colony of *H. glaber* there are marked differences in adult body size (Jarvis 1978, 1981, 1985; Brett, chap. 4; Jarvis et al., chap. 12), and only a single species, *H. glaber*, is presently recognized (see Honeycutt et al., chap. 2).

Burrow Systems, Burrow Dynamics, and Burrowing

All bathyergids are completely subterranean. They live underground in a system of burrows that consists mainly of shallow foraging tunnels running, in most species, at about the depth of their food belowground. Mole-rats are herbivorous, the majority subsisting primarily on subterranean portions of plants, especially corms, bulbs, and tubers (Jarvis and Sale 1971; Lovegrove and Jarvis 1986; Bennett 1988; Brett, chap. 5). In addition, species of the larger-sized genera *Bathyergus* and *Georychus* may supplement this diet with the aerial portions of plants (table 3-3); to obtain these, the animals remain underground, bite through the roots and stem base of the plant and then pull the aerial portions into the burrow. Occasionally individuals of both genera forage in the immediate vicinity of an open burrow, usually keeping their hind legs in the hole mouth and stretching out to gather portions of plants.

The rest of a mole-rat's tunnel system is deeper than the foraging burrows and comprises one or more nests, toilet areas, and (especially in *G. capensis* and in species of *Cryptomys*) food stores; deep, blind tunnels, into which mole-

Table 3-2
Early Collections and Descriptions of *Heterocephalus* 1842–1957

Publication Date	Author	Collector	Locality	Species
1842	Ruppell	Bretska	Shoa, Ethiopia	*H glaber*
1885a,b	Thomas	Phillips	Gergolbi, Somaliland	*H phillipsi*
1893	Parona & Cattaneo	Bottego, Expedition I	Err region, Somaliland	*H. glaber*
1895	Thomas	Bottego, Expedition I	Archeisa, Somaliland	*H. glaber*
1896	Bottego	Bottego, Expedition II	Italian Somaliland	*H glaber*
1896	Francaviglia	Bottego, Expedition II	Italian Somaliland	*H glaber*
1897	Thomas	Bottego, Expedition II	Lugh, Brava, Somalia	*H. glaber*
1903	Thomas	Bowen & Dulio	Mogadishu, Somaliland	*Fornaria phillipsi*
		Ansorge	Makindu country, Kenya colony	*H. ansorgei*
1904	Thomas	Thomas	Hargeisa, Somaliland	*H glaber*
			Wardairi, Gergolbi	*H ?, F phillipsi*
1909	Thomas	Dunn	Wardiari, Somaliland	*H. dunni*
1912	Allen	Allen	Neumann's Boma, Guaso Nyiro, Kenya	*H stygius*
1912	Lonnberg	Swedish Expedition	North Guaso Nyiro, Kenya	*H. g. progrediers*
1915	Senna	Stefanini & Paoli	Afgoi, Somaliland	*H. glaber*
1919	Hollister	Bottego, Heller, Percival	Archer's Post, Longaya Water, Kenya, etc	*H glaber*
1928	W. J Hamilton	Akeley	Kinna River, Kenya	*H glaber*
1934	De Beaux	De Beaux	Gardo, Somaliland	*H g scortecci*
1940	Thigpen	Anderson & Landauer	Kenya	*H glaber*
1946	Cei	Bottego, Expedition II	Italian Somaliland	*H. glaber*
1953, 1957	Hill et al	Hill et al	Isiolo, Dandu, Wajir Moyale, Kenya	*H glaber*
1957	Porter	Hill et al	Isiolo, Dandu, Wajir Moyale, Kenya	*H glaber*
1957	Stark	Stark	Harar region, Ethiopia	*H glaber*

NOTE Information from Brett (1986)

rats retreat for protection and possibly also for behavioral thermoregulation, may also be present (Genelly 1965; Jarvis and Sale 1971; Hickman 1979b; De Graaff 1981; Jarvis 1985; Davies and Jarvis 1986; Lovegrove and Jarvis 1986). Hickman (1979b) and Nanni (1988) suggested that these blind tunnels may also act as drainage sumps during flooding.

Although details vary, bathyergid mole-rats excavate their burrows in two main ways. Both of the *Bathyergus* species live in soft sand and dig with long, well-developed claws on their forefeet (De Graaff 1981; Davies and Jarvis 1986). The other four genera excavate the soil with large protruding incisors

Table 3-3

A Summary of the Major Characteristics of the Bathyergidae, Their Burrow Lengths, Diets, and Habitats

Species	2N	Sex	Mean Mass (g)	SD	N	Moles/ Burrow	Bio-mass/ Bur-row (g)	Mean Bur-row Length (m)	N	Mass Moles/ Meter (g/m)	Diet	Soil	Rainfall (mm)	Source
Bathyergus suillus	56	m	933	253.8	71	1	877	256	7	3.8*	over 60% aerial veg.; the rest are roots, small geophytes	soft windblown dune sands	380–500	1, 2
		f	635	135	73									
B. janetta	54	m	451	124	10	1	387	189	3	1.9*	aerial vegetation and small geophytes	soft windblown dune sands	100–250	3
		f	332	98	16									
Cryptomys hottentotus hottentotus	54	m	77	21.7	31	2–14	132–810	464	4	0.8*	entirely geophytes, usually small	compact soils occasionally sandy	200–500	1, 4, 5, 6
		f	57	10.9	19									
C. hottentotus natalensis	54	m	106	—	4	2–3	202	181	4	1.1*	geophytes and grass rhizomes	compact soils	400–600	7, 8
		f	88	—	4									
C. damarensis	74,78	m	104	28.6	37	12–25	1358–3893	—	—	—	geophytes, often large & dispersed	usually soft Kalahari sands	200–600	4, 9, 10, 11
		f	97	16.4	32									
Georychus capensis	54	m	181	73.3	51	1	169	48	14	4.5*	under 15% aerial; rest small geophytes	variety of soil, soft to more consolidated	350–380	5, 12, 13
		f	180	92.3	37									

Species									Food	Soil	Range	Ref.
Heliophobius argenteocinereus	60	160	—	4	1	100	47	4	2.1*	roots, medium-sized geophytes	variety of soil, often very compact	250–600	14, 15
Heterocephalus glaber (north Kenya)[†]	60	23	8.1	177	40–82	626–2111	595	1	2.1*	entirely geophytes & roots, sometimes large & dispersed	mainly hard soils	200–350	16
Heterocephalus glaber (south Kenya)[†]	60	33.9	3.4	715	25–97	511–3385	3027	1	1.1	entirely geophytes & roots, often large & dispersed	mainly hard soils	200–400	17

* Measurements obtained from excavated burrow systems in which the mole-rats had been captured and weighed. Other figures are estimates using telemetry (Brett 1986) or partly excavated systems. Karyotype data from Nevo et al. (1986) and George (1979).

[†] See Brett (chap. 4) for mass measurements on male and female *H. glaber.*

SOURCE: 1, Davies and Jarvis 1986; 2, Bevıss-Challiner and Broll, unpubl. data; 3, Jarvis, Griffin, Griffin, Hamerton, and Tharme, unpubl. data; 4, Bennett 1988; 5, Lovegrove and Jarvis 1986; 6, Broll, unpubl. data; 7, Hickman 1979b; 8, Nanni 1988; 9, Smithers 1983; 10, De Graaff 1981; 11, Lovegrove and Painting 1987; 12, Taylor et al. 1985, 13, Du Toit et al. 1985, 14, Jarvis and Sale 1971, 15, Kingdon 1974, 16, Jarvis 1985; 17, Brett 1986, chap. 5.

(Genelly 1965; Jarvis and Sale 1971). Once the soil has been broken up, all bathyergids push it beneath their bodies with their forefeet, collect it with their hind feet, and kick it behind them. When sufficient soil has accumulated behind the mole-rat, it moves backward along the burrow with the soil, goes up a side branch to the surface, and then pushes or kicks the soil through a hole, thus forming a mound or molehill. As the mole-rat pushes the soil along the burrow, it is compacted, and this core of soil is squeezed out of the hole onto the surface, like toothpaste out of a tube; the resulting mound is composed of these tubular cores (Genelly 1965; Jarvis and Sale 1971). With the exception of *Heterocephalus glaber*, once the initial hole to the surface has been made, there is always an earthen plug between the mole-rat and the surface opening during mound formation. The naked mole-rat differs from all the other bathyergids in that it kicks out a fine spray of soil through an open hole to form a mound resembling a minature volcano (Hill et al. 1957; Jarvis and Sale 1971; Brett, chap. 4; Braude, chap. 6). In all genera, once a mound has been made, the tunnel branch leading to the surface is packed with soil, sealing the system off from the surface and rendering it relatively airtight, watertight, and predator-proof.

Burrow length appears to be correlated with the biomass of the mole-rat(s) occupying a system, their diet, and the availability of food (table 3-3; Jarvis and Sale 1971; Hickman 1979b; Jarvis 1985; Davies and Jarvis 1986; Brett 1986, chap. 5). Species of *Cryptomys* and *H. glaber* have a significantly lower biomass per meter of burrow (1–2 g/m) than do *B. suillus* and *G. capensis* (3.5–4.5 g/m). Not all the soil that the mole-rats excavate is pushed out onto the surface; indeed, most species use it to plug old sections of the system (Davies and Jarvis 1986). This probably keeps burrow length within viable limits (i.e., limits that can be effectively defended and repaired by the inhabitants). Probably the viable burrow length is shorter for solitary mole-rats than for social ones; one mole-rat would have difficulty in speedily detecting and repairing damage to all sections of a very long burrow or in expeditiously detecting the entry of predators or foreign conspecifics.

Plugging old burrows with soil occurs most frequently at drier times of the year, when soil conditions are unfavorable for digging close to the surface. At these times, the soil near the surface is either extremely hard to bore through or is so soft and sandy that the burrows collapse. When it is very dry, there is thus little surface evidence of burrowing activity (Davies and Jarvis 1986; Brett, chap. 5; Jarvis, unpubl. data). Conversely, when the soil is easy to work there is increased mound-building activity. It is therefore difficult to assess seasonal trends in burrowing without the use of extensive radio-tracking and hand excavation.

Despite these problems, Genelly (1965) estimated that a *C. hottentotus darlingi* colony, containing an estimated 10 mole-rats, dug 135 m in a month. Mound production in *B. suillus* and *C. hottentotus hottentotus* was counted and plotted for a year, and entire burrow systems were excavated (Davies and Jarvis 1986). Average daily burrow extensions of 0.7 m were recorded for 6 solitary *B. suillus* and a colony of 3 *C. h. hottentotus*, and maximum monthly burrow extensions of 141 m for a single *B. suillus* and 47 m for the 3 *C. h. hottentotus*. Brett (1986, chap. 5) found that a colony of about 87 *H. glaber* extended their burrow system by 1 km during the short rains (March–May 1983). These figures, when converted into mass of soil moved, show that during peak burrowing periods, a single *B. suillus* and a colony of about 87 *H. glaber* individuals can excavate, move, and throw up as mounds more than 500 kg of soil per month.

There are physical limitations to the length of burrow that can be dug per unit of time. One of these is the rate at which the digging tools (teeth or claws) can grow. Although the species show differences in the morphology of their teeth, there must be a upper limit to the absolute rate at which teeth can grow, depending on the rate at which tooth tissue can proliferate, the availability of the minerals needed, and the energetic costs of tooth growth. The rates of tooth growth have not been measured in all the genera. However, the teeth of captive *B. suillus* and *G. capensis* can grow as much as 8–10 mm/wk and those of *H. glaber* by 4–6 mm/wk (Jarvis, unpubl. data). Brett (1986) noted that *H. glaber* has to stop frequently to sharpen the teeth when digging through hard soil. We have observed that even short periods of forced digging can result in extreme tooth wear. For example, after being pursued for 3–5 days, individuals of both *G. capensis* and *C. h. hottentotus* species had upper incisors worn to the gums. If wear and growth rates are similar in these sympatric species, the social *C. h. hottentotus* would have an advantage over the solitary *G. capensis* in being able to share the digging among several colony members. Indeed, only the last individuals to be captured in a colony of 10 *C. h. hottentotus* showed this extreme tooth wear.

The Subterranean Niche

MORPHOLOGY

Many morphological features of the Bathyergidae are a consequence of living underground. Mace et al. (1981) suggested that the subterranean way of life represents a simple sensory or perceptual niche, which may have led to the bathyergids' having the smallest relative brain size of any group of small mammals. Bathyergids have reduced eyes (Cei 1946; Eloff 1958; Rees 1968) and visual centers in the brain (Hill et al. 1957; Pilleri 1960). They apparently cannot see images, and it is unclear whether or not all species can perceive light (Eloff 1958; Poduschka 1978). When mole-rats of most species move about their burrows in the lab, their eyes are usually kept closed; the eyes are opened when the animals are alarmed or, in *Heterocephalus glaber*, when a flashlight is shined on them (Sherman, pers. comm.). Eloff (1958) suggested that the cornea may be used to detect air currents produced by breaks in the burrow system, but this suggestion was not supported by Rosevear (1969) and Poduschka (1978). In most mole-rat species, the ear pinnae are reduced, but hearing is nonetheless acute. The senses of touch and smell are well developed (Hill et al. 1957; De Graaff 1960; Jarvis, unpubl. data).

In all genera except *Georychus*, longer sensory hairs (vibrissae) stand out from the pelage of the head and body; these are the primary hairs remaining on the naked mole-rat (Francaviglia 1896; W. J. Hamilton, Jr. 1928; Thigpen, 1940). The pelage is thick in species occurring in mesic areas, but De Graaff (1981) indicated that it shows a tendency for reduction in species occurring in hot, arid regions with high burrow temperatures (e.g., *Bathyergus janetta*, *Cryptomys damarensis*); the skin of *H. glaber* is almost naked (see plate 3-1). Lovegrove (1986b, 1987a) showed that conductance in the thinner-coated *C. damarensis* was higher than in *G. capensis*, which has a thick coat. However, anomalies do occur, and Haim and Fairall (1986) also found high conductances in *C. hottentotus* from mesic areas. Conductance in *H. glaber* is also very high (McNab 1966; Withers and Jarvis 1980), and the skin is well vascularized; these two factors would facilitate rapid heat exchange between the animal and its environment. The skin of *H. glaber* lacks sweat glands but contains sebaceous glands associated with sensory patches on the head (bearing vibrissae) and the ano-genital area (Tucker 1981). Ectoparasites may sometimes occur in the fur of all the haired genera (De Graaff 1964b), although we have found few parasites on *C. damarensis*. Immmature stages of mites (Hypoderidae) occur subcutaneously around the groin and axilla regions of *H. glaber* (Fain 1968), and recently Braude has found several other species of mites associated with *H. glaber* (see Lacey et al., chap. 8).

Bathyergids have several other distinctive morphological attributes. They have short limbs, cylindrically shaped bodies, and very loose skin. These attributes probably facilitate moving and turning in the narrow confines of burrows. Prominent extrabuccal incisors are a feature of all the mole-rats. The lips meet behind the incisors and effectively separate them from the oral cavity, thereby keeping soil out of the mouth and digestive tract when the animals dig. The outer edges of the hind feet are fringed with stiff hairs that increase the surface area of the foot and may help contain the excavated soil as it is being swept or pushed back along the burrow and up a side branch to the surface. The relatively long tail of *H. glaber* (ca. half the body length) bears scattered sensory hairs and is used to guide the mole-rat as it moves backward through its burrows. Other genera have short tails fringed with stiff hairs; when moving backward with a load of soil, these mole-rats place their tails on top of the soil and thus keep it from falling onto their backs.

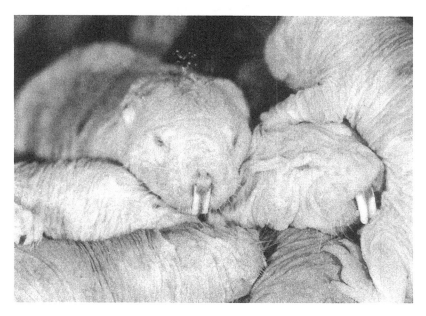

Plate 3-1. *Top*, A naked mole-rat (*Heterocephalus glaber*); *bottom*, a group of colony mates in the communal nest. Photos: C Springmann, R A. Mendez

PHYSIOLOGY

The microclimate in a burrow is buffered from large diel fluctuations in temperature (Kennerly 1964; McNab 1966; Jarvis 1978; Bennett et al. 1988). Burrows experience moderate seasonal temperature fluctuations in temperate areas and minimal fluctuations in tropical areas (especially deep underground; see Brett, chap. 5). Because the burrows are normally sealed from the surface, diffusion of oxygen and carbon dioxide takes place largely through the soil, and the burrow atmosphere can deviate greatly from that aboveground. Typically, the air in sealed burrows is characterized by higher relative humidities and levels of carbon dioxide and lower concentrations of oxygen than aboveground air (Arieli et al. 1977; Arieli 1979). Digging activity by the mole-rats is likely to enhance these differences and animals may encounter extreme conditions of hypoxia and hypercapnia at a time when their metabolic demands are also increased. Ar et al. (1977) found that the Mediterranean mole-rat, *Spalax ehrenbergi* (subfamily Spalacidae, table 3-1) exhibits a high tolerance to hypoxia and hypercapnia. Haim et al. (1985) and Haim and Fairall (1986) showed that some fossorial mammals, including *C. hottentotus*, excrete large quantities of bicarbonates and carbonates in their urine. They suggested that this may represent an important avenue for the loss of the carbon dioxide produced during exercise and also via fermentation processes in their ceca, which would otherwise be added to the already hypercapnic environment. However, this excretion of bicarbonates and carbonates could result in problems of acid-base imbalance, and this hypothesis needs further investigation. The blood of *H. glaber* has a high affinity for oxygen, which was viewed by Johansen et al. (1976) as a possible adaptation to subterranean life.

In the bathyergid genera, body size clearly tends to decrease with soil compaction and with food availability (table 3-3). Vleck (1979, 1981), Du Toit et al. (1985), and Lovegrove (1987b, 1989) have shown that the energetic cost involved in burrowing by gophers and mole-rats is high. Depending on soil density, compaction, and the size of the burrowing animal (and consequently the diameter of the burrow), digging can require 360 to 3,400 times as much energy as moving the same distance on the surface. These energetic considerations might explain why the large-sized mole-rats in the genus *Bathyergus* are found only in soft sandy soils, and why the largest species of the genus *Cryptomys* (*C. damarensis*) also occurs in sandy soils.

It could be argued that only large-sized mole-rats would be able to defend themselves effectively against predators in soft sand and that smaller mole-rats must rely on the "castles of clay" afforded by having burrows in harder soils. This argument seems to hold for *H. glaber* versus the other bathyergids, but it is not totally supported by the distribution of *B. suillus*, *G. capensis*, and *C. h. hottentotus* in parts of the southwestern Cape, South Africa. Where the soil is sandy, all three species may occur sympatrically, whereas sometimes as little as 100 m away, where the soil is more compact, only *G. capensis* and *C. h. hottentotus* are found (Lovegrove and Jarvis 1986, unpubl. data). Because the two smaller-sized species occur in both the soil types in this area and *B. suillus* does not, it implies that soil compaction is one of the determinants in the distribution of the larger *B. suillus*. The predation hypothesis predicts that the risk of predation must be lower for the smaller species in this particular area, allowing them to exist in the soft sand; however, we have no empirical data to substantiate this. Indeed, more potential predators of the smaller genera occur in the area than those of *B. suillus*. Furthermore, mole-snakes, cobras, herons, and small carnivores occurring in the area are known to prey upon all three species. In the Kalahari Gemsbok Park, body size in *C. damarensis* is appreciably larger in the soft dune valleys than in the more compact river beds (Lovegrove, unpubl. MS). Again there is no information on the comparative levels of predation in these two geographically close areas, but it seems unlikely that they differ markedly. Plasticity in body size in a single species, in relation to habitat and especially food resources, has been extensively documented in pocket gophers (Patton 1990) (Geomyidae, table 3-1), further demonstrating that a number of proximate factors play a role in determining the body mass of subterranean mammals. The reduction in body size culminates in *H. glaber*, which normally occurs in areas where the soil is extremely hard for most of the year, and where mean body masses for colony mates are about 32 g (see Brett, chap. 4). Where hard soil is accompanied by low food availability, mean body masses of *H. glaber* may be as little as 20 g (Jarvis 1985; Jarvis et al., chap. 12).

Other possible correlates of the gaseous environment within the burrow, the high metabolic cost of burrowing, and also habitat productivity and resource patchiness are (1) lower than expected metabolic rates and body temperatures and (2) higher than expected rates of thermal conductance; these lead to thermolability and poikilothermy in *H. glaber* (McNab 1966, 1968, 1979; Jarvis 1978; Withers and Jarvis 1980; Moon et al. 1981; Mustafa et al. 1981; see also Alexander, chap. 15). Yahav et al. (1989) demonstrated that when naked mole-rats are not huddling, there is a direct relationship between body temperature (T_b) and ambient temperature (T_a) throughout the ambient temperature ranges they tested ($15°$–$36°$C), indicating that these animals are essentially poikilothermic. Between T_a's of $15°$ and $28°$C, oxygen consumption ($\dot{V}O_2$) rose with T_a, thereby confirming this poikilothermic response. Above a T_a of $28°$C, changes in $\dot{V}O_2$ with increasing T_a followed the expected endothermic pattern. At these higher T_a's, oxygen consumption was minimal (0.79 ± 0.17 ml/g/h at $31.8° \pm 0.6°$C); thus, thermoneutrality falls within the normal range of burrow temperatures experienced by *H. glaber* ($28°$–$32°$C; see Brett, chap. 5). This resting metabolic rate is considerably higher than the 0.55–0.62 ml/g/h found by McNab (1966) and Withers and Jarvis (1980).

The body temperatures of freshly captured *H. glaber* lie within $2°$C of the burrow temperatures in which they were captured (28°–33°C). No animal ($n = 120$) was captured with a body temperature above 35°C even at the hottest time of the day (Jarvis, unpubl. data). Brett (1986) suggested that mole-rats "bask" in the early morning in superficial burrows lying directly under warm bare patches of soil, and observations in lab colonies confirm basking behavior (Lacey et al., chap. 8; Jarvis Appendix).

Yahav et al. (1989) found that at 31.8°C, evaporative water loss for individual naked mole-rats was significantly higher than for most other rodents. However, when 4–8 animals were huddling, savings in both $\dot{V}O_2$ and evaporative water loss were considerable. In addition, Yahav and Buffenstein (pers. comm.) have found that a huddling group of naked mole-rats can control their body temperature over the ambient temperature range normally encountered in their habitat (28°–32°C), whereas a lone animal or a pair of mole-rats cannot. This may be another factor favoring small groups of naked mole-rats dispersing as a unit, rather than as individuals or pairs. Lovegrove and Wissel (1988) stated that, irrespective of body mass, all bathyergids that have been investigated have similar metabolic rates per gram. Using the earlier values obtained by McNab (1966) for *H. glaber*, Lovegrove and Wissel (1988) showed that a 36-g *H. glaber* has practically the same mass-specific resting metabolic rate as a 1,000-g *B. suillus*. They suggested that low metabolic rates would lead to considerable energy saving, an important factor in mammals that have to expend considerable energy in digging through the soil (Lovegrove 1986a,b; Lovegrove and Wissel 1988).

Body growth rates of the Bathyergidae are also lower than expected for mammals of their size (Taylor et al. 1985; Bennett and Jarvis 1988a,b), but this is variable and may, in social genera, depend on the number of individuals already in the colony (Jarvis 1981; Bennett 1988; Bennett et al. 1990; Jarvis et al., chap. 12) and on the colony's social structure (Lacey and Sherman, chap. 10).

Many Bathyergidae consume a high proportion of fiber in their natural diet, and the adults of all species practice autocoprophagy. The cecal regions of *H. glaber* and *C. damarensis* contain large numbers of symbiotic protozoa, bacteria, and fungi (Porter 1957; Naumova 1974; Yahav and Buffenstein, pers. comm.), which together play an important role in the digestion of cellulose via the production of volatile fatty acids, CO_2, H_2 and CH_4. Yahav and Buffenstein (MS) have shown that in *H. glaber* the microbial organisms involved in cecal fermentation function optimally at 33°C; this is considerably lower than for other mammals but lies within the body temperature range normally experienced by naked mole-rats in their burrows. Yahav and Buffenstein (pers. comm.) have also shown that strong endogonic reactions are taking place in the cecum, such that cecal temperatures are consistently and significantly ($p < 0.001$) 1.5°C lower than body temperature. By contrast, the temperature in the fermentation chamber of ruminants is 2°–3°C higher than body temperature.

Yahav and Buffenstein (pers. comm.) suggest that these poikilothermic naked mole-rats cannot produce sufficient heat to counteract the endogonic reactions taking place in the cecum. The cecum therefore acts as a heat sink and, defying the laws of thermodynamics, remains at a lower temperature than the rest of the body.

Because of their subterranean existence, the Bathyergidae have little exposure to ultraviolet light and therefore are unable to metabolize vitamin D_3 (cholecalciferol) in their skins. Because they are herbivores, there is no obvious dietary source of vitamin D_3. In most mammals, vitamin-D-dependent calcium-binding proteins (calbindins) are responsible for active transport of calcium ions from the gut; relatively little calcium is taken up passively. Studies of calcium uptake by *H. glaber* indicate that it is passive and independent of vitamin D and the calbindins (Buffenstein et al. 1989). Furthermore, these studies have shown that the mole-rats are highly efficient in the uptake of calcium: they absorb more than 90% of the available calcium in the diet, whereas the efficiency of most mammals is below 60%. Naked mole-rats may have a high demand for calcium. In particular, the breeding female, who is pregnant and/or lactating throughout her life, would have high calcium demands. The ever-growing incisors, used in digging the burrows, also act as a calcium sink (Buffenstein and Yahav, MS). No data are available on the abundance of calcium in the natural diet of naked mole-rats. Their extremely efficient absorption of dietary calcium suggests strong selection to ensure that a sufficient supply is available. Brett (chap. 4) observed naked mole-rats in the field eating a bone and suggested that they were doing so to obtain additional calcium.

Social Structure

In solitary Bathyergidae (i.e., species in the genera *Bathyergus*, *Georychus*, and *Heliophobius*), each mole-rat occupies its own burrow system and for most of the year will defend its territory physically, fighting any conspecifics it encounters. Both sexes of *B. suillus*, *B. janetta*, and *G. capensis* drum on the ground with their hind feet (Bennett and Jarvis 1988a, unpubl. data), and this may function as a territorial advertisement. Nothing is known of the duration of burrow cohabitation by a pair of solitary mole-rats in the context of breeding. Observations of mating behavior in captive *G. capensis* (Bennett and Jarvis 1988a, this chapter) suggest that burrow cohabitation is brief (probably < 24 h), although multiple copulations occur while the pair are together. If the mother and weaned young of *G. capensis*, *B. suillus*, and *B. janetta* are kept together in the lab, levels of aggression build up, leading eventually (when the young are ca. 8-wk old) to serious fighting, injury, and death if the mole-rats are not separated (Bennett and Jarvis 1988a; Jarvis, unpubl. data). Much of the initial fighting seems to be over food, and the youngsters in particular steal from each other's food stores, even when there is an ad libitum supply. Aggression consists of shoving, biting (especially around the face and genitalia), tooth fencing, sparring, various vocalizations, and (in *G. capensis*) drumming with the hind feet.

There is confused and insufficient information on colony size and composition in many of the species and subspecies of *Cryptomys* (Genelly 1965; Hickman 1979b; Smithers 1983, Burda 1989), the only group-living genus aside from *Heterocephalus*. Interpretation of previous papers on *Cryptomys* are difficult because (1) few attempts have been made to trap entire colonies, and (2) researchers have typically assumed that the small mole-rats in a colony are younger than the larger ones. Recent studies on the growth rates and on the ages (using tooth wear or date of birth) of entire colonies of *C. hottentotus hottentotus* and *C. damarensis* (Bennett 1988; Bennett and Jarvis 1988b; Bennett et al. 1990) have shown that body size does not always reflect age, although the reproductive pair are among the largest and oldest animals in the colony.

Investigations into colony structure of wild-caught *C. h. hottentotus* from the Cape Peninsula of South Africa and *C. damarensis* from the Rehoboth district of Namibia indicate (Bennett 1988; Bennett and Jarvis 1988b) that colony sizes in the former are variable but smaller (range of 2–14, $n = 8$) than colony sizes in the latter (range of 8–25, $n = 9$). In each colony of both species, there is a single reproductive pair, usually the largest male and female in the colony; the female is usually one of the last animals to be captured. Post mortem examination of a complete wild-caught colony of 25 *C. damarensis* (14 females, 11 males) showed that only 1 female had placental scars but that spermatogenesis was occurring in the testes of many of the males (Bennett and Jarvis 1988b). This restriction of breeding among females (at least) and the difficulty in capturing the breeding female, might also explain why Smithers (1983) found only 5 gravid *C. h. darlingi* females in a sample of several hundred specimens he examined. Recently, Burda (1989, pers. comm.) has found indications of body-size and colony-size variability in what is probably another subspecies of *C. hottentotus*, from Zambia, in which up to 25 individuals may occur in a single colony.

Jarvis (1978, 1985), Brett (1986, chap. 5), and Lovegrove and Wissel (1988) proposed that because of the energetic costs of burrowing and the fact that the search for food is often blind, group size in mole-rats is a positive function of the mean distance between geophytes (bonanza-quality food resources). They argued that a cooperative effort is required to locate widely and randomly dispersed geophytes and thereby to reduce the risk of not finding food and starving to death. Where sufficient data on the size and distribution of geophytes are available to test this hypothesis (table 3-4), it certainly holds true (Brett 1986; Lovegrove and Painting 1987; Lovegrove and Wissel 1988;

Table 3-4

A Synopsis of Available Food, Its Dispersion, and Digestibility for Species from Five Bathyergid Genera

Species	Food Size (g wet wt.)	Available Energy (KJ/m²)	Digestible Energy (KJ/m²)	Mean % Digestible Energy (± SE)	Coefficient of Dispersion	Source
Bathyergus janetta	< 10	84 4	68 9	81 5 ± 7 2	1 6–5.0	1, 2 4
Cryptomys h hottentotus (mesic)	1–2	300 3	285 3	96 1 ± 1 8	14 8	3, 4
C. h hottentotus (arid)	1–4	181 0	159 0	92 1 ± 4 6	2 3–9 5	4
C damarensis	200–2000*	59 7, 610 0†	31 6, 323 0†	53 0 ± 9 8	15 4–24 9	1, 4
C damarensis	2–10‡	194 0, 345 0†	184 0, 330†	95 7 ± 2 2	10 6	4, 5
Georychus capensis	0 5–2	60 0	57 6	96 1 ± 1 8	2 6–11 2	4
Heterocephalus glaber	2–30,000	312 0	189 0–220§	60.0–70 0§	32 m between patches, 3 3 m between large tubers	6

NOTE For C h hottentotus, values are given for a mesic and an arid area, for C damarensis, values are given for an area with large tubers and another with small bulbs

* Monoculture of large tubers (Acanthosicyos)

† Because the geophytes are distributed in widely dispersed patches, the mean in and out of patches is given Similar data are not available for H glaber

‡ Monoculture of smaller bulbs (Dipcadi)

§ Digestibility is estimated for two species of tubers, Macrotyloma maranguense and Pyrenacantha kaurabassana (on the basis of digestible energy of Acanthosicyos for C damarensis)

SOURCE 1, Jarvis, unpubl data, 2, Bennett, unpubl data, 3, Jarvis and Lovegrove, unpubl data, 4, Bennett 1988, 5, Lovegrove 1987b, 6, Brett 1986

Jarvis and Bennett, unpubl. data). Thus, *C. damarensis*, probably the most social species in that genus (Bennett and Jarvis 1988b), and *H. glaber*, the most social of all the Bathyergidae, occur in arid areas with large widely dispersed geophytes, and where no solitary mole-rat genera occur. However, in equally arid regions of the northwestern Cape, where the geophytes are small and grow close together, four colonies of *C. h. hottentotus* contained only five to seven animals (Jarvis and Bennett 1990), suggesting to us that large colony sizes are not favored in these regions; unfortunately, predator pressures in these regions have not been quantified (see also Alexander, chap. 15).

With the possible exception of the octodontid *Spalacopus cyanus* (Reig 1970) the family Bathyergidae contains the only known social subterranean mammals. Hypotheses on the evolution of sociality in this family must examine all the genera and must also address the question of why other families of completely subterranean mammals are not social. Furthermore, explanations must be found as to why solitary bathyergids are absent from arid areas where food consists of widely dispersed tubers but can occur in very arid regions where the food is not dispersed. Thus, *B. janetta*, the only solitary bathyergid inhabiting very arid regions, only occurs where there are small, closely spaced tubers (table 3-4; Jarvis, M. Griffin and E. Griffin, unpubl. data).

Social bathyergids occur in a much wider range of habitats than do the solitary genera, in soft (*C. damarensis*) and hard, compact soils (*H. glaber*), where food is in the form of large, widely dispersed tubers (*C. damarensis* and *H. glaber*) or small, closely spaced tubers (*C. h. hottentotus*). We believe that predator avoidance (Alexander et al., chap. 1; Alexander, chap. 15) may be a contributing factor if cooperative digging decreases the total time necessary to excavate foraging burrows and dispose of the soil on the surface. Predation (by snakes) is most likely to occur during digging and mound formation (Fitzsimmons 1962; Brett, chap. 4; Braude, chap. 6). Time could be saved (and exposure to predators reduced) if the food located by the digging team were large, providing the colony with enough to enable them to temporarily halt their foraging activities. We believe it is significant that, to our knowledge, Africa is the only continent where many species of plants adapted to arid habitats have large subterranean storage organs; it is therefore a place where benefits can be gained from cooperative foraging by subterranean rodents using habitats inaccessible to solitary species. Interestingly, with one exception (the African striped weasel, *Poecilogale albinucha*), Africa lacks burrow-dwelling mammalian predators such as stoats and weasels.

Apart from restricted reproduction (i.e., only one pair reproduces), division of labor and definite behavioral roles are poorly developed in *C. h. hottentotus* but do occur in *C. damarensis* (Bennett and Jarvis 1988b; Bennett 1988, 1989). The smaller-sized mole-rats in *C. damarensis* colonies perform significantly more maintenance work (digging, sweeping, carrying food and nesting material) than do the larger animals (Bennett and Jarvis 1988b; Bennett 1990). Bennett (1988) has found that *C. damarensis* "workers" are often, but not invariably, the younger animals.

Bennett (1988) and D. Jacobs, Bennett, Jarvis, and T. Crowe (MS) have analyzed interactive behaviors in three *C. damarensis* colonies and have determined their dominance indices (Aspey 1977). Dominance hierarchies (Chase 1964) within these colonies and their linearity were tested using Landau's Index (Bekoff 1977) and confirmed by *Q*-type factor analysis. By these methods, Bennett (1988) and Jacobs et al. (MS) have shown that dominance in *C. damarensis* colonies is linear, with the breeding male at the top of the hierarchy followed by the breeding female. In the rest of the colony, males are dominant over females, and dominance is related to body mass. A similar linear-dominance hierarchy has also been described by Rosenthal (1989) for a colony of *C. h. hottentotus*.

At present, we have only limited information on how a *C. damarensis* breeding pair suppresses breeding in the other members of their colony. There appear to be frequent overt behavioral interactions, especially when dominant and subordinate individuals meet in the burrow or communal nest. In the burrow, these agonistic behaviors include vocalizations, tail raising, tail pulling, smelling genitalia, and shoving subordinates (in *H. glaber*, shoving, a mild form of aggression, is also frequently initiated by the breeding female; Reeve and Sherman, chap. 11). In the communal nest, *C. damarensis* juveniles allogroom dominant animals, especially the breeding pair. The breeding female grooms the juveniles from her previous litter, and this continues until her next litter is born (Bennett 1988, 1990). Juvenile mole-rats of both *Cryptomys* species engage in sparring bouts, in which individuals lock incisors and push and pull each other; this may serve to determine status within the colony, but too few long-term data on captive colonies are available to test this hypothesis.

Colony size and structure in *H. glaber* has been described elsewhere (Jarvis 1978, 1981, 1985; Brett 1986, chap. 4). In uncultivated habitats, colonies containing approximately 100 naked mole-rats have been captured, and the average colony size is 70–80. In eight complete and almost complete colonies captured there has never been more than one reproductive female (Jarvis 1985; Brett, chap. 4). The breeding female is usually the longest-bodied and one of the heaviest mole-rats in the colony (Jarvis et al., chap. 12). She is usually the last or one of the last to be captured. Within the colony, there is division of labor (Jarvis 1981; Lacey and Sherman, chap. 10; Jarvis et al., chap. 12). The breeding female appears to maintain her status within her colony through a mixture of behavioral and semiochemical control (Jarvis 1981, 1982, 1984; Abbott et al., 1989; Reeve and Sherman, chap. 11; Faulkes et al., chap. 14, in press). Often there are two to three reproductive males in a colony (Jarvis, chap. 13; Lacey and Sherman, chap. 10). Interestingly, there is however little aggression between the males and both (all) will mate with the breeding female when she is in estrus.

From when the pups are weaned (ca. 4 wk) until they are about 2-yr old, they tooth fence, spar, and wrestle with each other (Lacey et al., chap. 8; Jarvis, chap. 13); sometimes four to five individuals will team up against another group or an individual. As with *Cryptomys* species (Bennett and Jarvis 1988b; Bennett 1988, 1990), sparring in *H. glaber* may be important in determining the eventual position of individuals within the colony hierarchy; however, no quantitative data are available about this.

In the social mole-rats (e.g., *H. glaber*, *C. damarensis*, and *C. h. hottentotus*), colony members share a communal nest (e.g., plate 3-1) and toilet area and also share the food found by the colony. On entering the toilet, naked mole-rats frequently sweep with their hind feet and then groom their body, face, and mouth with their soiled hind feet, thereby spreading excretory products of the colony, and perhaps more importantly of the breeding female, onto themselves (Lacey et al., chap. 8). These products may constitute a colony odor and also may directly affect the recipient mole-rat; the individual also acts as a vector carrying the products out of the toilet into the burrow system and nest. In two species of *Cryptomys* (Bennett 1988), individuals may sweep back with their hind feet while in the toilet, but grooming is restricted to doubling up and cleaning the fur on their postero-ventral surface and head with the forefeet and mouth. On leaving the toilet area, *H. glaber* as well as species of *Cryptomys* may drag their ano-genital area along the floor of the burrow (see Lacey et al., chap. 8); in *Cryptomys*, at least, other individuals smell these marked areas.

Reproduction

Data on breeding seasons, litter sizes, and gestation periods of mole-rats are incomplete (table 3-5). Most species in the family Bathyergidae are seasonal breeders, with the exception of *Heterocephalus glaber* and *Cryptomys damarensis*. Environmental factors acting as cues for breeding are unknown, but because mole-rats live almost entirely in the dark, these factors are more probably rainfall or temperature fluctuations than changes in photoperiod. Quay (1981) found that the pineal gland of *H. glaber* is atrophic and the smallest in absolute size of any rodent so far examined. This is consistent with their subterranean existence and suggests that photoperiodic cues do not influence reproductive cyclicity. The pineals of other Bathyergidae have not been examined.

Within the Georychinae, there is a progressive change in courtship and mating behavior as sociality becomes more pronounced. In the solitary *Georychus capensis*, the initial contact between a pair is through drumming with the hind feet (Bennett and Jarvis 1988a; P. Narins, Reichman, Jarvis and E. Lewis, unpubl. data). Since drumming sounds carry through soil, this signalling probably first occurs when the two burrow systems are still discrete, and it might conceivably convey information on the sex and receptivity of neighboring mole-rats. In captivity, male *G. capensis* initiate courtship with bursts of rapid drumming, and the female responds with much slower drumming. After the pair meet, their drumming is intermingled with vocalizations and locking of incisors. The mounting and mating sequence is very brief: the male simply seizes and mounts the female, who raises her tail. Mating often terminates with the female turning around and chasing the male. The impression gained by an observer is that (in captive animals, at least) the male seizes his opportunity, mates, and runs. Once apart, both animals groom their genitalia. In captivity, the mating sequence may be repeated over a 2-mo period (June and July), with the male hiding from the female between matings (Bennett and Jarvis 1988a);

Table 3-5

Reproductive Data for Selected Species in the Family Bathyergidae

Species	Pregnant and Lactating Females	No. of Litters Annually	Litter Size Range	Litter Size x̄	Gestation Duration (days)	Birth Mass (g)	Pups First Leave Nest*	Eyes Open*	First Eat Solids*	Weaned*†	Begin to Spar*	Disperse*	Source
Bathyergus suillus	Jul–Oct (winter)	1–2	1–4	2.4	ca. 52	34	4–5	10	10–13	21	12–13	60–65	1, 2, 3
Bathyergus janetta	Aug–Dec	1–2	1–7	3.5	—	15.4	9	15	13	28	11–16	60–65	4
Cryptomys h. hottentotus	Oct–Dec	1–2?	1–6	6	59–66	8–9	5	13	10	28	10–14	social	5
Cryptomys damarensis	all year	1–4	1–5	3	78–92	8–10	1–3	18	6	28	18–25	social	6
Georychus capensis	Aug–Dec (spring/summer)	1–2	4–10	6	44	5–12	7	9	17	28	35	55–60	7
Heliophobius argenteocinereus	Apr–Jun	?	2–4	—	ca. 87	7	—	—	—	—	—	—	1, 8
Heterocephalus glaber (captive)	all year	4–5	1–27	13	66–74	1.8	14	30	14	21	21	social	9
Heterocephalus glaber (wild)	all year	4–5	4–15	10								social	10, 11

* Data are given in days after birth.   † Pups are considered weaned when most of the diet is solid foods; they may still suckle infrequently.

SOURCE: 1, Jarvis 1969, 2, Jarvis, unpubl. data; 3, van der Horst 1970; 4, Jarvis, D. Hamerton, M. Griffin, and E. Griffin, unpubl. data, 5, Bennett 1990, 6, Bennett and Jarvis 1988b; 7, Bennett and Jarvis 1988a; 8, Kingdon 1974; 9, Jarvis, chap. 13; 10, Jarvis 1985, 11, Brett, chap. 4.

mating may also occur postpartum (ca. 10 days after the first but not the second litter) or after the young of the first litter have dispersed.

In a captive *Bathyergus suillus* pair housed in adjacent tunnel systems separated by a wire-mesh partition, the male drummed for 2 wk before the female became perforate and began to drum back. This was in June, the beginning of the breeding season of wild *B. suillus* (Jarvis 1969; van der Horst 1970). Initially, the male drummed with single beats and the female responded with a two-beat drum; however as this "concert" progressed, the tempo of the drumming by the male increased until the sounds were indistinguishable. When the partition between the two tunnel systems was removed, the male followed the female, smelling her genital region. At length she would stop walking, squeal loudly, raise her tail, and he would then attempt to mount. Unfortunately, the female would never allow courtship to progress further than this, and after several attempts to mate, the animals would begin to fight and were separated to prevent injury (E. McDaid and Jarvis, unpubl. data).

Courtship in the social *C. hottentotus hottentotus* is initiated by the male when he encounters the estrous reproductive female in the burrow. Courtship is more elaborate than in *G. capensis*. Courtship behaviors include vocalizations, the female raising her tail, the male smelling her genital region, the male taking the female's rump into his mouth and chewing it gently, and the male stroking the female's sides with his head (Hickman 1982; Bennett 1988, 1989; Rosenthal 1989). This is followed by mounting and mating. The mating male is not subsequently chased by the female.

Cryptomys damarensis differs from *C. h. hottentotus* in that the female now initiates the courtship sequence. In captivity, on encountering the reproductive male in the burrow, the *C. damarensis* reproductive female vocalizes, briefly drums with her hind feet, and then sometimes mounts the male's head. The pair then enter a chamber and chase each other, head to tail, in a tight circle before the female pauses, raises her tail, is smelled, mounted and mated; there is no post-mating chasing (Bennett and Jarvis 1988b; Bennett 1990). This sequence is repeated quite frequently for about 10 days and then ceases.

In *H. glaber* the female also initiates courtship and mating, usually by vocalizing and then crouching in front of a breeding male and adopting the lordosis position, whereupon the male mounts (Jarvis, chap. 13). Mutual ano-genital nuzzling and sniffing may also occur as a prelude to or during courtship (Lacey et al., chap. 8). Mating rarely continues for more than a few hours. Behavioral estrus in *H. glaber* usually occurs about 10 days postpartum (Lacey and Sherman, chap. 10; Jarvis, chap. 13).

In all the colonial (i.e., social) mole-rat species, courtship and mating is restricted to a single reproductive female and one to three reproductive males; the other males in the group show no apparent interest in the estrous female. There is seldom any fighting between males for mating opportunities. In captivity, pair-bonding between the reproductive animals outside the mating period appears to be very weak in *C. h. hottentotus*, to be largely restricted to the male grooming the female in *C. damarensis*, and is strongly maintained with mutual ano-genital nuzzling and sniffing in *H. glaber* (see Jarvis, chap. 13).

No sexual dimorphism exists in the solitary *G. capensis*, whereas it is marked in *B. suillus* and *B. janetta* (table 3-3). This may suggest that, as with *Thomomys bottae* (Reichman et al. 1982), only the largest and most aggressive *Bathyergus* males mate; they may well be polygynous. *Bathyergus suillus* males have extremely thick skin (up to 10 mm) on the ventral surface of the neck (Davies and Jarvis 1986), and this may also indicate that males fight each other. Indeed, two badly injured *B. suillus* males were found by us in the field. They had deep wounds on the head and broken incisors; the nature of the wounds was consistent with punctures made by the incisors of another mole-rat. The largest animal in a *Cryptomys* colony is usually the reproductive male. Since other males in the colony may be smaller than the reproductive female, the existence of sexual dimorphism may be masked by the interrelationship between social status and the body size of colony members. However, morphometric studies (whole-body measurements and skull dimensions) on one colony of *C. h. hottentotus* (n = 13 individuals) and one of *C. damarensis* (n = 25), both sacrificed soon after capture, revealed no sexual dimorphism for the former species, but clear dimorphism for the latter (Bennett 1988; Bennett et al. 1990). Hagen (1985) found no sexual dimorphism in *H. glaber*. In naked mole-rats, the highest levels of aggression occur between the largest females in the colony, often involving shoving encounters between the animals (Lacey and Sherman, chap. 10; Reeve and Sherman, chap. 11; Jarvis, chap. 13). In *C. damarensis*, both the breeding female and the male pull and push colony mates along the burrows and also take food away from other colony members (Bennett 1988; Jacobs et al., MS).

With the exception of *H. glaber*, litter sizes among bathyergids tend to be small (1–6 pups, table 3-5) with one and maximally two litters born annually (Jarvis 1969; De Graaff 1981; Smithers 1983; Bennett and Jarvis 1988a,b; Bennett 1988; Burda 1989). *Georychus capensis* may occasionally have up to 10 pups in a litter (mean = 6) (Taylor et al. 1985; Bennett and Jarvis 1988a). Although the maximum litter size of captive *H. glaber* exceeds that recorded for any other mammal (up to 27 young/litter; Jarvis, chap. 13), Brett (1986, chap. 4) suggests that litter size in the wild rarely exceeds 12. However, the high litter size shown by captive *H. glaber* indicates that they have the potential to bear larger litters under optimal conditions. Perhaps this enables *H. glaber* to capitalize on the reproductive opportunities offered during years with exceptionally high rainfall or food availability. If this is so, it is intriguing that *C. damarensis* has such small litters (i.e., a maximum of 5 pups). In captivity, the *H. glaber* breeding female gives birth about every 70–80 days (four to five litters/year; Jarvis, chap. 13; Lacey and Sherman, chap. 10). If the colony is regarded as one reproductive unit, containing 40–100 animals (Jarvis 1985; Brett, chap. 4), a maximum of 54 young born to the single reproductive female in the colony each year is equivalent to a recruitment of 1.0 to 2.7 pups by each female in the colony (if they all bred). The other social bathyergids, *C. h. hottentotus* and *C. damarensis*, have a single breeding female per colony, and she gives birth to 1–5 pups per litter (Smithers 1983; Bennett 1988; Bennett and Jarvis 1988b). *Cryptomys h. hottentotus* probably breeds seasonally, but recent work on *C. damarensis* (Bennett and Jarvis, unpubl. data) indicates a lack of seasonality and up to four litters annually. These data again indicate very low annual recruitment rates for entire colonies in this genus. By contrast, in the solitary genera (e.g., *Georychus*), where each adult female in the population probably has one or more litters per year, annual recruitment may be 5–10 pups per female.

Although little is known of gestation periods (table 3-5), these appear to be longer in bathyergids than expected for rodents of their size. Phylogenetically, this may reflect their hystricomorph affinities (Weir 1974; Honeycutt et al., chap. 2). At birth, the pups of all genera are hairless or nearly so and have limited powers of locomotion, although they are able to move around the nest to find their mother. The pups of social genera stay or are kept by the other mole-rats in the communal nest and, if alarmed, nonbreeding colony members carry the pups out of the nest and later assist in returning them (Jarvis, chap. 13). Mole-rat pups become independent fairly rapidly. *Cryptomys damarensis* pups are the most precocial, and they walk along the burrows and begin to eat solids within a week of birth. *Heterocephalus glaber*, *G. capensis*, and *B. suillus* pups are more altricial and take 2–3 wk to reach a comparable degree of independence. Coprophagy is practiced by pups of all the Bathyergidae. At weaning, *G. capensis* juveniles visit the toilet area for feces, which they then eat, whereas *H. glaber* and *C. damarensis* pups beg feces from adults; they may also visit the communal toilet area for feces.

The young of the solitary genera disperse soon after weaning (i.e., 7–8 wk of age in *G. capensis*, *B. suillus*, *B. janetta*: Bennett and Jarvis 1988a; Jarvis, unpubl. data). Natal dispersal is preceded by increasingly violent sparring, incisor fencing, and biting, and foot drumming between littermates. Juveniles often disperse by extending portions of their mother's burrow and establishing satellite systems in the immediate vicinity; sometimes dispersal is aboveground, and new systems are established some distance away from the natal burrow (i.e., more than 50 m in some cases). Low vagility is a characteristic of many subterranean mammals (Patton 1990).

Longevity, Predation, and Other Causes of Mortality

Little is known of longevity in most bathyergids. In captivity, 3 *Cryptomys hottentotus hottentotus* lived for 5 yr after capture, and 1 *C. damarensis* has been in captivity for 10 yr (as of October 1989); these mole-rats therefore live longer than other rodents of similiar body size (e.g., *Rattus norvegicus*). Exceptional longevities have been found in captive *Heterocephalus glaber*. Indeed, 17 adult (i.e., probaby 1 yr or older) naked mole-rats captured in July 1974 are still alive, and 2 reproductive females were still breeding in Jarvis's lab in October 1989, making them 16 yr or older. Moreover, 60–70 adults captured by the editors of this volume in September 1977 and December 1979 are still alive as of this writing (i.e., 11–13 yr later).

In general, as with other subterranean rodents, bathyergid mole-rats are protected from predators by living underground in burrow systems that are thoroughly sealed from the surface. Apart from snakes, only two predators in Africa are small enough to enter burrows and strong enough to catch any species of mole-rat. The burrows of *Bathyergus suillus* are large enough to accommodate the African skunk (*Ictonyx striatus*), and this animal has occasionally been caught in traps set for *B. suillus* (Jarvis, unpubl. data). The widespread (but rare) striped weasel, only weighs 260–270 g at the maximum. Captive specimens will eat *Cryptomys*, but have difficulty in killing a half-grown *Rattus rattus* (Smithers 1983), suggesting that they may not generally be formidable mole-rat predators. We know of no predators able to dig fast enough to catch mole-rats by excavating their burrows, although some can dig and catch mole-rats as they form their mounds. Once a side branch to the surface has been thoroughly sealed, and provided that the burrows are not damaged, all the bathyergids would seem to live in safe environments. Mole-rats are most vulnerable when working close to the surface during the forming of mounds and when foraging for shallow roots and bulbs. In all the genera except *H. glaber*, the working mole-rat is partly protected by the plug of soil separating it from the surface. Nevertheless, this plug offers far less protection than does the thoroughly packed side branch that is below a completed mound. It seems likely that only during mound formation and feeding would an aboveground predator know where a mole-rat was.

Snakes may be the primary predators on mole-rats. Observations in the field (Jarvis, unpubl. data) of the behavior of mole snakes (*Pseudapsis cana*) suggest that they are attracted by the smell of freshly turned soil. Broadley (1983) stated that mole snakes push the forepart of their bodies through a molehill into the burrow below and wait in this position for the return of the mole-rat. Penetration by mole snakes into burrows would be easiest through the loose soil above an incomplete mound; furthermore, the snake would not have to wait long at an incomplete mound for the return of the mole-rat with its next load of soil. The mole snake is known to prey on many species of southern African mole-rats, including *B. suillus* (FitzSimons 1962; De Graaff 1964a; F. Duckitt, pers. comm.). Because it is a widespread species also occurring in eastern Africa, it may also prey on *H. glaber*. De Graaff (1981) trapped a shield-nosed snake (*Aspidelaps scutatus*) in a *C. hottentotus* burrow and found that it contained the hindquarters of a mole-rat. Similarly, the editors of this volume, and R. L. Honeycutt (pers comm.), dissected three rufous-beaked snakes (*Rhamphiophis oxyrhynchus rostratus*) captured in naked mole-rat burrows in Kenya and found them to contain *H. glaber* carcasses. Brett (1986, chap. 4) observed the rufous-beaked snake, the file snake (*Mehelya capensis savorgnani*), and the herald or white-lipped snake (*Crotaphopeltis hotamboeia*) catching *H. glaber* workers as they kicked soil out of an open hole onto the surface; Braude (chap. 6) found a sand boa (*Eryx colubrinus*) with a naked mole-rat in its stomach.

A number of species of birds (herons, storks, and barn owls, as well as other birds of prey) and carnivores (wildcats, caracals, jackals, mongooses, and even hyenas) also capture the haired mole-rats as they push soil to the surface (De Graaff 1981). Although we know of no direct evidence, it seems possible that birds with long beaks (e.g., storks and herons) could also catch *H. glaber* during "volcanoing;" we have observed *C. h. hottentotus* taken this way and *H. glaber* is clearly visible to a predator as it kicks out soil (Braude, chap. 6).

Georhychus capensis periodically abandon the burrow system and move on the surface; thus, areas worked by mole-rats are suddenly abandoned, and fresh workings appear some distance away in previously unoccupied areas. In the suburbs of Cape Town, mounds of *G. capensis* appear on grassed traffic islands completely surrounded by highways; the only access to these areas would be if the mole-rat came aboveground and crossed the highway (Jarvis, unpubl. data). During the move aboveground, the mole-rat would be especially vulnerable to predation by many animals that would normally be unable to catch them. Roberts (1951) reported aboveground movement by *C. hottentotus*.

Data from *B. suillus* (Davies and Jarvis 1986), *C. damarensis* (Jarvis and Bennett, unpubl. data) and *H. glaber* (Brett 1986, chap. 5) suggest that these mole-rats remain resident in an area for long periods, working and reworking the same home range; they are probably exposed to lower predation pressures than *G. capensis* because they remain primarily underground. Of these three species, *B. suillus* comes aboveground most often, and farmers claim that wandering on the surface commonly occurs before rainfall (M. Jarvis, pers. comm.). Furthermore, a recent survey of prey remains from black eagle (*Aquila verreauxi*) nests from the fynbos (areas with mediterranean chapparal-like plant communities favored by *B. suillus*) showed that *B. suillus* formed a major part of their diet (M. Jarvis, pers. comm.)

In all areas except those with an exceptionally high plant biomass, *H. glaber* colonies are spaced far apart, and new small-sized colonies are rare (Jarvis 1985; Brett 1986, chap. 4, 5); this may imply that established colonies grow very slowly and that recruitment approximately balances predation (see also Jarvis, chap. 13). Brett (1986, chap. 4) has calculated that, even in the absence of predation, a colony containing 90 animals and 1 reproductive female would have taken at least 2 yr to attain this size; even moderate amounts of predation (e.g., 2 mole-rats/mo) would double the time taken for the colony to grow to 90 animals, and predation of 4–5 animals a month would halt colony growth. Even lower levels of predation would halt the growth of colonies of *C. damarensis.*

In the southwestern Cape, many of the lowland areas favored by mole-rats are subject to seasonal flooding; this also occurs in parts of the Karroo (Roberts 1951) and probably in other areas. Mole-rats (*C. h. hottentotus* and *H. glaber*) can swim for brief periods (W. J. Hamilton, Jr. 1928; Hickman 1978, 1983). However, during flooding, the southern African species at least are often forced to find refuge on high ground and also come out onto the surface where they frequently die of cold or are taken by predators (Roberts 1951; Jarvis and Lovegrove, unpubl. data; F. Duckitt, pers. comm.).

Burrow Spacing

The burrow systems of the solitary mole-rats frequently occur in groups or clusters separated from other groups by vacant ground. Neighboring burrow systems may interdigitate and their side branches may come within 10 cm of each other (Davies and Jarvis 1986; E. McDaid, pers. comm.). Open spaces between clusters of burrows may become filled as the population density increases. Each burrow system is a separate unit containing, for most of the year, a single mole-rat. No data are available on the degree of relatedness of mole-rats within a cluster but, because of the way the young disperse, clusters may contain relatives in many instances.

Data on burrow spacing in the social mole-rats are limited. *Heterocephalus glaber* colonies are usually spaced 400–1,000 m apart (Jarvis 1985; Brett 1986, chap. 4, 5). It is difficult to determine the spacing of *Cryptomys h. hottentotus* and *C. damarensis* burrows without excavating entire systems. One *C. h. hottentotus* burrow that was dug up contained 14 mole-rats and was 1,000 m long (Jarvis and Lovegrove, unpubl. data). It is therefore hardly surprising that few systems have been excavated in their entirety (Genelly 1965; Hickman 1979; Davies and Jarvis 1986; Lovegrove and Jarvis 1986, unpubl. data). Five neighboring colonies of *C. damarensis* have been captured at Dordabis, Namibia in an area of approximately 200 × 200 m. The nearest tunnels of two of these systems were less than 10 m apart.

As with many other subterranean mammals (Patton 1990), the distribution of all the genera and species of mole-rats is patchy (see Honeycutt et al., chap. 2); large tracts of land may be entirely devoid of mole-rats and then when a slight change in soil, rainfall, or vegetation occurs, the mole-rats will be present again (Shortridge 1934; Roberts 1951; Jarvis, pers. obs.). This patchy distribution, as well as the relatively poor mobility of mole-rats, may have important implications for speciation in the Bathyergidae (Nevo et al. 1986, 1987; Honeycutt et al. 1987; chap. 2).

Discussion

Most subterranean mammals live alone (Nevo 1979). The Bathyergidae are exceptional in having colonial genera. Only one other subterranean rodent, the octodontid *Spalacopus cyanus*, appears to live in colonies (Reig 1970); however, little is known of its social organization. Within the Bathyergidae can be found the entire range of sociality, from completely solitary to colonies with a division of labor (Jarvis 1981; Brett 1986; Bennett and Jarvis 1988b; chap. 10–14 in this volume), opening the way for a study of the determinants of sociality. This is currently being investigated along a broad front, incorporating the ethology, ecology, and physiology of the Bathyergidae; the discussion that follows should therefore be considered as partly speculative. Various contributors to this volume present a number of different hypotheses about the evolution of sociality in mole-rats; here we present our own interpretations. We do not address why the Bathyergidae originally became subterranean, but we speculate on why two genera are social whereas the other three are solitary. The question of the evolution of eusociality in one of the two social genera is discussed in chapters 1, 10, and 15.

It is beyond our scope to explore in depth whether the first Bathyergidae were solitary or social. However, two factors seem to indicate that the move was from solitary to social. First, few if any other completely subterranean mammals are social. Second, the largest living Bathyergidae are solitary (e.g., *Bathyergus* and *Georychus*; see table 3-2); this holds even for *B. janetta*, which lives in very arid regions. The earliest known fossil Bathyergoidea were large animals, and if they were indeed completely subterranean, this might suggest that they also were solitary. However, since mtDNA analyses (Honeycutt et al. 1987; chap. 2) indicate that the split between the solitary and social Bathyergidae occurred early in the evolutionary history of this family, this issue cannot be resolved until more data are available.

From the past and present distribution of the Bathyergidae, it is apparent that most species of solitary genera occur in more mesic areas than do species of social genera. In mesic areas, the rainfall is higher and more predictable than in the arid areas, the soil is easily worked for more days in the year, and the mole-rats eat the aerial parts of plants as well as small corms and bulbs with a high nutritional value and high digestibility (table 3-4; Du Toit et al. 1985; Lovegrove and Jarvis 1986; Bennett 1988; Jarvis and Bennett, unpubl. data). In these habitats, shorter burrows can meet the energetic requirements of a solitary mole-rat. The arid areas are characterized by low, and perhaps more important, unpredictable rainfall and by high soil surface and burrow temperatures for at least some of the year. In the case of *Heterocephalus glaber*, the deeper burrows maintain a practically steady temperature throughout the year (Brett, chap. 5). The food in arid areas tends to be more fibrous and of a poorer quality (in terms of its digestibility) than in mesic areas (table 3-4; Bennett 1988), and individual items of food are often widely and patchily distributed, but of greater mass (Brett 1986; Jarvis and Bennett, unpubl. data; Lovegrove and Painting 1987).

In species adapted to more arid habitats, burrow systems are longer per gram of mole-rat (table 3-3; Davies and Jarvis 1986). Unpredictable rainfall and high evaporation rates mean that in many months the soil is dry and difficult to work, and digging conditions are optimal for only short periods. We have already suggested that even under optimal conditions, there are limits to the distance that can be dug each day by a single mole-rat. Two of these constraints are (1) the energetic cost of digging and (2) the rate at which the incisors can grow each day. Because of these limitations, the harder the soil, the shorter the distance one mole-rat can dig in a day. Unlike the aboveground vegetation, the biomass of tubers, bulbs, and roots is seasonally relatively constant, although the cost incurred by a mole-rat to obtain it varies seasonally and regionally.

Mole-rats in arid environments also tend to be small. Small mole-rats require a burrow system with a smaller diameter (and therefore need to move less soil/m dug); in a thermally stable environment, they also need less food than larger mole-rats. Considerable energy and water is also conserved when mole-rats huddle together at rest (Withers and Jarvis 1980; Yahav et al. 1989, unpubl. MS).

The burrow microclimate, and probably also the pressures due to the high energetic cost of burrowing and to food availability (McNab 1966; Vleck 1979, 1981), may favor the evolution of mammals with lower than expected basal metabolic rates; the lowest rates are in mole-rat species occurring in hot arid areas. These species adapted to arid habitats also often have high conductances (McNab 1966, 1980; Jarvis 1978; Withers and Jarvis 1980; Lovegrove 1986a,b, 1987b) and may exhibit poikilothermy. Consequences of a low metabolic rate are low growth rates, slow maturation, and long life spans, features usually found in larger mammals with low rates of extrinsic mortality, but not in most small mammals in which thermoregulatory constraints usually require high metabolic rates. The thermally stable burrow environment of mole-rats perhaps frees them from these constraints.

When all these factors are considered together, a picture emerges of multiple interlinking features, all contributing to the evolution of sociality in *Cryptomys* and *Heterocephalus*. We suggest that, with increasing aridity, mole-rats must dig longer burrow systems for two main reasons: (1) to find their patchy food supply, and (2) to find sufficient food to last them through a period (of considerable and uncertain length) during which it is difficult and more costly to dig for food. Lovegrove and Wissel (1988) have shown mathematically that in regions where food is widely and randomly spaced, the risks of a solitary mole-rat not finding food are very high and that these risks are greatly decreased when several mole-rats search for the food (see also Brett, chap. 5). In these (arid) regions, survival therefore depends on having a work force that reduces the risks of not finding food and that shares the cost of burrowing, of repairing the tunnel system, and also of defending it against predators. Obviously, the benefits accrued from this sharing only pay off if sufficient food is then found to feed all the mole-rats sharing the system; here the size of the tuber or tuber patch is of critical importance. The latter factor may be the key as to why *B. janetta* is solitary: where this mole-rat occurs, rainfall is lower than in the habitats of *C. damarensis* and *H. glaber*, but the geophytes are small in size and not widely dispersed (table 3-4). Under these conditions, no advantage would accrue from being social. Like the social *C. damarensis*, the solitary *B. janetta* occurs in soft sands.

Because of the low mobility of most mole-rats and their vulnerability to predation when moving on the surface, the most likely way to increase the number of individuals in a system would be for a family to remain together. A highly organized colony in which some members never reproduce would be more likely to have evolved if colony members were related and the risks of dispersal were high. For a more complete discussion of this point and explanations of alternative hypotheses, see Alexander et al. (chap. 1), Lacey and Sherman (chap. 10), and Alexander (chap. 15).

It has been argued that the energetic benefits of sociality can be fully realized only if the total energy expenditure of the colony is minimized (Jarvis 1978; Lovegrove and Wissel 1988). Adaptations that reduce energetic costs include reducing body size, mass-specific metabolic rates, and thermoregulatory costs, and having low rates of body growth and low recruitment to colonies. Consequences of these measures as well as of reduced extrinsic mortality are increased longevity and a slowly expanding colony living together for an extended period of time (at least several years), something quite exceptional among small-sized rodents, in which huge annual population fluctuations and short life spans are the norm.

The close confines under which mole-rats live together (e.g., narrow burrows and a communal nest chamber) would enhance the ability, already inherent in the Rodentia (Lee and Boot 1955; Whitten 1959; McClintock 1983a,b), of the dominant animal(s) to restrict the reproduction of subordinates, until eventually reproduction is confined to a single pair. Division of labor and the protection from predation of the breeding pair and other potential reproductives would follow. We suggest (but cannot substantiate) that during most years predation would more or less balance recruitment, and colony size would remain little changed. It would only be after one, or perhaps more, exeptionally good years that colony growth would lead to budding and the formation of new colonies. This may explain why only one small (apparently recently established) colony has ever been found, despite extensive searching (Brett 1986, chap. 4; Jarvis, unpubl. data).

We also suggest that although group living by mole-rats could have evolved as a strategy for sharing costs and risks, the evolution of a highly structured colony, such as found in *H. glaber*, was only possible when individuals remained together for extended periods of time. Furthermore, because other families of subterranean rodents have not become social, there must have been special features and relationships inherent within the Bathyergidae that preadapted them for sociality. These features still need to be clearly defined, but some of them may be a consequence of their hystricomorph affinities (Honeycutt et al., chap. 2). For example, the hystricomorph-type of ovary (Tam 1974; Weir and Rowlands 1974) could well have preadapted the Bathyergidae for hormonal control of colony reproduction by one female with highly secretory ovaries (Kayanja and Jarvis 1971). Preadaptations within the Bathyergidae and the propensity in Africa for plants adapted to arid habitats to have large subterranean storage organs thus may have set the stage for the evolution of sociality.

Summary

The naked mole-rat is one representative of the family Bathyergidae, a group of 12 African subterranean rodents in which individuals of three genera are solitary and two are social. When compared to the other Bathyergidae, the naked mole-rat represents one end of a continuum from a solitary life-style to eusociality. It also exhibits features, such as hairlessness and poikilothermy, not found in any other subterranean rodent. In order to better understand the proximate and ultimate factors leading to the evolution of eusociality in some, but not all bathyergids, this chapter presents an overview of the ecology and behavior of the entire family.

All the Bathyergidae are completely subterranean and live in a sealed system of burrows. Unless they are disposing of excavated soil onto the surface or foraging in superficial burrows, the mole-rats live in an environment that is well protected from predators and from extremes in temperature. They feed mainly on roots and subterranean storage organs of plants, which they find by digging foraging tunnels. In arid parts of Africa, there are many plants with large underground storage organs (e.g., 1–50 kg in size), and these geophytes are widely dispersed. None of the solitary mole-rats occur in such areas, possibly because of the high risks of not finding a tuber; the risk of starvation is greatly reduced when several individuals in a colony burrow in search of food.

The colonies of the two social genera, *Heterocephalus* and *Cryptomys*, contain a single reproductive female and one to three reproductive males; the remainder of the individuals do not breed. Young born to the reproductive female do not disperse, and colonies show an overlap of generations. Colonies of *Heterocephalus glaber* range widely in size (averaging 70–80 individuals), and those of *C. damarensis* contain up to 25 mole-rats; both these species show division of labor among the nonreproductive animals. By contrast, the solitary genera (*Georychus*, *Bathyergus*, and *Heliophobius*) are strongly territorial, and except in the breeding season, each individual occupies and defends its own burrow system. The pups of the solitary genera disperse when about 2-mo old. Territorial advertisement in the form of drumming with their hind feet is a feature of the solitary genera but apparently not of the social genera. The major differences in the reproduction of social and solitary bathyergids are reviewed in this chapter.

Acknowledgments

We thank P. W. Sherman, O. J. Reichman, R. Buffenstein, and two anonymous reviewers for helpful comments on the manuscript. The Council for Scientific and Industrial Research (South Africa), the University of Cape Town, and the National Geographic Society provided the funds to collect animals and to maintain them in captivity. We are grateful to numerous people who accompanied us on field trips, and provided us with hospitality and help; we especially thank M. Griffin, E. Griffin, M.-L. Penrith, M. Penrith, P. Bally, H. Bally, and R. Bally, I. Aggundey, Mr. and Mrs. F. Duckitt, Mr. and Mrs. P. Luhl, Mr. and Mrs. F. Kennedy, and J. Booysen. We also thank the Namibia Department of Nature Conservation for permission to work in Namibia.

4 The Population Structure of Naked Mole-Rat Colonies

Robert A. Brett

Studies of naked mole-rats in captivity have shown that they are eusocial (Jarvis 1981). A division of labor in a colony into reproductives and nonreproductives, and size-linked differences in behavior have been recorded (Lacey and Sherman, chap. 10; Jarvis et al., chap. 12; Faulkes et al., chap. 14). However, little is known about the structure and composition of wild naked mole-rat colonies and still less about the long-term behavior and demography of wild populations. No life-history data have been available hitherto for judging the validity of the results obtained from studies of the demography and behavior of captive animals housed in artificial tunnel systems. For these reasons, my dissertation research (Brett 1986), which is summarized in the next two chapters, was designed to obtain field data on the sizes and life-history patterns of naked mole-rat colonies, their burrow structure, food preferences, mortality sources, timing of breeding, litter sizes, individual growth rates, and the structure and manner of formation of new colonies. Evidence was also sought for a division of labor within a wild colony that might complement the laboratory observations cited above.

Only relatively recently was it noted that naked mole-rats live in large colonies within a single burrow system. Thomas (1903) observed, without further comment, that a certain Dr. Dulio reported naked mole-rats living in colonies of 50–100 animals near Mogadishu, Somalia. Hill et al. (1957) reported that the burrow systems of *Heterocephalus glaber* in northern Kenya did not contain more than 20 animals. However, it became apparent that naked mole-rats live in much larger groups once complete colonies were captured and entire burrow systems were excavated (e.g., by Jarvis 1969; Jarvis and Sale 1971; see Jarvis and Bennett, chap. 3).

Few data on the breeding biology of wild *H. glaber* were available when I began my research. Jarvis (1969) did not capture any pregnant animals over a complete year of sampling, but she did capture several juveniles that were probably born during February through April (the months of rainfall maxima in Kenya). Hill et al. (1957) found two pregnant females during capture in May. These observations led Jarvis (1969) to suggest that breeding in *H. glaber* is confined to the long rains (in the spring), with the possibility of producing a second litter during the short rains (winter), if conditions were good. Subsequently Jarvis (1984, chap. 13) and Lacey and Sherman (chap. 10) reported that the gestation period of breeding females in captive colonies is 70–80 days. Kayanja and Jarvis (1971) determined from the histology of the ovary of an estrous female that 4–6 mature follicles were produced, and Jarvis (1969) caught a wild female that bore five pups. In captivity, breeding females may have up to 27 young in one litter (Jarvis 1984, chap. 13).

Here I present information obtained during a 2-yr field study (1982–1983) of free-living *H. glaber* colonies in Kenya. The results are divided into six sections: the distribution of colonies, colony size and composition (including size variations within and between colonies, and colony growth rates), differences between the sexes, breeding, and the formation of new colonies. The final section deals with evidence, such as can be obtained from captures alone, for the existence for a division of labor in *H. glaber* colonies. This section also includes data on the monopoly of reproduction by particular animals and other evidence obtained during capture that permitted assessment of behavioral differences among colony members. My data are presented along with all previous results on the population structure of wild colonies; the majority of the latter were gathered on two *H. glaber* populations located about 300 km apart: one at Lerata (near Isiolo) in north-central Kenya (studied by Jarvis 1985, unpubl. data), and the other near Mtito Andei in southeastern Kenya (studied by me and by the editors in 1979). A map showing all these study localities is given as figure 7-1 in Honeycutt et al. (chap. 7); the sites are depicted in plate 4-1.

Methods

CAPTURE OF NAKED MOLE-RAT COLONIES

All dirt roads and trails in the areas around Kamboyo (inside Tsavo West National Park) and Kathkani (an agricultural area), near Mtito Andei (fig. 4-1), as well as the area 500 m to either side of these tracks, were surveyed for naked mole-rat colonies. Colonies were located by detection of discrete groups of molehills (or "volcanos;" see plate 4-2). Many colonies were found burrowing near or underneath dirt roads (Brett, chap. 5). The locations of all colonies that I found were recorded on a 1:50,000 scale map of the region. The habitat and ecology of colonies in the two study localities are described in detail in chapter 5.

I captured naked mole-rats using the methods described previously (Hickman 1979a; Brett 1986). These techniques are summarized by Jarvis (Appendix) and will not be reiterated here. It should be noted that at the end of a day's capture of animals from a colony's burrow system (using either soil traps or shutter traps), I stuffed pieces of sweet potato into the ends of opened burrows, which were then sealed with loose soil. The mole-rats did not block off these

Plate 4-1. *Top,* A view of the Lerata area in north-central Kenya, where four naked mole-rat colonies described in this chapter were collected. *Bottom,* A view of the Mtito Andei area (specifically the Kamboyo-1 colony site) in southeastern Kenya The majority of the work on *Heterocephalus glaber* has been done in the Mtito Andei vicinity. Photos: J.U.M. Jarvis, R. A. Brett.

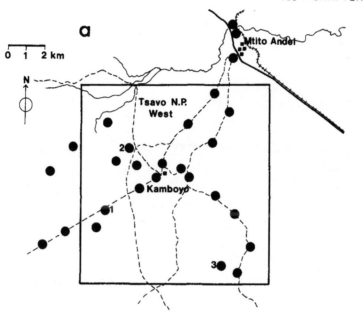

Fig. 4-1. a, Map of the Kamboyo area of Tsavo West National Park.

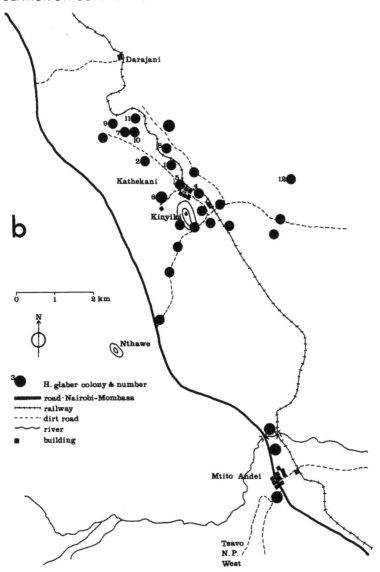

Fig. 4-1 b, the Kathekani area, showing the locations of naked mole-rat colonies. Colonies Kamboyo 1-3 and Kathekani 1-12 are numbered.

Plate 4-2. *Top*, A molehill "erupting" as loose soil, unearthed during digging by naked mole-rats, is kicked out onto the surface. *Bottom*, A group of fresh molehills indicating recent excavations by naked mole-rats. Photos· R. A. Brett, J.U.M. Jarvis.

burrows, and in the morning, tooth marks on the sweet potatoes indicated the burrows that were still occupied. When no more blocking off of burrows or feeding at the sweet potatoes was found for 2 days after the last animal was trapped, I assumed that the complete colony had been captured.

On several occasions, capture of mole-rats from certain colonies had to be abandoned, either because the hardness of the soil made further digging impossible or because of interference from humans (at Kathekani) or lions (in the Kamboyo area). In such cases, the entire colony was not captured; for data analyses, these colonies were classified as "almost complete," if many animals had been captured and the capture rate of those remaining in the burrow system was very low (e.g., < 5 per day), or "incomplete," if only a few animals had been captured and the capture rate was still high (e.g., 10–15 per day).

DATA COLLECTION

The following information was taken on a total of 715 animals I captured from 14 different *Heterocephalus glaber* colonies: sex, body weight (g), body length from nose to tip of tail (mm), reproductive condition (e.g., a perforate vagina and teats distinguished breeding females), any other distinguishing features (e.g., scars, deformities), and the order in the daily capture sequence of animals from each colony.

In one colony (Kathekani 9; n = 97 mole-rats), for which almost all of animals were caught over 4 days after heavy short rains in December 1983, many of the mole-rats' incisor teeth were clogged with soil as a result of their recent digging activities. Each animal in this colony was scored at capture according to whether its incisors were clean, dirty (light covering of dry soil), or clogged (covered by moist soil). These data were used to infer the amount of digging performed by different animals in the colony just before capture.

In November and December 1982, 31 animals were captured from my main study colony in Tsavo West National Park, which I called Kamboyo 1 (fig. 4-1a). All 31 mole-rats were individually toe-clipped and returned to their burrow system. An additional 38 animals were captured, marked and released in Kamboyo-1 colony during January to November 1983. Recaptures of marked animals were used to assess growth rates and demographic parameters.

Results

COLONY DISPERSION

Nineteen *Heterocephalus glaber* colonies were discovered within an area of 100 km² circumscribed around Kamboyo Hill in Tsavo West during 1982–

1983 (fig. 4-1a). The total area used by individual colonies, bounded by peripheral groups of molehills, was usually at least 1 ha. Measuring from the center of a colony area, the mean nearest-neighbor distance for the 19 colonies was 933 ± 308 m.

Most of the colonies detected were found on or within about 1 km of roads. It is likely that the distribution presented is not complete because finding colonies near roads may be easier than finding colonies away from them. Thus, 933 m may overestimate the true mean distance between neighboring colonies.

Outside the National Park, in the agricultural fields around the village of Kathekani, the areas covered by each of the colonies that I studied are harder to describe. Colonies were much closer together and were difficult to distinguish, particularly where the molehills of one colony were widely separated. Around Kathekani it is even possible that one huge burrow system stretched out underground but contained several recent or incipient colonies that were blocked off from each other. Where colony areas could be separated by the positions of molehills, distances between colonies were often less than 150 m (fig. 4-1b). Measuring from the center of each putative colony's area, the mean nearest-neighbor distance of the 22 colonies in the Kathekani area was 473 ± 97 m.

THE STRUCTURE OF COLONIES

A comparison of the demographic and body-size data from the 14 naked mole-rat colonies that I captured in the Mtito Andei area (3 at Kamboyo, 11 at Kathekani), and similar data from 12 colonies sampled by previous workers in the same and other parts of Kenya (table 4-1) shows that considerable variation exists in the sizes, sex ratios, biomasses, and mean body weights of colonies and colony fragments.

COLONY SIZE

Four colonies captured at Kathekani (2, 7, 8, and 10: table 4-1) were judged to be complete, using the criteria described above; four other colonies (Kathekani 4, 9, 11, Kamboyo 1) were judged to be almost complete when no more animals could be caught (the breeding females in these colonies were not captured). At the small end of the colony-size spectrum, Kathekani-10 colony comprised only 25 animals, including 1 breeding female; this colony probably had been formed recently (see below). At the large end of the colony-size spectrum, Kathekani 9 comprised at least 97 individuals when it was unearthed in December 1983. When no further digging was possible, several burrows were still being blocked by mole-rats, sweet potato lures were still being nibbled at, and no breeding female had been caught.

In October 1984, a colony comprising 295 mole-rats was captured at Ka-

Table 4-1
Data from Twenty-Six Naked Mole-Rat Colonies Captured in Northern and
Southern Kenya from 1977 to 1983

Location/ Colony		Sample Size			Sex Ratio	Body Weight (g)		Bio-mass (g)	Breeding Female BW(g)	Capture Date	Source
		Total	M	F		$\bar{x}$	SD				
Northern Kenya											
Lerata											Jarvis 1985
1	√	60	35	25	1 40	20 9	5.2	1254		Jul 80	
2	√	40	21	19	1.11	17.7	4.3	624	28	Aug 80	
3		22	12	10	1.20	18.8	4.2	361	22	Aug 80	
4	√	82	45	37	1 22	27.9	8 9	2111	52	Aug 80	
Totals		204	113	91	1.24	21.3	4 0				
Southern Kenya											
Kathekani											Brett 1986
1		23	11	12	0 92	31.0	8.0	714		May 82	
2	√	52	26	26	1.00	39.1	10.1	2034	61	May 82	
4		53	30	23	1 30	40 0	11.7	2118		Jun 82	
5		30	17	13	1.31	30.7	6 8	922		Aug 82	
6		29	15	14	1.07	31.9	11.6	924		Aug 82	
7	√	93	46	47	0.98	30 7	7.8	2857	54	Apr 83	
8	√	70	38	32	1.19	33.1	9.4	2315	59	Dec 83	
9		97	54	43	1.26	34 9	9 9	3385		Dec 83	
10	√	25	10	15	0.67	20 4	9 8	511	41	Apr 83	
11		72	36	36	1 00	33 1	9.3	2372		May 83	
12		28	13	15	0.87	41.2	12.1	1153		Jul 82	
Kamboyo											
1		69	39	30	1.30	31.0	8.3	2140		May 82	
2		16	10	6	1.67	32 9	7 7	526		May 82	
3		39	20	19	1.05	37.9	11 0	1479		Oct 82	
Misc. others		19	11	8	1.38	—	—	—		—	
Totals		715	376	339	1.11	33.9	3 4				
Mtito Andei											
I		15	8	7	1.14	32.5	13.2	471		Oct 77	Withers & Jarvis
II		40	23	17	1.35	31.9	13 0	1206		Oct 77	1980
2		35	25	10	2.50	32.1	13 5	1100		Dec 79	Alexander,
4		23	12	11	1.09	20.0	12 6	460		Dec 79	unpubl data
A		26	13	13	1.00	28.9	8.2	749		Dec 79	Sherman,
B		34	18	16	1.13	21.7	10.6	739	38	Dec 79	unpubl. data
Athi River		13	8	5	1.63	24.2	7.2	314		Jul 80	Jarvis 1985
Ithumba		27	14	13	1 08	35.4	10.4	955		Sep 80	Jarvis 1985
Total		213	121	92	1.37	28.3	6.8				

NOTE· √, complete colony; BW, body weight Animal weights under 10 g were excluded from the mean body weights of colonies captured by Jarvis (1985) Only one animal weighing under 10 g was captured by Brett (1986).

thekani in a 2-ha field of cultivated sweet potatoes. Most remarkably, no breeding female was captured; this huge colony thus must have been even larger. Though the 295-member colony was by far the largest, the burrow systems of three other large but partially captured colonies at Kathekani were situated in fields containing root crops. The implication is that the largest colonies occur where food is most abundant.

A mark-recapture study of Kamboyo 1 enabled me to estimate the total size of this colony. Between November 1982 and November 1983, 69 animals were captured and marked; 52 recaptures of marked animals were made between April and November 1983. I assumed that there was no emigration or immigration, that recruitment equaled mortality, and that all animals (except for the breeding female) were equally catchable. I employed the method of De Feu et al. (1983) for estimating the size of bird populations from mark-recapture data. This method permitted the use of data from single capture sessions (e.g., 2–4 days when conditions were suitable) spread over a long period. De Feu et al.'s equation is

$$[1-(n/p)] = [1-(1/p)]^{(n+r)},$$

where p is the estimate of the total population size, n the number of new animals captured within one capture period, and r the number of recaptures within one capture period. Three estimates were made, based on the whole study period and on two subsamples (table 4-2); by the three calculations, the size of Kamboyo 1 in 1983 was estimated at 87 animals.

Considering the three complete colonies containing more than 50 animals, plus the estimate of the total number of mole-rats in Kamboyo-1 colony, the mean colony size was 78 ± 22 ($n = 4$). Considering these four colonies and the four almost complete colonies that I captured (table 4-1), the mean colony size was 74 ± 18 mole-rats. In only three of nine additional colonies excavated by others in Kenya were all colony members thought to be captured (Jarvis 1985): these populations numbered 60, 40, and 82 (table 4-1). The implication of these data is that "mature" *Heterocephalus glaber* colonies are variable in size but average 70–80 animals.

COLONY BIOMASS

The mean biomass of my three complete colonies numbering more than 50 individuals (Kathekani 2, 7, and 8) was $2,402 \pm 418$ g. The mean biomass of seven colonies larger than 50 individuals (three complete, four almost complete) was $2,460 \pm 490$ g. The biomasses of the complete colonies from southern Kenya were considerably greater than those from Lerata in northern Kenya (table 4-1; Jarvis 1985). This is attributable mainly to the lower individual body weights of the Lerata mole-rats and, to a lesser extent, to the larger sizes of colonies recorded in southern Kenya.

Table 4-2
Estimates of the Size of Kamboyo-1 Colony from Mark-Recapture Data Using the Method
of De Feu et al. (1983)

Capture Period	Total	New Animals (n)	Recaptures (r)	Colony-Size Estimate (P)	Standard Error
Whole study					
19 Nov 82–30 Nov 83	116	64	52	86	7.6
1983 only					
18 Mar 83–30 Nov 83	88	56	32	89	11 2
Late 1983					
11 Aug 83–30 Nov 83	71	49	22	88	14 4

SIZE DISTRIBUTION OF MOLE-RATS

In captures of colonies at Kamboyo, Kathekani, and Mtito Andei, animals with a body weight (BW) of less than 21 g were classified as subadults; clustering of body weights of animals from the same litter were apparent below this weight. Juveniles were defined as those weighing less than 17 g. These classifications are inappropriate for animals captured in northern Kenya (Lerata), where average body weight was significantly lower (table 4-1) and was probably influenced by the more arid environment (Jarvis 1985; Jarvis et al., chap. 12; Brett, chap. 5).

Body weight is clearly a good indicator of body size for all age classes in wild *H. glaber* colonies. In the largest colonies (i.e., those with numerous data), there were highly significant regressions between body weight and body length (e.g., Kathekani 9: $y = 1.069x + 99.54$; $r = 0.949$, $n = 97$). Only 1 animal of the 715 captured over the study period (a male) was noticeably thin and underweight. This male deviated significantly from the regression line in its own colony (Kathekani 9: $t = 4.608$, $p < 0.05$) and was very shrunken and slow in its movements, but it did not have skin lesions or any external abnormalities. This animal may have been the former breeding male (see Jarvis et al., chap. 12). None of the four breeding females that I captured was pregnant; neither did they deviate significantly from the regression of weight versus length in their colonies.

SIZE DISTRIBUTIONS OF INDIVIDUALS WITHIN COLONIES

The mean body weights of adult mole-rats (i.e., BW > 21g) from seven complete or almost complete colonies captured at Kathekani was 33.9 ± 4.9 g ($n = 651$). There were small peaks at the heavy end of the distributions of some of these colonies (fig. 4-2), which might indicate subdivision of colonies into age classes, or breeders and nonbreeders, on the basis of size. Peaks at the light end probably comprise litters growing toward the mean adult body size.

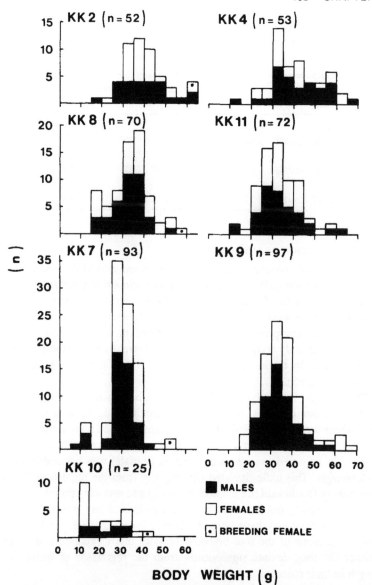

Fig. 4-2. Body-weight distributions of naked mole-rats from seven complete or almost complete colonies captured at Kathekani (KK).

The size distributions of large colonies were tested for normality to look for any significant biomodalities or subdivisions into body size groups. Only two of the six colonies with over 50 members had body-weight distributions that differed significantly from a normal distribution (Kathekani 7 and 9; see fig. 4-2). The size distributions of different colonies would be expected to vary seasonally because of reproduction, and Kathekani-7 colony had a recent litter of six that was separate from the main distribution (Kolmogorov-Smirnov test: $D = 0.114$, $p < 0.01$, $n = 93$). Kathekani 9 contained a subgroup of animals that were significantly heavier than the others ($D = 0.116$, $p < 0.01$, $n = 97$).

VARIATION IN SIZE DISTRIBUTION BETWEEN COLONIES

Considering only colonies of more than 50 mole-rats ($n = 7$), the mean body weights of adults were greater in smaller colonies ($r = -0.97$, $p < 0.01$). For example, members of Kathekani 2 ($n = 52$) and Kathekani 4 ($n = 53$) had significantly greater mean body weights than did colony members of Kathekani 8 ($n = 70$), Kathekani 11 ($n = 72$), and Kamboyo 1 ($n = 87$) (see table 4-1). The dates of capture of colonies of different sizes were sufficiently spaced throughout the year to dismiss any seasonal effect.

Larger colonies may have lower mean body weights because of (1) the recent birth of several litters, (2) lower growth rates of individuals within these colonies (see Jarvis et al., chap. 12), (3) more competition for food, or (4) a lack of any particularly heavy animals (BW > 49 g). Since growth beyond a body weight of approximately 35 g is relatively slow (Jarvis 1978; Lacey and Sherman, chap. 10), the absolute number of animals attaining very large size may indicate the age or stage of maturity of a particular colony.

There is considerable variation between colonies in the proportion of animals weighing more than 49 g (table 4-3). Ignoring the small colony Kathekani 10 (half of which was subadults), there were relatively fewer heavy animals in large colonies ($r = -0.73$, $p < 0.05$, $n = 6$; e.g., compare colonies Kathekani 2 and 4 with the others). The proportion of subadults, which presumably represents recent recruitment to the colony, shows no relationship to colony size (table 4-3; $r = 0.26$, $p > 0.05$).

The three colonies with the highest mean body weights (Kathekani 2, 4, and 12) were captured in areas adjoining agricultural land. The food biomass available on or close to agricultural land was probably greater than that available in Tsavo West National Park. There may also be an effect of higher food availability on colony size; the largest colony ($n > 295$) was captured on agricultural land. Moreover, some of the larger colonies appeared to have had particularly high recruitment (e.g., Kathekani 7, 49% of animals weighed less than 30 g at capture). Nevertheless, no significant differences were found between the mean body weight of all adults (BW > 21g) captured inside (Kamboyo, $n = 111$) and outside the National Park (Kathekani, $n = 521$) ($t = 0.31$, $p > 0.05$).

Table 4-3

Proportions of Naked Mole-Rats in Different Size Classes in Eight Colonies
Captured at Kathekani (KK) and Kamboyo (KA)

Colony	n		% Subadults (BW < 21 g)	% Adults (21 g < BW < 50 g)	% Large Adults (BW > 50 g)	Breeding Female BW (g)	Heaviest Male BW (g)
KK2	52	√	1.9	82.7	15 4	61	60, 59
KK4	53		3.8	71 7	24.5		65
KK7	93	√	6.5	90.3	3.2	54	42, 40
KK8	70	√	15.7	78.6	5.7	59	50
KK9	97		3.1	88 7	8 2		57, 54
KK10	25	√	56.0	44.0	0 0	41	35
KK11	72		4.2	90 3	5.5		60, 59
KA1	69		14.5	82.5	2.9		51

NOTE· √, complete colony, BW, body weight

This contrasts with the dramatic difference in mean body weights of naked mole-rats from four colonies sampled by Jarvis (1985) at Lerata (table 4-1; see also chap. 12).

GROWTH RATES

A total of 52 recaptures were made from the Kamboyo-1 colony during the study period. Of these, 15 (5 adults and 10 subadults, including 3 juveniles) were marked in November 1982 and recaptured in November 1983. Thus, it was possible to control for any seasonal effects on growth of these individuals. The body-weight changes over a year were expressed as a power curve (fig. 4-3a). There is a highly significant negative regression of initial body weight on the percentage of change in body weight ($r = 0.94$, df $= 13$, $p < 0.0001$). The growth rates calculated here for wild mole-rats are approximately three times higher than rates reported by Jarvis (1981) for 29 captive individuals (see also Jarvis et al., chap. 12).

No significant deviations in growth rate from the curve (fig. 4-3a) for animals measured over 12 mo in Kamboyo 1 were found. However, if growth rates over the year are expressed as the percentage of change in body weight per day (fig. 4-3b), a wide variation in adult growth rates is revealed. There is some evidence of a seasonal effect on growth rates in these data. In addition, some adults that had similar initial body weights grew considerably faster than others over the same period. For example, between September and November 1983, female 236 grew over three times faster than female 201, and males 209 and 222 grew three to five times faster than males 210 and 212. Differences in growth rates between the sexes and between different litters are discussed below and by Jarvis et al. (chap. 12).

Table 4-4
Seasonal Growth of Naked Mole-Rats in Kamboyo-1 Colony

Time Period	Rainfall (mm)	Growth Rate (% Change in BW per Day)		No. of Animals
		$\bar{x}$	SD	
Nov 82 – Sep 83	321.6	0.0524	± 0.0413	7
Nov 82 – May 83	304.3	0.0386	± 0.0370	5
May 83 – Sep 83	17.3	0.0384	± 0.0345	6
Sep 83 – Nov 83	101.6	0 1279	± 0.1086	12
May 83 – Nov 83	118.9	0.0616	± 0.0260	10

For all adults captured in the Kamboyo and Kathekani areas (n = 651), females (mean BW = 36.5 ± 4.5 g) were significantly heavier than males (mean BW = 34.1 ± 3.7) (t = 3.056, df = 650, p < 0.01). Females were significantly heavier in six colonies: Kathekani 7, 8, 9 and 12 and Kamboyo 1 and 3 (complete details in Brett 1986). In seven out of nine colonies captured by other workers in Kenya (table 4-1), females were, on the average, heavier than males; however, only in Lerata 1 was the difference between the sexes significant (t = 2.51, df = 58, p < 0.01). Females gained weight significantly faster than males over 12 mo in Kamboyo 1 (t = 2.25, df = 13, p < 0.05). The growth rates of two females were significantly greater than those of two males, paired on the basis of having the same initial body weight (t = 3.12, p < 0.05).

SEASONAL GROWTH

Mean growth rates (expressed as the percentage of change in body weight per day) were calculated for all animals captured during 1982–1983 at the beginning and end of time periods with varying amounts of rainfall (table 4-4). From May to September 1983, it was dry; during the following period (September–November 1983), roughly half of the rainfall of the 1983 short rains accumulated. I suspected that growth rates might be higher in wet months than dry months because of increased food availability and quality. If tubers were more accessible and more succulent in wet seasons (see Brett, chap. 5), it would at least be expected that mole-rats would increase in body weight because of hydration. Indeed, weight gains averaged across all recaptures in the wet 1983 period were about three times higher than in the dry periods of November 1982–May 1983 and May–September 1983.

The growth rates of animals depended on their initial size (fig. 4-3a). Thus, average seasonal growth rates (table 4-4) might depend on which mole-rats were recaptured (i.e., whether or not the sample contained the same proportions of small and large animals). To control for this, five animals were captured before and after the dry period (May–September 1983) and again after

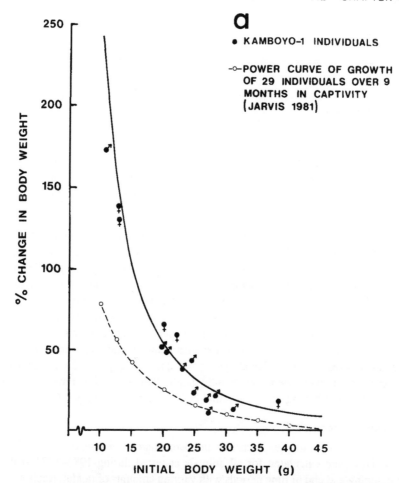

Fig. 4-3. Growth rates of naked mole-rats from Kamboyo-1 colony over 12 mo (November 1982–November 1983). a, The growth curve for 15 wild animals (y = 53587 1x–2.35). *Dashed line*, A growth curve for animals in captivity (from Jarvis 1981).

the rainy period (September–November 1983). Thus, their growth rates in both periods could be compared. These five mole-rats showed significantly higher mean weight gains in the 1983 rainy period compared with the preceding dry season (paired *t* test: *t* = 5.33, df = 4, *p* < 0.01).

SEX RATIOS

Considering the combined sex ratio of the 715 animals captured and sexed over the study period, there was a slight, but not statistically significant bias

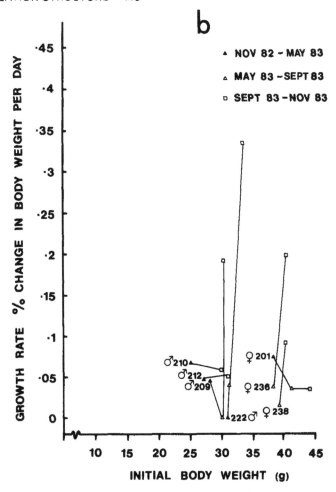

Fig. 4-3b. Changes in the weight-gain trajectories of seven wild adult naked mole-rats in three successive time periods over 12 mo. Lines connect successive growth rates of the same animal, plotted against the body weight at the start of each period (*filled triangle*, November 1982–May 1983; *open triangle*, May–September 1983; *open square*, September–November 1983).

toward males (1.11:1.0); one of the four complete colonies that I captured was male-biased. However, when sex-ratio data from other sources (table 4-1) are considered, it becomes clear that there is a general bias toward males across all colonies. For all animals captured in the Mtito Andei area by previous workers, the bias toward males is significant ($n = 217, \chi^2 = 4.70, p < 0.05$). The sex-ratio data of adult naked mole-rats captured by Jarvis (1985) also show a significant male bias ($n = 249, \chi^2 = 3.94, p < 0.05$); all three of the complete colonies that she captured at Lerata were male-biased. When all previous data are pooled with the data from Kamboyo and Kathekani, the bias toward males is highly significant (n = 1136, χ^2 = 7.78, p < 0.01). Jarvis (chap. 13) also presents data showing male-biased sex ratios in *Heterocephalus glaber* colonies.

The male bias was not as pronounced among subadults. At Kamboyo and Kathekani, the sex ratio of the 64 subadults captured was almost parity (1.03:1.0), and the sex ratio of 26 juveniles was exactly 1:1 (however, see Jarvis, chap. 13). This suggests that, among adults, the survival rate for females may be lower than that for males. If this were true, then in mature colonies with many large animals, the sex ratio ought to be most male biased, assuming that competition and fighting among females for breeding opportunities lead to female mortality. The lack of association between colony sex ratio and colony size (r_s = 0.092, df = 13, p > 0.05) or between sex ratio and mean body weights (rs = 0.149, df = 13, p > 0.05) does not support this hypothesis, however. In fact the reverse was true in Kathekani 8 (n = 70), where large adults (BW > 40 g) were mostly females, and small adults (BW, 22–40 g) were mostly males (χ^2 = 5.083, p < 0.05). For the 113 individuals weighing more than 40 g in the seven almost complete colonies captured (fig. 4-2), the sex ratio was biased slightly toward females (1:0.92); the bias toward females was greater in the 39 animals weighing more than 49 g (1:0.77). Thus, the overall sex-ratio bias toward males in *H. glaber* colonies cannot be attributed unequivocally to female-female aggression, although this hypothesis seems more likely than the alternative that female emigration exceeds male emigration.

BREEDING DATA

LITTER SIZES

In the complete or almost complete Kamboyo and Kathekani colonies, several clearly distinguishable litter clusters were found. These were groups of animals, with some or all members weighing less than 22 g, showing an obvious aggregation of body weights within a range of ± 3 g. Individuals whose body weights were over 3 g higher or lower than the mean for each cluster were excluded as members of the same litter. Considering the youngest litter in each colony, the mean number of weaned young per litter was 9.71 ± 2.81 (n = 7) (table 4-5). Normal litter size at birth in wild colonies may be slightly higher. It is interesting that the smallest juvenile caught weighed only 9 g but was highly mobile.

Putative litters have been captured from 11 colonies in other field studies. These data yield an average of 4.3 ± 2.3 pups per apparent litter (table 4-5). It is uncertain whether or not any of these were complete litters. However, Lerata 1 was almost a complete colony (the breeding female was not caught), and

Table 4-5
Breeding Data from Naked Mole-Rat Colonies Captured in Northern
and Southern Kenya, 1977–1984

Location/ Colony	Month of Capture	Estimated Mo. of Birth	Number of Litters (Litter Size)	Source
Mtito Andei	1967–68	Feb–Apr	several (BW < 17 g)	1
Mtito Andei	Oct 77	Sep 77	2 (2, 5)	2
MA 2	Dec 79	Oct 79	1 (3)	3
MA 3	Dec 79	Nov 79	1 (2 newborns)	4
MA 4	Dec 79	Sep 79	1 (7)	3
MA B	Dec 79	Sep 79	1 (4)	4
Athi River	Jul 80	May 80	1 (1)	5
LR 1	Aug 80	—	1 (1)	5
LR 2	Aug 80	—	1 (7)	5
LR 3	Aug 80	Aug 80	1 (5, unweaned)	5
LR 4	Aug 80	Jul 80	1 (8)	5
KK n = 295+	Oct 84	Sep 84	1 (13)	6
KK 7	Apr 83	Feb 83	1 (6)	6
KK 8	Dec 83	Jul 83	2 (3, 8)	6
KK 9	Dec 83	Aug 83	1 (13)	6
KK 10	Apr 83	Mar 83	> 2 (2, 10)	6
KK 12	Jul 82	Mar 82	1 (11)	6
KA 1	May 82	May 82	2 (7, 3)	6

NOTE· MA, Mtito Andei, LR, Lerata, KK, Kathekani; KA, Kamboyo A 75-day gestation period is
assumed.
SOURCE· 1, Jarvis 1969, 2, Withers and Jarvis 1980; 3, Alexander, unpubl data; 4, Sherman, unpubl data, 5, Jarvis 1985, 6, Brett 1986

Lerata 2 and 4 were complete. Thus, the sizes of weaned litters in the Lerata area seem to have been slightly smaller than those at Kamboyo and Kathekani.

GROWTH RATES OF YOUNG

Jarvis (1978) calculated mean growth rates of young mole-rats born in captivity. Four litters born to Kathekani-10 colony in captivity (in 1984) increased in weight over the period of 0–2 mo old at rates similar to those Jarvis reported. The mean body weight of 2-mo-old young was 13.4 ± 1.09 g (19 animals, 4 litters). The consistent growth rates of litters born to colonies in captivity were used for predicting the growth of wild-born young up to 2 mo of age. These data and those on growth rates of the three juveniles from the mark-recapture study (ca. 2–14 mo of age), allowed the calculation of a growth-rate curve for animals weighing up to 30 g (fig. 4-4). The three wild juveniles grew faster than Jarvis' (1978) juveniles after 2 mo of age.

Using the estimated growth curve, I extrapolated the body weights at capture of the 64 subadults caught at Kambyo and Kathekani to yield estimated

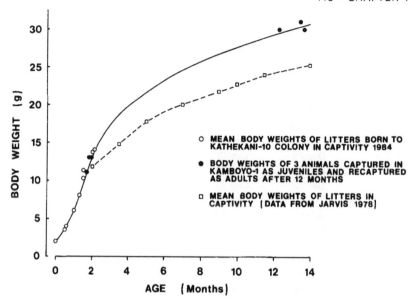

Fig 4-4. Estimated growth curve for juvenile naked mole-rats in wild colonies.

dates of birth for each litter. Dates of capture were sufficiently spaced to allow reasonable estimation of dates of birth throughout the year and, assuming a 75-day gestation period (Jarvis, chap. 13), estimation of dates of conception.

TIMING OF LITTERS

A total of 23 surviving litters was born in my 14 study colonies over 22 mo in 1982–1983. There was no significant correlation (r_s = -0.03, df = 18, p > 0.05) between monthly rainfall measured at Kamboyo and the number of surviving litters (Kamboyo and Kathekani, 1982–1983) estimated to have been born in each month (fig. 4-5). Runs tests also revealed no significant difference from a random distribution of litters over 12 or 22 mo. Even when only juveniles (BW < 17 g) were considered, thereby increasing the accuracy of birthdate estimates, litters were spread over the year and showed no association with monthly rainfall.

It is possible that peaks of rainfall only affect the timing of breeding after some time lag, perhaps the time taken for rainfall to affect the softness of subsoil layers (ca. 0.5–1.0 m) or tuber quality and succulence. However, monthly rainfall totals were not significantly correlated with the number of litters born per month at any possible time lag (1–12 mo from rainfall to estimated birth dates). The monthly numbers of litters estimated to have been born and conceived in 1982–1983 were grouped into a 12-mo period. No significant correlations were found between times of the conception or births of surviving litters and mean monthly rainfall at Kamboyo (1973–1983) for

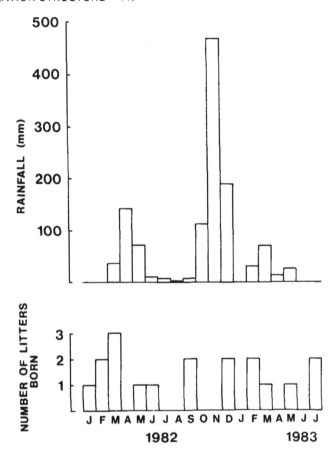

any possible time lag (1–12 mo, including the assumed 75-day gestation period).

Four nonpregnant breeding females were captured at Kathekani during rainy seasons; in three other almost-complete colonies, no breeding female had turned up when catching had to be terminated. It is possible that breeding females were not captured in the latter three colonies because they were very pregnant and near term and thus unable to get out of a nest chamber or to penetrate the small-diameter burrows close to the surface where most captures were achieved.

Estimates of birth dates derived from previous captures of juveniles from colonies in Kenya (table 4-5) seem to indicate birth peaks in the winter and spring (the rainy seasons); however, these data are not representative because most studies were done in the latter half of the year. Estimates of birth dates of colonies at Lerata using the growth curves of Mtito Andei mole-rats (fig. 4-4) were probably erroneous because of the smaller size of mole-rats at Lerata. However, given the lack of breeding seasonality found in the Mtito Andei area, it is interesting that one unweaned litter was captured by Jarvis at Lerata in August (the dry season).

LITTER INTERVALS

In eight colonies captured at Kathekani and Kamboyo, I found juveniles that appeared to be from two or more successive litters. A mean inter-litter interval of 94 ± 25 days for these colonies was calculated using the growth curve (fig. 4-4; also Brett 1986). One colony (Kathekani 10) produced three litters in a year, with intervals between litters of 71 and 91 days. Thus, naked mole-rat colonies in nature can produce pups every 2–3 mo. The interval between apparent litters in the field was similar to the 76–90 day inter-litter interval in captivity (see Jarvis, chap. 13; Lacey and Sherman, chap. 10).

THE FORMATION OF NEW COLONIES

In April 1983, one small but complete colony (Kathekani 10) was caught; it comprised only 25 individuals (table 4-6). The colony contained 11 adults, including 1 breeding female and 1 breeding male (subsequently identified in captivity), 4 subadults (from two litters) and 10 juveniles (all about the same size and probably littermates).

The nearest neighboring colony to Kathekani 10 was Kathekani 7; the latter contained 93 animals. The nearest molehills between these colonies were 60 m apart. Kathekani 7 and 10 were probably connected at one time because there was evidence of old workings on the surface between the two colony locations. However, the burrow systems were definitely sealed off from each other: all Kathekani-10 burrows leading toward Kathekani 7 had blind endings. No blocking off or feeding at planted sweet potatoes was found for several days after the last Kathekani-10 animal had been removed, confirming a complete colony capture.

It is likely that Kathekani 10 was a newly founded colony, perhaps budded off from Kathekani 7 at least 7 mo before capture. Assuming that the largest subadults (21–26 g) were born in the new colony area, and the new colony was formed one gestation period before their birth date (i.e., the time when reproductive suppression by the Kathekani-7 breeding female was relaxed), the projected date of colony fission was mid-June 1982. Assuming that only the adults I captured formed the group that split from the parent colony, the minimum number of mole-rats that left Kathekani 7 to form Kathekani 10 was 11 individuals.

Table 4-6
Composition of Kathekani-10 Colony

Sex	Body Weight (g)	Capture Order	Sex	Body Weight (g)	Capture Order
Female	24	1	Male	21	11
Female	30	2	Female	21	13
Female	29	3			
Male	26	4	Male	17	14
Female	30	5	Male	15	15
Male	28	6			
Male	30	7	10 pups*	10–11	16–25
Female	30	8			
Male	35[†]	9			
Male	30	10			
Female	41[‡]	12			

* A litter containing 8 females and 2 males
[†] Breeding male
[‡] Breeding female

It is also possible that Kathekani 10 was founded by the breeding pair (order of capture: 9 and 12; table 4-6), and that the first litter born contained seven to nine animals (order of capture: 1, 2, 3, 4, 5, 6, 7, 8, 10), followed by three further litters. All of the adults (except the breeding pair) were small (BW < 31 g), and the breeders were lighter than those found in other colonies. Colonies are easily founded from breeding pairs in captivity (Jarvis, Appendix), and recent observations of *H. glaber* individuals walking aboveground at night outside of rainy seasons (S. Braude, pers. comm.; L. Nutter, pers. comm.) suggest that the founding of new colonies by breeding pairs is at least a possibility.

DIVISION OF LABOR

BREEDERS

As indicated previously, I captured four breeding females, one from each of the four complete colonies collected at Kathekani (table 4-3). In all colonies captured by myself and by others, either no breeding females were found or only one was captured. Although captive colonies sometimes contain more than one breeding female for long periods of time (Jarvis et al., chap. 12), there is no evidence that wild *Heterocephalus glaber* colonies ever have more than one breeding female.

Evidence from captive studies (Lacey and Sherman, chap. 10; Jarvis, chap. 13) indicates that each *H. glaber* colony has one to three breeding males; these animals often started out among the largest males in a colony. In five large or complete colonies I captured (Kathekani 4, 7, 8, 9 and 11), one or two males were significantly heavier than the rest (table 4-3). Perhaps these large males were the breeders; they could not be distinguished morphologically (e.g., by protuberant penises) other than by size. However, in Kathekani 10, the identity of the extra-large male (capture order, 9; table 4-6) as the breeder was confirmed in 1984 while the colony was in captivity in London (Faulkes et al., chap. 14).

WORK: RECENT DIGGING

The degrees of clogging of the incisor teeth of 97 individuals captured over a 4-day period from Kathekani 9 during the short rains of 1983 were examined as a function of body weight (fig. 4-6). No significant differences were found between the mean body weights of animals in the three different categories of digging effort, as defined by the amount of earth adhering to their incisor teeth. Considering all animals with body weights over 30 g (i.e., only adults), the percentage of animals with clean incisors correlated positively and significantly with increasing body weight (in 5-g increments; $r_s = 0.81$, df = 10, $p < 0.01$). Although the larger size classes had fewer animals, this suggests that smaller adults did more digging. The frequency of clogged incisors in small adults (BW = 22–40 g; $n = 72$) differed significantly from that in large adults (BW > 40 g; $n = 22$) ($G = 4.36$, $p < 0.05$); small adults more often had clogged incisors. The difference was more significant if larger animals (BW > 45 g; $n = 12$) were compared to small adults ($G = 5.28$, $p < 0.01$). Significantly more large animals (BW > 40 g) had clean incisors than expected on the basis of randomness ($\chi^2 = 6.4$, $p < 0.02$).

The influence of sex on the degree of incisor clogging in Kathekani 9 was investigated using three-way G-tests (Sokal and Rohlf 1981). It was found that the three factors—sex, clogging, and size—were not independent ($G = 28.27$, df = 6, $p < 0.001$), and that each factor taken in turn depended significantly on the combined influence of the other two. Conditional independence (independence of two factors at given level of the third) was found in all cases. The most important interaction of sex and size was found in the non-digging category (clean incisors): small non-digging animals were mostly males, and large non-digging animals were mostly females.

Assuming that effective cleaning of incisors occurred randomly regardless of size, these data indicate that digging decreased with increasing body size. As described below, the association of "nonworking" behavior and large body size may be connected with defensive behavior. Interestingly, the first six animals caught in this colony had clean incisors and weighed an average of 51.7 ± 10.7 g, significantly heavier ($p < 0.01$) than the mean colony body weight of 34.9 ± 9.9 g.

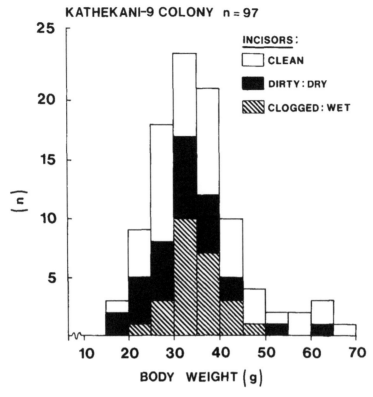

Fig. 4-6. The body weights of naked mole-rats from Kathekani-9 colony, show-ing the distribution of clogging of the incisor teeth of individuals (presumably an indicator of recent digging).

COLONY DEFENSE

During rainy seasons, I witnessed five attacks by rufous-beaked snakes (*Rhamphiophis oxyrhynchus rostratus*), which are poisonous, on naked mole-rats that were "volcanoing" (i.e., kicking out soil onto the surface to form molehills; see plate 4-3) (Brett 1986). Other mole-rats that were digging and kicking soil in burrows close to the hole where a colony mate was caught by the snake gave alarm grunts and sealed off the snake and the bitten mole-rat with soil, apparently to prevent entry of the predator into deeper burrows. One attack at a molehill by a snake was interrupted by the observer, and the freshly bitten animal, which died after about 5 min, was found to weigh 31 g. This is below the average body weight for this locality, but because this colony was not captured, the size of the dead animal relative to other colony members is unknown (Braude has shown that volcanoers are often among the largest indi-viduals in their colony; chap. 6).

Plate 4-3 *Top,* The rufous-beaked snake (*Rhamphiophis oxyrhynchus rostratus*), a known predator on naked mole-rats. *Bottom,* The beaked snake attacks a naked mole-rat colony by entering an erupting molehill Photos: J.U.M. Jarvis, R. A Brett.

Although blocking-off behavior may be effective in excluding snakes that attack at molehills, *H. glaber* is also highly aggressive to snakes found in burrows. Violent attacks on snakes introduced into captive colonies have been recorded (Lacey and Sherman, chap. 10; Alexander, chap. 15), albeit in a situation in which neither the mole-rats nor the snakes could escape. In captivity, aggressive behavior toward introduced snakes was performed mainly by large animals; only the largest nonbreeding animals confronted the snake, though they weighed only 30–40 g. Lacey and Sherman (chap. 10) also found that threatening behavior toward foreign mole-rats was positively correlated with body weight and was performed primarily by large nonbreeders.

The laboratory observations cited above led me to predict that in a wild colony there might be size-related differences between individuals in the capture order and the recapture frequency. For example, large, nonworking animals might behave defensively as a "soldier" caste, or "praetorian guard" (Jarvis 1981, 1984), particularly in response to attacks by snakes at surface diggings or invasions of mole-rats from foreign colonies.

CAPTURE ORDER

Significant negative correlations between body weight and capture order (fig. 4–7) were found in 5 of the 6 complete or nearly complete colonies that I investigated: Kathekani 4 ($r = -.033$, $p < 0.01$), Kathekani 7 ($r = -0.27$, $p < 0.001$), Kathekani 8 ($r = -0.44$, $p < 0.0001$), Kathekani 9 ($r = -0.56$, $p < 0001$), and Kathekani 10 ($r = -0.81$, $p < 0.0001$). Negative correlations were also found in 4 of 12 colonies captured previously by other workers: Mtito Andei I ($r = -0.63$, $n = 15$, $p < 0.01$), Mtito Andei II ($r = -0.34$, $n = 40$, $p < 0.01$; Withers and Jarvis 1980); Mtito Andei B ($r = -0.38$, n=34, $p < 0.001$; Sherman, unpubl. data); and Lerata 1 ($r = -0.33$, $n = 60$, $p < 0.001$; Jarvis 1985). In no colony was there a significant positive relationship between body weight and capture order. Apparently, the first mole-rats to visit breaks in tunnel systems, and thus to be captured, are the larger individuals.

I also compared the weights of the first 5 or 10 animals caught on each day in colonies Kathekani 2, 4, 7, 8, 9, and 10 (fig. 4-7) with the mean colony body weight and the mean body weight of animals captured on that day. No significant differences were found except in Kathekani 9, where the first 5 animals ($t = 3.23$, df = 100, $p < 0.01$) and the first 10 animals ($t = 3.77$, df = 105, $p < 0.001$) weighed significantly more than the mean colony body weight. Ratios of animals smaller and larger than the mean colony body weight were examined in groups of 10 in order of capture (in colonies shown in fig. 4-7). Again, significant deviations from equal numbers of large and small animals in each group of 10 were found in Kathekani 9 for the whole capture period ($\chi^2 = 44.24$, df = 8, $p < 0.001$). Finally, runs tests were performed on the capture data from 10 colonies at Kathekani (2, 4, 7, 8, 9, 10, 11) and Kamboyo (1, 2, 3). These tests revealed a highly significant nonrandomness (aggregation) of large and small animals (above and below mean colony body weight) in 6 of the 10 colonies. There was significant aggregation of animals larger or smaller than mean size in all the large ($n > 50$) colonies except Kathekani 4 (analyses detailed in Brett 1986).

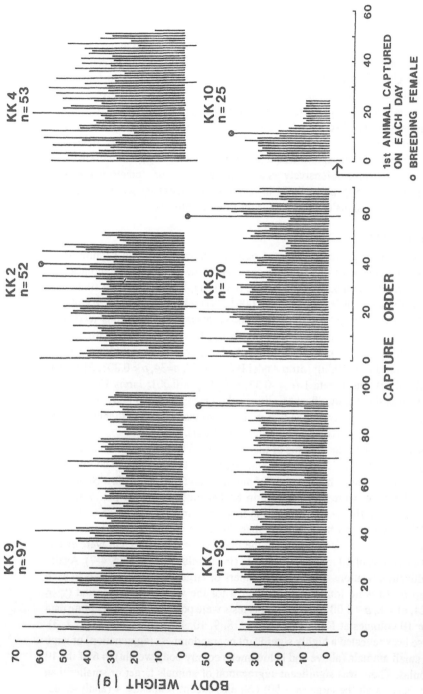

Fig. 4-7 The order of capture and body weights of naked mole-rats from six complete or almost complete colonies from Kathekani (KK).

These results indicate that in wild colonies, sampled animals steadily get smaller as capturing proceeds. However, not all of the particularly large animals are captured first; for example, the breeding males and females are typically caught late in the sequence. Accompanying the decrease in body size, there was clustering of animals of similar size. This is seen most clearly in the capture data for Kathekani 10, where adults were captured first, followed by juveniles and then pups (table 4-6; fig. 4-7).

RECAPTURE FREQUENCY

For 17 animals first captured in November 1982 and recaptured in September–November 1983, in Kamboyo 1, significant positive correlations were found between recapture frequency and body weight at first capture ($r_s = 0.74$, $p < 0.01$) and body weight at final capture ($r_s = 0.48, p < 0.05$). Similarly, when the initial body weights of all 32 of the 1983 captures were considered, significant positive correlations were found between recapture frequency and initial and final body weight (Brett 1986). However, no significant differences were found between mean body weights of animals that had been recaptured the same number of times and groups of animals that had been recaptured more often or less often; thus, there was evidence for a trend toward increasing catchability with increasing body size. The implication is that large animals indeed visit and revisit breaks in burrows most frequently.

Discussion

COLONY SIZE AND COMPOSITION

The 715 naked mole-rats captured from 14 colonies at Kamboyo and Kathekani during 1982–1983 confirmed that the numbers of animals inhabiting the same burrow system in *Heterocephalus glaber* are the largest found in any fossorial mammal. The next largest group (colony) sizes are found in the related bathyergids from southern Africa, *Cryptomys hottentotus hottentotus* (maximum $n = 14$), and *C. damarensis* (maximum $n = 25$) (Davies and Jarvis 1986; Jarvis and Bennett, chap. 3), and the octodont rodent *Spalacopus cyanus* (maximum $n = 15$) from Chile (Reig 1970). Little is known of the breeding characteristics of *Spalacopus* and the other highly social South American octodont rodent, the tuco-tuco (*Ctenomys sp.*). Breeding in both the *Cryptomys* species is monopolized by a single pair, and *Cryptomys damarensis* shows a division of labor somewhat similar to, though simpler than, that found in *Heterocephalus glaber* (Jarvis and Bennett, chap. 3). Several species of birds and non-fossorial mammals are highly social (review in Lacey and Sherman, chap. 10; Emlen 1984). None of these species live in integrated cooperatively breeding groups as large as those found in *Heterocephalus glaber*. The dwarf mongoose (*Helogale parvula*) has the largest group size in other mammals in which breeding is monopolized by a single dominant pair (Rood 1983, 1986).

The variable colony size found in naked mole-rats (25–295+ individuals, mean = 70–80) is probably a reflection of the amount of food available to a colony, as well as its age. High survival rates of very large litters in areas with abundant food (similar to litter sizes and survival rates in the laboratory where food is provided ad libitum) apparently result in very large colony sizes. High birthrates may have caused some colonies to have a majority of small animals (e.g., Kathekani 7) and body sizes that were not normally distributed, in contrast to more "normal" or mature colonies with a higher proportion of large animals (e.g., Kathekani 2 and 4).

Although there was no significant effect of food availability on the mean body weights in colonies sampled at Kathekani and at Kamboyo (in the National Park; see table 4-1), the same was not true in other parts of the *H. glaber* range. As discussed by Jarvis et al. (chap. 12), Lerata 4 colony, excavated in northern Kenya, was found in an area that was more lush than its three neighboring colonies (Lerata 1, 2, 3). Both the number of animals and the mean body size in Lerata 4 were higher than in the three nearby colonies. The Mtito Andei area has higher rainfall than *H. glaber* habitats in northern Kenya. The quality and distribution of food items is also different: southern areas apparently have more food plants with large tubers (Jarvis 1985). This might contribute to the remarkable difference in mean body sizes between mole-rats captured at Lerata and at the Mtito Andei area (table 4-1). However, the lack of difference between mean body weights of the southern colonies suggests that *H. glaber* has a maximum body size and that this is achieved around Mtito Andei.

It is probable that the body-weight gains found in Kamboyo 1 were high during the rainy season (table 4-4, fig. 4-3b) in comparison to the dry season because of the changes in the food supply. Less energy would have to be channeled into digging soft soil (Brett 1986, chap. 5; Lovegrove 1987a, 1989), and more varied and more succulent food resources would be available (Brett, chap. 5). Large differences were found between the growth rates of some adult animals (30–40 g) monitored over the same wet and dry periods, and these may have been the result of social influences on growth rather than ecological factors (fig. 4-3b; see also Jarvis et al., chap. 12). With my small sample it was not possible to determine the causes of these differences, whether associated with food availability, sex, or behavioral status.

Wild juveniles grew faster (fig. 4-4) than juveniles reared in captive colonies (Jarvis 1978), possibly because the food supply in the wild is more conducive to growth. Moreover, in the confines of the laboratory, suppression of growth seems to increase with numbers of animals in a colony; Jarvis et al. (chap. 12) have found that the mean growth rates and final weights of the first two litters born to breeding pairs in captivity are significantly greater than those of subsequent litters.

If a mole-rat colony of average size produces four litters of 10 young a year, it could increase by 40 animals per year. If the largest colony captured ($n > 295$) was produced by a single breeding female, it would have taken a minimum of 7 yr to build up to its size at capture (assuming no mortality). The net recruitment rate averaged over all females in a colony (1.9 young/female/yr) is low compared with other rodents (Jarvis 1973) and is considerably lower than the per capita recruitment rates for many other fossorial mammals (e.g., *Spalax ehrenbergi*, Nevo 1961; *Tachyoryctes splendens*, Jarvis 1973). The restriction of reproduction in *H. glaber* colonies to a single female obviously accounts for the low per capita recruitment rate.

Survival rates and mortality of wild-caught individuals (or colonies) are unknown (but are currently being investigated by S. H. Braude [pers. comm.]). In captivity, longevity can exceed 16 yr (Jarvis, chap. 13), a remarkably long time for a mammal the size of *H. glaber*. Alexander et al. (chap. 1) discuss the ultimate or evolutionary reasons for long life span in *H. glaber*. On a proximate level, long life spans are probably linked to the animals' very low metabolic rate, their relative safety from predators, and perhaps, their slow rate of growth and reproductive development (Jarvis and Bennett, chap. 3). It is also known that a breeding female may reproduce for over 12 yr in captivity (Jarvis, chap. 13). In the Kathekani colonies, the breeding female was typically one of the last animals to be captured (see fig. 4-7). The only exception was Kathekani 10, where the breeding female was the last adult captured, followed by 13 juveniles and pups. Jarvis (1985) recorded a similar phenomenon at Lerata: the eighty-second (last) animal captured from Lerata 4 colony and the last adult captured from Lerata 2 were the breeding females. This suggests that the breeding female is reluctant to investigate disturbances to the burrow system and is thus most protected from attacks by predators. Lacey and Sherman (chap. 10) observed that in laboratory trials with live snakes and foreign colonies, breeding females usually kept their distance from the disturbance and often remained in or near the nest.

Differences between the Sexes

The sex-ratio bias toward males within and among colonies (table 4-1) may be due in part to greater female mortality, perhaps resulting from competition for breeding status. Evidence from captive colonies suggests that the breeding tenure of males is shorter than that of females (Jarvis, chap. 13), and therefore the probability of an individual male's breeding during its lifetime may be higher than that of an individual female. If so, males may have less incentive to compete strongly for breeding status once a breeding animal dies or is removed. It seems unlikely that female mortality is higher because of greater predation on adult females. In this study, digging (as indicated by dirt-clogged incisors) was not associated with one sex more than the other, and Lacey and Sherman (chap. 10) found no consistent differences between males and females in any colony-maintenance or defensive behaviors. A discussion of evolutionary reasons for the prevalence of male-biased sex ratios in cooperatively breeding vertebrates is given by Alexander et al. (chap. 1).

It is interesting that the only animal of 715 mole-rats captured at Kathekani and Kamboyo that was significantly underweight for its length and had a thin and shrunken appearance was a male. This animal may have been a former breeder. Jarvis et al. (chap. 12), reported that, in their captive colonies, a large breeding male became physically shrunken and eventually died after being supplanted by another male. Jarvis et al. also reported that breeding males do not live as long as breeding females. *Heterocephalus glaber* breeding males have the highest levels of testosterone among males in the colony, which may be the proximate mechanism influencing their survival rates; breeding females also have high levels of testosterone, as well as high levels of estrogen (B. Broll, pers. comm.; see also Faulkes et al., chap. 14).

Although females grew faster than males on the average in Kamboyo 1, the most rapid weight increases in wild mole-rats were not confined to females (fig. 4-3b). These variations in growth rates were not nearly as great as those recorded by Jarvis (1981) for captive animals, in which differences were most marked when a new breeding female was establishing herself. Such fast-growth phases may be seen in the wild only when females are growing during behavioral and/or pheromonal competition for dominance. Thus, the reason for the significantly greater weights of adult females may be the fast growth of several former candidates for breeding-female status (Jarvis 1981, chap. 13; Lacey and Sherman, chap. 10), though only one female ultimately breeds. The higher weight of females may also be a side effect of females' living longer than males on average and thus having longer to grow.

BREEDING

The mean size of weaned litters in the Kamboyo and Kathekani colonies (ca. 10; table 4-5) is large compared with other bathyergids except *Georychus capensis* (Bennett and Jarvis 1988a; Jarvis and Bennett, chap. 3). Inter-litter intervals in the wild were similar to those observed in captivity, suggesting that wild colonies are as capable as captive colonies of producing four to five litters per year. Dates of capture of juveniles indicate that breeding in the Mtito Andei area was not seasonal during 1981–1983 (fig. 4-5). In contrast, breeding appears to be seasonal in all the solitary bathyergids (Jarvis 1969; De Graaff 1981; Smithers 1983; Bennett and Jarvis 1988).

Year-round breeding in *H. glaber*, and possibly in *Cryptomys damarensis*, suggests that the supply of food is consistent. If there is any effect of very dry years on breeding output in *H. glaber*, rainfall may be the only seasonal cue

that can stimulate breeding. Quay (1981) has found that the pineal gland of
H. glaber is atrophic and the smallest in absolute size of any so far described
in rodents. This is consistent with independence of the animal from photoperiodic cues that might influence physiological or reproductive cyclicity and
suggests that, if there are seasonal cues, they do not come from the light cycle.

If there is high recruitment into colonies because of year-round production
of large litters and because of high survival rates when food is succulent and
predation is low, irregular or opportunistic breeding may help explain the variation in total size of colonies and the varied weight distributions of colony
members (table 4-1, fig. 4-2). The frequency of colony fissioning, as well as
the time of its last occurrence, will also affect colony size and body-weight
distributions, as will the differential growth of individuals (fig. 4-3b) and any
increase in mortality in one of the size cohorts.

DISPERSAL AND THE FORMATION OF NEW COLONIES

The circumstantial evidence presented on dispersal is the first indication that
new colonies may form through fissioning. The founders of Kathekani 10
probably split off from Kathekani 7. This is supported by the probable high
recruitment to Kathekani 7 before the split (i.e., Kathekani 7 had large numbers of animals weighing ca. 21 g; see fig. 4-7). The group that left was probably composed of small animals of both sexes and may have been a single
breeding pair (see table 4-6).

It is likely that the formation of a new colony by a small group would
succeed only if there were a sufficient work force to find food for the new
breeding female and the first litter produced. A dispersing breeding pair of
mole-rats might survive only if they had a substantial amount of food available
(perhaps in the form of a very large tuber) to supply them until their first litter
was old enough to start working. It is not clear when colony fission occurs, but
I suspect that it would be most common during rainy seasons, when digging is
least costly (Brett 1986, chap. 5).

When considering the limits that food supply, rain, and the size of a burrow
system place on the size and stability of naked mole-rat colonies, I hypothesize
that dispersal takes place through any of four main routes.

1. Consider a situation in which a colony steadily depletes its food resources. As a consequence of the continual search for more food, the burrow
system expands, both in total length within a circumscribed area and outward
(increasing the total area). This may be accompanied by a breakdown of suppression by breeders of fertility in nonbreeding colony members (Jarvis, chap.
13) caused by less frequent behavioral and chemical contact (see Reeve and
Sherman, chap. 11; Faulkes et al., chap. 14). If this occurs, several females
may come into estrus and fight for the opportunity to breed. In captive colonies, estrous females may fight violently, often resulting in the death of one or
several competitors (Jarvis 1978, chap. 13; Lacey and Sherman, chap. 10). In

the wild, this aggression could result in dispersal of rival females, as well as serious injuries and mortality. Perhaps the mole-rats that spend the most time away from the nest, the breeding female, and the toilet may develop reproductively and may disperse within the length of burrow system that already exists. Potential reproductives might block themselves off within a portion of it, rather than risk fights with breeding animals.

Dispersal may also occur because of the effects of colony size. As colony size increases, each individual has a lower probability of reproducing personally (see Alexander et al., chap. 1). Therefore, at large colony sizes, the relative advantages to individuals of staying in the natal colony to help versus dispersing and trying to breed independently may shift toward dispersal. Additionally, there appears to be an asymptotic relationship between pup survival and colony size, such that when colonies are large each individual's assistance is worth less (to pup survival) than when colonies are very small. Together these two factors make attempts to reproduce independently more likely at larger colony sizes. Newly potent mole-rats (perhaps accompanied by some worker animals) might thus "escape" and occupy new burrow systems blocked off from the parent colony.

2. In areas with rich and numerous food reserves that are available year-round, a burrow system may become overpopulated. High fecundity resulting in a large number of young adults (e.g., Kathekani 7) and high survival rates in good conditions may be recognized by nonreproductives as the best time to try to disperse (or block themselves off within the existing burrow system) and form new colonies. In areas of plentiful food, the work force needed to locate the food would be relatively small (Brett, chap. 5), and thus the minimum size for newly founded colonies would be smaller than otherwise, perhaps even enabling pairs of animals to found new colonies (e.g., Kathekani 10, perhaps).

3. The death of a breeding female may stimulate colony fission. In captive colonies, the removal of a breeding female and her suppressive influence often results in several females' growing rapidly, reaching puberty, and coming into estrus (Jarvis 1981, chap. 13; Lacey and Sherman, chap. 10). In the wild, with unlimited digging opportunities (especially during rainy periods), potential breeding pairs may disperse as soon as aggression within the colony rises after the death of a breeding female.

4. Colony division or dispersal may also be a chance event caused by external factors. Alexander et al. (chap. 1) have hypothesized that predation by snakes might stimulate dispersal. Blocking-off behavior (Brett 1986) is a common antipredator response, and it may result in some animals being sealed off from colony mates. Although naked mole-rats are good at rejoining severed burrows (Brett, chap. 5), it may be disadvantageous for some animals to link up severed systems if not doing so allows them to achieve breeding status. If enough time elapses for potential breeding females to come into estrus, the separated animals may never rejoin the divided portions of the burrow system.

Division of a colony through burrow collapse, caused by large mammals walking overhead (e.g., African elephants, *Loxodonta africana*) or by flood erosion, might have the same effect. Dispersal caused by such events would be most likely to occur in rainy seasons. During the rains, molehill openings most readily admit snakes; furthermore, torrential rains make flooding likely, and burrows in softer, moister soils collapse more readily. Mole-rats working at the periphery of a colony may be cut off from the main burrow system by a flood, or perhaps through the digging and blocking activities of other animals in response to the entry of a snake (Brett 1986). If smaller mole-rats indeed do more digging (see fig. 4-6), and are found more often at the periphery of a burrow system or in burrows close to the surface, animals dispersed "by accident" are probably of small body size. If, however, large animals do most of the digging (see Lacey and Sherman, chap. 10), then such accidental dispersal should result in colonies initially composed of large mole-rats.

DISPERSION OF COLONIES

It is interesting that distances between colonies were so great in Tsavo West National Park (ca. 1 km; fig. 4-1a) compared with the nearby agricultural area (fig. 4-1b). Locations were found in the National Park where there was evidence of very old *H. glaber* workings (molehills), but these were long distances away from the nearest colony with fresh molehills. These old workings may have been made by colonies that had since died out. Alternatively, whole colonies may have moved in one direction over a long period. However, when four of the colonies monitored at Kamboyo in 1982–1983 were inspected intermittently between 1984 and 1989, none were found to have moved at all, judging by the locations of groups of molehills in exactly the same areas as I saw them in 1982–1983 (Brett, chap. 5).

Geographic and geologic barriers (e.g., formations of impenetrable soil or rock) may influence colony distribution in the National Park more than in agricultural areas. The latter are suitable both for cultivation and for *H. glaber* burrow systems, where high food availability and soil suitability might stimulate frequent, successful dispersal events, resulting in the smaller mean distances between colonies. It is also possible that just as many new colonies form in the National Park, but that fewer survive, perhaps because of a higher number of snake predators (in areas without humans), a sparser and less predictable food distribution, or both.

MORTALITY FACTORS

Frequent observations of predation attempts on naked mole-rats by various snakes (e.g., beaked snakes *Rhamphiophis sp.*; plate 4-3) suggest that snakes are an important mortality factor (see also Braude, chap. 6; Jarvis and Bennett, chap. 3). Snakes represent a threat to all size and age classes of naked mole-rats. Radiotelemetric studies of 20 individuals in Kamboyo 1 (chap. 5) indicated that small animals did more digging and food carrying in superficial burrows, near the soil surface, than did large animals (see also fig. 4-6). It seems likely that small mole-rats may frequently meet up with snakes while foraging at the periphery of colonies. However, Braude (chap. 6) reported that larger mole-rats were the most frequent participants in "volcanoing." Thus, if snakes do manage to enter burrow systems, they may be confronted by larger colony members, some of which will undoubtedly be killed.

Since naked mole-rats are particularly subject to snake attacks when there are openings into the burrow system (i.e., when molehills are produced), predation by snakes should be correlated with rainy seasons (Brett, chap. 5). All five of my observations of snake predation were made during rainy seasons. Predation on other bathyergids that inhabit softer soils and come out of their burrows to feed on the aerial portions of plants is likely to be more severe than predation on *H. glaber*. For example, *Tachyoryctes splendens*, which lives in softer soils and at higher altitudes in Kenya, is a major food item for many small carnivores and raptors (Jarvis 1973). I suspect that naked mole-rats may suffer less predation or none at all during dry seasons, when the soil is virtually impenetrable and burrow systems are completely sealed for long periods.

Naked mole-rats may die of starvation if colonies are located in areas without much food, or if they consume their food resources (e.g., patches of tubers) and are unable to locate more. A critical minimum number of animals must be required to locate and exploit any given distribution of food resources (Lovegrove 1987a; Brett, chap. 5). Thus, if food availability is limited and sufficient animals die to reduce the number in a colony below the critical threshold, the whole colony might starve. Similarly, a newly formed colony that does not have sufficient numbers to locate widely scattered food may die out, whereas larger colonies living in a area with a similar food distribution could survive. This would place a lower limit on the number of animals necessary to ensure survival of a newly formed colony. Therefore, I predict that groups dispersing successfully in areas of low food availability or wide dispersion will be large compared to successfully dispersing groups in areas of high food availability.

If any colony member were to contract an infectious disease, the whole colony could be afflicted very quickly, owing to the frequent contact between colony members. Close genetic similarity between members of a colony (Honeycutt et al., chap. 7; Reeve et al. 1990) would also increase the vulnerability of a colony to disease. Obligate auto- and allo-coprophagy spreads pathogens rapidly through a colony, and intestinal infections have resulted in the extinction of several captive colonies (Jarvis, Appendix). Thus, colony extinction is a potential consequence of the initial infection of a single animal. However a burrow system closed to the surface would seem to be an effective barrier to parasites or diseases that are airborne or carried by other mammal species. Furthermore, it should be noted that genetic monomorphism does not necessarily doom naked mole-rat colonies or populations: some free-ranging large mammals (e.g., cheetah, *Acinonyx jubatus*) are virtually monomorphic over large portions of their range (O'Brien et al. 1983), showing that such populations are able to survive.

Apart from old age, the only other significant mortality factor for *H. glaber* is likely to be intraspecific aggression. Mole-rats are xenophobic; if freshly captured animals from neighboring colonies are placed together, they fight vigorously (Jarvis, Appendix). Similarly, animals from different colonies maintained in captivity will fight and defend their tunnel systems against intruders (Lacey and Sherman, chap. 10; Jarvis, chap. 13). Intracolonial aggression occurs when females are competing for breeding status, and this may result in deaths in the field as it does in captivity.

DEFENSIVE BEHAVIOR

Several other group-living mammals exhibit division of labor in defense. For example, in banded mongoose (*Mungos mungo*) packs, subordinate males guard the young when other members of the pack are foraging; adult males perform 75% of all guarding, whereas lactating females do not guard (Rood 1974). Similar guarding of young or sentry behavior occurs among subordinate males in dwarf mongooses; such behavior is particularly important in avoiding predation by snakes (Rood 1978, 1980; Rasa 1976). In wild dogs (*Lycaon pictus*), subordinate males stay near the den and guard young while other pack members are hunting (Lawick and Lawick-Goodall 1970).

The capture-order data of whole *H. glaber* colonies show that large animals tend to be captured first (fig. 4-7) and recaptured most often. In general, the body size of animals decreased gradually with order in the capture sequence, rather than there being distinct groups of large animals caught first. Any defense against intruding snakes or mole-rats would probably be most useful to the survival and reproduction of the colony if it were initiated as far away from the nest as possible. In captivity, larger colony members tend to lie near exits from nest chambers, to rush out into burrows first when there is a disturbance, and to defend their colony against snakes and foreign mole-rats (Lacey and Sherman, chap. 10). From my capture data, it may be inferred that defensive behaviors in the field are also performed by larger individuals at some distance from the nest. Observations of predation in nature are as yet insufficient to prove whether or not large colony members are the most frequent victims; preliminary data (e.g., Braude, chap. 6) suggest that they might be.

WORKER STATUS: CAPTURE DATA

The body-weight data from complete or almost complete colonies do not show convincing biomodalities or trimodalities, as might be expected if *H. glaber* colonies were divided into discrete morphological castes (as originally suggested in Jarvis 1981). However, some colonies do contain a few extra-large individuals (fig. 4-2). The lack of consistency in these size distributions does not suggest that they are associated with distinct behavioral castes.

The data collected on tooth clogging among freshly captured animals from Kathekani 9 suggested differences in digging effort within a colony: the teeth of small animals were proportionally more clogged with soil than those of large animals (fig. 4-6). Divisions within a colony into distinct working and nonworking groups based on weight (Jarvis 1981) were again not seen in this test. The character used in judging worker status may have been affected by any variation in the frequency with which animals in a colony wiped their incisors clean. It is also possible that, over the 4 days it took to capture the colony, the tooth clogging I observed resulted mainly from escape attempts. If so, however, only 26% of the colony had recently been digging (trying to escape). In summary, the scoring of the condition of incisor teeth in this colony did not allow a critical test of worker behavior, since it was not clear that I was studying worker behavior. Even if it were worker behavior that I was studying, it was only one of several worker behaviors (see Lacey and Sherman, chap. 10; Faulkes et al., chap. 14).

MONOPOLIZATION OF REPRODUCTION

The most conclusive result was that there was never more than 1 breeding female in any colony captured by me or by anyone else. Jarvis (1985) found one breeding female in each of three complete colonies captured at Lerata and was the first to observe that captive colonies have only a single breeding female (Jarvis 1978, 1981, chap. 13). These data demonstrate that breeding among females is monopolized by a single animal in both wild and captive *H. glaber* colonies. Breeding among males is restricted to one to three large animals (Lacey and Sherman, chap. 10; Jarvis, chap. 13). The data strongly support the hypothesis that adults in a naked mole-rat colony are divided into two groups, reproductives and nonreproductives.

Lacking data on longevity of naked mole-rats in the wild, and with no information on how long females and males retain breeding status, I cannot quantify the probability of an individual mole-rat breeding during its lifetime. Nevertheless, we may calculate the approximate chances of breeding by females, given a few assumptions.

Consider a small colony founded by, say, 4 males, 4 females and a new breeding pair (a group similar to the possible founders of Kathekani 10). Mean longevity might be as much as 10 yr. A breeding female dies and is replaced about every 5 yr in captivity (Jarvis, pers. comm.). Net recruitment (including births, juvenile mortality, predation, etc.) to the colony is 10 animals per year and is not sex-biased. Even assuming that the ability to become the breeding female does not diminish with increasing age or body size and that breeding females are "chosen" at random, a female in this hypothetical colony has only a 1 in 34 chance of becoming the breeder when the first breeding female dies, and only a 1 in 54 chance when the second breeding female dies. At that point, the colony has grown to 108 individuals, and the founders are all dead.

The assumptions made above are conservative, and the figures for longevity and turnover time of breeders are informed guesses: net recruitment to a colony is probably much higher, reproductive females may fail to breed, and individuals may live for much longer. In addition, it is often females who start out small that achieve breeding status when a breeding female dies (Jarvis 1981; Lacey and Sherman, chap. 10; Jarvis et al., chap. 12), suggesting that there may be a time window within a female's life when she is most likely to become a breeder if the opportunity arises. I conclude that for females the numerical probability of breeding is low but not insignificant (perhaps less than 1 in 100 in a colony of average size).

We still know nothing about how breeding animals are selected, only that competition between rival estrous females occurs and can result in escalated aggression, perhaps preceded by a period of behavioral and/or pheromonal contests in which neither animal can gain supremacy. It is interesting that aggressive interactions have rarely been observed between breeding males and/ or potential breeding males (Lacey and Sherman, chap. 10; Jarvis, chap. 13). Perhaps the dominant breeding female selects her own mates, making fighting superfluous. Less competition between males may also be indicative of the increased probability of their breeding because of rapid turnover of male breeders.

Summary

I studied the size and structure of wild *Heterocephalus glaber* colonies (715 animals captured from 14 colonies) in the vicinity of Tsavo West National Park in southeastern Kenya. Colony size was quite variable, ranging from 25 (for a newly formed colony) to over 295 (a colony captured in a cultivated field). The mean colony size was 75 animals. It was found that litters averaging about 10 young were produced year-round. The projected birth dates of litters showed no association with seasonal rainfall. In colonies where successive litters were captured, inter-litter intervals of 70–90 days, similar to those observed in captive colonies, were found; mole-rat colonies can thus produce four to five litters per year.

The spacing between neighboring colonies on agricultural land (ca. 500 m) was about half the nearest-neighbor distance in the National Park (ca. 1 km). This was probably related to higher food availability in the cultivated areas. Growth rates of all animals decreased with increasing body size, and animals took more than 1 yr to reach mean adult size (ca. 34 g).

The first indication of the manner of dispersal was found when an apparently newly formed colony, consisting of 11 adults and 14 subadults, was unearthed. This suggests that pairs, or more likely small groups of adults of both sexes, disperse with a potential reproductive pair.

The body-size distributions of colonies were not conclusively bimodal or trimodal, as would be expected if there were a discrete physical caste system. However, in one captured colony, small animals showed more evidence of having recently dug than did large animals. Large animals were caught first and recaptured more often, suggesting that they have a defensive role in the colony and that defensive behavior may increase with size. Finally, the capture of eight complete or almost complete colonies, each with a single breeding female, confirmed the evidence obtained from captive colonies and wild-colony fragments that breeding is monopolized by one female per colony.

Acknowledgments

I thank the Office of the President, Nairobi, Kenya for permission to conduct research, and the Wildlife Conservation and Management Department for permission to work in Tsavo West. I am grateful to J.U.M. Jarvis, B.C.R. Bertram, F. W. Woodley and G.M.O. Maloiy for their help and advice, and N. Mathuku, S. Komu Nzioki and D. Natta for their digging and trapping skills. The manuscript benefited greatly from the comments of P. W. Sherman, J.U.M. Jarvis and two anonymous referees. The work was supported by a research studentship from the Natural Environment and Research Council (Great Britain).

5 The Ecology of Naked Mole-Rat Colonies: Burrowing, Food, and Limiting Factors

Robert A. Brett

The naked mole-rat occupies a very specialized niche in the underground ecology of the East African savannah (Jarvis and Bennett, chap. 3). Because eusociality is unusual among vertebrate species (Andersson 1984; Alexander et al., chap. 1; Lacey and Sherman, chap. 10), the extrinsic or ecological factors that might have encouraged the evolution of this social organization in *Heterocephalus glaber* are of considerable interest.

In this chapter I describe recent findings on the ecology of wild colonies, gained from field work conducted by myself and others in northern and southern Kenya. The main source of information is my 2-yr study (1982–1983; Brett 1986) of naked mole-rat colonies around the village of Mtito Andei. A primary objective of the study was to establish the size and structure of a complete *H. glaber* burrow system. By radio-tracking individuals from one colony, I attempted to determine which behaviors (e.g., digging, foraging, inactivity, etc.) were associated with different parts of the burrow system. I also aimed to quantify the amount of burrowing going on underground through observations of molehill production on the surface and by excavatiion of multiple colonies in and around Tsavo West National Park, Kenya (Brett, chap. 4) to explore the different types of burrows made by naked mole-rats.

Details of the preferred food plants and foraging patterns of *H. glaber* are also presented in this chapter. The relationship of foraging patterns to edible plant biomass and distribution are investigated, and the possible relationship of burrowing patterns to the exploitation of particular food items is examined.

The results are drawn together in a discussion of the major ecological factors affecting naked mole-rats individually and collectively, particularly those related to location and exploitation of their food resources. The issue of how ecological limits have influenced the evolution of group living and the high degree of social cooperation in *H. glaber* is also discussed (see also Alexander et al., chap. 1; Jarvis and Bennett, chap. 3; Alexander, chap. 15). Before proceeding, however, it may be useful to to introduce the reader to the results of previous studies on *H. glaber* distribution, burrowing patterns, and food preferences.

GEOGRAPHIC DISTRIBUTION AND FAVORED HABITAT

Naked mole-rats are found only in the arid belt of the horn of Africa. This comprises the Ogaden region of Ethiopia and most of Somalia (Drake-Brockmann 1910) and the Northern Frontier District of Kenya, extending southward to Tsavo National Park. A map of the distribution of the naked mole-rat in Kenya, showing the location of study sites mentioned in this chapter is presented as figure 7-1 in Honeycutt et al. (chap. 7, p. 197; see also p. 56).

Heterocephalus glaber favors fine, sandy soils, though some colonies are found in areas with large amounts of quartz or large crystalline material (e.g., the Yatta Plateau and Tsavo River areas of Tsavo National Park), which probably hinder digging and the construction of permanent burrow systems. Naked mole-rats are also found in pure gypsum soils (in northern Somalia) as well as in the red, lateritic loams of the Tsavo area. Generally, *H. glaber* localities have fine soil, which becomes particularly hard in dry seasons and only softens appreciably after rain. Inhabited areas are characterized by low annual rainfall (< 400 mm), low altitude (< 1000 m), little diurnal fluctuation in temperature, and high radiation and evaporative water loss (see Jarvis and Bennett, chap. 3).

MOLEHILLS AND BURROW SYSTEMS

The only surface indication of the presence, size, and distribution of wild *H. glaber* colonies is the periodic appearance of molehills. These are usually produced in small groups and have a characteristic volcanolike shape (see Braude, chap. 6). Molehills are usually composed of fine, dry soil, kicked out in a pulsing spray from small surface openings of burrows (see plate 4-2 in Brett, chap. 4). The manner of molehill production and their size and shape have been described previously (Hill et al. 1957; Jarvis and Sale 1971). Hill et al. deduced that teamwork was involved in digging, because soil was kicked onto the surface in a continuous stream for a relatively long period of time. Cooperative digging and the formation of digging teams, where lines of animals shift soil up to mole-rats kicking soil onto the surface, was subsequently described in captive colonies by Jarvis and Sale (1971) and diagrammed by Jarvis (1984).

There is considerable variation in the reported distribution of molehills and the areas covered by colonies in different *H. glaber* localities. Stark (1957) reported that in the Harar region of Ethiopia, colonies were about 500 m apart and that groups of molehills from one colony covered 20 × 50 m. Hill et al. (1957) recorded the distribution of molehills before excavating burrow systems at Isiolo in northern Kenya. They reported that molehills occurred in large groups, scattered over an area of 25–35 m, and that adjacent groups were roughly 80 m apart. They excavated the burrow systems running beneath these groups of molehills, and discovered that adjacent groups were often connected underground. Jarvis and Sale (1971) found that molehills from one colony at Mtito Andei were aggregated into groups of about 20, which were 22.0 ± 12.5 m apart.

Hill et al. (1957) and Jarvis and Sale (1971) noted that the production of molehills seemed to be associated with rain showers; they suggested that this occurred because the efficiency of burrowing increases with softness or workability of soil after rains. Digging trials in the field (Brett 1986) and laboratory (Lovegrove 1989) have confirmed that naked mole-rats can more quickly excavate soft, moist soils than hard soils. Hill et al. also observed that the openings of burrows were securely blocked during heavy rains, perhaps to prevent flooding, and also that there were bursts of molehill production just before the onset of rainy seasons.

Previous studies of *H. glaber* burrow systems involved the laborious excavation of every tunnel, usually in extremely hard soil. These were undertaken at Isiolo (Hill et al. 1957) and Lerata (Jarvis 1985) in northern Kenya, and around Mtito Andei (Jarvis and Sale 1971; Sherman, Jarvis, and Alexander, unpubl. data) in southern Kenya. In no instances could all tunnel branches be followed; thus, the limits of burrows were virtually unknown. However, the impression gained by all these workers was that burrow systems are extensive, with multiple layers, often covering at least 100 m in total length.

FOOD PLANTS

The arid areas inhabited by *H. glaber* show a preponderance of geophytic plants, which possess swollen and often succulent subterranean roots, tubers, and bulbs; naked mole-rats appear to feed primarily on these. Hill et al. (1957) recorded species that were possible naked mole-rat food plants at Isiolo (e.g., herbs with large bulbs: *Ammocharis*, *Barleria*, and *Indigofera* spp., and a creeper with tuberous roots: *Cissus aphyllantha*). They also suggested that the roots of several trees and shrubs were eaten (e.g., *Erythrina*, *Hydrora*, *Amarantica* spp. and several *Acacia* and *Commifera* spp.). Hill et al. presented a large inventory of vertebrate and invertebrate fauna found in *H. glaber* burrows, including toads, centipedes, and scorpions, though none of these were mentioned as definite food items.

Jarvis and Sale (1971) recorded that tubers of an unidentified *Vigna* sp. (a legume) were eaten by a colony near Mtito Andei, and they found the remains of the husks of a *Mariscus* sp. (a sedge) in nest chambers that they excavated. As well as excavating *Vigna* tubers that were eaten in situ (but which were still sprouting), Jarvis and Sale observed that foraging burrows appeared to branch extensively in patches of *Vigna* plants, and that this might be an adaptation to enhance the rate of their discovery.

At Lerata, Jarvis (1985) determined that an *Asparagus* sp., a *Convolvulus* sp. with thick fleshy roots, and the swollen roots of an unidentified vine with a green, jointed stem were eaten by *H. glaber*. W. J. Hamilton, Jr. (1928) quoted some observations of one Mrs. Delia J. Akeley who found that two naked mole-rats captured near the Kinna River in northern Kenya had ant remains in their stomachs. Hamilton investigated the stomach contents of these specimens further and found "the remains of a beetle, a pupal case, a few seeds, several small rootlets and grasses and a small quantity of fine quartz pebbles" (1928, p. 182). However, Jarvis (pers. comm.) examined the stomach contents of more than 100 naked mole-rats and found no invertebrates or green plant material.

The evidence gathered to date thus indicates that *H. glaber* is primarily herbivorous. Since many invertebrates are found in mole-rat burrows, it is possible that some of these are occasionally eaten. Sherman (pers. comm.) found the head capsules of termites in the feces of a few recently captured animals at Mtito Andei and discovered that captive mole-rats readily consume mealworms (but not earthworms). Jarvis (chap. 13) has recorded instances of captive animals feeding on dead pups as well as adult conspecifics. The primary feces of naked mole-rats in the field are often milky and consist mainly of the endosymbionts that aid their digestion by breaking down cellulose (Porter 1957).

Methods

STUDY SITES

One naked mole-rat colony in Tsavo National Park (West) at Kamboyo (38° 07' E, 02° 45' S; 10 km southeast of Mtito Andei) was studied intensively. This colony, which I called Kamboyo 1, comprised about 87 mole-rats (Brett, chap. 4) and was located in an area of wooded grassland dominated by *Acacia* spp. and *Combretum aculeatum* scrub and the grasses *Chloris roxburghiana* and *Digitaria macroblephara*. In addition, colonies were studied (by excavation) near the village of Kathekani, 6 km north of Mtito Andei (see chap. 4).

The Tsavo area has two distinct rainy seasons, the long rains of March-May, and the short rains in November–December (fig. 5-1); the duration of each is relatively unpredictable, and rains may fail completely (Jarvis and Sale 1971). The mean annual rainfall for Kamboyo (659 mm, 1953–1983) is high compared with most other *H. glaber* localities (Jarvis 1985), and the Mtito Andei region is lush for a greater portion of the year than is the more arid Northern Frontier District of Kenya. A graph of soil temperature at Kamboyo (fig. 5-2) shows the large diurnal fluctuations found near the surface. At a depth of 60 cm, the soil temperature is stable diurnally at about 27.5°C, and varies by ± 2°C seasonally. The mole-rat's burrows extend to depths of 1–2 m (below),

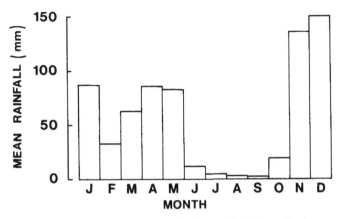

Fig. 5-1. Mean monthly rainfall totals for 1953–1983 at Kamboyo weather station, Tsavo National Park (West), Kenya.

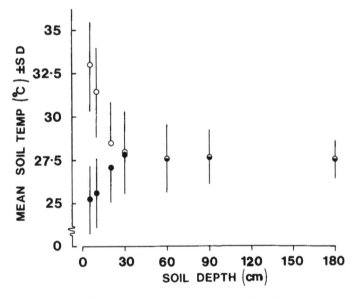

Fig. 5-2. Daily soil temperatures (± standard deviation) of 712 readings over 2 yr, 1982–1983, at depths of 10 cm – 180 cm, measured at Kamboyo weather station, Tsavo National Park (West), Kenya. *Open circles,* mean maximum temperature at 3·00 P M ; *filled circles,* mean minimum at 9:00 A M ; *half-filled circles,* no difference in maximum and minimum soil temperatures.

where the seasonal variation in temperature is only ± 1°C. Humidity in *H. glaber* burrows is uniformly close to 90% (McNab 1966; Jarvis 1978). Thus, conditions in naked mole-rat burrows are remarkably stable.

MAPPING OF THE MAIN STUDY COLONY

The approximate extent of the area covered by the burrow system of the Kamboyo-1 colony was determined initially from the position of peripheral groups of molehills. Two reference axes, defined by lines of wooden pegs placed at right angles to each other, were laid out across the colony area, with their origin at the approximate center of the colony. Pegs were hammered 20 cm into the ground at 2-m intervals. All positional data (e.g., the location of molehills, burrow paths, and food plants) were recorded on graph paper as distances from the axes, using a tape measure, lengths of string, or pacing. Additional lines of pegs were placed in areas used frequently by the mole-rats, especially those at some distance (> 50 m) from the main axes. The positions of all points were recorded as Cartesian coordinates, to an accuracy of ± 0.5 m. In areas of complex branching of burrows, greater accuracy of coordinates was required, and a grid of pegs located 1 m apart was placed over these areas.

MOLEHILLS

The area surrounding Kamboyo 1 was inspected at least weekly, and the positions of all newly formed groups of molehills were mapped. The volume and weight of a representative sample of molehills produced over 22 mo (March 1982–December 1983) was recorded. Weights and volumes of individual molehills were measured by gathering up the soil, firmly packing it into a wooden box of known weight and volume (8,000 cm³), and then weighing the box. Only molehills composed of dry soil (not moist or sticky), and produced outside rainy seasons, were weighed and measured. In total, 112 molehills were weighed, and the volumes of 95 molehills were determined.

MAPPING AND DESCRIPTION OF BURROWS

Twenty animals were captured for radio-tracking from Kamboyo 1 between September 1982 and November 1983, using the trapping techniques described by Jarvis (Appendix). There was a two-fold range in body weights of animals that I radio-tagged (27 g–51 g); the mean body weight of these animals was 35.7 ± 7.9 g, slightly higher than the mean body weight in the whole colony. Individuals were fitted with one of two types of radio tags (SR-1 transmitters, 173.2 MHz; manufactured by Biotrack Ltd., Wareham, Dorset, U.K.).

1. *Collar transmitters* were attached by shortening and crimping the loop aerial (insulated thin steel cable) around the mole-rats' neck to a smaller diam-

eter than that of the head, such that free rotation of the collar around the neck was still possible. Collars were attached soon after capture, and no immobilizing drugs were used. Radio-tagged mole-rats were kept in an artificial burrow system for a few hours, to check for possible hindrance or adverse reaction, before releasing them at the site of capture. Collar transmitters weighed 2.0 g, which was 5%–6% of the animals' mean body weight.

2. *Implant transmitters* were covered in a thin layer of silicone sealant, sterilized, and then implanted subcutaneously in the neck region of animals immobilized with 1–2 mg of ketamine hydrochloride (Ketalar, Parke-Davis, Ltd.). The skin is quite loose in the neck region of a naked mole-rat, and the implants were placed over the clavicle, where they would least impede movement. Mole-rats with implants were kept under observation for 24 h before release at the site of capture. Implant transmitters weighed 1.5 g, or 4%–5% of the animals' mean body weight.

Ten implants and 10 collar transmitters were employed, and animals captured at the same time were fitted with the same transmitter type, released, and tracked together in groups of two to five. Collars performed better, having a greater range (60–80 m) than the implants (20–40 m). The only negative was that in three cases, collar units were apparently chewed into by other mole-rats, judging by the change in transmitted tone from a bleep to a continuous whine before they failed. The range obtainable on both transmitter types decreased with the depth of the burrow. Mean transmitter lifetime was 15 ± 5 days (range: 7–29 days). Collars had no obvious effect on the behavior of the tagged animals; it seems likely that the implant transmitters caused slight debility. One collared animal was recaptured (with its collar still attached) more than 3 mo after the batteries had expired; this animal had gained 2 g in weight.

Radio-tagged mole-rats were located initially by walking over the colony area with a three-element yagi antenna attached to the receiver, which gave a directional fix on the animal. Once the area occupied by the mole-rat was located, it was approached very quietly, so as not to disturb or interrupt behavior. I found that normal walking near tagged animals interrupted foraging or digging activities, although walking near nest sites (see below) did not cause the frenzied exit of animals from nests that is often observed in captive colonies (Jarvis, Appendix; pers. obs.). Once the tagged animal had been located within an area of about 5 × 5 m, a smaller loop antenna was fitted to the receiver. This allowed more accurate location of the animal down to an area of 5 × 5 cm in burrows near the surface, and 20 × 20 cm in deeper burrows.

Animals were located at hourly intervals in daylight hours (6:00 A.M.–6:00 P.M.), and the positions of all burrows along which radio-tagged mole-rats moved were plotted. Animals were tracked over a period of 15 mo, for a total of 1,918 h, including 105 complete days of monitoring. Movement of the radio signal or fluctuations in its strength indicated activity. Subterranean nest sites were located by noting the positions where several inactive (stationary) tagged animals were repeatedly found together. The main use of a burrow could be inferred if the same behavior occurred repeatedly in it (e.g., foraging). Thus, I was also able to study differences in the frequency with which certain behaviors were performed by animals of different size (Brett 1986).

Radio-tagged mole-rats often moved rapidly without pause for distances greater than 50 m, along straight burrows leading from nest sites to peripheral burrows, or between nest sites. If an active animal was located under the aerial portions of a known food plant, and the sound of chewing at a root or tuber could also be heard, the mole-rat was clearly foraging. Sounds of chewing at food plants were often associated with animated squeaking (or "chirping") from the tagged animal or its immediate companions (the significance of this behavior is discussed by Pepper et al., chap. 9). If an animal was located at a site where there was audible chewing at the soil, or kicking soil along burrows, I interpreted that the burrows were being extended (or reopened). It was clear from the noise (or location) when radio-tagged animals were digging, shifting soil, or kicking soil out onto the surface to form molehills. Thus, in addition to distinguishing activity from inactivity and locating nest sites, I could infer whether digging, earth moving, or foraging behavior were occurring in different burrows.

In the process of capturing mole-rats from Kamboyo 1, from 2 other colonies in the Kamboyo area (Kamboyo 2 and 3), and, especially, from 11 colonies in the Kathekani area (see chap. 4), parts of burrow systems were excavated, some to depths of 2 m. This permitted characterization of various burrow types, according to their paths, diameters, depths, contents, circumstances, and likely uses.

IDENTIFICATION OF FOOD PLANTS

Many food plants of *H. glaber* were discovered during excavation of burrow systems. Foraging was indicated when swollen roots or tubers were found to have been chewed or when hollowed-out tubers were found at the termini of blind-ending burrows. Several tuber species that had clearly been encountered during burrowing were circumvented; these were not chewed and had evidently been rejected.

The positions of large, well-defined patches of geophytes, whether edible or inedible were recorded over the whole area of Kamboyo 1. Food plants discovered during excavation of colonies at Kathekani were collected for identification if they differed from those around Kamboyo. In the areas of study colonies, all species of grasses, herbs, and shrubs that had swollen roots, tubers, rhizomes, or bulbs were excavated. The aerial portions were photographed in situ and pressed for later identification. Roots were photographed, weighed, and measured. Most of the plants in Tsavo West National Park had been identified by Greenway (1969). The identity of my plant specimens was confirmed by botanists at the Nairobi Herbarium and at the Royal Botanic Gardens at Edinburgh and Kew.

VEGETATION PLOTS

The area of Kamboyo 1 was far too large to permit a complete survey of the distribution of individual food plants. Therefore, once the colony was mapped via radio-tracking, three areas (plots) were selected for special scrutiny. Plot A, an area of 3,000 m², was divided into 30 quadrats (10 × 10 m). It contained a complete patch of *Macrotyloma maranguense*, an important food-plant species, and an extensively used (in 1983) foraging branch of the burrow system. The positions of foraging burrows were plotted using radio fixes and, occasionally, by excavation (accuracy, ± 10 cm). Plot A was used to assess the interaction of foraging-burrow patterns with food distribution and biomass. Plots B and C were each 1,200 m² in size and were divided into 12 quadrats (10 × 10 m). These plots were located in the center of the area covered by the colony but had minimal encroachment by *H. glaber* foraging burrows.

Within every quadrat on plots A–C, each square meter was marked with wooden pegs and examined in detail. The locations of each potential food plant and those plant species assumed to be unpalatable were plotted on graph paper. Specimens of each geophyte species found on the plots were excavated to establish the average weight of their subterranean portions. These vegetation surveys were conducted at the end of two wet seasons (plots B and C, December 1982; plot A, December 1983), when the aerial portions of grasses, herbs, and vines had fully sprouted and/or had flowered, thus facilitating identification.

PALATABILITY, FEEDING RATES, AND ENERGY CONTENT

In addition to recording which plants were chewed on in the Kamboyo-1 burrow system, I performed an experiment to determine which food plants were palatable and which were avoided (and likely to be unpalatable). Pieces of roots and tubers of various species found in the area of Kamboyo 1 were offered, one species at a time, to naked mole-rats housed in an artificial tunnel system, and palatability was noted.

During August 1983 (dry season), weighed pieces of roots or tubers of the four most important food plants in the Kamboyo-1 area (in terms of biomass in plot A), *Pyrenacantha*, *Macrotyloma*, *Dactyliandra*, and *Vigna* spp., were placed at the end of burrows of Kamboyo 1. Food items were offered one at a time and were weighed before and after five 24-h intervals. Pieces of sweet potato, which is highly palatable, were also offered to the field colony and to captive animals in order to compare feeding rates.

The energy content of tubers of the four main food species on plot A were determined by bomb calorimetry. The mean energy content per 100 g of tuber (dry weight) of each species (n = 5 samples) was measured in the Nuffield Laboratories of Comparative Medicine in the Institute of Zoology, London. For these determinations, only samples of material from the homogeneous interior of tubers were used.

Results

THE PRODUCTION OF MOLEHILLS

WEIGHT AND VOLUME OF SOIL

The mean weight (± SE) of 112 molehills produced by Kamboyo-1 colony during 1982–1983 was 9.05 ± 0.37 kg, and their mean volume was 6,648 ± 346 cm³. The mean diameter of holes at the base of molehills was 24.9 ± 4.1 mm (n = 14). Assuming a mean burrow diameter of 3.0 cm (the average for burrows at all depths), I calculated that one molehill of average size represented the excavation of 940 cm of tunnels (1 kg of soil, with a water content of 10%, represented the amount of material removed from 60 cm of tunnel).

In 1982–1983, Kamboyo 1 produced 400–500 molehills, equivalent to 3,600–4,500 kg of soil, and 2.3–2.9 km of new burrows excavated per year. Molehills usually appeared in groups of four to eight, and the time for production of a group of six molehills was about 4 h. This represents a flow to the surface of about 13.5 kg of soil per hour, or about 380 times the mean body weight of a naked mole-rat.

DISTRIBUTION

The positions of all groups of molehills produced by Kamboyo 1 over 22 mo were plotted (fig. 5-3). The burrow system, as indicated by the molehill positions, covered an area of about 440 × 240 m. I was certain that all the molehills were produced by one colony because (1) the distances to molehills in adjacent colonies were more than 600 m, whereas distances between groups of molehills in the colony area were less than 40 m, and (2) radio-tracking confirmed that all the molehills were visited by individuals occupying a single burrow system (see below).

A dirt track ran across the area of Kamboyo 1, and there were several areas of bare ground above the colony (plate 5-1). Of all the molehills produced during the study period, 46% appeared in these areas of bare soil, whereas bare ground comprised only 9% of the whole colony area. There was thus a distinct preference for producing molehills on open, hard-packed ground rather than on grass-covered areas.

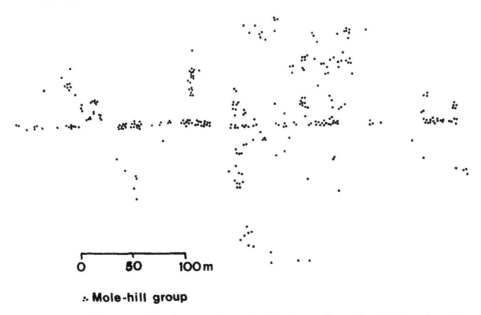

∴ Mole-hill group

Fig. 5-3. Map of the area of Kamboyo-1 colony, showing the positions of molehills produced from March 1, 1982 to December 31, 1983

Many molehills were also produced in areas of foraging activity, particularly in patches of tubers, and groups of molehills were sometimes produced simultaneously in several different areas (separated by > 100 m). New molehills were also produced in the same positions as molehills that were months or perhaps years old, suggesting that this colony may have reexcavated parts of its burrow system that had been blocked with soil and even may have revisited previously exploited food resources.

TIMING AND SEASONALITY OF PRODUCTION

In Kamboyo 1, the majority of molehills (70%, $n = 120$) were started late at night (2:00 A.M.–5:00 A.M.) and most were completed in the first hours of daylight (6:00 A.M.–8:00 A.M.). Most of the remaining 30% of molehills were started in the early morning; only 17% were begun during the period 1:00 P.M.–4:00 P.M. (the hottest part of the day).

There was a significant positive correlation between the number of molehills produced monthly and rainfall (fig. 5-4; $r = 0.48$, df $= 20$, $p < 0.02$). Over the 22 mo, 73% of all molehills were produced in the rainy months of March–May and October–December (i.e., < half the year). A significant positive correlation was also found between monthly molehill production and the soil

Plate 5-1 *Top,* The dirt track that bisected Kamboyo 1, the main study colony in Tsavo National Park (West). Many molehills were thrown up on and along this track *Bottom,* A naked mole-rat volcano "erupting" in the middle of a dirt track Photos R A. Brett, M. Griffin.

temperature at depths of 180 cm ($r = 0.59$, df = 20, $p < 0.01$) and 10 cm (9:00 A.M. (min): $r = 0.61$, df = 20, $p < 0.01$; 3:00 P.M. (max): $r = 0.47$, df = 20, $p < 0.05$).

Rainfall has a marked influence on soil temperature (fig. 5-4); mean soil temperatures dropped shortly after the start of rainy seasons. Soil temperatures dropped first near the surface before a less dramatic reduction occurred deeper down. Thus, the soil temperature at 10 cm underground showed strong daily as well as seasonal variation, but the temperature at 180 cm showed seasonal variation only. Digging of molehills commenced immediately after the first showers of each rainy period, and even small amounts of rain were enough to stimulate huge bursts of molehill production. Molehill production ceased after continued heavy daily rain, when the surface soil was saturated and muddy. Indeed, during heavy rains, the entrances to burrows from molehills were firmly plugged with soil.

THE BURROW SYSTEM

DESCRIPTION OF BURROWS

In all colonies excavated in the Mtito Andei area, five burrow types were observed:

1. Superficial burrows
2. Slanted connecting burrows
3. Highway burrows
4. Nest chambers
5. Toilet chambers

Immediately below all molehills, at depths of 0–20 cm, complex networks of small-diameter *superficial burrows* (2.5–3.0 cm) were found (plate 5-2). These were generally wide enough to hold only one mole-rat at a time. Superficial burrows were often dug while molehills were being produced, as evidenced by the sounds of animals digging in the immediate vicinity of erupting molehills; these networks were temporary and were blocked off with soil soon after use. The superficial burrows branched frequently, often on several levels, particularly in association with patches of small tubers (e.g., *Macrotyloma*). The superficial networks below molehills rarely covered more than an area of 5 × 5 m. In tuber patches, however, networks extended over a larger area. Superficial networks usually had one or two routes to deeper burrows.

Slanted connecting burrows (plate 5-3) led from superficial networks deeper underground (20–50 cm). They branched little and were characteristically grooved, being much higher than they were wide (3.5–4.0-cm wide, 4–6-cm high). It is likely that the shape of these burrows was caused by the repeated action of digging teams (Jarvis and Sale 1971) wearing away the

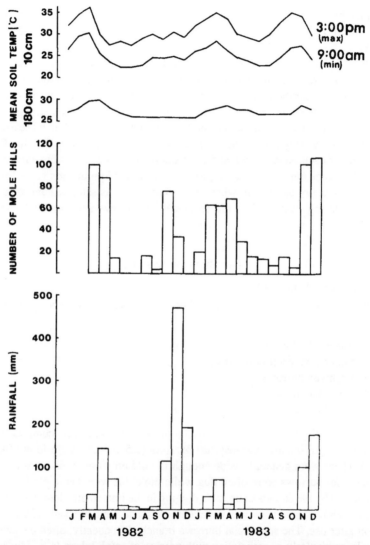

Fig. 5-4. The number of molehills produced per month from March 1982 to December 1983 by Kamboyo-1 colony, with monthly rainfall totals and mean monthly soil temperatures (measured daily at 9:00 A M [minimum] and 3·00 P M [maximum] at soil depths of 10 cm and 180 cm)

Plate 5-2. A network of superficial burrows excavated underneath a group of molehills. Photo· R. A. Brett.

Plate 5-3. Excavation of burrows beneath a group of molehills, showing a slanted connecting burrow leading down from the superficial network to a deeper highway burrow Photo: R. A. Brett.

floor and roof of the burrow while shifting soil upward from lower levels for expulsion onto the surface. The slanted burrows led down to the next level of tunnels.

Highway burrows (plate 5-3) were round, 4–5 cm in diameter, and located 50 cm or deeper underground. Mole-rats could probably pass freely through them without straddling each other. Highway burrows ran for long distances without much branching, connecting nests with foraging areas; they included short (2–5 cm) blind-ended burrows, which were probably places where animals could pull in and pass in different directions or turn around (see p. 222).

Highway burrows were kept permanently open, and if they were experimentally bisected, the mole-rats almost always joined them up again, even if sections as long as 1–2 m were removed. Damaged highway burrows were rejoined by first blocking off the two openings, then digging new burrows under, or occasionally around the side of the pit I had dug to gain access to the burrow. It thus appeared that animals located in both parts of the burrow were able to dig directly toward each other. Solitary bathyergids, geomyids, and rhizomyids are also apparently able to join up burrows in this way (Jarvis, unpubl. data; Reichman, unpubl. data); how they accomplish this is unknown, but I suspect that they hear each other and burrow toward the sounds.

To avoid disturbance, *nest chambers* were not excavated in Kamboyo 1, but nests were found in the excavation of numerous other colonies (table 5-1). The dimensions and depths of nests varied, but in general they were found deep in the ground (> 50 cm), at the level of highway burrows. Nest chambers appeared to be burrows widened sufficiently to allow several animals to rest together. Nest chambers usually contained husks and bulblike bases of the sedge *Cyperus niveus* and the grass *Chloris roxburghiana*, as well as the epidermes of tubers, particularly the small (and portable) roots of *Macrotyloma maranguense*.

The burrows leading out of nest chambers were exceedingly complex, branched frequently, and ran on several levels. At Kathekani, I found that there were usually two to five burrows exiting from nest chambers; Sherman, Jarvis, and Alexander (unpubl. data) also found multiple exits from nest chambers in colonies at Mtito Andei. However, Jarvis (unpubl.) has found some nests with just one exit. Most nests also had two to three *bolt holes* leading from them. These were blind-ending burrows that descended very steeply to deep soil, and did not end in another nest; bolt holes are also a feature of the burrow systems of other social and solitary mole-rats (Jarvis and Sale 1971; Davies and Jarvis 1986; Lovegrove and Painting 1987; Jarvis and Bennett, chap. 3).

Only one *toilet chamber* was excavated (at Kathekani). It was a blind-ending chamber measuring 5 × 8 cm in cross section, but it was presumably much larger once because the blind end was filled with dry, hard-packed feces; the feces were moister toward the entrance. This chamber was located 2.5 m from a large nest chamber and 55 cm underground. Jarvis (1985) also found a

Table 5-1

Sizes of Nest Chambers Excavated in Naked Mole-Rat Colonies
in the Field (all data in cm)

Colony	Depth	Height	Length	Width	Source
Mtito Andei	30 0	10.0	11.0	11.0	1
Mtito Andei	30 0	18.0	38.0	41.0	1
Mtito Andei	30 0	15.0	40 0	40 0	1
Mtito Andei	45.0	7 5	18 0	23.0	1
Mtito 2A	33.1	45.8	22.9	15.3	2
Mtito 2B	22.9	30.5	25.5	7.6	2
	63.6	35.6	30 5	10.2	2
Mtito 3	33 1	8.9	—	—	2
Athi River	47.0	7.0	11 0	12.0	3
Rhino Corner	44 0	9.0	43.0	22.0	3
Ithumba	67 0	small	—	—	3
Lerata 1	82.0	28.0	15.0	33 0	3
Lerata 2	41.0	9.0	15.0	10.0	3
Lerata 3	90.0	—	—	—	
Kathekani 2	64 0	9.0	17.0	12.0	4
Kathekani 4	77.0	9.0	22.0	9 0	4
Range	67 1	38 8	32 0	33 4	
Mean	50.0	17.3	23.8	18.9	
±SD	21.1	12 5	11.0	11.9	

NOTE. *Depth* is distance underground

SOURCE 1, Jarvis and Sale 1971; 2, Sherman, Jarvis, and Alexander, unpubl data;
3, Jarvis, unpubl. data; 4, Brett 1986

toilet area in the Lerata-1 colony. It was simply a blind-ending tunnel located less than 1 m from a nest chamber and packed with hard, slightly moldy
feces.

THE SIZE AND STRUCTURE OF A BURROW SYSTEM

A complete map of all the burrows used by Kamboyo 1 (fig. 5-5) was constructed from the accumulated radio-fixed locations of the 20 animals tracked
from September 2, 1982 to December 17, 1983. The majority of mapped burrows were used throughout this period, particularly the highway burrows that
connected nest chambers and served as routes to major foraging areas.

The total length of Kamboyo 1's burrow system (fig. 5-5) was 3.02 km. I
estimate that the total burrow length used by the colony over the study period
was 3.6–4.0 km, 20%–30% longer than the system that I mapped. My estimate
is based on calculations of the shortest distances from the mapped burrow
system to areas clearly used by the mole-rats (e.g., active molehill production),
although not actually visited by radio-tagged mole-rats. It is striking that the
major highway burrow in the system followed a straight path; it was located

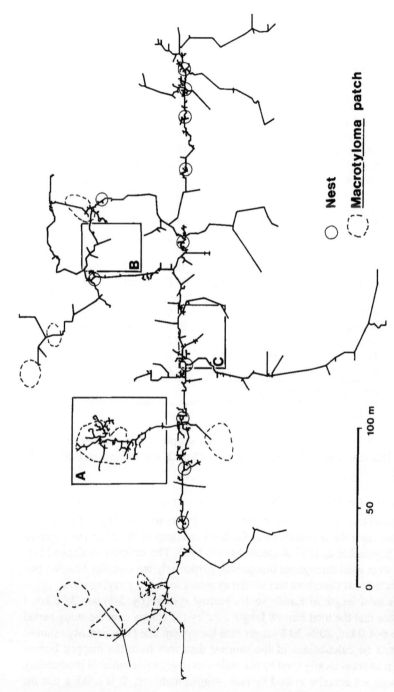

Fig. 5-5. Map of the burrow system of Kamboyo-1 colony as indicated by the positions of radio-tagged animals in 1982–1983, showing the sites of nests, and the positions of vegetation plots A, B, and C, and patches of *Macrotyloma maranguense*.

Nest ○

Macrotyloma patch ⬭

directly beneath the dirt track mentioned earlier. Molehills were produced preferentially on bare areas, particularly this track.

Eleven different nest sites were identified from radio-tracking in Kamboyo 1. These were used by the mole-rats over the tracking period, but never all at once. A few nests were used more than others, and nest use was related to the areas of the burrow system where most foraging was taking place (Brett 1986). Sometimes more than one nest was in use by the colony, but no more than two to three nests were ever used simultaneously.

FOOD PLANTS

IDENTIFICATION AND DESCRIPTION

A list of all food-plant species identified from the area of Kamboyo-1 colony is presented as table 5-2, which also includes three plant species that were evidently avoided; these same three species were also judged unpalatable on the basis of the feeding trials. I list three additional food plants (including the sweet potato) found only during excavations of colonies near Kathekani.

The two most important food-plant species associated with the burrow system of Kamboyo 1 (highest percentage of biomass in plots A and B) were *Pyrenacantha kaurabassana* and *Macrotyloma maranguense*.

Pyrenacantha kaurabassana. The geophyte *Pyrenacantha kaurabassana* (plate 5-4) produces long, leafy vines, and orange, grape-sized fruits in wet seasons. It has a very large tuber, often football-sized or larger, with yellow-orange, succulent flesh of a woolly consistency and a thin (< 1-mm thick) cream-colored epidermis (table 5-3). Tubers often have two parts, an upper tuber from which the vine sprouts, connected by a stout stalk to a larger lower tuber. Although the *P. kaurabassana* tubers around Kamboyo 1 did not exceed 13 kg, they are known to reach 30 kg in weight (Jarvis 1985; Brett 1986; Sherman, Jarvis, and Alexander, unpubl. data). At Kamboyo 1, these tubers were found at depths of 25–50 cm.

The mole-rats seemed to concentrate their foraging on the flesh of *P. kaurabassana* tubers, and they left the thin epidermis alone. Indeed, the mole-rats bored into these tubers rather than eating away the sides. One tuber excavated on plot A had been burrowed into by the mole-rats and then blocked up with soil. Subsequent radio-tracking revealed that the burrows leading to the tuber were reopened, and that the tuber was consumed once more. Several other *P. kaurabassana* specimens found by me and by Sherman, Jarvis, and Alexander (unpubl. data) had been burrowed into by mole-rats but were still sprouting healthily (plate 5-4). It therefore appears that these tubers are not destroyed, but nibbled at in situ, and later blocked up with soil. Interestingly, this may permit some regeneration.

Table 5-2

Plant Species Consumed by Naked Mole-Rats in Kamboyo-1 Colony
and in the the Burrows of Colonies at Kathekanı,
and Species that the Animals Avoided

Colony/Species	Family
CONSUMED	
Kamboyo	
Pyrenacantha kaurabassana Baillon	Icacinaceae
Macrotyloma maranguense (Taubert) Verdcourt	Leguminosae
Vigna friesiorum Harms var. *angustifolia* Verdcourt	Leguminosae
Vigna membranacea A. Richard subspecies *caesia* (Chiovenda) Verdcourt, or *Vigna frutescens* A. Richard	Leguminosae
Dactyliandra stefaninii (Chiovenda) Jeffrey (syn.: *Trochmeria stefaninii*)	Cucurbitaceae
Coccinia microphylla Gilg	Cucurbitaceae
Anthericum venulosum Baker (probably)	Iridaceae
Talinum caffrum (Thunberg) Ecklon & Zeyher	Portulacaceae
Stylochiton salaamicus N E. Brown (probably)	Araceae
Cyperus niveus Retzius	Cyperaceae
Chloris roxburgiana	Graminae
Kathekani	
Unidentified species A	Asclepiadaceae
Thunbergia guerkeana	Acanthaceae
Ipomoea batatus (sweet potato)	Convolvulaceae
NOT CONSUMED	
Kamboyo and Kathekani	
Neorautanenia mitis (A. Richard) Verdcourt	Leguminosae
Cyphostemma orondo	Vitaceae
Cissampelos paviera	Merispermaceae

Table 5-3

Mean (± SD) Weight and Dimensions of *Pyrenacantha kaurabassana* Tubers and
Macrotyloma maranguense Roots and the Depth at Which They Were Found

Characteristics	Pyrenacantha kaurabassana			Macrotyloma maranguense		
	$\bar{X}$	SD	n	$\bar{X}$	SD	n
Weight (g)	5280	3560	16	13.6	4 4 (w)	40
				8 6	2 2 (f)	10
Diameter (cm)	23	2 7	5	2 1	0 5	7
Height (cm)	24	13	7	6 7	2.2	7
Depth (cm)	25 6	12	13	0–15	—	—

NOTE w, whole root, f, flesh only, n, sample size

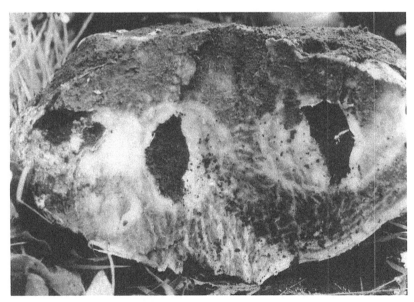

Plate 5-4. *Top,* A *Pyrenacantha kaurabassana* tuber; this is the primary food source of *Heterocephalus glaber* in the Tsavo area. *Bottom,* A *P. kaurabassana* tuber that has been burrowed through and eaten by the mole-rats, then blocked up with soil; such tubers often regenerate. Photos: R. A. Brett, P. W. Sherman

Plate 5-5. *Top,* A *Macrotyloma maranguense* plant showing the relatively small size and shallow depth of the tuber belowground. *Bottom, M maranguense* tubers come in various sizes and shapes. Photos R. A Brett

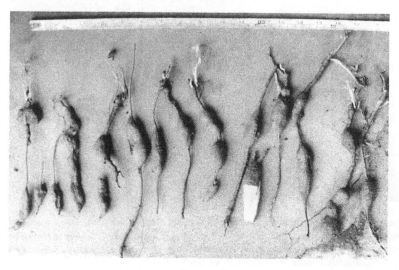

The degree of dependence of naked mole-rats on these tubers in the Mtito Andei area is suggested by a survey conducted by the Sherman, Jarvis, and Alexander. They drove 26 km of dirt roads and tracks in December 1979 (after a rainy period), noting molehills and *P. kaurabassana* vines. Although they found some *P. kaurabassana* plants more than 500 m from the nearest molehills, they never saw molehills more than 500 m from the nearest *P. kaurabassana* plant (Sherman, pers. comm.).

Macrotyloma maranguense. The legume *Macrotyloma maranguense* (plate 5-5) occurs in dense vegetative patches, and individual plants are connected by thin underground rhizomes. Foliage only appears during wet seasons. The roots, which are no more than 10-cm long and found in the surface 15 cm of soil, have swollen portions (up to 4-cm wide), often two or three on one plant; these swollen roots contain white, succulent, and slightly stringy flesh. The brown root epidermis is thick and fibrous, making up 37% of the total root weight (table 5-3).

In areas where mole-rats had been foraging on *M. maranguense* plants, excavations revealed that the swollen portions of the roots were often completely cut out and removed; these pieces of root would be small enough to be transported along the mole-rats' burrows. Swollen root portions were also found to have been chewed on in situ without separating the upper plant from the tap root. In these cases, only the flesh of the swollen portions had been consumed. Pieces of the epidermis of this species were found in nest chambers, suggesting that only the fleshy portion of the roots was eaten. The roots and tubers of other food-plant species did not have thick epidermes and were apparently totally consumed by the mole-rats.

PALATABILITY, FEEDING RATES, AND ENERGY CONTENT

Naked mole-rats housed in an artificial tunnel system ate root or tuber material of 11 of the 14 species offered to them. The tubers of the three rejected species (*Neorautanenia*, *Cyphostemma*, and *Cissampelos* spp.) were also not eaten by animals in the burrows of Kamboyo 1 (table 5-2). Mole-rats investigating such food items retreated quickly after sniffing them, further indicating their unpalatability.

For the approximately 87 mole-rats of Kamboyo 1, the mean feeding rate on a sweet potato was 48.6 ± 22.8 g/24 h (equivalent to 0.56 g/animal/day; $n = 13$ trials). *Dactyliandra steffaninii* tubers were eaten at a similar rate (40–45 g/24 h), and *P. kaurabassana* tubers were eaten at 4–10 g/24 h. The small roots of *M. maranguense* and *Vigna* spp. that I placed at the ends of Kamboyo-1 burrows were excavated and quickly carried away by the mole-rats; thus, the feeding rates on these plants could not be determined. From the amounts consumed by the captive animals, *Macrotyloma* and *Vigna* were judged to be as palatable as sweet potato and more palatable than *P. kaurabassana*.

It is unlikely that the plant parts that I offered to Kamboyo 1 were the only food sources for the colony during this time. Thus, I cannot infer the exact daily food requirements of this colony. I can, however, predict the rate of consumption of a single food item similar in palatability to a sweet potato. For example, if a *M. maranguense* patch of average size (mean area = 229 ± 96 m², *n* = 9; this is equivalent to 8.62 kg of food) was depleted by a colony of 87 animals at a rate of 48.6 g per day, it would last 177 days, assuming that all roots were located. A single *P. kaurabassana* tuber of average size (table 5-3) eaten at a rate of 10 g per day would last the same colony 528 days, assuming in both cases that feeding rates do not vary as a function of time of year.

Three captive mole-rats consumed sweet potato (their only food) at an average rate of 11.8 ± 4.0 g in 24 h (equivalent to 3.93 g/animal/day; *n* = 5 trials). All other things being equal, a colony of about 87 individuals (i.e., Kamboyo 1) would consume 342 g/day. If fed upon exclusively, and at this rate, a *M. maranguense* patch would last the colony about 25 days (assuming that all tubers were located), and a *P. kaurabassana* tuber would last about 75 days. The energy demands of isolated individuals may, however, be considerably greater than their demands might be if they were members of a large colony. Jarvis (1978) has shown that the energy requirement of a colony is reduced by huddling, and this would probably have greater influence in larger groups. Thus, there may not be a linear relationship between the number of animals in a colony and its total food requirement. The situation is also complicated by seasonal and individual variation in the amount of energy expended by mole-rats in wild colonies because they dig to find their food.

Considering all edible plant species together, the mean energy content of *Heterocephalus glaber* food sources in the area of Kamboyo 1 was 1,887 ± 30.7 kJ/100 g of root or tuber (Brett 1986). The energy content of the two primary food-plant species were 1,837 kJ/100 g (*P. kaurabassana* tubers), and 1,900 kJ/100 g (*M. maranguense* roots). Assuming a 70% water content by weight, an average *P. kaurabassana* tuber would thus contain 29,098 kJ or 59% of the estimated energy content of an *M. maranguense* patch of average size (49,134 kJ; below).

FOOD-PLANT DISTRIBUTION

Macrotyloma maranguense was the only major food-plant species that grew in well-defined patches. Nine such patches were found (fig. 5-5), and the mean nearest-neighbor distance between them was 32.9 ± 21.5 m. The mean area of these patches was 229.3 ± 96.4 m². The area actually covered by *M. maranguense* plants made up only 3.5% of the total area circumscribing them (58,800 m²).

The distribution patterns of the major palatable geophytes and of those that were not eaten were mapped in plots A–C (for details, see Brett 1986), and analyzed using nearest-neighbor techniques (Southwood 1966) to detect departures from randomness (table 5-4). *Pyrenacantha kaurabassana* tubers

Table 5-4
Biomass and Distribution Analyses of Food Plants, and Plants not Eaten (unpalatable),
in Vegetation Plots in the Area of Mole-Rat Colony Kamboyo 1

Plant Species	n	Density (n/m²)	Biomass (g/m²) $\bar{x}$	Biomass (g/m²) SD	Mean Nearest-Neighbor Distance (m)	χ^2	df	Distribution	p
			PLOT A						
Plants eaten									
Pyrenacantha	63	0.021	110.9	±74 8	3 28	55 48	50	random	
Macrotyloma									
Whole plot	1919								
Patch only	1912	4.406	5 7	±19.4(w)	0.24	—	—	patchy	
			37.6	±12.2(f)					
Center of patch	585	5.850	79 3	±25.8(w)	0 21	—	—		
(10 × 10 m²)			50.1	±12 9(f)					
Vigna spp.	14	0.005	0.05		8.69	24.03	18	random	
Cucurbitaceae	25	0.008	1.60	±0.61	5.50	47.55	36	random	
Dactyliandra	18	0.006	1 16	±0.44					
Coccinia	7	0.002	2.26						
Talinum	21	0.007	0.99	±0.12	2.68	10.29	36	patchy	<0.005
Stylochiton	3	0.001	0 05		15 19	3.38	4	random	
Plants not eaten									
Neorautanenia	12	0.004	23 00	±15.56	9.24	58 88	18	spaced	<0.001
Cyphostemma	21	0.007	1.93	±0.25	6.80	44.92	28	spaced	<0.025
Cissampelos	28	0.009	18.67		3 29	69.17	40	patchy	<0.005
			PLOT B						
Plants eaten									
Pyrenacantha	134	0.112	589.60	±397.5	1 11	38 56	50	random	
Macrotyloma	256	4.406	37.6	±12.20	0.24	—	—	patchy	
Dactyliandra	12	0.010	1 93	±0.73	4.91	24.97	16	random	
Plants not eaten									
Neorautanenia	32	0.027	153 30	±103 70	2 97	77.04	48	spaced	<0.01
Cyphostemma	10	0.008	80.21	±0.29	4.79	10.28	14	random	
Cissampelos	38	0.032	63 33		1.54	32.11	56	patchy	<0.005
			PLOT C						
Plants eaten									
Pyrenacantha	53	0.044	233.20	±157.20	2.27	44.98	50	random	
Dactyliandra	9	0.008	1.44	±0.55	5.10	9.57	12	random	
Plants not eaten									
Neorautanenia	32	0 027	153.30	±103.70	2 89	33 55	40	random	
Cyphostemma	10	0 008	80.21	±0 29	7 38	30 39	16	spaced	<0.025

NOTE. For plot A, the area is 3,000 m²; w, whole tuber, f, flesh only Plots B and C are control plots of 1,200 m² each
The sequence of species indicates their importance in the diet of the mole-rates of Kamboyo 1

Table 5-5
Proportion of Palatable Biomass Accounted for by Each Food-Plant Species
on Plot A (see fig. 5-6)

Food-Plant Species	% Biomass Including Macrotyloma	Excluding Macrotyloma	In Macrotyloma Patch Only
Pyrenacantha kaurabassana	91.48	96 0	63 55
Macrotyloma maranguense	4 50	—	32.77
Vigna spp.	0 04	0.04	0 04
Dactyliandra stefaninii	0 95	1.03	1 55
Coccinia microphylla	2 16	2.00	0.00
Talinum caffrum	0 82	0 88	1 99
Stylochiton salaamicus	0 04	0.05	0 10

NOTE The sequence of species indicates their importance in the diet of the mole-rats at Kamboyo 1

were randomly distributed within the areas of the vegetation plots. However, there did not appear to be an even distribution in plot A, and it is possible that this species does have a clumped distribution on a larger scale (i.e., over the whole area of the Kamboyo-1 colony). Certainly the distances between *P. kaurabassana* tubers were far greater than those between *M. maranguense* plants within a patch. The distribution of all plants palatable to mole-rats, except *M. maranguense* and *Talinum caffrum,* was random (table 5-4), even on a microgeographic scale. There may have been a negative influence of the patch of *M. maranguense* on the growth of other geophyte species in plot A; however, *T. caffrum* did not appear to be affected, there being an aggregation of its tubers in the *M. maranguense* patch.

BIOMASS

Density and biomass figures for all major food plants in plots A, B, and C, and those not eaten (unpalatable) are given in table 5-4. Only the fleshy (edible) portions of *M. maranguense* roots were used for estimates of food biomass available to Kamboyo 1. It was assumed that the entirety of all the other palatable underground plants were eaten, since no other species had a thick bark. To assess the effects of patchiness of *M. maranguense* on the food availability in small areas (e.g., the size of plot A), the figures are divided into percentages for biomass outside and inside *M. maranguense* patches.

Pyrenacantha kaurabassana was by far the major food-plant species available in plots A (table 5-5), B, and C (> 90% of the total palatable biomass). *Macrotyloma maranguense* made up less than 5% of the total palatable biomass. However, within the area of the *M. maranguense* patch in plot A, its proportion of the edible biomass was, of course, much higher. The whole patch was estimated to have 25.9 ± 8.4 kg of swollen *M. maranguense* roots, containing 16.3 ± 5.3 kg of food (excluding husks).

Table 5-6
Total Underground Biomass Densities in Plots A, B, and C

Biomass Density	Plot A (3,000 m²)	Plot B (1,200 m²)	Plot C (1,200 m²)
Palatable (excluding Macrotyloma maranguense)	113	591	235
Unpalatable	43	297	234
Total	156	888	469

NOTE All figures are expressed in grams per square meter

Assuming that *M. maranguense* plants were at similar densities in each patch, the mean biomass of food in an average-sized patch was 8.62 kg Thus, although the food density of *M. maranguense* roots within patches was 50.1 ± 2.9 g/m², when the whole area occupied by the colony was considered, the density of *M. maranguense* food was only 1.32 g/m². Considering the mean density of *P. kaurabassana* biomass from the three vegetation plots, and the density of *M. maranguense* biomass in the total area bounding the nine patches located, *M. maranguense* biomass was only 0.42% of the biomass of *P. kaurabassana* tubers and only 4.42% of the total palatable biomass available. In contrast, *P. kaurabassana* tubers made up 92% of the total palatable biomass.

Data on the density and mean weight of subterranean plant material were analyzed to yield figures for the total palatable and unpalatable food-biomass density in each vegetation plot (table 5-6) and a mean biomass density for the whole area of Kamboyo 1. A considerable amount (28%–50%) of the total underground plant material was apparently unavailable to the mole-rats (unpalatable). Averaging the mean palatable biomass densities of food in the three plots (table 5-6), the total food density available to Kamboyo 1 was estimated at 313 ± 203 g/m². The total food biomass within the whole colony area (240 × 440 m) was calculated to be 37,712 kg.

OTHER FOOD ITEMS

Invertebrates. Invertebrates collected in burrow systems were presented to a small captive *H. glaber* colony (n = 3 individuals). Invertebrates offered included myriapods, spiders, ants, termites, and beetles, including a species of tenebrionid beetle (*Paoligena* sp.) found exclusively in the burrow systems of *H. glaber* (Penrith 1982). None of the invertebrates was eaten. Indeed only moving specimens elicited any reaction, usually a bite with the incisor teeth.

Bones. In August 1983, three radio-tagged mole-rats from Kamboyo 1 used a branch of burrows and nearby nest site located on vegetation plot A (fig. 5-5) almost exclusively. One individual was tracked to a new position in this burrow branch near the surface, where the burrow ran directly beneath a 20-cm piece of an old bone (the radius or ulna of an equid, presumably a zebra, *Equus burchelli*). A rasping sound was heard from underneath the bone for about 2 min while the radio-tagged mole-rat was there. When the animal had left the area, I turned over the bone and found a blind-ended burrow running directly underneath. A substantial amount of one end of the bone had been chewed away, and there were a great number of chewing marks along the length of the underside of the bone. All three mole-rats tracked during this period made visits to this bone, usually spending 2–3 min chewing at it. Since few fragments of bone were found on the floor of the burrow underneath, it was assumed that the bone was being consumed.

FOOD DISTRIBUTION AND BURROWING

Jarvis and Sale (1971) observed that *H. glaber* burrows appeared to branch frequently in patches of *Vigna* spp. They suggested that this might increase the rate of discovery of such food items. The presence of a foraging branch of the Kamboyo-1 burrow system on part of plot A (fig. 5-5) and the extent of the plot (3,000 m^2) allowed me to assess the effects on burrowing of the distribution and biomass of *P. kaurabassana* and *M. maranguense*.

Figure 5-6 shows the distribution of *P. kaurabassana* tubers, the area covered by the *M. maranguense* patch in plot A, and the positions of burrows mapped in my radiotelemetric study. Burrows are not obviously associated with the position of large tubers. There is very little branching of the burrow in plot A in response to the higher density of *P. kaurabassana* tubers in the area near the main burrow network (fig. 5-5); the first *P. kaurabassana* tuber was encountered after 20 m or more of burrowing out from the main highway burrow (fig. 5-6), after many tubers located fairly close to the burrow (< 1 m) had been "missed." In contrast, branching of the burrow system was obviously associated with the patch of *M. maranguense* plants in plot A. It appears that the density of most plant species apart from *M. maranguense* was actually reduced in the area of burrowing. These impressions were tested statistically in three ways.

First, presence or absence of mole-rat burrows and various plant species (listed in table 5-2) was established for the 30 quadrats of 10 × 10 m and the associations of plant species and burrows and pairs of plant species were examined statistically. No significant associations between burrows and plant species or pairs of plant species were found, except for *M. maranguense*, which exhibited a highly significant positive association with burrows ($\chi^2 =$ 13.27, df = 28, $p < 0.005$).

Second, the total length of burrow in each quadrat was calculated by summing the distances between successive "fixes" of radio-tracked mole-rats traveling along them. I tested for correlations between these lengths and the numbers of plants of each species across the 30 quadrats. A highly significant positive correlation was found between the length of the burrow and the number of *M. maranguense* plants ($r = 0.79$, df = 30, $p < 0.001$). No significant

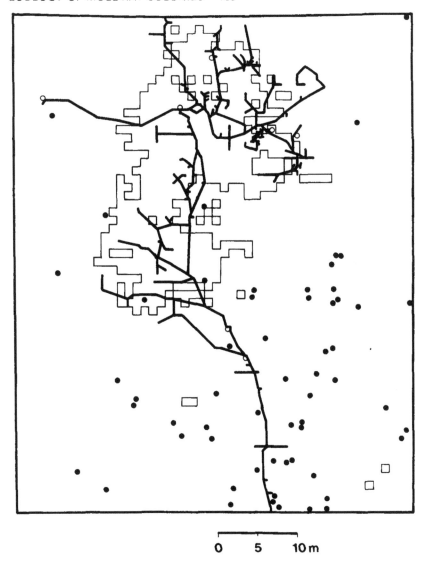

0   5   10 m

Fig. 5-6. Map of the burrow system of Kamboyo-1 colony in vegetation plot A, show-ing the distribution of *Pyrenacantha kaurabassana* tubers and the limit of the patch of *Macrotyloma maranguense*, which is plotted by presence or absence in quadrats of 1 m². *Heavy lines*, naked mole-rat burrows; *thin lines*, limit of *M. maranguense* patch; *filled circles*, uneaten *P. kaurabassana* tuber; *open circles*, eaten *P. kaurabassana* tuber.

correlations were found with any other plant species, and with all species ex-cept *T. caffrum*, the correlations were negative.

Third, the number of branches per meter of burrow in plot A was compared inside and outside the *M. maranguense* patch. Outside the patch, 89 m of burrow contained 26 dichotomous branches (0.29 branches/m); inside the patch, 82 m of burrow contained 79 dichotomous branches (0.96 branches/ m). Thus, branching per meter of burrow was more than three times as great within the *M. maranguense* patch as outside it. From this, I conclude that *M. maranguense* tubers are worthwhile enough food items (or the soil around them is soft enough) that mole-rats will make the extra digging efforts necessary to obtain them, at least in terms of relatively more excavation (burrow branching) within a limited area (tuber patch).

Burrow length was negatively associated with the number and distribution of all tubers other than *M. maranguense* and *T. caffrum*. This might have been because they were eaten out by the mole-rats in their search for *M. maranguense* tubers. Apart from enhancing the chances of finding other plants of the same species, increased branching of burrows probably also increases the rate of discovery of other species of geophyte within *M. maranguense* patches. Thus, 4 of 8 *P. kaurabassana* tubers located in the *M. maranguense* patch were found by mole-rats and eaten (fig. 5-6); in contrast, only 2 of more than 30 *P. kaurabassana* tubers located outside an *M. maranguense* patch were found and eaten. Thus, the food potential of an *M. maranguense* patch is actually somewhat greater than the estimate given (tables 5-4, 5-5).

Discussion

LIMITING FACTORS

MOLEHILL PRODUCTION

The measures of soil transport that I presented provide an indication of the prodigious digging abilities of naked mole-rats. In the 1983 rainy season alone, the approximately 87 animals of Kamboyo 1 excavated about 1 km of new burrows; the colony excavated 2.3–2.9 km of tunnels in 1982–1983. It must be remembered that these figures exclude the amount of soil actually shifted around inside the burrow system (i.e., not removed to the surface). It is possible that some of the soil appearing as molehills in a rainy season is loose soil accumulated mainly from food searches and burrowing activity in the previous dry season. New tunnels may be constructed in the dry seasons, even though few molehills are produced, and soil may be shifted from the newly excavated burrows into old ones rather than being disposed of on the surface. However, the amount of soil thrown out onto the surface must represent the total area of burrow reopened or newly excavated, unless many old burrows are used as permanent repositories for excavated soil.

Kamboyo 1 covered an area much larger (ca. 105,000 m^2) than that of any colony previously described (on the basis of excavations). Digging in superficial burrows took place simultaneously in several different places separated by more than 100 m. This gave the appearance of workings from two different colonies and may be the reason for the comparatively low estimates of the size of colony areas previously published (Hill et al. 1957; Stark 1957; Jarvis and Sale 1971). Although the present data were collected primarily from one colony, judging from the distribution of molehills, other colonies in Tsavo National Park occupied at least as large an area as did Kamboyo 1. For example, a neighboring colony at Kamboyo measured about 900 m from one side to the other.

The tendency of naked mole-rat colonies to produce molehills on roads and tracks (i.e., areas of bare soil without grass cover) seen in my study was also noted by Hill et al. (1957), Braude (chap. 6), and Sherman, Jarvis, and Alexander (unpubl. data). There are two possible reasons for this preference for bare soil. First, Hill et al. reported that attempts at molehill excavation in a grassy patch at the side of a track were soon abandoned, perhaps because the grass cover impeded the expulsion of soil. Second, and more importantly, the hard-packed soil under roads and tracks may be difficult for predators to excavate and thus the safest place to open the burrow system.

Most of the nest sites and main highway burrows of Kamboyo-1 colony were also located beneath a hard-packed dirt track. This track was put in by the National Park authorities around 1950; it seems unlikely that the straight track followed the path of molehills made by a colony that was already in place. Several reasons may account for the location of nest sites in these bare areas.

1. Bare areas of ground absorb more heat than areas with vegetation cover, thus providing favorable areas for huddling and heat absorption. Radio-tagged mole-rats in Kamboyo 1 appeared to come to superficial burrows in bare areas to warm up. Excess body heat may also be reradiated faster in burrows near a bare surface when temperatures drop (e.g., at night). These factors would increase the efficiency of ectothermy in bare areas (see McNab 1966; Jarvis 1978; Withers and Jarvis 1980).

2. In bare areas, the soil is baked and packed hard, perhaps reducing the likelihood of a damaging collapse of burrows and particularly of nest chambers. These areas may thus be good locations for large underground cavities. The hardness of the soil in these areas would also make it exceedingly difficult for any predators to excavate nests.

3. Dirt tracks in the National Park (such as the one passing over Kamboyo 1) were graded such that the sides were lower than the center to encourage runoff. This may result in the center being the place that is least likely to flood, and thus the best place to locate a nest (as well as to open the burrow system); the water runoff also created rich foraging areas at the sides of tracks (i.e., near nests).

Molehill production was most noticeable in the early morning and late evening (Hill et al. 1957; Jarvis and Sale 1971; Braude, chap. 6). This temporal pattern may be an adaptation for avoiding snake predation (e.g., *Rhamphiophis* spp., *Eryx* spp.; see chap. 3, 4, 6). Snakes were most active during the rainy season (Pitman 1974) when molehill production also peaked. By producing molehills when reptiles are most likely to be torpid, the mole-rats may reduce the likelihood of attack. During the early morning (2:00 A.M.–5:00 A.M.), most foraging snakes lose much of the heat absorbed through daytime basking (Heatwole 1976), but the heat retained by the soil at the greater depths inhabited by the mole-rats probably allows them to remain active all day and night, and certainly for short periods in the early morning when air temperatures are at their lowest (e.g., fig. 5-2). In addition, burrowing activity will raise a mole-rat's body temperature.

It is likely that seasonality in molehill production is a consequence of seasonal excavation of soil belowground because of the expansion of the burrow system with foraging. Genelly (1965) found a clear link between soil moisture and the digging of foraging burrows by *Cryptomys hottentotus hottentotus*. The strong correlation between rainfall and molehill production at Kamboyo 1 (fig. 5-4) suggests that the expansion of burrow systems and probably most of the foraging occurs when soils are softest and most easily worked and when excavation is least energetically costly.

THE BURROW SYSTEM

Kamboyo 1 inhabited what is certainly the largest and longest continuous system of burrows ever discovered for a fossorial mammal, and considerably greater in size than other *H. glaber* burrow systems excavated (Jarvis and Sale 1971, 165 m; Jarvis 1985, 595 m; Sherman, Jarvis, and Alexander, unpubl. data, 237 m). The burrow system of Kamboyo 1 was also far longer than burrow lengths measured for two other social bathyergids, *Cryptomys hottentotus* (Davies and Jarvis 1986, 387 m; Hickman 1979b, 340 m) and *C. damarensis* (Lovegrove and Painting 1987; they excavated a primary burrow, perhaps equivalent to a highway burrow in *H. glaber*, that was 82 m long). At any one time, the length of open and unblocked burrows in the Kamboyo-1 system was undoubtedly less than the total length of burrow used by the tagged mole-rats over the 15-mo tracking period, because some burrow blockage probably took place. However, I believe the figure for the length of the burrow system (3.02 km) underestimates the actual total by 20%–30% because a substantial length of burrow was probably undetected by the radio-tracking.

The Kamboyo-1 burrow system did not change in total extent or move in any direction over the 2-yr study period. The colony has been revisited intermittently, but at least once yearly, since 1983. In 1989, 7 yr after the start of observations, Kamboyo 1 still occupied exactly the same area, with groups of molehills thrown up in many of the same spots as those shown in figure 5-3. Other colonies in the Kamboyo area also appeared, from the position of molehills, not to have moved, expanded, or fragmented over this 7-yr period. Colonies in this area are thus apparently able to survive on a sustainable yield of subterranean tubers.

DIET AND FORAGING PATTERNS

The diet of naked mole-rats in the Mtito Andei area was composed primarily of underground plant material. A huge biomass of subterranean geophytes was available, but one-fourth to one-half of it was unpalatable (tables 5-4, 5-6). Some edible tubers were preferred over others; in general, the small fleshy tubers were more palatable than the large fibrous ones.

The rate of food supply to an *H. glaber* colony is primarily related to the rate of discovery of the large and deep *P. kaurabassana* tubers (averaging 5.3 kg of food per tuber), which made up more than 90% of the palatable biomass in each of the three study plots. Assuming that naked mole-rats find and harvest most of the individual plants as they burrow within the widely dispersed patches of *M. maranguense* (averaging 8.6 kg of food per patch), patches of this species represent the next largest food resource available.

No evidence was found that mole-rats in the study area ate any of the invertebrate fauna found in their burrows. The discovery of mole-rats eating a bone lying on the surface was intriguing. It is possible that this bone was used just for honing the incisors. However, tooth sharpening is performed by resting mole-rats by shearing the upper and lower sets of incisors against each other (Lacey et al., chap. 8); moreover, large portions of the bone had been removed. Perhaps bones are an important supplement to the diet, providing calcium and phosphorus salts, which may be rare in tuberous food. Because the incisor teeth of *H. glaber* grow and wear rapidly, it is likely that a regular source of minerals is necessary for this rate of tooth replacement; animal bones might well be this source. How the subterranean mole-rats locate bones lying on the surface is unknown.

Heterocephalus glaber foraging behavior can clearly be nonrandom. Mole-rats may increase their rate of discovery of food by changing their burrowing behavior in response to particular food plants. Analyses of the burrowing patterns in plot A confirm the suggestion of Jarvis and Sale (1971) that increased branching of burrows occurs in patches of vegetative tubers (in this case, *M. maranguense*). Since the burrowing patterns inside and outside a patch were clearly different, it appears that this foraging behavior enables the animals to harvest one particular type of food plant. No increase in branching was found where burrows met larger, randomly distributed tubers (e.g., *P. kaurabassana*). The change in burrowing pattern from a straight, relatively unbranching burrow to more turning and branching in patches of small, dispersed, but abundant food items may be an example of area-restricted searching (Krebs 1978) as an optimal-foraging strategy. It is analogous to the paths taken by blackbirds (*Turdus merula*) in foraging on a patchy food (worms) distribution on grass lawns (Smith 1974).

The larger tubers eaten by the mole-rats were usually found far underground, and it is probable that the deep highway burrows encounter these. However, I found no evidence that the highway burrows in Kamboyo 1 were directed toward *P. kaurabassana* tubers; rather, they appeared to follow a dirt track closely. Burrowing also could not be adequately characterized as random, because burrows did radiate out from the central highway burrow "trunk" (fig. 5-5), and there were few back loops or circular paths (as might be expected if burrowing were truly random).

I suggest that most food searching is conducted by burrowing in a blind but unidirectional fashion at different levels. Blind search could be a reasonable foraging behavior for discovering the large *P. kaurabassana* tubers, which, although they were quite dense in some areas (e.g., plot B), were still randomly distributed. Patches of vegetative tubers (e.g., *M. maranguense*) were located far apart (mean interpatch distance was 33 m), and patches were probably also encountered by chance. Once a large patch has been encountered, the mole-rats could maximize the rate of discovery of these smaller, portable food items by using area-restricted searching.

It seems unlikely that naked mole-rats can sense the presence of tubers over long distances through the soil (e.g., chemically); they did not appear to direct their burrowing straight toward tubers. As Kamboyo-1 colony members were excavating the branch of burrows in plot A (figs. 5-5, 5-6), they evidently burrowed for over 20 m, past six *P. kaurabassana* tubes positioned less than 1 m from the burrow line, before locating a tuber and consuming it. The mean nearest-neighbor distance of *P. kaurabassana* tubers in plot A was 3.3 m (table 5-4). These tubers and those of other species are probably only detected by foraging mole-rats at very short distances (e.g., < 10 cm), perhaps by smell. Thus, foraging behavior in *H. glaber* probably does not conform to the assumptions of Pyke's (1978) optimal-foraging model, which requires sensory input to be the most important factor in maximizing food intake. When responding to the location of food patches and when foraging within them, the mole-rats probably use information gained from the location of the last food item(s) encountered.

Although area-restricted searching was indicated for Kamboyo 1, at Lerata in northern Kenya, Jarvis (1985) found no relation between the distribution of food plants and the burrow configuration nor any evidence that naked mole-rats followed roots as cues to the location of further plants. The Lerata area differed from the Mtito Andei area in that there were few, if any, tubers; at Lerata, the mole-rats ate long, thin fleshy roots. Area-restricted searching may be advantageous only where food is at least sometimes concentrated into patches, as at Kamboyo 1.

Heterocephalus glaber does not always destroy larger tubers (e.g., *P. kau-rabassana*) but typically consumes only part of them and uses soil to fill the internal cavity created while foraging; the animals then block off the partially eaten tuber, allowing it to regenerate. I suggest that the filling in of partially consumed tubers by foraging mole-rats may be a type of farming of such food resources in order to maintain the colony's food supply within one area. The ease with which burrows blocked by mole-rats could be reopened by hand during excavations of burrow systems suggests that it would not be difficult for individuals to relocate previously exploited food resources, even if the burrows leading to them had been plugged with soil. The (likely) reexploitation of regenerated food resources by *H. glaber* should be contrasted with the removal and storage of small food items (e.g., bulbs and corms) by the solitary mole-rats *Georychus capensis* (Du Toit et al. 1985) and *Spalax ehrenbergi* (Nevo 1961). Food stores are built up by these mole-rats in wet seasons for use during dry seasons when less food is available aboveground and when the soil is hard to excavate.

Kamboyo 1 has remained in essentially the same spot from 1982 to 1989; if colonies typically live in one area for many years, as appears to be the case, judicious use of local food resources is a possibility (given the similarity in the reproductive interests of colony members; Alexander et al., chap. 1). It is not known how foraging naked mole-rats affect the growth of vegetative patches of food tubers. Because of the complex rhizomatous network in *M. maran-guense* patches, the mole-rats may rarely "eat out" or destroy a food patch completely. However, an area of the *M. maranguense* patch in plot A did appear to be destroyed by intense burrowing and exploitation. Interestingly, in terms of the farming hypothesis, many foraging burrows in *M. maranguense* patches were blocked off with soil after use, and the plants continued to grow healthily and spread in these areas.

LIMITING FACTORS: SOIL HARDNESS, TEMPERATURE, AND RAINFALL

Variation in hardness of the soil, particularly near the surface, depends on the timing and quantity of rainfall and associated soil temperatures. Because the softness of soil has a profound influence on the efficiency of digging by naked mole-rats (Brett 1986; Lovegrove 1989), the accessibility of food depends on the timing and quantity of rains, as well as on temperature variations. The efficiency of digging must have a considerable influence on the mean and minimum energy costs of digging between major food resources (see below), especially if, as I suspect, they are located blind.

Rainfall has a dramatic effect on the production of molehills (fig. 5-4). I suggest that the extreme hardness of the soil 5–10 cm below the surface during dry seasons is the reason why molehills are not often produced at this time. It is in this region of the soil profile that the largest changes in soil moisture content and soil penetrability occur with rainfall (Brett 1986). Animals extended tunnels almost five times faster when digging in moist soil as they did in soil with the same penetrability as the very hard surface crust in dry seasons. Although neither energy expenditure nor the amount of wear on the incisor teeth of digging mole-rats has yet been quantified, excavation in very hard soil is costly in terms of both factors. It should be noted for completeness that digging in soft, moist soil may also involve costs: incisors get clogged with soil and need frequent grooming, and there may be thermoregulatory problems if an animal gets too wet.

Surface burrow temperatures in the middle of the day during the dry season often reach levels far above the tolerance limit for *H. glaber* (e.g., $40°$–$45°$C; Jarvis 1978). Rainfall reduces the amplitude of diurnal temperature variation near the surface (fig. 5-4); with less extreme temperatures, there is more time in a day when superficial burrows are tolerable to the mole-rats. Thus, rainfall not only makes digging more efficient, but it increases the length of time over which it can occur. Furthermore, rainfall probably also improves the nutritional quality and energy content of subterranean tubers. Rainfall peaks may also stimulate dispersal of mole-rats to form new colonies through (1) reducing the energetic cost of digging new burrows, (2) increasing the amount of food available, which may improve the chances that a small group can survive, and (3) physically separating and thus isolating groups (e.g., as a result of flash floods or burrow cave-ins) that can potentially start new colonies (see Brett, chap. 4).

If molehill production is positively correlated with the total amount of soil excavation going on belowground, the rate of location of food sources is also probably influenced by the rainfall. However, high levels of foraging behavior and general activity are maintained by the mole-rats in dry seasons (Brett 1986). Thus, although molehill production is distinctly seasonal, activity (time spent out of nest chambers) is not, and it is thus likely that food is continuously harvested year-round. If there is indeed a stable supply of food to the colony, it might explain the lack of seasonality of breeding in the Mtito Andei area (Brett, chap. 4).

Since the surface soil gets very hard in dry seasons, food located in this layer (such as patches of *M. maranguense* tubers) must cost more to dig for and exploit at these times. Consequently, I suggest that foraging for large tubers, located deeper but in softer soil, provides the stable food supply that the mole-rats depend on. During the dry seasons, this may be the only source of new food that it is energetically feasible to exploit. The penetrability of soil at the mean depth of *P. kaurabassana* tubers (38 cm) in the Mtito Andei area is sufficiently high to allow efficient foraging for these tubers year-round (Brett 1986).

The thermal stability of soil at depths below 30 cm, compared to wide diurnal and seasonal variation nearer to the surface (figs. 5-2, 5-4) allows *H. glaber* to be ectothermic and must lower the energy consumption of a colony considerably. Furthermore, Jarvis (1978) has shown that huddling reduces the metabolic requirements of a given number of mole-rats. In burrow systems that are completely sealed for long periods, with reduced O_2 levels and elevated CO_2 levels, a low metabolic rate and a high oxygen affinity of red blood cells (Johansen et al. 1976) have obvious adaptive value for naked mole-rats (Lovegrove 1987a). Nakedness clearly facilitates efficient heat transfer among mole-rats, and from burrow to mole-rat (and vice versa; McNab 1966; Withers and Jarvis 1980); there may also be some energetic saving made by not growing fur in the first place. Tucker (1981) believes that lack of fur may also be an advantage when digging in very hard soils and when traveling backward. Further discussion of the evolution of hairlessness in *H. glaber* is given by Alexander (chap. 15).

THE COSTS OF BURROWING

The costs to individual mole-rats of locating new food resources by digging may be reduced, and the rate of locating food may be increased, by group foraging. I now ask whether it is possible for solitary mole-rats to survive on food resources of the biomass and distribution determined in the area of the Kamboyo-1 colony, and how foraging is constrained by the energy required to find or exploit new food resources or by other costs (e.g., the wear on incisor teeth).

Using data on the distribution, biomass, and energy content of food (this chapter) and on the digging performance of *H. glaber* (Brett 1986), I constructed a hypothetical energy budget for a single mole-rat. The methods used were similar to those employed in studies of burrowing energetics in the solitary bathyergid *G. capensis* (Du Toit et al. 1985) and the pocket gopher *Thomomys bottae* (Vleck 1979, 1981). First, the distribution of *P. kaurabassana* tubers in plot A (fig. 5-6) was used to estimate the average distance that an individual would have to burrow from one tuber to the next if tubers were encountered by chance. The following expression (Brett 1986) was used:

$$D_c = \frac{(A - ta)/t}{W} ,$$

where D_c is the mean distance burrowed between encounters with tubers, A is the total area of the plot, t is the number of tubers in the plot, a is the horizontal area of a tuber of mean diameter (assuming that tubers are spherical), and W is the search width of a foraging mole-rat (the distance on either side of a burrow at which a tuber could be detected, plus the diameter of the burrow itself). Two estimates of W were used: one based on the assumption that *H. glaber* cannot detect tubers through soil, and a second based on the assumption that *H. glaber* can detect tubers through the soil 10 cm or less from the center of the burrow.

The estimates assumed that the tubers were randomly distributed through the plot and that consumption of one tuber did not reduce the average probability of encountering another within the same average inter-tuber distance. Burrowing in the vertical plane was ignored in these calculations.

The mean burrowing rates of individual mole-rats in hard (4% moisture content) and soft (14% moisture content) soils have been determined (Brett 1986). These were used to calculate the average time required by a single individual to burrow from one tuber to the next (T_c). Mean burrowing rates of *H. glaber* individuals were measured over a period of 10 min; a mole-rat would probably have to rest and recuperate between longer digging bouts. The time required for incisor growth would also decrease the digging rate; the magnitude of this factor is unknown and was not taken into account in my model. However, in my digging experiments, the rate of wear on incisors was an important constraint on burrowing rate, particularly in hard soils, where animals had to stop frequently to hone their teeth in order to continue burrowing.

The mean metabolic rate of single mole-rats at rest at $30°C$ was calculated by McNab (1966) and Jarvis (1978). Assuming that, as was found in *G. capensis* by Du Toit et al. (1985), the metabolic rate of a burrowing animal is three times that of an animal at rest, the metabolic rate of a single burrowing naked mole-rat was estimated to be 3.42 cm³ O_2/g/h. Lovegrove (1989) measured a digging metabolic rate in *H. glaber* of 3.36 ± 0.25 cm³ O_2/g/h in damp sand and 2.78 ± 0.25 cm³ O_2/g/h in dry sand. The rate of energy consumption of a burrowing mole-rat was calculated using the conversion equation of Du Toit et al. (1985),

$$E_s = [20.08 \, (BW)] \, \dot{V}O_2,$$

where E_s is the rate of energy consumption in burrowing (J/h), $\dot{V}O_2$ is the metabolic rate of an animal during burrowing (cm³ O_2/g/h), and BW is the body weight of the burrowing animal (g). The mean body weight of adult *H. glaber* (> 21 g) in the Mtito Andei area was 34 g (recorded by Brett 1986, chap. 4). Thus, the energy (E_c) in kilojoules consumed in travel over a distance of D_c meters (in a time of T_c hours) was calculated. This was compared with the average energy obtainable from a single *P. kaurabassana* tuber, calculated using energy-content data presented earlier. A *P. kaurabassana* tuber of average size weighed 5.28 kg (wet weight) or 1.58 kg (dry weight) and had a diameter of 23 cm. It was assumed that *P. kaurabassana* tubers, like the geophytes eaten by *G. capensis* (Du Toit et al. 1985), contained 70% water by weight but had a digestibility of 50% (I assume that *H. glaber* food is more fibrous than *G. capensis* food; see below). Thus, an estimated average of 14,552 kJ of energy was available in each *P. kaurabassana* tuber.

Estimates of D_c, T_c, and E_c were calculated for a single mole-rat of average size burrowing in hard and soft soils with ($W = 20$ cm) and without ($W = 3$ cm)

Table 5-7

Estimates of the Distance (D_c) and Time (T_c) of Travel for a Solitary Naked Mole-Rat between *Pyrenacantha kaurabassana* Tubers and *Macrotyloma maranguense* Patches, and the Energy Expended in Burrowing between Them (E_c), and in Lifting Soil Vertically to the Surface (E_L) from Burrows of Length D_c between Tubers and Patches

| Parameters | P. kaurabassana | | M. maranguense | |
| | Able to Detect Tubers through Soil | | Able to Detect Tubers through Soil | |
	No	Yes	No	Yes
W (search width, cm)	3	20	3	20
D_c (m)	1586	237	210.1	31.5
T_c (hours)				
Hard soil	1468	2198	227.0 yr	34.2 yr
Soft soil	3147	471	48.8 yr	7.3 yr
E_c (kJ)				
Hard soil	34,250	5127	4,539,000	680,800
Soft soil	7340	1099	972,600	145,900
No. of molehills	168	25	223,344	3352
E_L (kJ)	29	4	992	149

the ability to detect tubers through the soil (table 5-7). A 34-g "blind" foraging mole-rat would use 2.4 times the energy available in one *P. kaurabassana* tuber while digging between tubers in hard soil and 50% of the energy available while digging between tubers in soft soil.

Estimates of the cost of burrowing between patches of *M. maranguense* plants were also calculated (table 5-7). There were nine *M. maranguense* patches (combined area = 2064 m²) in the approximately 105,000-m² area of Kamboyo 1. Each patch contained, on the average, 2.59 kg of food (dry weight), which, assuming an energy content of 19 kJ/g dry weight and a digestibility of 97% (as for *G. capensis* corms; the flesh inside *M. maranguense* roots is not fibrous), would contain 47,660 kJ of energy. The estimates of D_c, and especially T_c (table 5-7), reveal that a solitary mole-rat could not exist on a diet of *M. maranguense* alone. Even if a mole-rat burrowing in soft soil could detect patches at distances of up to 10 cm through the soil, the energy expended between patches would be three times that available in a single patch of average size, even assuming all roots within the patch were discovered (e.g., using area-restricted searching).

A solitary mole-rat might survive by foraging on *P. kaurabassana* tubers alone if it could detect the tubers through the soil and/or burrowed in soil that was constantly soft. However, the energy expenditures listed in table 5-7 are for excavation only; energy required to remove soil from burrows and to transport it to the surface was not included. It is improbable that a mole-rat could burrow between tubers without removing soil; indeed it would have to eject soil to the surface to keep a pathway open, since a mole-rat could not store the energy needed to burrow to the next tuber in the form of fat without revisiting of the first tuber (assuming 1 g of fat contains 28 kJ of energy), and since loosened soil occupies a greater volume than compacted soil.

Du Toit et al. (1985) estimated the energy expended by *G. capensis* in transporting soil to the surface and expelling it using Vleck's (1981) expression,

$$E_L = 5 \ (9.81) \ (M) \ (D),$$

where E_L is the energy cost (J) of lifting soil to the surface through distance D (m). The number 5 is a factor to account for the 20% efficiency of work done against gravity by fossorial rodents, M is the mass of soil (kg), and 9.81 is the acceleration due to gravity (m/s^2). *Heterocephalus glaber* molehills have a mean weight of 9.05 kg, and a packed volume of 6,648 cm^3. Each molehill would be equivalent to a length of 9.4 m of burrow of a mean diameter of 3 cm. The vertical distances (D) used were 0.376 m for burrows at the depth of *P. kaurabassana* tubers and 0.100 m for burrows at the depth of *M. maranguense* patches (table 5-3).

The energies required in lifting to the surface the volume of soil needed to excavate a clear burrow between two *P. kaurabassana* tubers or two *M. maranguense* patches (table 5-7) are negligible compared with the energy expended by *H. glaber* in burrowing (E_C), but they include neither the energy expended in shifting soil along burrows horizontally to the surface, nor the energy expended in the animal's particularly vigorous manner (compared to that of other fossorial mammals) of throwing soil onto the surface. When ejecting soil, *H. glaber* may kick the soil as much as 20 cm above ground level.

The success of a lone mole-rat would be strongly affected by whether it could detect tubers through soil. My data and observations suggest that *H. glaber* was unable to detect *P. kaurabassana* even when these tubers were less than 1 m from burrows. Lovegrove (1987a) found that the other colonial bathyergid *Cryptomys damarensis* passed within 2–3 cm of undetected tubers during foraging.

If it were feeding solely on *P. kaurabassana* tubers, a mole-rat of average size burrowing in soft soil with a maximum tuber-detection distance of 10 cm, would have to burrow continuously for about 20 days before discovering another tuber (not including the time taken to remove soil to the surface). About 8% of the energy available from a single tuber would be used in this burrowing. Assuming that there is a limit to the hardness of teeth, even if there is strong selection acting on this trait, it is possible that the rate of tooth regeneration, as well as the amount of energy available in the tubers, would limit the survival of a single mole-rat.

Unless a mole-rat lived continuously in soft soil, it would be unable to survive on *P. kaurabassana* tubers alone without cooperation in foraging from other mole-rats, if group foraging would increase the rate of discovery of tubers. Cooperation would not decrease the total energy required to locate individual food resources, but it would decrease the average time between their discovery and hence save on both energy (maintenance costs) and tooth wear. Whereas an individual mole-rat could not afford to expend the energy to find new tubers, tubers are large enough to support the shared efforts of several mole-rats. Most importantly, cooperation with others would, theoretically, reduce the variance in the time between discoveries of food resources. Cooperation would also enable the colony to exploit fully the times when soil conditions were optimal for digging. Much of the food needed for later use could be located during these times. Consequently, a group or colony of individuals would be less likely than an individual to starve because of failure to locate a new food resource on the amount of energy supplied by the last item.

Lovegrove (1987a) developed a detailed model to estimate the probability of unproductive foraging in several bathyergids, including *H. glaber* (see Jarvis and Bennett, chap. 3). He found that a solitary forager in a habitat with low geophyte density, such as the area of Kamboyo 1, would have a 50% risk of not encountering a geophyte after burrowing a distance of 84 m. Lovegrove (1987a) concluded that solitary foraging is precluded in arid habitats where geophytes are widely dispersed, and a strategy of group living and cooperative foraging becomes advantageous.

THE ADVANTAGES OF LIVING IN COLONIES

From the preceding discussion, I suggest that an important limit on survival and successful breeding of mole-rat colonies is the rate of discovery of large tubers. This rate is influenced by the constraint of soil hardness, either in terms of the burrowing energy required to travel between food resources or the cost in wear to the digging incisor teeth. A mole-rat species living in arid areas, where the food resources are widely dispersed, costly to locate, and only seasonally accessible, may be predisposed to sociality because of the advantages of group foraging (Jarvis and Bennett, chap. 3), and possibly of group defense against predators (see Alexander et al., chap. 1; Alexander, chap. 15). These advantages would outweigh the cost of group living that the greater food requirements of a larger number of animals would impose.

A large tuber or a patch of small ones would only contain enough food to support, say, a pair of mole-rats for a limited time. Any offspring produced by the pair would increase the food demand but also the work force, and because of their smaller body size, the young could excavate burrows and locate food more efficiently than their parents. Body size is clearly important in influencing the cost of digging a burrow, because a small increase in the diameter of a burrow being excavated carries a greatly increased energetic cost (Vleck 1979). If the costs of digging burrows of increasing diameter go up with the second power of the radius of a mole-rat's body, having small individuals do the digging may be cost effective, especially since the metabolic rate of mole-rats is about 40% below that predicted by their body size alone (McNab 1966). Whether or not the small individuals indeed do most of the excavating is unclear, despite several recent studies (see Lacey and Sherman, chap. 10; Jarvis et al., chap. 12).

The widely spaced bonanza-quality food resources I found around Mtito Andei are, in general, rich enough for many small animals to feed on. However, because of the spacing and hard soil in which these plants occur (and their apparently chance location by mole-rats), the total cost in terms of the time and energy needed to locate and harvest them is more than could be afforded by an individual, or a pair of mole-rats, particularly if they were large animals. The suggested ecological limits on the size of *H. glaber* colonies and the area covered by them closely resemble those of the resource-dispersion hypothesis (Macdonald 1983) for carnivore territoriality. In carnivores, territory size is also determined by the distribution and predictability of the food supply. Assuming a year-round food supply, the size of the group occupying any given territory is then determined by the abundance of food in that area.

I agree with Jarvis and Bennett (chap. 3) that group living in *H. glaber* may be a consequence of becoming exclusively subterranean and of feeding primarily on large, widely dispersed geophytes. Specializing on subterranean food is probably selected for in open, arid areas where plants have evolved very large underground food reserves for surviving through difficult times. In these areas, the little vegetation that is sporadically available aboveground may be already fully exploited by more efficient harvesters (e.g., large ungulates). Escape from predators is another probable selective force in stimulating development of a totally subterranean mode of life (see Alexander, chap. 15). Defense against predators within a burrow system excavated in very hard soil would likely be more effective by groups than by singletons: groups might prevent entry of predators when the burrow system is broken into (e.g., by blocking behavior in superficial burrows; see chap. 4) and they also could more effectively attack intruders en masse (see Lacey and Sherman, chap. 10).

A COMPARATIVE VIEW

THE BATHYERGIDAE

The hypothesis that aridity somehow influences the advantages of group living in the Bathyergidae is supported by a comparative study of social organization (Jarvis and Bennett, chap. 3). Body size in the five bathyergid genera decreases with the increasing aridity of habitat, although in *Cryptomys dama-*

rensis this is somewhat offset by the soft soil in which they burrow. The general decrease in body size is accompanied by an increase in sociality from the solitary genera *Bathyergus suillus*, *Georychus capensis*, and *Heliophobius argenteocinereus* (Jarvis and Sale 1971; Du Toit et al. 1985), through *Cryptomys* (which ranges in group size from 2 to 22, depending on the species; Genelly 1965; Hickman 1979b; Davies and Jarvis 1986; Bennett and Jarvis 1988b) to *Heterocephalus glaber*. The social genera have less hair; *Cryptomys* spp. have a thin covering, and *Heterocephalus glaber* is virtually nude.

The social *C. hottentotus* and *C. damarensis* are specialist feeders on geophytes and grass rhizomes in arid areas of southern Africa (Genelly 1965; Hickman 1979b; Davies and Jarvis 1986; Lovegrove and Painting 1987) and rarely, if ever, come aboveground. The solitary genera live in more mesic habitats and can be characterized as more generalist feeders in that they will eat the roots of a plant from below and then pull down aerial portions into their burrows for consumption (e.g., *Heliophobius argenteocinereus*, *B. suillus*, and *G. capensis*; Jarvis and Sale 1971; Jarvis and Bennett, chap. 3; Beviss-Challinor et al., unpubl. data, quoted by Du Toit et al. 1985).

It is instructive to compare the food biomasses available to *H. glaber* and to the solitary bathyergid *G. capensis*, particularly in relation to the possible relative success of solitary and group foraging in two different habitats: the Tsavo area for *H. glaber* and the southwestern Cape area of South Africa for *G. capensis* (Du Toit et al. 1985). The biomass of food per unit of area (biomass density) available to *H. glaber* around Kamboyo 1 was almost three times that of the four small corm species available to *G. capensis*. The mean biomass density of *P. kaurabassana* in plots A, B, and C was 312 g/m^2, and for *M. maranguense* in plot A, it was 163 ± 53 g/m^2; this compares with 116 ± 33 g/m^2 for the *G. capensis* corms. However, the density of *P. kaurabassana* tubers (average density in plots A, B, and C was 0.06 plants/m^2) was far less than the density of corms eaten by *G. capensis* (< 500/m^2). Lovegrove (1987a) has shown that the probability of finding widely dispersed food resources using random digging is far lower for *H. glaber* than for *G. capensis*.

The absolute energy content figures for *H. glaber* food (Brett 1986) are 16% higher than the mean energy content of the corm species eaten by *G. capensis* (Du Toit et al. 1985). However, the energy that can be extracted by *H. glaber* from its foods is probably lower (i.e., lower assimilation efficiency) because the food is more fibrous and in general harder to digest. Bennett and Jarvis (pers. comm.) have found that the assimilation efficiency of mole-rats feeding on a high-fiber diet is 30%–40% lower than that of mole-rats feeding on sweet potatoes or corms.

It is certain that the cost of finding and harvesting food is higher in the harder soil of *H. glaber* localities. The biomass of *G. capensis* food is much more evenly distributed than that of *H. glaber* food, whereas the amount of food in a *P. kaurabassana* tuber (5.3 kg) is vastly greater than that in an aver-

Table 5-8
Burrow Lengths and Volumes of Species in Three Genera
in the Family Bathyergidae

Species	Burrow Length (m)	Burrow Volume (m³)	Source
Georychus capensis	130	1.02	Du Toit et al 1985
Cryptomys hottentotus	58–340	0.11–0.67	Hickman 1979
Heterocephalus glaber			
Kamboyo 1	3022	2.14	Brett 1986
Lerata 1	595	0.42	Jarvis 1985

age *G. capensis* corm (0.23 g; Lovegrove 1987a). This difference in the dispersion and item size of food must affect the costs and benefits to mole-rats of group living and of cooperating in exploiting food in the two areas.

A comparison of the lengths and volumes of individual burrow systems of three bathyergids (table 5-8; the solitary *G. capensis*, the semicolonial *C. hottentotus*, and the colonial *H. glaber*) shows that the burrows of social mole-rats are relatively longer than those of solitary mole-rats. If the biomass of mole-rats in a burrow system is similar, this increase in burrow length with increasing sociality reflects the greater distances that the social genera must burrow for the typically widely dispersed or patchy food resources that they exploit. In terms of mole-rat biomass per unit of burrow length, Davies and Jarvis (1986) found that burrow systems of *C. hottentotus* (0.8 g of mole-rats/m of burrow) are very similar to the Kamboyo-1 system of *H. glaber* (0.89 g/m).

OTHER HERBIVOROUS AND INSECTIVOROUS MOLES

The rhizomyid mole-rat *Tachyoryctes splendens* lives in soft soils (at higher altitudes in Kenya than *H. glaber*) and is a generalist feeder on shallow roots and bulbs, also coming aboveground to gather aerial vegetation (Jarvis 1973a,b). The blind spalacid mole-rat *Spalax ehrenbergi* lives in the eastern Mediterranean region, where it inhabits a variety of soft soils, ranging from wet loams to drier sandy soils. It is also entirely vegetarian, feeding on bulbs and rhizomes (which are stored) and on seeds collected aboveground (Nevo 1961). The insectivorous moles of the families Chrysochloridae and Talpidae are also confined to soft soils. The South African chrysochlorid *Amblysomus hottentotus* mostly inhabits dry sand, through which it burrows with a swimming motion (Kuyper 1979, 1985). The Ruwenzori golden mole (*Chrysochloris stuhlmanni*) lives in swampy, peaty soils, which are light and easily worked; it rummages aboveground in the thick mat of vegetation covering the soil (Jarvis 1974) and forages just under the soil surface for snails, earthworms, and dipteran larvae. The European mole (*Talpa europaea*) is confined to soft, moist soils. Animals occupy individual territories and feed on earthworms, beetles, and fly larvae (Godfrey 1955; Godfrey and Crowcroft 1960).

Table 5-9

Biomass, Mean Body Weight (± SD) and Burrow Length of Two Naked Mole-Rat Colonies, One from Northern and the Other from Southern Kenya

Locality	No. in Colony	Body Weight (g)		Biomass (kg)	Burrow Length (m)	Biomass/ Burrow Length	Source
		x̄	SD				
Kamboyo 1 (S Kenya)	87	33.3	±6.5	2.65	3022	0.96	Brett 1986
Lerata 1 (N Kenya)	60	20.9	±5 2	1.25	595	2.11	Jarvis 1985

Radio-tracking studies have shown that *T. europaea* makes lengthy excursions aboveground (up to1 km), often in search of water or nest material (Stone and Gorman 1985).

These herbivore and insectivore moles have social organizations similar to those of the solitary bathyergid genera. Individuals, or at most pairs, occupy burrow systems and are territorial. Neighboring animals of opposite sex apparently meet up only for mating, after having located each other using auditory or chemical signals. The young disperse from the maternal burrow soon after birth, often doing so aboveground (e.g., *T. europaea*). Development of solitary life in moles may be a consequence of living on food items that are small but uniformly distributed in soils that are easily worked, just as sociality may be a consequence of the need for cooperative exploitation of patchy but rich food items located in hard soil. There is apparently an antipredator advantage to living in hard soils. This is further emphasized by the evidently increased risk of digging burrows in soft soil, through which a variety of small carnivores and raptors can dig and/or grasp with talons or claws (e.g., in *T. splendes*; Jarvis 1973a,b).

BIOMASS AND BURROW LENGTH

The energy requirements of a naked mole-rat colony may be lower if the total biomass of animals is reduced. This could be accomplished while holding the number of animals (the potential work force) constant by reducing the body weight of each animal. In areas where food resources are more widely distributed, a longer burrow system would be needed to exploit them.

To examine this, I compared the biomass, mean body weight, and burrow length of two *H. glaber* colonies, one from southern Kenya (near Mtito Andei) and the other from northern Kenya (near Lerata; table 5-9). The mean body weight of Lerata-1 animals was 39% lower than that of Kamboyo-1 colony members, and there were 20% fewer animals in Lerata 1. The Mtito Andei site had a much higher available food biomass because of the sporadic occurrence of large *P. kaurabassana* tubers, which were largely absent at Lerata. Jarvis

(1985; Jarvis et al., chap. 12) also found that one Lerata colony, in an area that she judged to be more lush than the areas of three neighboring colonies, had a higher mean body weight (see Brett, chap. 4). Jarvis suggested that *H. glaber* responds to low food availability, not by reducing the work-force size but by having a relatively reduced mean body size within each colony. This presumably occurs because animals can grow bigger in areas where food is more abundant or where burrowing constraints are less severe.

The ratios of biomass to burrow length are similar (1–2 g of mole-rat per meter of burrow) in both Kamboyo 1 and Lerata 1 (table 5-9). However, the ratios of numbers of individuals to burrow length differ considerably between the two localities, given the difference in average mole-rat body weight between them: Lerata 1 had 9.9 m of burrow per mole-rat, and Kamboyo 1 had 33.6 m per mole-rat.

I suggest that in areas of lower rainfall, poorer food quality, and reduced food availability (e.g., Lerata), *H. glaber* body size is limited by low energy intake, but colony size remains fairly constant to provide a sufficient work force to locate and harvest food. Jarvis (1985) recorded complete colony sizes of 40, 60, and 82 near Lerata (see Brett, chap. 4). In areas with higher rainfall and better food (e.g., Mtito Andei), body size reaches a maximum, controlled only through limits set by chemical or behavioral suppression of growth (Jarvis et al., chap. 12). The sizes of Mtito Andei colonies were in general similar to those at Lerata, but were occasionally much larger (e.g., > 295) in spots with superabundant, easily accessible food. The suggestion that body size is limited by food availability is supported by the fact that animals captured in northern Kenya, normally very small in the wild, reached the same average body size as animals from southern Kenya after a period of time in captivity where food was provided ad libitum (Jarvis et al., chap. 12).

The findings of this chapter seem to support Jarvis's (1978) suggestion that *H. glaber* colonies are limited by the patchiness of their food resources and the costs of digging for them, as much as by the size (and quality) of these resources. Thus, large differences in the distribution and quality of food between localities may influence the total number of animals in naked mole-rat colonies, in addition to the mean body size of their members.

Summary

The production of molehills by a naked mole-rat colony (n = ca. 87 animals) in Tsavo West National Park, Kenya (Kamboyo 1) was monitored over a period of 22 mo, and its burrow system was mapped over 15 mo using radio-telemetry. Excavations provided information on the characteristics of burrows. The identity, distribution, and biomass of plants eaten were determined. Feeding rates and energy contents of food were also measured.

Distinct seasonal peaks of molehill production coincided with rainfall maxima. However, surface digging activities ceased after saturation of the soil by prolonged torrential rain, suggesting that there is an optimum soil softness at which digging is most efficient. Three-quarters of all molehills were produced in the rainy seasons of March–May and October–December. Kamboyo-1 colony threw up 400–500 molehills per year, equivalent to 3,600–4,500 kg of soil or 2.3–2.9 km of burrow excavated per year. The mapped burrow system covered an area of roughly 105,000 m^2 and was 3.02 km in total length at the minimum. Up to 11 nest sites were used over the study period. The majority of molehills were started in the early morning and these, as well as nest sites, were located preferentially on (under) areas of bare ground, particularly dirt tracks; possible reasons for this are discussed.

Naked mole-rats consumed subterranean food from a wide variety of geophyte species. The only non-plant material consumed was a bone lying on the surface. The major food item was a large, deep tuber from the plant *Pyrenacantha kaurabassana* (Icacinaceae); these tubers were randomly distributed, and foraging mole-rats apparently located them by chance. An average-sized *P. kaurabassana* tuber could by itself support an average-sized *Heterocephalus glaber* colony for about 75 days. The second most important food resources in terms of biomass were large, dense patches of *Macrotyloma maranguense* (Leguminosae); these plants had small swollen roots that were very palatable to mole-rats. The foraging burrows within these well-defined patches branched and turned much more than burrows outside the patches. Foraging *H. glaber* thus used area-restricted searching behavior, which may increase the rate of discovery of roots within a patch. Because it is unlikely that *H. glaber* can detect tubers through the soil, its foraging behavior apparently involves using information from the location of previously encountered food items, rather than sensory input from items not yet located. An average-sized *M. maranguense* patch could by itself support an average-sized *H. glaber* colony for about 25 days.

The major ecological factors limiting naked mole-rat individuals and colonies are discussed. It is concluded that group living and cooperative foraging are probably necessary for survival when exploiting widely dispersed, high-quality food resources in hard soil, because of the high costs of digging for them (in terms of energy and wear on the teeth) and the animals' inability to detect them through the soil.

Acknowledgments

I thank the Office of the President, Nairobi for permission to conduct research in Kenya and the Wildlife Conservation and Management Department for permission to work in Tsavo West National Park. I am indebted to wardens

G. Lolkiniei and P. Muange for the use of Park accommodation over the study period and for the use of data collected at the weather station at Kamboyo. I am grateful to G. Williams at the Institute of Zoology for his bomb calorimetry of mole-rat tubers, and to J.C.M. Alexander of the Royal Botanic Garden, Edinburgh, C.H.S. Kabuye and J. B. Gillett of the East African Herbarium, Nairobi, and C. Jeffrey, G. P. Lewis, R. M. Polhill and B. Verdcourt of the Royal Botanic Gardens, Kew, who identified food plants and other vegetation.

I thank F. W. Woodley, G.M.O. Maloiy, and N. Mathuku for their help in Kenya, and J.U.M. Jarvis, B.C.R. Bertram, J. B. Wood and G. M. Mace for their assistance and advice. P. W. Sherman, J.U.M. Jarvis, and two anonymous referees provided useful comments on earlier drafts of this manuscript. The work was supported by a research studentship from the Natural Environment Research Council (U.K.), with additional funding from the Royal Society (for radiotelemetric equipment) and from the Central Research Fund of London University.

6 Which Naked Mole-Rats Volcano?

Stanton H. Braude

As is clear from previous chapters in this volume, naked mole-rats are fossorial and apparently spend their entire lives underground. *Heterocephalus glaber* inhabits large colonies (75–80 individuals on the average); its burrow systems are extensive, in some cases reaching about 3 km in total length (Brett 1986, chap. 5). Burrows are typically dug through hard, packed soil. The hard soil overhead protects naked mole-rats from most terrestrial predators. Unfortunately, it also makes their behavior difficult to study in the field. Indeed, division of labor in naked mole-rat colonies has been studied thus far primarily in the laboratory (Jarvis 1981; Payne 1982; Braude 1983; Isil 1983; Lacey and Sherman, chap. 10; Jarvis et al., chap. 12; Faulkes et al., chap. 14).

The only mole-rat behavior that is directly observable in nature is the production of molehills, also known as "volcanoing" (see W. J. Hamilton, Jr., 1928; Hill et al. 1957; Brett, chap. 5; Lacey et al., chap. 8). Volcanoing occurs when a mole-rat uses its hind feet to sweep loose dirt and debris out of a burrow leading vertically to the surface. The earth being removed is the product of subterranean excavations. The 10–15 cm tall geysers of soil particles and the conical mounds of earth that accumulate around the holes (20–30 cm in diameter) look like miniature volcanos (plate 6-1; see also plate 4-2).

Volcanoing is most commonly observed at dusk and dawn (Brett, chap. 5), although I have seen animals volcano at all hours of the day. Naked mole-rats certainly volcano at night because large new mounds are frequently found at dawn; however, I have never actually seen volcanoing at night because the policy of the National Park where I worked was to prohibit outdoor activity from dusk to dawn. Brett (chap. 5) has suggested that nocturnal or crepuscular volcanoing serves an anti-predator function: the mole-rats open their burrows when it is cool and snakes are most likely to be torpid. Naked mole-rats may volcano during any time of the year, but this activity is most common during or just after the October–December or March–April rains (Brett, chap. 5). Volcanoing is apparently stimulated by light rain showers and is curtailed by torrential rains. In a large colony (e.g., 100–200 mole-rats), 15 or more volcanos may be active at one time, but 5–8 simultaneously active volcanos per colony is more typical.

Volcanoing would seem to be one of the riskiest behaviors naked mole-rats perform, because the volcanoer's posterior is directly exposed to predators, and volcanoers are the first individuals a predator would encounter. Brett

Plate 6-1. *Top*, A naked mole-rat molehill or "volcano" erupting on the edge of a road *Bottom*, Looking down into the crater of an active molehill, one can occasionally see the tail or, in this case the head, of the animal doing the volcanoing Photos J U M Jarvis.

(1986), Sherman, Jarvis, and Alexander (unpubl. data), and I have all found the remains of partially digested *H. glaber* in the stomachs of snakes, and Brett (chap. 4) observed rufous-beaked snakes (*Rhamphiophis oxyrhynchus rostratus*) enter volcanos on five occasions. None of us, however, knows if the ingested individuals had been volcanoing just before being eaten.

Jarvis and Sale (1971) observed captive mole-rats dig through soil in a glass enclosure. They suggested that there is a division of labor in this task with one animal remaining near the volcano hole to kick out the soil brought to it by colony mates. A diagram of this "digging chain" is presented by Jarvis (1984). Given the importance of volcanoing for tunnel construction and the potential dangers associated with it, I set out to identify the individual mole-rats involved in volcanoing in three free-living colonies in Kenya. In this chapter I attempt to answer the following questions: (1) Do all naked mole-rats in a wild colony perform volcanoing behavior with equal frequency? (2) If not, which colony members seem to perform this behavior most often? (3) Is volcanoing a behavior that might endanger the lives of the animals doing it?

Methods and Results

Volcanoing was investigated among wild *Heterocephalus glaber* colonies inhabiting Meru National Park, Kenya during 1986–1989. Meru is about 150 km northeast of Nairobi, where the eastern base of the Mt. Kenya plateau meets the equator. My three primary study colonies were located between the Bwatherongi and Mulika Rivers in the northern sector of the Park ($0°5'$–$0°15'$N $38°10'$–$38°20'$E). The terrain is flat, with patches of open savannah and bush. Colonies B and C were located at opposite ends of the Bwatherongi River Campsite (ca. 150 m apart); colony T was located 4 km southeast of colonies B and C. Colonies D, Y, X, S, and M were located along the road that runs parallel to and between the Mulika and Bwatherongi Rivers.

COLONY B

This colony was at the eastern end of the Bwatherongi Campsite. Colony members ($n = 36$) were captured, toe-clipped for permanent identification, and immediately released during the period December 22–30, 1986. I attempted to recapture the whole colony in January 1987, June 1987, and June 1988, but apparently because many animals had been captured three to four times in December 1986, they became trap-shy and would block the traps I used (described by Jarvis in the Appendix) with soil after only 5–10 animals had been caught. However, I am confident that I had marked the entire colony because no unmarked animals were captured in June 1987 or June 1988.

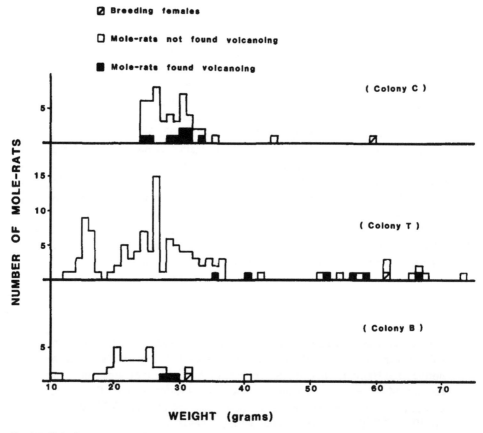

Fig 6-1 Naked mole-rats marked as volcanoers in colony B (*lower panel*) and colony T (*middle panel*) were significantly heavier than colony mates who were not observed volcanoing, volcanoers in colony C (*upper panel*) were heavier than non-volcanoers, but this difference was not significant (see table 6-1)

On the morning of December 29, 1986, three new volcanos were being thrown up by colony B. I captured three animals engaged in volcanoing at two different holes by grabbing their tails with forceps and pulling them out (plate 6-1). I had previously trapped and marked two of these mole-rats; the third individual had not been captured before. The three volcanoers weighed 28 g, 29 g, and 30 g at capture; two were females, and one was a male. Among all individuals in the colony, the volcanoers were the fourth, fifth and sixth heaviest (fig. 6-1, bottom); excluding the breeding female, they were the third, fourth, and fifth largest individuals in the colony. Volcanoers were significantly larger than non-volcanoing mole-rats in the colony, regardless of whether the breeding female was included or excluded from the analysis (table 6-1).

Table 6-1

Weights of Naked Mole-Rats That Participated in Volcanoing in Three Field Colonies in Comparison to Weights of Colony Mates That Were not Seen Volcanoing during the Same One-Hour Sampling Periods

							Body Weights of Volcanoers versus Non-volcanoers (Mann-Whitney Tests)					
							Including the Breeding Female as a Non-volcanoer		Exluding the Breeding Female from the Analysis		Excluding the Breeding Female and All Juveniles ≤ 18 g from the Analysis	
	Body Weight(g)											
	Non-volcanoer			Volcanoer								
Colony	x̄	SD	n	x̄	SD	n	Z	p	Z	p	Z	p
C	29.7	± 6.3	38	30.3	± 2.9	10	1.434	<0.1515	1.599	<0.1099	1 599	<0.1099
T	28.9	±12 7	106	52 0	±11.6	6	-3.437	<0.0006	-3 495	<0 0005	-3.316	<0 0009
B	23.6	± 5.4	33	29 0	± 1 0	3	-2.261	<0.0238	-2.416	<0 0157	-2 360	<0.0183

COLONY T

This colony was located 800 m from the Mulika River Bridge, about 4 km from the Bwatherongi Campsite. On the morning of May 5, 1987 I observed volcanoing at colony T. My assistant and I marked the hindquarters of the mole-rats that were volcanoing between 7:20 A.M. and 8:20 A.M. by touching them with a cotton swab dipped in either blue india ink or gentian violet solution. We marked the volcanoers at every opportunity during that hour; at least 10 different volcanos were simultaneously active during the period.

We began to trap the mole-rats in colony T at 8:25 A.M. that same morning and had captured 64 individuals when we quit at 5:00 P.M. Six of the captured individuals had blue or purple hindquarters indicating that they had been marked as volcanoers; 3 were males and 3 females. The captured animals were kept at the field camp in a tin trunk with soil and newspaper until the entire colony was captured, which took an additional 9 days. The breeding female of colony T was captured on the seventh day. The colony consisted of 112 mole-rats including 48 females, 61 males, and 3 individuals of undetermined sex. In colony T the volcanoers ranked fourth, ninth, eleventh, thirteenth, sixteenth, and twentieth in size (fig. 6-1, middle). They were significantly larger than the non-volcanoing mole-rats in the colony (table 6-1).

I estimated the number of mole-rats left in colony T's burrow system after every day of trapping by using Brett's (chap. 4) technique of plugging each open hole with a sweet potato at night and checking the amount eaten the next morning. After May 14 (i.e., the ninth day of trapping) no sweet potato was eaten. To further ensure that I had captured every member of colony T, I left one tunnel unplugged. When mole-rats remain in a burrow system, they find such an opening and block it up with soil (Jarvis, Appendix). In this case, the open tunnel was not blocked with soil after May 14. (The unplugged tunnel was covered by a 75-liter basin with its edges sealed down using rocks and soil to keep snakes out of the burrow. The basin did not prevent mole-rats in other colonies from recognizing that the tunnel was open and behaving accordingly.)

COLONY C

Colony C was located on the western end of the Bwatherongi Campsite, about 150 m from colony B. The colony was captured in April 1987 and contained 87 mole-rats. I ensured that all members were trapped using the procedures described above. When the tunnel I left open was no longer blocked at night, I knew I had captured the entire colony. I also re-trapped 18 members of the colony in June 1987; all 18 had been marked previously.

On the morning of June 26, 1988, I began to trap colony C. Seven mole-rats were captured between 7:50 A.M. and 8:30 A.M.; their body weights were 32, 26, 27, 26, 27, 26, and 31 g. After trapping, I noticed about 12 new volcanos being kicked up 50 m east of where I was trapping. From 8:45 to 9:45 A.M., my assistant and I squirted gentian violet stain on the hindquarters of the vol-canoers at all of these holes, using plastic squirt guns. At 9:40 A.M., I set traps in the tunnels where I had marked the volcanoers, and by 6:25 P.M., I had caught 46 of the 48 mole-rats remaining in colony C. Of these, 10 were marked with gentian violet; 4 of the volcanoers were females, and 6 were males. The volcanoers in colony C were not among the very largest in the colony (fig. 6-1, top); however, their mean weights were greater than those of the non-vol-canoers (this difference was not statistically significant; see table 6-1).

The breeding female and a large male (possibly her mate) were the only members of colony C captured after June 26. They were caught on July 8 and 9, respectively. On July 13, the breeding female gave birth to a litter of 9 pups in the tin trunk at camp. On August 6, these pups weighed about 5 g, and the entire colony of 20 female and 28 male adults plus the pups was returned to its burrow system.

All 55 adults in this sample were definitely from the same colony, because they had been trapped together and toe-clipped in 1987. Interestingly, 32 individuals that had been captured from colony C in 1987 were not recaptured in 1988; what happened to them is unknown. It is possible that they may have budded off to form a new colony (see Brett, chap. 4), but there was no evidence of any young colony within a 100-m radius of colony C.

COLONIES D, Y, X, AND S

During July 1989, volcanos were observed at four different colonies in the early morning (6:30 A.M. to 9:00 A.M.). None of these colonies had been re-

Table 6-2

Time That Naked Mole-Rat Volcanos Were Active (i e , spewing forth earth) and Intervals between Activity at Four Wild *Heterocephalus glaber* Colonies

Col- ony	Date (1989)	Bouts of Vol- canoing	Total Time Observed (min)	Time Volcanoing (seconds)					Intervals between Bouts of Volcanoing (seconds)				
				n	$\bar{x}$	SD	Sum	Range	n	$\bar{x}$	SD	Sum	Range
D	3 Jul	72	34	72	6 85	2 88	493	1–14	71	21 78	19 38	1546	3–123
Y	9 Jul	18	25	18	59 39	100 38	1069	3–436	17	23 53	27 27	400	5–57
X	17 Jul	34	19	34	18 44	10 93	627	2–43	33	16 27	17 79	537	7–103
S	30 Jul	17	7	17	15 24	7 65	259	8–42	16	10 81	3 87	173	5–16

cently disturbed by trapping. I parked my vehicle at least 50 m away from the volcanos and sat no closer than 20 m. I observed only one volcano at each colony and recorded the exact times that soil began to spray out and exactly when soil ejection ceased. Observations continued until volcanoing at that hole ended for the morning or, in the case of colony S, when a tourist vehicle drove over the volcano.

The frequency and duration of volcanoing bouts and the intervals between bouts for the four colonies are summarized in table 6-2. There was considerable variation in the time during which soil was being kicked out onto the surface. In all four colonies, however, there were long periods of time when soil was not being expelled from the open hole (i.e., in colony D, no soil was kicked out for 76% of the volcanoing "bout"; in colony Y, 27%; in colony X, 46%; and in colony S, 40%).

I do not know if the lack of soil spraying out indicated that no mole-rat (i.e., the volcanoer) was waiting at the hole entrance, because I did not disturb the colonies by walking up to them. However, while marking the volcanoers in colonies T and C (above), I noticed that when soil was not being expelled, no mole-rat was visibly guarding the opening. Of course, a guard could have been waiting farther back in the tunnel, but there was no visible evidence of one.

COLONY M

On July 23, 1988 a sand boa (*Eryx colubrinus*) was caught leaving a naked mole-rat volcano at colony M. I dissected the snake and found that it had one large naked mole-rat in its stomach. The dead animal was female 483, which had been marked on May 25, 1987. At 37 g, she was the thirteenth largest animal in that colony of 98 animals (twelfth largest excluding the breeding female).

Discussion

Divisions of labor in naked mole-rat colonies have been studied primarily in the laboratory. Jarvis (1981) described the largest mole-rats in a colony as "nonworkers" because they seldom participated in colony maintenance activities (which she defined as digging and transporting). Volcanoing behavior was not included in Jarvis's analysis. Payne (1982) and Isil (1983) both observed that larger mole-rats worked less frequently, but neither could identify distinct nonworking groups or castes. Braude (1983) and Lacey and Sherman (chap. 10) found that the largest mole-rats in laboratory colonies performed the majority of the digging (but see Jarvis et al., chap. 12), but they did not study volcanoing; Lacey and Sherman (chap. 10) also presented evidence that the largest mole-rats defend the colony against snakes and intrusions from foreign colonies. Brett (chap. 5) found large naked mole-rats to be generally less active than small ones in his radiotelemetric studies of 20 animals; Brett did not specifically investigate volcanoing behavior. In his digging-efficiency experiments, Brett (1986) reported that large mole-rats displaced smaller ones from excavation sites and performed the majority of the digging.

My investigation of division of labor in wild colonies focused specifically on volcanoing. In colonies B and T, the volcanoers were significantly larger than their colony mates; in colony C, the volcanoers were also larger, but not significantly so (table 6-1). I cannot be certain why colony C differed. It might have been the result of a recent change in the colony composition because of emigration or mortality. The intrusive presence of my traps at a distance of 50 m or less and the capture of seven colony mates may also have affected the volcanoing behavior in colony C. The difference could also have been due to an urgent need to find more food because of the impending birth of a new litter of pups. This would explain the relatively large number of volcanos (12) for the size of the colony . The volcanoing in colony B may also have been influenced by trapping and releasing of naked mole-rats during the preceding week, perhaps weakening the trend seen in this colony. In colony T, where the relationship between size and volcanoing was the strongest, there was no interference with the colony for more than 1 mo before the marking of the volcanoers.

My data indicate that although volcanoing was not restricted to the very largest mole-rats in a colony, the behavior was performed nonrandomly (fig. 6-1), with larger individuals, but not the breeding female, participating most often. There appeared to be no sex bias in the performance of volcanoing. Thus, participation in volcanoing seemed to follow a pattern similar to what Lacey and Sherman (chap. 10) reported for digging and various behaviors associated with colony defense. In other words, larger colony members, which Jarvis (1981) classified as "nonworkers," did seem to perform an important task within my study colonies.

Volcanoing is probably one of the riskiest behaviors of naked mole-rats because (1) the animal's hindquarters are exposed, (2) predators can see (and hear) the small geysers of soil from some distance away, and (3) this is the only time when the tunnel system is opened and when predators can gain easy access. The 1989 observations of the time volcanos were active and inactive indicated that snakes would have had ample opportunity to enter naked mole-rat tunnels when earth was not being spewed out (and, perhaps, the hole was unguarded; see table 6-2).

Although I saw no predation on volcanoers, there is no doubt that a snake or other predator could catch a volcanoer by the tail, because I was able to do so. In fact, natural predators would find catching volcanoers considerably easier than I did because they are obviously not concerned with avoiding injury to the mole-rat. The truncated tails of some wild naked mole-rats (ca. 1% of the 1,550 mole-rats I have caught) could conceivably be the result of attempted predation while those animals were volcanoing. With regard to predation by rufous-beaked snakes, Brett (1986, p. 308) suggested that "it is probable that the mole-rats were struck at by the snake from above while kicking soil out at the base of a molehill." Brett did not report the sizes of mole-rats he found in the stomachs of snakes but the one that I found was among the largest in its colony. This suggests that at least sometimes larger individuals, rather than small colony members or juveniles, fall prey to snakes.

Summary

Molehill construction or "volcanoing" behavior was observed in seven wild naked mole-rat colonies. In three of these, the volcanoers were identified. In all three colonies, the volcanoers were larger than their colony mates, and in two of the three colonies significantly so. There was no evidence of sex-bias in volcanoing. Volcanoing is probably a particularly risky behavior because the volcanoer is exposed to predators; one mole-rat was recovered from the stomach of a snake, and it was among the heaviest 15% of animals in its colony. In the four colonies where I observed volcanoing closely, the volcanos were found to be inactive for considerable periods of time, allowing ample opportunities for snakes to enter the apparently unguarded burrow.

Acknowledgments

I thank R. D. Alexander for help and advice in all aspects of the work. I also thank the President of the Republic of Kenya, Daniel Arap Moi, and the Wildlife Conservation and Management Department of Kenya for allowing me to work in Meru National Park. Valuable comments on the manuscript were provided by J.U.M. Jarvis, P. W. Sherman, and two anonymous reviewers. I am grateful to the following agencies for financial support: The American Society of Mammalogists, Sigma Xi, The Explorers Club of New York, and the Rackham Graduate School, Museum of Zoology (Hinsdale Fund) and the Biology Department at the University of Michigan.

7

Genetic Variation within and among Populations of the Naked Mole-Rat: Evidence from Nuclear and Mitochondrial Genomes

Rodney L. Honeycutt, Kimberlyn Nelson, Duane A. Schlitter, and Paul W. Sherman

Populations of naked mole-rats are subdivided into colonies that are separated spatially (Brett 1986; chap. 4, 5). Although females can produce rather large litters every 2–3 mo (Brett, chap. 4; Jarvis, chap. 13), the restriction of breeding within colonies should result in small effective population sizes. What predictions can be made about genetic variation within and among *Heterocephalus glaber* colonies? On the one hand, if levels of genetic variation within and between colonies are similar, then several explanations are possible: (1) considerable migration is occurring, (2) small effective population sizes within colonies coupled with frequent population bottlenecks have reduced variation throughout the species, or (3) there is a correlation between niche width and genetic variation, as proposed by Nevo et al. (1984). On the other hand, if the rate of migration between colonies is low, as seems likely, then intracolonial genetic variation ought to be lower than intercolonial variation because of genetic drift reinforced by isolation (Wright 1940; Varvio et al. 1986).

Here we present the results of three studies on genetic variation within and among naked mole-rat colonies. Two of these investigations are independent analyses of nuclear gene variation using standard gel electrophoresis; the third study is a geographic survey of mitochondrial DNA (mtDNA) variation in *H. glaber* throughout Kenya. Our data allowed us to compare the population structure of *H. glaber* with that of other fossorial rodents and to discuss the likely factors affecting genetic variation in these monophyletic groups.

Materials and Methods

ELECTROPHORESIS

A total of 130 naked mole-rats was collected from 21 wild colonies in Kenya, East Africa. These samples represented collecting efforts by three groups: (1) R. L. Honeycutt and colleagues from Harvard University in 1985, (2) P. W. Sherman, R. D. Alexander, and J.U.M. Jarvis in 1979, and (3) J.U.M. Jarvis in 1977. The animals were collected from two geographic localities separated by about 300 km (fig. 7-1): Mtito Andei (17 colonies, 107 animals) and Lerata (4 colonies, 23 animals). Colonies were defined on the basis of molehill patterns (Brett, chap. 5); molehills that were close together (< 50 m) were assumed to be connected underground.

Naked mole-rats were captured alive using a technique developed by Kenyan farmers (see Brett, chap. 4). Briefly, active "volcanos" were located (see Braude, chap. 6), and an opening was made into the burrow systems associated with the mound. Mole-rats attempting to plug the opening were caught by blocking the burrow behind the opening with a hoe or shovel (Jarvis, Appendix). Animals in the Harvard sample were removed from the burrow and returned to a field laboratory where they were sacrificed. The hearts, livers, and kidneys of these mole-rats were then frozen in liquid nitrogen (-196°C); the samples were analyzed within 6 mo from when they were collected. Animals in the Cornell and Cape Town samples were returned to the lab alive, and a few drops of blood were obtained from the tail of each. Blood samples were frozen on dry ice or in a −70°C freezer; all the blood samples were analyzed in within 2 mo from when they were obtained.

Independent surveys of allozyme variation were conducted at Harvard University and Cornell University. Heart, liver, and kidney tissues (Harvard) or blood (Cornell) from each individual were studied. Individuals were screened for allozyme variation using standard starch-gel and polyacrylamide electrophoresis (Shaw and Prasad 1970; Selander et al. 1971; Harris and Hopkinson 1976). Each locus was examined across a range of pH and gel buffers to detect potential hidden alleles. The Cornell study was completed in August 1983, and the Harvard study took place in January 1986. Data obtained at Cornell were not revealed to the Harvard team until the latter's electrophoretic runs were completed. Thus, data were obtained blind with regard to the other group's work, and conclusions from one study did not bias the results of the other study.

MITOCHONDRIAL DNA

Fifty-four naked mole-rats collected from 16 colonies by the Harvard group were used to examine mtDNA variation. The sample consisted of 42 individu-

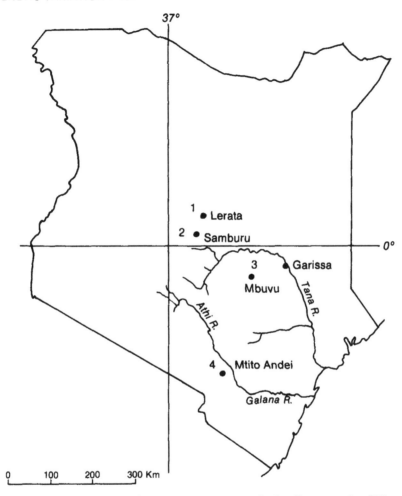

Fig. 7-1 Sampling localities for naked mole-rats used in the allozyme and mtDNA studies. Specific localities are: 1, Lerata; 2, Samburu; 3, Mbuvu; 4, Mtito Andei.

als (in nine colonies) from Mtito Andei, 5 individuals (in two colonies) from the Samburu District, Buffalo Springs, and 7 individuals (in three colonies) from the Kitui District, Mbuvu (fig. 7-1).

Mitochondrial DNA was isolated from frozen liver using CsCl/propidium iodide gradient centrifugation (Densmore et al. 1985). Purified mtDNA from each individual was digested with restriction endonucleases according to the recommendations of the supplier (New England Biolabs). The restriction fragments were end-labeled with d^{-32P} dNTP's using the large (Klenow) fragment DNA polyermase I (W. M. Brown 1980), and separated by gel electrophoresis on 1.2% agarose or 3.5% polyacrylamide. Internal size standards, HindIII digested λ DNA and SV40 digested with *Hin*fI, were included on each gel. The digestion patterns were visualized by autoradiography.

The following 11 restriction endonucleases were used to digest all mtDNAs: *Acc*I, *Bst*EII, *Dde*II, *Eco*RI, *Hin*dIII, *Hin*PI, *Hpa*II, *Mbo*I, *Sau*96I, *Sty*I, and *Taq*I. The individuals from Mtito Andei (*n* = 42) were also examined using *Alu*I, *Dra*I, *Hae*III, *Hin*fI, and *Rsa*I. In addition to the above enzymes, a few individuals from several of the Mtito Andei colonies were examined with 14 other restriction endonucleases. These latter enzymes produced few restriction fragments and revealed no variation.

Results

ELECTROPHORETIC VARIATION

Twenty-nine proteins representing 49 presumed loci were examined (see the Appendix to this chapter). Because different tissues were used, and data were collected independently at the two universities, some loci were not examined for all individuals. However, each mole-rat was examined at a minimum of 34 loci.

The 107 individuals from 17 colonies collected at Mtito Andei were monomorphic and fixed for the same allele at all loci. This absence of variation was apparent despite the fact that samples were collected at three different times, 1977, 1979, and 1985, and analyzed in two independent labs.

One sample of 23 individuals representing four colonies was collected at Lerata in north-central Kenya (by Jarvis) and analyzed at Cornell. Of the 34 loci examined in the Lerata sample, 33 were monomorphic. At the IDH-1 locus, two alleles were segregating—IDH-1^{100} and IDH-1^{43}—with 11 individuals being homozygous IDH-1$^{100/100}$ and 12 being heterozygous IDH-1$^{100/43}$ (table 7-1). No homozygous IDH-1$^{43/43}$ individuals were found. The frequency of IDH-1^{100} was 0.74 and IDH-1^{43} was 0.26, and the average heterozygosity for Lerata individuals at IDH-1 was 0.015. There was a significant deficit in homozygous IDH-1$^{43/43}$ individuals (table 7-1).

Variation between colonies from Lerata and Mtito Andei was found at two loci, XDH and IDH-1. For XDH, all individuals from Mtito Andei were fixed for XDH^{100}, whereas all individuals from Lerata were fixed for an alternative allele, XDH^{90}. For IDH-1, one allele (IDH-1^{100}) was shared between the two localities, and a unique allele (IDH-1^{43}) was found at Lerata (above). Variation at the IDH-1 locus was further examined at Harvard using 12 specimens collected in 1986 from three colonies in the Samburu District, just south of Lerata (fig. 7-1). All 12 individuals were fixed for the same IDH-1^{100} allele seen at Mtito Andei.

Table 7-1

Frequency of IDH-1 Genotypes in Naked Mole-Rat Colonies from Lerata, Kenya
(Jarvis's 1977 sample, analyzed at Cornell) and Tests of Goodness of Fit to
Expected Genotype Proportions

Colony No.	n*	IDH-1 Genotypes			3:1 Ratio		1.2 1 Ratio	
		100/100	100/43	43/43	G	P	G	P
1	2	2	0	0	NA†		NA†	
2	6	2	4	0	0.68	0 58	3.45	0 176
3	7	2	5	0	1 32	0 25	4.10	0.127
4	8	5	3	0	0 51	0.52	7 437	0.024
Pooled	23	11	12	0	0 04	0 83	15 29	0 001

* Number of individuals sampled
† NA, not applicable.

MITOCHONDRIAL DNA VARIATION

VARIATION WITHIN LOCALITIES

Colonies around Mtito Andei were examined most extensively for mtDNA variation. Most of the 16 restriction endonucleases used to assess mtDNA variation had four-base recognition sites and therefore produced digestion profiles consisting of numerous fragments. A total of 259 restriction sites representing about 6.7% of the mitochondrial genome was revealed, and two haplotypes differing by a single restriction site defined by the restriction endonuclease *Hpa*II were found. The average nucleotide sequence divergence (δ) (Nei and Li 1979) between these two haplotypes was 0.055%. The distribution of this haplotype variation among the nine study colonies was: (1) six colonies (22 individuals) had haplotype A only; (2) one colony (3 individuals) had haplotype B only; and (3) all 17 individuals in two colonies were heteroplasmic for both the A and B haplotypes (see fig. 7-2). Heteroplasmy refers to the occurrence of more than one mitochondrial genome within an individual. Site heteroplasmy, as seen in *Heterocephalus glaber*, is rare relative to the more common length heteroplasmy (Honeycutt and Wheeler 1990).

Densitometric scans were conducted on *Hpa*II digests from individuals that were heteroplasmic to determine the relative proportion of the two haplotypes (mitochondrial genomes) in each individual. Within most heteroplasmic individuals, haplotype A was the most common, but in two animals, haplotype B represented 51% and 52% of the genomes present. This suggests that there was turnover occurring within individuals relative to the frequency of each haplotype. The higher frequency of haplotype A within heteroplasmic individuals was to be expected, because the A haplotype was the most common among all individuals from Mtito Andei. The B haplotype was more restricted and there-

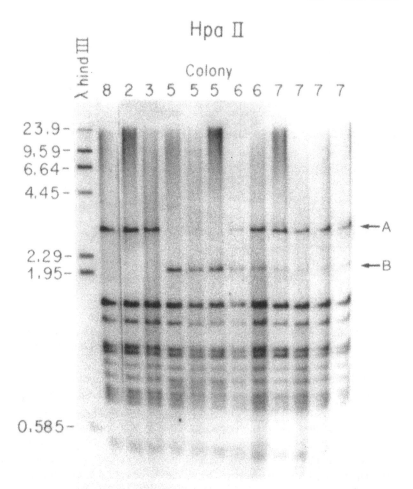

Fig 7-2. *Hpa*II digestion patterns of mitochondrial DNA samples collected from naked mole-rat colonies near Mtito Andei, Kenya. Colonies 2, 3, and 8 have the A haplotype, colony 5 the B haplotype, and colonies 6 and 7 are heteroplasmic for A and B.

fore may have been more recently derived, possibly from a heteroplasmic intermediate. The close geographic proximity (< 1 km) of the colony with haplotype B and the heteroplasmic colonies (haplotype AB) supports the idea that B may have been derived from AB.

Intracolonial mtDNA variation was absent at the Mbuvu and Samburu localities, where 164 restriction sites representing 4.3% of the genome were examined. The two colonies at Samburu, however, differed by a minimum of two restriction sites. The average nucleotide sequence divergence (δ) between

Table 7-2

The Average Nucleotide Sequence Divergence (δ) between mtDNA Haplotypes
of Naked Mole-Rats within and between Sampled Localities

	Sample Size		Within	Between
Locality	Individuals	Colonies	Locality (%)	Localities (%)
Mtito Andei	42	9	0.055	
Samburu	5	2	0.10	
Mbuvu	7	3	0	
Mtito Andei/Samburu				5.40
Mbuvu/Samburu				5 40
Mtito Andei/Mbuvu				0.40

NOTE· See figure 7-1 The divergence estimates were calculated from restriction-fragment data according to the procedure of Nei and Li (1979)

the two haplotypes at Samburu was 0.10%. These two colonies were collected 4.0 km apart.

VARIATION BETWEEN LOCALITIES

Considerable geographic variation was seen among the Samburu, Mbuvu, and Mtito Andei localities (table 7-2). For example, mole-rats from the Samburu locality (fig. 7-1) differed from those at Mbuvu and Mtito Andei by a δ of 5.4%. The Mbuvu and Mtito Andei samples were more similar to one another (δ = 0.4%). In terms of proximity, however, Mbuvu is closer to Samburu, and no apparent geological or physiographic barriers account for these observed patterns of variation.

Discussion

Genetic diversity in naked mole-rats was mainly restricted to divergence between disjunct populations in the samples we analyzed. Genetic diversity within and between colonies in a local population or geographic area was low. In particular, intra- and intercolonial variation in allozymes and mtDNA was extremely low around Mtito Andei. DNA "fingerprinting" (i.e., the use of minisatellite DNA variation to generate individual-specific restriction-fragment patterns) has also shown that colonies in this region are extremely monomorphic (Reeve et al. 1990). Indeed, only monozygotic twins and strains of mice that have mated siblings for about 60 generations have DNA fingerprints as similar as the Mtito Andei naked mole-rats, thus suggesting a high level of inbreeding. Likewise, the mtDNA studies of Mtito Andei animals revealed three haplotype combinations (A, B, AB), with each colony containing only one combination.

Although genetic variation in local populations of naked mole-rats was low, what little variation existed may provide some insight into the structure of naked mole-rat populations. First, DNA fingerprints derived from minisatellite probes revealed the highest inbreeding coefficient ($F = 0.45 \pm 0.18$) yet recorded among free-living mammals (Reeve et al. 1990). Second, although the sample sizes were small, variation at the IDH-1 locus for the Lerata colonies that were heterozygous revealed a lack of homozygotes for the IDH-1[43] allele (table 7-1), suggestive of deviations from random mating. Third, the distribution of mtDNA haplotypes defined by the restriction endonuclease *Hpa*II, coupled with the observed variation relative to the frequency of haplotypes A and B in the heteroplasmic individuals, suggests the possibility of new colony formation from preexisting colonies (see also Brett, chap. 4). If common equals "primitive," then the A haplotype seems to be ancestral, with the B haplotype being a more recent derivative. Given the close proximity of the B haplotype colony to the heteroplasmic colony (possessing AB), colony B's inception from the AB lineage is a reasonable hypothesis.

Genetic divergence between colonies separated by larger geographic distances is evident. The δ value of 5.4% between mtDNA haplotypes at Samburu relative to Mbuvu and Mtito Andei is as high as that found among subspecies of *Peromyscus maniculatus* ($\delta = 2.0\%$; Lansman et al. 1983), and populations of *Rattus rattus* found on different continents ($\delta = 4.0\%$; Brown and Simpson 1981). The observed discontinuity in haplotype divergence between Samburu and Mtito Andei-Mbuvu suggests a long-term, extrinsic barrier to gene flow between these two regions. Avise et al. (1987) indicated that this level of intraspecific divergence is most commonly encountered in species with discontinuous distributions. In fact, another fossorial rodent, *Geomys pinetis*, also demonstrates considerable regional divergence (Avise et al. 1979). Allozymes and chromosomes also reveal geographic variation in *Heterocephalus glaber*. The Lerata allozyme sample, which came from an area north of the Samburu samples used for mtDNA analysis, differed from Mtito Andei at two loci, one of which was fixed for alternative alleles (XDH[100] vs. XDH[90]). As indicated in chapter 2, variation in the size of chromosome arms has also been observed between naked mole-rats from Somalia and Kenya.

Factors causing the patterns of genetic variation in *H. glaber* can be evaluated by comparison to organisms with similar life histories. *Heterocephalus glaber* shares many characteristics with other fossorial mammals, such as a sealed burrow system with a stable microhabitat, adaptations for burrowing, a patchy distribution throughout their range, and population or colony subdivision. These traits can influence patterns of genetic variation (Nevo 1979; Patton 1985), and two main explanations for the patterns of variation in fossorial mammals have been proposed.

On the one hand, Nevo et al. (1984) considered fossorial mammals to have lower levels of genetic polymorphism ($\bar{P}$, percentage of loci that are polymorphic) and heterozygosity ($\bar{H}$, percentage of loci that are heterozygous) per population than non-fossorial mammals. Nevo and his colleagues (Nevo and Shaw 1972; Nevo and Cleve 1978; Nevo et al. 1984; Nevo et al. 1987) have suggested that such patterns support the niche-width–genetic-variation hypothesis. This hypothesis suggests a correlation between reduced genetic variation in fossorial mammals and the presumed low levels of environmental diversity in the "narrow" (i.e., stable, constant, and predictable) subterranean niche. On the other hand, this idea has been challenged by several authors (Patton and Yang 1977; Patton and Feder 1978; Schnell and Selander 1981; Nei and Graur 1984; Sage et al. 1986) who suggested, alternatively, that levels of genetic variation in fossorial mammals may indicate historical and stochastic processes involving fluctuations in effective population size, degree of gene flow, and genetic drift.

As can be seen in table 7-3, the range of electrophoretic variation in fossorial rodents (overall average $H = 0\%–9.3\%$; $P = 0\%–50\%$) is large; indeed, it encompasses the total range of variation observed across all mammalian species, including terrestrial, arboreal, and flying mammals ($\bar{H} = 4.1\%$; Nevo et al. 1984). Variation in *H. glaber* falls at the lower end of this range (especially relative to other bathyergids). Nevertheless, the variation depicted in table 7-3 seems to offer little overall support for the niche-width–genetic-variation hypothesis.

Because few studies of fossorial mammals have examined mtDNA variation, it is difficult to compare the patterns seen in *H. glaber* relative to those in mammals with a similar natural history. Nevertheless, there was a trend toward increasing levels of divergence as more disjunct populations were compared, and these levels of divergence were comparable to those seen in other animals with disjunct or patchy distributions (Avise et al. 1987).

Selander and Whittam (1983) suggested that interpreting differentiation in terms of evolutionary factors is difficult, especially when one attempts to do so in the absence of historical demographic information. Although we agree on this point, we feel that it is useful to present a historical and demographic hypothesis that might help explain the patterns of variation in *H. glaber* and that could serve as a working basis for further research.

The naked mole-rat is an arid-adapted species restricted to regions of East Africa that are characterized by low annual temperature fluctuations and annual rainfall less than 700 mm (Jarvis and Bennett, chap. 3). Colonies are distributed patchily with local colony size and distribution resulting in part from the locations of favored food resources (Jarvis 1978; Lovegrove and Wissel 1988; Brett, chap. 5). Aside from the advantages of group foraging (Brett, chap. 5), naked mole-rats probably also gain advantages in terms of predator avoidance by living in groups (see Alexander, chap. 15). Individual mole-rat colonies consist of long-lived, overlapping generations, with a reproductive division of labor and recruitment of both sexes into the natal group (Brett 1986, chap. 4; Jarvis et al., chap. 12). Finally, the formation of new colonies apparently occurs at least sometimes by budding or fissioning from preexisting

Table 7-3
Genic and mtDNA Variation in Fossorial Rodents

Taxa	$\bar{H}$ (range %)	$\bar{P}$ (%)	δ (%)	Source
Geomyidae				
Geomys bursarius	3 8	23 0		1
Geomys tropicalis	0	0		1
Geomys pinetis	2 5	40 0	0 4 (3.4)	2, 3
Geomys personatus	4 4	19 0		1
Geomys arenarius	5 0	12.0		1
Thomomys talpoides	4 7 (0 8–8.5)	23 5 (9 7–35 5)		4
Thomomys bottae	9 3	33.4		5
Thomomys umbrinus	5 9 (2 2–10 0)	29.0 (13.0–44 0)		6, 7
Cratogeomys zinseri	0	0		8
Cratogeomys fumosus	0	0		8
Cratogeomys gymnurus	0	4 5		8
Zygogeomys trichopus	0.8	9.7		9
Orthogeomys heterodus	2 0	6 4		10
Orthogeomys underwoodi	2 0	3 3		10
Orthogeomys cherriei	1 0	3 2		10
Orthogeomys cavator	0	6 5		10
Orthogeomy dariensis	1.0	3.2		10
Spalacidae				
Spalax ehrenbergi	3 9	20 0	0.9 (5.5)	11
Ctenomyide				
Ctenomys haigi	9 0	18.0		12
Ctenomys mendocinos	7.6	17.0		12
Ctenomys dorbignyi	6 0	12 0		12
Bathyergidae				
Heterocephalus glaber	1.5	6 0	0.07 (2.9)	13
Cryptomys hottentotus	4.4	45.0	0.78 (1 8)	14
Cryptomys natalensis	1 7	42.0	0 (1.7)	14
Georychus capensis	5.0	50 0	0 44 (7.1)	14

NOTE $\bar{H}$, indicates the percentage of loci that were heterozygous per population, $\bar{P}$, the percentage of loci that were polymorphic, and δ, the average sequence divergence between mtDNA haplotypes For δ, numbers in parentheses represent interregional comparisons With the exception of naked mole-rats, estimates of δ were not made with four-base restriction endonucleases

SOURCE 1, Selander et al 1974, 2, Avise et al 1979, 3, Laerm et al 1982; 4, Nevo et al 1974, 5, Patton and Yang 1977, 6, Patton and Feder 1978, 7, Hafner et al 1987, 8, Honeycutt and Williams 1982, 9, Hafner and Barkley 1984, 10, Hafner, unpubl data, 11, Honeycutt, unpubl data; 12, Sage et al 1986, 13, this chapter, 14, Honeycutt et al 1987 colonies (Brett, chap. 4). All these behavioral and ecological characteristics suggest that the cost of leaving home and finding or founding a new colony may be higher than the cost of inbreeding within a colony (Reeve et al. 1990).

A lack of migration between colonies coupled with a small number of breeding individuals (one female, one to three males; see Brett, chap. 4; Lacey and Sherman, chap. 10; Jarvis, chap. 13) leads to a reduction of the effective population size, with the result that genetic variation within a colony is diminished. Reduced variation within colonies can be extended to low levels of variation between colonies with a consideration of possible past events, such as population bottlenecks caused by regional or local extinctions. Population bottlenecks can influence overall levels of genetic variation, in proportion to the size of population decrease and its recovery rate subsequent to the bottleneck (Nei et al. 1975).

There is some indication of the extinction of local naked mole-rat populations (i.e., bottlenecks) in historical times. *Heterocephalus glaber* has a fossil record in eastern Africa extending back approximately 3 million yr (i.e., to Plio-Pleistocene times; Van Couvering 1980). There is fossil evidence suggesting that *H. glaber* occupied several regions of eastern Africa in the Rift Valley (e.g., Tanzania, and Omo in southern Ethiopia) where the species is not found today (Wessleman 1984; Denys 1985). Paleoclimatic data indicate several alternations of climate during the Pleistocene in eastern Africa as a result of glacial and interglacial periods (Cooke 1972; Livingston 1975; Hamilton 1982; Kappelman 1984). Because *H. glaber* is a desert-adapted mammal with a patchy distribution, shifts in climate in the recent past could greatly affect the survival and distribution of colonies throughout the species' range.

These historical events, resulting in population fragmentation, might help account for the divergent nature of the Samburu populations relative to Mbuvu and Mtito Andei. Local extinctions and recolonizations in more recent times could also account for the low levels of variation seen at Mtito Andei and other localities (see Maruyama and Kimura 1980). For instance, intense coloniality may make naked mole-rats particularly susceptible to local extinctions (e.g., due to disease; Reeve et al. 1990). Following the extinction of a colony, rapid recolonization of the vacated area or even the old burrow system could occur (e.g., via fissioning and dispersal by other local colonies). Thus, the Mtito Andei, Lerata, and Samburu colonies may each be of very recent common ancestry, yet be quite divergent from one another as a result of long-term interpopulation isolation and local extinctions and recolonizations.

Summary

Genetic variation within and among 26 naked mole-rat colonies representing 142 individuals was investigated using genetic markers from both the nuclear and mitochondrial genomes. Variation within a locality was extremely low among colonies. In terms of *nuclear genes, a total of 49 presumptive loci were* studied independently at Harvard and Cornell using electrophoresis. Among these, IDH-1 was the only locus with any intracolonial variation; colonies at one locality (Lerata) contained either IDH-1$^{100/100}$ or both IDH-1$^{100/100}$ and IDH-1$^{100/43}$ genotypes (no IDH 43/43 individuals were found). Mitochondrial DNA variation within colonies was limited to restriction-site heteroplasmy detected in one colony at a second site. Overall, the patterns of intra- and intercolonial variation seen within small geographic regions suggested that colonies represent separate breeding units and that there was considerably more variation between colonies than within colonies.

Genetic divergence was higher between geographically separated colonies. Samples from Samburu and Lerata in northern Kenya differed from samples from Mbuvu and Mtito Andei in the south. The Lerata locality had two allozyme alleles, XDH^{90} and IDH-1^{43}, not seen in samples at Mtito Andei. Although Lerata was not examined for mtDNA variation, the Samburu sample (from just south of Lerata) differed by 5.4% from the southeastern localities; this difference is as large as the difference between subspecies of some other rodents.

Patterns of genetic variation in naked mole-rats were compared with patterns in other fossorial rodents. These comparisons suggested that *Heterocephalus glaber* populations are comparatively monomorphic and homozygous. Historical and demographic elements might help explain the patterns of variation seen in *H. glaber*. Historical factors such as population bottlenecks resulting from climatic changes and local colony extinctions/recolonizations coupled with unique life-history characteristics (i.e., extreme restriction of breeding within semi-isolated colonies) may be responsible for the patterns of genetic variation observed in naked mole-rats.

Acknowledgments

We thank R. D. Alexander, L. K. Alexander, M. W. Allard, N. Chondo, P. Gathinji, C. Kagarise Sherman, J.U.M. Jarvis, and M. B. Qumsiyeh for assistance in the collection of specimens, J.U.M. Jarvis for mole-rat blood samples and advice, and J. C. Avise, M. S. Hafner, T. Lamb, and an anonymous reviewer for comments on earlier drafts of the manuscript. R. E. Leakey, I. Aggundey and colleagues at the National Museum of Kenya made this research possible through support and field assistance. We also thank the Office of the President and National Parks and Wildlife, Nairobi, for issuing research clearance and collecting permits. Fieldwork was supported by the Barbour and Richmond Funds from the Museum of Comparative Zoology at Harvard University, the Netting Research Fund, Cordelia Scaife May Charitable Trust, and the O'Neil Fund of The Carnegie Museum of Natural History, and the National Geographic Society. Electrophoretic work at Cornell was done in the Laboratory of Ecological and Evolutionary Genetics, under the direction of B. P. May. Research was supported by a National Institutes of Health Predoctoral Traineeship in Genetics to K.N., by National Science Foundation grants to R.L.H., K.N., and P.W.S., and by Harvard and Cornell Universities.

Appendix

Presumed Loci Surveyed for Genetic Variation Using Electrophoresis
in Twenty-One Colonies of Naked Mole-Rats ($n = 130$ individuals)
from Two Localities in Kenya

Protein	Locus	No. of Alleles	Cornell Sample	Harvard Sample
Acid phosphatase	AP	1		X
Aconitase	ACON-1	1		X
	ACON-2	1		X
Adenosine deaminase	ADA	1	X	
Aldolase	ALD A	1		X
	ALD B	1		X
Aspartate aminotransferase	AAT-1	1		X
	AAT-2	1	X	X
Catalase	CAT	1	X	X
Diaphorase	DIA-1	1	X	X
	DIA-3	1	X	X
Esterase	EST-1	1		X
	EST-2	1		X
	EST-4	1	X	
General protein	GP-1	1	X	
	GP-2	1	X	
	GP-3	1	X	
	GP-4	1	X	
	GP-5	1	X	
General protein	GR-1	1	X	X
	GR-2	1	X	X
	GR-3	1	X	
Glucosephosphate isomerase	GP1	1	X	X
Glucose-6-phosphate dehydrogenase	Gd	1	X	X
Glutamate pyruvate transaminase	GPT	1	X	
Glyceraldehyde-phosphate dehydrogenase	GAPDH	1	X	X

Appendix, cont

Protein	Locus	No of Alleles	Cornell Sample	Harvard Sample
Isocitrate dehydrogenase	IDH-1	2	X	X
	IDH-2	1		X
Lactate dehydrogenase	LDH-1	1	X	X
	LDH-2	1	X	X
Leucine aminopeptidase	LAP	1	X	X
Malate dehydrogenase	MDH-1	1		X
	MDH-2	1	X	X
Malic enzyme	ME	1	X	X
Mannose phosphate isomerase	MPI	1	X	
Nucleoside phosphorylase	NP	1		X
Phosphoglucomutase	PGM-1	1		X
	PGM-2	1	X	X
	PGM-3	1		X
Phosphogluconate dehydrogenase	PGD	1		X
Phosphoglycerate kinase	PGK	1	X	
Peptidase	PEP-A	1	X	X
	PEP-B	1	X	X
	PEP-C	1	X	
Peroxidase	PER	1		X
Sorbitol dehydrogenase	SORDH	1	X	
Superoxide dismutase	SOD-1	1	X	X
	SOD-2	1	X	X
Xanthine dehydrogenase	XDH	2	X	X

NOTE The Cornell sample comprised 88 mole-rats from 12 colonies, and the Harvard sample 42 individuals from 9 colonies An X indicates that the locus was surveyed and the number of alleles discovered is listed (monomorphic loci had one allele)

8 An Ethogram for the Naked Mole-Rat: Nonvocal Behaviors

Eileen A. Lacey, Richard D. Alexander,
Stanton H. Braude, Paul W. Sherman,
and Jennifer U. M. Jarvis

Ethograms are species-specific catalogs of behavior that describe an animal's actions to individuals not familiar with that organism. Ethograms also provide standardized labels for behaviors that can be used by different investigators, thereby increasing the consistency and repeatability of behavioral studies. Ideally, biologists would like to determine the reproductive significance of each aspect of an organism's behavior, since natural selection acts to shape behavior through differences in reproductive success. However, the reproductive consequences of individual behaviors are only rarely understood. As a result, ethologists are faced with a choice: they can characterize behaviors in ways that imply function even when reproductive significance has not been demonstrated, or they can characterize behaviors in ways that are primarily descriptive (but that provide little information about function) even when some evidence of adaptive significance is available.

Here we present an ethogram for the naked mole-rat. Nonvocal behaviors are described in this chapter; vocal behaviors are addressed in the next. Although some readers may find these chapters less theoretically stimulating than others, we note that while the unusual social organization of *Heterocephalus glaber* has been well documented (e.g., Jarvis 1981; this volume), the behavioral repertoire of this species has not been formally characterized. The most appropriate place for such a characterization is in this volume, particularly because many subsequent chapters rely on the terminology and descriptions of behaviors that are presented here.

In constructing this ethogram, we compromised between the functional and descriptive approaches outlined above; whereas the labels and explanatory text used to characterize each behavior are descriptive and lack functional connotation, the order in which the behaviors are presented is intended to suggest functional similarities between activities. Specifically, the 72 nonvocal behaviors of naked mole-rats described here have been grouped into 17 categories based on presumed functional similarity; these categories are intended to serve as hypotheses regarding the adaptive significance of the behaviors described. As additional data regarding reproductive significance become available and the presumed adaptive functions of behaviors become clearer, the behaviors in question can simply be moved from one category to another; the labels and explanatory text associated with these activities need not be altered. We believe that our approach represents the most effective means of presenting adaptive hypotheses without compromising the utility of the descriptive labels and explanations used to characterize behaviors.

Methods

The following ethogram was developed from observations of naked mole-rat behavior conducted by the authors. An initial list of behaviors was compiled during discussions at Ann Arbor, Michigan in July 1985 (see the Preface). Quantitative data regarding specific behaviors were later obtained from observations of colonies housed in translucent (acrylic plastic [i.e., plexiglass] or glass) tunnel systems in laboratories at Cornell University ($n = 5$ colonies), the University of Michigan ($n = 3$ colonies), and the University of Cape Town ($n > 20$ colonies). The methods used to house and maintain these colonies are described by Jarvis in the Appendix.

Each behavior has been given a descriptive, mnemonic label; for behaviors not identifiable by label alone, a more detailed description of the actions comprising that behavior has been developed. Both labels and associated explanations are descriptive. Photographs of selected behaviors are included (see also Pennisi 1986), along with line drawings to highlight particular details. Whenever possible, the subset of colony members performing a particular behavior and the context in which the behavior occurs have also been characterized. To suggest adaptive functions for specific activities, behaviors have been grouped into categories based on presumed functional similarities. Behaviors were categorized on the basis of (1) evidence of adaptive significance and (2) apparent functional congruence with similar behaviors in other species. This ethogram is intended to serve both as a guide to the behaviors discussed elsewhere in this volume and as a reference for future studies of naked mole-rat behavior.

Descriptions of Behaviors

GROOMING

Naked mole-rats exhibit the following behaviors associated with grooming their own bodies (i.e., autogrooming):

1. Cleaning the forefeet or hind feet with the incisors
2. Cleaning the incisors with the forefeet

3. Wiping the face and muzzle with both forefeet
4. Scratching the flank, underarm, mouth, or head regions with the hind foot (fig. 8-1)
5. Grooming the tail with the teeth or forefeet
6. Grooming the genitals with the tongue or incisors
7. Honing the incisors

When *honing the incisors*, mole-rats rub together the ends of the incisors, typically with the bottom incisors extending beyond the top ones such that grinding occurs on the inner surface of the lower teeth and the outer surface of the upper teeth. Incisor honing often occurs when the animals are reclining (below).

Autogrooming occurs at a low but consistent frequency; 1.3% ± 0.1% of data points per individual collected during scan sampling consisted of observations of autogrooming (10-min scan intervals; ≥ 125 scans per animal; $n = 65$ animals in three Cornell colonies). In contrast, allogrooming is almost never observed among adult naked mole-rats. Although adults do groom pups (e.g., licking or nibbling pups, see "Neonate Tending," p. 231), similar behaviors are not observed among adults.

Ectoparasites of *Heterocephalus glaber* include the subcutaneous hypoderatid mite *Acotylopus canestrinii* (Parona 1895), the chigger *Euschongastia bottegi* (Parona 1895), and an undescribed species of mite in the genus *Androlaelaps* (B. M. OConnor, pers. comm.; collected by S.H.B.). However, the virtual absence of allogrooming in *H. glaber* adults suggests that (1) these parasites are not common in the laboratory, (2) they do not cause significant irritation to the animals, or (3) individuals are able to deal with their own parasites more effectively than another animal could. Clearly, allogrooming in *H. glaber* has not developed the complex secondary social functions that it has in other highly social mammals, most notably primates such as bonnet macaques (*Macaca radiata*, Silk 1982) and savannah baboons (*Papio cynocephalus*, Saunders 1988; see also reviews in Sparks 1967; Seyfarth 1977, 1983).

Fig 8-1. A naked mole-rat scratching its face with a hind foot Scratching occurs either as a part of autogrooming or immediately following urination. Scratching that occurs in the latter context is much more vigorous and exaggerated than scratching that is not associated with urination.

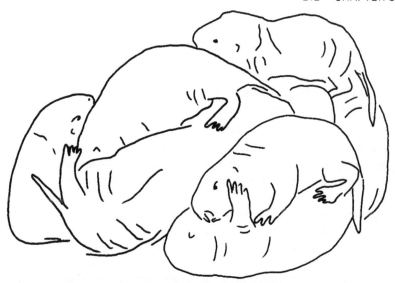

Fig 8-2 A group of naked mole-rats reclining in a nest box Reclining is characterized by the relaxed position of the animals, as well as by the absence of any locomotor activity The mole-rats may recline alone but are typically observed reclining in physical contact with other colony members (e g , in the nest)

RESTING

Resting behaviors are characterized by a lack of motor activity. Although it is difficult to demonstrate that an animal is actually resting or sleeping, we can say that it is not engaged in any other identifiable activity. Behaviors categorized as resting are:

1. Yawning
2. Dozing
3. Reclining

When *dozing* an animal stands motionless with its head drooped, giving the impression that it is asleep on its feet. Dozing is typically observed in the tunnels, frequently at sites where lamps provide a localized source of heat. Animals may doze singly or in small groups of two to four.

Also called huddling (e.g., Withers and Jarvis 1980; Jarvis 1981), *reclining* occurs when an animal lies on its side, back, or stomach with its eyes closed, apparently asleep; often the limbs twitch slightly. Mole-rats may recline singly or in groups; animals reclining in groups are in physical contact with one another, often with individuals lying on top of, as well as beside others (fig. 8-2).

Reclining represents a substantial portion of an individual's daily activity. For nonbreeding animals (*n* = 56) in three Cornell colonies, 56%–61% of data

Plate 8-1. *Top*, A group of naked mole-rats reclining (or huddling) in the colony's nest; the breeding female (nipples visible) is in the center of the pile, and two 2–3-day-old pups are in the foreground. *Bottom*, A close-up of a similar nest huddle showing how tightly the mole-rats pack themselves. Photos: J.U M. Jarvis, R. A. Mendez.

points per individual collected during scan sampling consisted of observations of reclining (10-min scan interval; $n \geq 125$ scans/animal). Breeding males in these colonies ($n = 5$) behaved similarly; 58%–65% of data points for these animals consisted of observations of reclining. In contrast, breeding females in these colonies ($n = 4$) were observed reclining during only 34%–46% of data points from activity scans (see also Reeve and Sherman, chap. 11).

Although naked mole-rats may recline singly, individuals are most frequently observed reclining in large groups (> 10 animals) in the colony's nest box (plate 8-1). It was our strong impression (confirmed by Jarvis, chap. 13), that the tendency of individuals to recline in groups in the nest increased shortly before a litter of pups was born into a colony; this suggests that reclining in groups may have either thermoregulatory or antipredator significance. *Yawning* is depicted in plate 2-1 (p. 57).

THERMOREGULATION

The thermoregulatory physiology of naked mole-rats has been studied by numerous investigators (e.g., Johansen et. al. 1976; Mustafa et. al. 1981; see Jarvis and Bennett, chap. 3). Naked mole-rats are unusual in that they are virtually hairless and essentially ectothermic. In captivity, the body temperatures of individuals vary directly with ambient temperature (McNab 1966). In the field, however, naked mole-rats are thought to be functionally homeothermic because of the thermal stability of their subterranean environment (Jarvis 1978) and behavioral mechanisms of thermoregulation (Withers and Jarvis 1980). In the laboratory, the following behaviors appear to be associated with thermoregulation:

1. Crouching
2. Shivering
3. Basking

A *crouching* animal stands in a hunched posture, in physical contact with other colony members. Individuals often shiver while crouching. Crouching differs from reclining in that animals are standing, rather than lying down. Crouching differs from dozing in that dozing animals are relaxed, whereas crouching mole-rats assume a hunched posture. Reclining typically occurs in nest boxes, but crouching and dozing are most commonly observed in the tunnels.

Shivering is a very rapid shaking or quivering of the body and extremeties. An animal stands in a hunched posture, with the extremities held under the body. Unlike the jerking motion associated with the loud chirp vocalization (Pepper et. al., chap. 9), shivering is characterized by individual convulsions of small amplitude, too rapid to be distinguishable from one another.

A *basking* animal stands with its back or side against a tunnel wall, typically under electric lamps that provide localized heat sources. Basking differs from dozing in that the animal's head does not droop, but, along with the body, is pressed tightly against the warm tunnel wall.

Naked mole-rats in our lab colonies appeared to respond behaviorally to changes in ambient temperature. Specifically, colony members spent more time crouching when the ambient temperature was low (e.g., < 23°C; also Withers and Jarvis 1980). Furthermore, the locations at which basking typically occured corresponded to the locations of electric lamps, and altering the locations of these lamps could induce colonies to switch nest boxes and basking sites.

Individuals were also observed to shiver when the ambient temperature was low. In addition, shivering was observed in two contexts not necessarily associated with thermoregulation: shivering by pregnant females while crouching or reclining in the nest, and shivering by injured or diseased animals. Shivering in these contexts may reflect physiological stress in addition to attempts to thermoregulate.

FEEDING

In the field, naked mole-rats feed on large subterranean tubers (e.g., *Pyrenacantha kaurabassana*; Brett, chap. 5). The animals tunnel through the soil to reach these tubers and then apparently consume them there by gradually hollowing out the tuberous portion of the plant (plate 8-2). The animals also feed on smaller corms and roots (e.g., the legume *Macrotyloma maranguense*; Brett, chap. 5) that are apparently transported to the nest for consumption. Five feeding behaviors have been observed in the laboratory:

1. Brushing food
2. Licking food with the tongue
3. Gnawing
4. Nibbling
5. Chewing

Brushing food involves rapidly moving the forefeet up and down along the sides of a food item that is held in the incisors. Brushing is typically observed before an animal begins feeding on small wet or dirt-covered food items; brushing is not observed while animals are gnawing on large food items.

When *gnawing*, an animal stands with its legs braced against the sides or floor of a tunnel. The incisors are closed across the surface of a large food item (e.g., a tuber), thereby scraping off small bits of food (fig. 8-3; plate 8-2, top).

A *nibbling* animal holds a food item with both forepaws (often while reclining) and consumes small portions of food using the tongue and incisors (fig. 8-4).

Plate 8-2. *Top*, A naked mole-rat gnawing on a subterranean tuber *Bottom*, A mole-rat has gnawed through the tuber The animals consume large tubers by removing small pieces; sometimes these are carried back to the nest. Photos· R A. Mendez

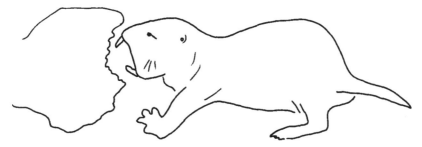

Fig. 8-3 A naked mole-rat gnawing at a large tuber. Gnawing is characteristic of animals feeding on large, hard food items The legs are braced against the tunnel walls, and the incisors are used to scrape small slivers of food from the tuber

Fig. 8-4. A naked mole-rat nibbling on a piece of food Nibbling is characteristic of animals feeding on small or soft food items The food item is held between the forepaws and consumed using the tongue and incisors.

The type of feeding behavior observed appears to depend on the size and consistency of the food item being consumed. Gnawing typically occurs when an animal attempts to consume a large, hard, unmovable food item, whereas nibbling occurs when animals consume small or soft (e.g., bananas) pieces of food that can be held with the forefeet. In captivity, feeding behaviors are observed in all portions of the tunnel system, although these behaviors are most common at the site where food is introduced and in the nest box. The types of food items consumed by laboratory colonies are discussed by Jarvis in the Appendix. Behaviors associated with transporting food items through the tunnel system are discussed below.

ELIMINATION

Naked mole-rats usually urinate and defecate in specific areas of the tunnel system, typically in dead-end tunnels or in boxes with only a single tunnel access ("toilet" boxes). All colony members appear to use the same toilet area or areas. Elimination behaviors include:

Fig 8-5. A naked mole-rat urinating The ano-genital area is extended toward the substrate, with the tail held above the body

1. Defecating
2. Urinating (fig. 8-5)
3. Urinating with crotch dragging
4. Urinating with scratching
5. Wallowing

Defecating mole-rats expel two types of feces: moist, soft, light-colored feces that are often immediately reingested by the defecating animal (or a colony mate) and that may function as caecotrophs, and dry, hard, dark-colored feces that are not consumed directly (but may occasionally be retrieved from toilet areas and eaten). Animals excreting the first type of feces assume the doubled-up posture typical of autocrophagy (see below); in contrast, animals excreting the second type of feces stand in a hunched posture with the tail raised away from the substrate.

In *urinating with crotch dragging*, an animal urinates with the hind legs splayed and the ano-genital area extended toward the substrate (fig. 8-5). Immediately following urination, the animal drags its ano-genital area along the floor of the toilet area.

In *urinating with scratching*, an animal scratches its head, shoulders, or open mouth with a hind foot immediately after urinating (as in fig. 8-1). This type of scratching is much more vigorous and exaggerated than the scratching associated with grooming and occurs exclusively after urination while the animal is in the toilet box or in the tunnel leading to it. Scratching associated with urination may be either preceded or followed by wallowing.

A *wallowing* animal rubs its shoulders or flanks against the bottom or sides of the toilet box or a nearby tunnel immediately after urinating or defecating; sometimes a wallowing animal rolls onto its back (plate 8-3). Wallowing often immediately precedes or follows scratching of the head and shoulders. Wallowing typically occurs in the toilet area; the occurrence of wallowing may be related to the wetness of the substrate in the toilet (O'Riain, pers. comm.).

Elimination behaviors appear to fall into two categories: those associated only with the physiological need to excrete wastes (e.g., 1 and 2), and those

Plate 8-3. A naked mole-rat wallowing in the colony's communal toilet. Wallowing occurs after urination or defecation and probably serves to coat colony members with a common odor. Photo: R. A. Mendez

that involve more elaborate behavior patterns with apparent social functions (e.g., 3–5). The urine of naked mole-rats may contain semiochemicals which function in nest-mate recognition. Wallowing and scratching are two ways that colony members can cover themselves with urine and feces, thereby acquiring or reinforcing their own distinctive colony odor.

COPROPHAGY

Adult naked mole-rats consume both their own feces and the feces of other colony members. Three behaviors are associated with coprophagy:

1. Autocoprophagy
2. Allocoprophagy
3. Begging

In *autocoprophagy*, an animal consumes its own feces, typically while doubled over so that it is sitting on its hindquarters with its mouth in contact with its anus (fig. 8-6).

Allocoprophagy is the consumption of the feces of another colony member. Allocoprophagy is typically preceded by begging for feces.

A *begging* animal uses its muzzle to nudge and tug at the anal area of a colony mate. The begging animal gives a special vocalization (Pepper et al.,

Fig 8-6 Autocoprophagy Naked mole-rats consume their own feces by doubling over so that the mouth is in contact with the ano-genital area

Fig. 8-7 A naked mole-rat pup begging for feces from an adult colony mate The pup nudges and tugs at the ano-genital area of the adult with its muzzle The animal providing feces either lies on its back or assumes a doubled over posture with its muzzle near its ano-genital area

chap. 9). The individual providing feces either lies on its back or assumes the doubled-up posture typical of autocoprophagy (fig. 8-7).

Coprophagy is an essential part of the nutrition of some lagomorphs (see Thacker and Brandt 1955), rodents (see Barnes et al. 1963), and termites (Waller and La Fage 1987). It seems likely that naked mole-rats also obtain nutrients by consuming feces (their own and those of colony mates). Pups beg for feces as they approach weaning (plate 8-4), and feces appear to provide a transitional source of food as pups cease to nurse and begin to consume solid food. Consumption of feces may also provide pups with endosymbiotic gut flora (Porter 1957), as in many termites (La Fage and Nutting 1978). In lab colonies studied by Jarvis (chap. 13), only nonbreeders provided feces to pups.

Among adults, begging is apparently exhibited only by the breeding female. Of 21 incidents of begging by adults observed at Cornell, all were by breeding

Plate 8-4. A naked mole-rat pup begging feces from a nonbreeding adult (sitting up, doubled over); the pup in this photo is about 4-wk old and nearly weaned. Photo J.U.M. Jarvis.

females. In 14 (66%) of these incidents, begging was directed toward a breeding male. Thus, although breeding females did beg and receive feces from nonbreeding colony members, allocoprophagy among adults occurred primarily between two breeding animals.

LOCOMOTION

Naked mole-rats have exhibited several forms of locomotor activities in the artificial burrow systems in our laboratories:

1. Walking (forward or backward)
2. Running (forward or backward)
3. Splayed walking
4. Crouch advancing
5. Darting
6. Swimming
7. Passing
8. Turning

In *splayed walking*, an animal walks with its legs held out to the sides of its body (resembling the stance of a salamander) and its trunk held close to the substrate.

Fig 8-8. Two naked mole-rats passing each other in a tunnel One animal, usually the smaller one, crouches against the bottom of the tunnel (*top*) and the other (usually larger) individual crawls over the crouching animal (*bottom*). Mole-rats can also pass side by side while moving through a tunnel

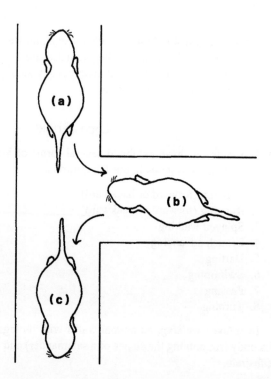

Fig 8-9 A naked mole-rat reversing its direction of locomotion by using a T-shaped tunnel junction to execute a three-point turn. The animal first passes the T junction (a), then backs into the branch of the junction that runs perpendicular to the direction of motion (b) The mole-rat then turns and resumes locomotion (but in the opposite direction) down the stretch of tunnel from which it had initially come (c).

Table 8-1

The Relative Frequencies of Occurrence of the Four
Most Commonly Observed Types of Locomotion

Type of Locomotion	Times Observed	
	No.	%
Walking forward	32	17
Walking backward	15	8
Running forward	104	55
Running backward	38	20
Total	189	100

NOTE Data are from a random sample of 189 instances in
which nonbreeding animals moved a distance of 20 cm or
more A total of 42 animals in two colonies were observed at
Cornell during 1986, the movements of each animal were
sampled three to eight times

In the behavior called *crouch advancing*, an animal moves forward a few
steps with its legs bent and its body held close to the substrate; the animal
pauses, and then moves forward again. This form of locomotion is most frequently observed following some type of colony disturbance (e.g., opening the
tunnel system; see "Alarm Reactions," p. 240).

A *darting* animal very rapidly moves a short distance (one to two body
lengths) forward or backward.

Naked mole-rats do have some *swimming* ability. W. J. Hamilton, Jr. (1928)
and Hickman (1983) reported that naked mole-rats are able to keep themselves
afloat for periods of up to 2 h by simultaneously dog-paddling with the hind
feet and sculling with the tail.

Two animals *passing* one another while moving through the tunnel system
may pass side by side, or, more commonly, one over the other (fig. 8-8). Individuals may pass one another while moving either forward or backward.

Turning animals can reverse the direction of locomotion by completing a
forward somersault followed by a 180° twist, such that the ventral portion of
the body faces the substrate. Alternatively, individuals can reverse direction by
completing a three-point turn in which the animal passes a T-shaped junction
in the tunnel system (in nature, a small outpocketing), backs into the branch of
the T perpendicular to the direction of movement, turns 90°, and resumes locomotion (but in the opposite direction) back down the stretch of tunnel that it
came from (fig. 8-9).

Naked mole-rats appear to move forward or backward with equal rapidity.
In a randomly chosen sample of 189 instances of locomotion (data from two
colonies at Cornell, $n = 42$ animals observed), 75% were observations of running, and 25% were of walking (table 8-1). The animals moved most rapidly
when traversing long (3–5 m), straight stretches of tunnel.

The type of locomotion exhibited by naked mole-rats changes if the animals are removed from the tunnel system. When they are placed in an open, flat area (e.g., a table top), locomotion consists primarily of splayed walking. The animals exhibit no apparent perception of "edges;" locomotion does not slow or otherwise visibly change as the animals approach drop-offs. Furthermore, the mole-rats will repeatedly fall off such edges, giving no indication that they learn to detect and respond to edges.

When mole-rats encounter one another in the tunnel system, one individual may pass over the other or both may pass side by side. It was our impression that passing side by side occurred most frequently when similar-sized small mole-rats encountered one another. Among members of three Cornell colonies ($n = 65$ animals), the tendency to pass over other individuals correlated significantly and positively with body weight (all $r > 0.56$, all $p < 0.01$). Although breeding females were among the largest individuals in the colonies, they passed over colony mates no more frequently than predicted on the basis of body size. Thus, it may simply be easier for a small mole-rat to squat and let a large one step over it than vice versa.

ORIENTATION

The sensory capabilities of naked mole-rats have not been quantitatively examined. However, the animals' morphology, in particular the minute size of the eyes, suggests that vision is a less important sensory modality than touch, olfaction (see Tucker 1981), or hearing (Pepper et al., chap. 9). Several orientation behaviors have been observed in our laboratory colonies:

1. Sniffing
2. Head pressing
3. Exit darting
4. Tail sweeping

During *head pressing*, an animal stands motionless with the top of its head pressed against the ceiling of a tunnel.

In *exit darting*, an animal in a nest box darts (see "Locomotion," above) forward until its nose contacts the sides of the nest box. The animal then darts backward several steps before repeating the forward motion. This pattern is repeated several times until the animal arrives at a tunnel opening leading out of the nest box.

An animal engaged in *tail sweeping* moves its tail from side to side (the tail often contacts the tunnel walls) while walking, running, or sweeping (e.g., nesting material; below); tail sweeping is especially pronounced when animals are moving backward through their burrow systems.

Orientation behaviors are particularly apparent when a colony is disturbed (e.g., by the introduction of a new tunnel or an unfamiliar conspecific), or when a colony is hungry and numerous individuals are searching for food. Individuals encountering a foreign mole-rat or a novel object alternately approach and retreat from the area, exhibiting both the crouch advance and frequent head presses. The animals also sniff intently at any newly introduced object, typically approaching it, sniffing briefly, and then retreating rapidly (one to two body lengths) before advancing again. The vibrissae and the stiff hairs on the tail are also likely used to obtain (tactile) information concerning the environment (Hill et al. 1957; Tucker 1981).

Naked mole-rats appear to learn tunnel configurations, as indicated by an increase in the speed of backward and forward locomotion once the animals have made several passes through a novel tunnel configuration and frequent navigational mistakes (bumping into walls) following changes in tunnel system configuration. In addition, the animals often walk through the tunnel system with their eyes apparently closed. Interestingly, in the field, mole-rat colonies whose tunnel systems have been experimentally bisected are able to orient the direction of tunnel excavation so as to exactly reconnect the severed tunnels (Brett 1986, chap. 5); how the two halves of a colony orient toward each other is unknown.

TRANSPORT

In the laboratory, naked mole-rats transport a variety of materials (e.g., pieces of food, nesting material, chunks of dirt, small stones; see plates 10-2, 10-3) through the tunnel system. In the field, both corms and the husks of bulbs are found in nests, indicating that these items are transported through the burrows (Brett, chap. 5). In both the field and lab, the tunnels are free of food remnants and debris, suggesting that the burrows are being cleaned frequently. Four methods of transport have been noted:

1. Mouth carrying
2. Dragging
3. Sweeping
4. Backward kicking

A mole-rat engaged in *mouth carrying* holds an item between the incisors and lifts it off the tunnel floor. Items may be carried in this way while the animal is moving either forward or backward (fig. 8-10; plate 8-5).

In *dragging*, an item is held with the incisors and dragged along the tunnel floor, usually while the animal is moving backward through the tunnel system.

A *sweeping* mole-rat kicks loose items (e.g., sand, dirt, wood shavings) behind itself while moving backward through a tunnel. The hind feet are rotated outward and the legs are synchronously lifted and kicked backward several times as the animal balances on its forelegs. The animal then moves a step or two backward before repeating this kicking motion.

Fig 8-10. A naked mole-rat carrying a piece of food in its mouth The
piece of food is held between the incisors and lifted off the substrate
This method of transport is typically associated with small objects,
large or heavy items are dragged

Plate 8-5. A nonbreeding naked mole-rat carrying a piece of food it has just excised from
a large tuber Photo: R A. Mendez

Backward kicking occurs when the back legs are synchronously lifted and
kicked upward, propelling material above and behind the animal (fig. 8-11). In
contrast to sweeping, material is moved vertically (i.e., upward) as well as
horizontally, and the animal does not move backward between successive
kicks.

Whether objects are carried or dragged appears to depend on their size:
smaller items (e.g., bits of food) are usually carried (plate 8-5), whereas larger
items (e.g., rocks) are typically dragged. Items that are apparently either too
small or too numerous (e.g., wood shavings used as nesting material) to be
easily picked up with the incisors are swept through the tunnels. Although all

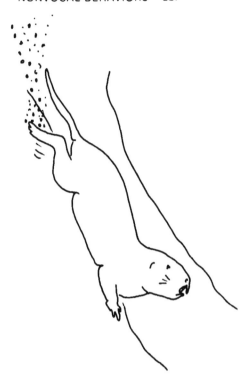

Fig 8-11 A naked mole-rat kicking backward. The hind feet are synchronously lifted and kicked, propelling soil above and behind the animal. This behavior is associated with creating molehills or "volcanoing" (diagram modified from Jarvis 1984).

nonbreeding colony members transport food, nesting material, and dirt through the tunnel system, transport behaviors are most commonly exhibited by smaller nonbreeders (Lacey and Sherman, chap. 10; Jarvis et al., chap. 12; Faulkes et al., chap. 14).

In the field, the backward kick is used to expel loose soil from the tunnel system (Jarvis and Sale 1971; Brett, chap. 5; Braude, chap. 6). In the laboratory, the backward kick is often used to move loose dirt into an empty nest box.

DIGGING

Naked mole-rats are fossorial, and digging clearly represents an important part of the animals' behavioral repertoire. In the field, *H. glaber* colonies are most easily found by locating the molehills (Brett, chap. 5) or "volcanos" (Braude, chap. 6) formed as the animals eject soil loosened by digging from their burrow systems. The digging behavior of naked mole-rats is a cooperative activity in which chains of individuals help excavate dirt and transport loose soil to burrow entrances (Jarvis and Sale 1971; see also Jarvis et al., chap. 12; Jarvis 1984). In the laboratory, the animals readily dig when provided with an appro-

priate substrate such as dirt (Lacey and Sherman, chap. 10; Jarvis, chap. 13) or
cork (Pepper et al., chap. 9). Three digging behaviors have been observed in
the lab:

1. Gnawing
2. Backshoveling
3. Foreleg digging

In *gnawing*, the incisors are scraped along a dirt face, loosening chunks
of soil (plate 8-6). This is the predominant mode of excavation, especially
when animals are working on hard surfaces.

Backshoveling means that the forelegs are synchronously lifted, moved forward and down onto a pile of dirt, and then moved backward under the body,
propelling the dirt underneath and slightly behind the animal.

Foreleg digging occurs when one foreleg is brought forward and the foot is
scraped along a dirt face, removing small pieces of dirt; the foreleg is then
returned to a position under the body. The action is usually repeated rapidly
using alternate forelegs. Foreleg digging and gnawing often occur in succession; an animal first gnaws at the earthen face for 15–30 s, then moves loosened dirt out of the way using the foreleg dig and backshovel.

Digging is one of the most important behaviors of naked mole-rats. Because
observing the details of tunnel excavation in the field is impossible, we must

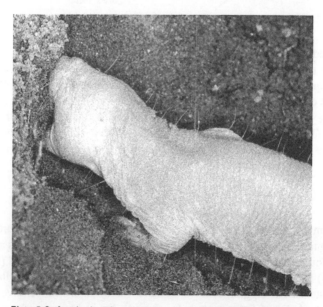

Plate 8-6. A naked mole-rat gnawing at a dirt face This is the predominant mode of tunnel excavation in this species Photo J U.M
Jarvis.

rely on inferences from laboratory observations. In captivity, all nonbreeding animals perform the three digging behaviors described above. However, it is currently somewhat unclear which colony members participate most frequently in these activities. Both positive (Braude 1983; Lacey and Sherman, chap. 10) and negative (Jarvis et al., chapt. 12) relationships between body weight and gnawing and digging frequency have been reported. The reader is referred to the relevant data and discussions in chapters 5, 10, and 12 of this volume.

MATING

The most distinctive feature of the mating system of naked mole-rats is that sexual activity is usually limited to a single female and one to three males per colony (Lacey and Sherman, chap. 10; Jarvis, chap. 13). There are three mating behaviors:

1. Backing
2. Mounting
3. Copulating

In *backing* behavior, the breeding female backs up to a male, exhibiting a lordosis-like posture that exposes her genitalia. During backing, the breeding female often gives a distinctive trilling vocalization (Pepper et al., chap. 9).

Mounting the breeding female from behind, the male typically braces his front legs against the sides of the tunnel system. He exhibits a pedaling motion of the hind legs as he attempts to bring his genitals into contact with those of the female (fig. 8-12).

Copulating is defined by contact between the genitalia of the male and female. Pelvic thrusting by the male is observed just before and during copulation. Ejaculation is not obvious. Copulations last less than 15 s, after which the male dismounts.

Female naked mole-rats exhibit behavioral estrus for 24 h or less. During this period, breeding animals remain in close proximity and move frequently around the tunnel system. Numerous mounting attempts occur but do not result in copulations. Aggression among breeding males does not occur during the breeding female's estrus, although aggression may be observed several days before mating (Lacey and Sherman, chap. 10; Jarvis, chap. 13).

Copulations and attempted copulations appear to occur whenever the estrous female encounters a breeding male in the tunnel system. Of 38 copulations observed at Cornell, 26 (68%) were initiated when the breeding female approached and backed up to a breeding male. The remaining 12 copulations (32%) were initiated when a breeding male approached, nuzzled, and mounted the breeding female. Mounting by males was also frequently observed during late pregnancy; the function of these mountings is unknown. Mounting and

Fig 8-12 A male naked mole-rat mounts the breeding female before copulation The male rests on the female's back as he pedals his hind feet against the female's flanks and attempts intromission

other sexual behaviors are not exhibited by nonbreeding colony members. Mating behavior is described in greater detail in chapter 13.

Reproductive behavior in laboratory colonies is often associated with a fourth behavior, ano-genital nuzzling. Because ano-genital nuzzling between breeding animals is observed during all phases of the female's reproductive cycle and because ano-genital nuzzling is occasionally observed between nonbreeding animals (Jarvis, chap. 13), this activity has been categorized as an interactive behavior (see p. 235) rather than a mating behavior.

Birthing

Birthing consists of a single behavior, *parturition*. A female that is about to give birth becomes very active, running through the tunnel system with frequent pauses to nudge or lick her genital area. She may also autogroom intensely during this period. Pups are typically born head first, and the head of a pup may be seen protruding from the female's vagina as she moves through the tunnel system. As the pup is finally expelled from the female's reproductive tract, the female often doubles over so that she is sitting on her hindquarters, with her muzzle in close proximity to her genital region (e.g., see figs. 8-6, 8-7); the female sometimes pulls at the pup (or afterbirth) with her incisors, apparently helping to remove the pup from her reproductive tract.

Data gathered at Cornell indicate that pups are born primarily in the tunnels; 30 of 42 pups (71%) whose births were observed were born in the tunnels, rather than in the nest box. At Michigan, 6 of 8 pups (75%) whose births were witnessed were born in the tunnels. During parturition, the breeding female is not obviously assisted by other colony members, although colony mates that encounter the female in the tunnels often stop to sniff at her before moving away. The newly born young are not immediately attended by either the breeding female or the breeding males. Instead, pups are cleaned and carried to the nest by nonbreeding colony members (see "Neonate Tending," below). Parturition typically lasts 1–3 h (depending on litter size), with pups born every 10–30 min. Parturition and associated behaviors are also described in chapter 13.

Neonate Tending

Naked mole-rat pups are cared for by adult and juvenile colony members during the first 3–4 wk after birth. Care of pups consists of six behaviors:

1. Carrying pups
2. Grooming pups
3. Nudging pups
4. Pushing
5. Nursing
6. Sweeping pups

Carrying pups occurs when an adult grasps a pup with its incisors and either carries or drags (see "Transport," p. 225) the pup through the tunnel system. Pups are typically grasped by the skin at the nape of the neck, although the belly or back may also be held (fig. 8-13).

When *grooming pups*, an adult grasps a pup with both forepaws and either licks the pup with its tongue or nibbles at the pup with its incisors. Grooming is most frequently directed toward the ano-genital area of the pup.

Nudging pups is a general contact with pups in which an adult either noses a pup with its muzzle or manipulates a pup with its forepaws.

When an adult shoves the blunt, anterior end of its muzzle (with the mouth closed) against the body of a pup, it is said to be *pushing*. The adult's head vibrates rapidly side to side, and its body jerks forward with each push. As a result of contact with the adult's muzzle, the pup is knocked several centimeters away from the adult, sometimes with such force that the pup is lifted into the air. Pushes are usually repeated in rapid succession such that the pup is often moved 15–20 cm. Pushing typically occurs in the nest box, although it is sometimes observed in the tunnels.

Fig. 8-13 An adult naked mole-rat carrying a pup. The pup is held between the adult's incisors and lifted completely off the substrate while being carried through the tunnel system.

Fig 8-14. A breeding female nursing her pups Nursing begins when the female sits up or rolls onto her back, exposing her teats to the pups, nursing ends when the female pushes the pups away and rolls onto her stomach, thus preventing access to her teats.

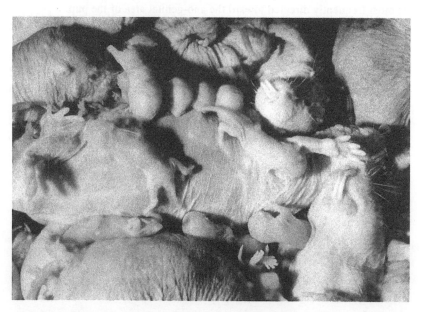

Plate 8-7. A breeding female naked mole-rat nursing her pups in the center of a crowded nest box; these pups are about 3-days old Photo J.U M Jarvis.

Nursing of the pups is done only by the breeding female (fig. 8-14), although nonbreeding males and females sometimes develop enlarged teats just before parturition (Jarvis, chap. 13). Nursing begins when a female rolls onto her back or side, making her teats accessible. Pups apparently actively seek out and approach the female, and several pups nurse simultaneously. Nursing often occurs in the nest, with the breeding female and her pups surrounded by or even lain on by nonbreeding colony members (plate 8-7). Nursing ends when the female either rolls onto her stomach or pushes pups away with her forefeet.

Sweeping pups involves an adult kicking a pup with its hind feet, using a motion that appears to be identical to the sweeping behavior observed during the transport of nesting material or debris (see "Transport," p. 225). A pup may be kicked once or several times in rapid succession; repeated kicks usually result in the pup's being tumbled rapidly along a stretch of tunnel.

After birth, pups are cleaned and carried to the nest by nonbreeding colony members. The breeding female nurses the pups, and lactation continues for about 3–4 wk. During lactation, the breeding female and the colony's breeding males are the most frequent participants per capita in grooming, nudging, and pushing pups (Lacey and Sherman, chap. 10); nonbreeders, especially juveniles, sometimes perform these behaviors (see Jarvis, chap. 13).

An unusual aspect of pup care in *H. glaber* is the occurrence of pushing. This behavior occurs frequently when pups are present in the nest, and almost appears to be a form of aggression. However, because it is so regularly performed by breeding animals (i.e., the parents of pups), it seems unlikely that it is detrimental to the young (see Lacey and Sherman, chap. 10). Pushing is directed exclusively toward pups and young animals (< 1-yr old). Pushing may function (1) to move pups quickly to the periphery of the nest, thereby preventing them from being trampled by other colony members, (2) to move pups quickly out of the way of predators, (3) to enhance the peristaltic gut action of pups, or (4) to help enforce social dominance among colony members (see also Reeve and Sherman, chap. 11). None of these hypotheses has been quantitatively examined.

The rate of successful weaning of litters is discussed in chapter 13. Differences in the care given to successful (i.e., reared to weaning) and unsuccessful (i.e., not reared to weaning) litters are striking, such that the fate of a litter can usually be predicted within 24 h of birth. Pups in successful litters are regularly nursed by the breeding female and are kept in the nest box where adult colony members spend considerable time crouching and reclining with the young; these pups are quickly retrieved and returned to the nest if they fall or are knocked out of the nest box. In contrast, pups in unsuccessful litters are nursed infrequently and are often left in tunnels outside the nest box; in the tunnels, pups in these litters are often stepped on or swept back and forth by nonbreeders. We do not know why only some litters are successful, although one hypothesis is presented in chapter 13 (also see the Appendix). Once a pup dies, cannibalism is frequently observed; adults also sometimes kill live pups (Jarvis, chap. 13).

JUVENILE-SPECIFIC BEHAVIORS

The ontogeny of pup behaviors and morphological development are described in chapter 13. Several behaviors were exhibited primarily by juvenile naked mole-rats or were exhibited by juveniles in unique contexts:

1. Wrestling
2. Dragging

Two or more *wrestling* pups roll and tumble together, often while batting or incisor fencing (plate 8-8). Wrestling can involve two to four animals.

In *dragging*, one pup grabs another (typically by the back of the neck, hindquarters, or tail) with its incisors and drags that individual through the tunnel system.

Juvenile-specific behaviors are first observed at about the time pups are weaned (ca. 3–4 wk after birth). These behaviors are common among littermates for the next 1–2 mo, after which their frequency gradually declines. Juvenile-specific behaviors cease entirely by the time the animals are about 2-yr old.

In addition to wrestling and dragging, juveniles also exhibit *incisor fencing* and *batting*; these behaviors resemble certain agonistic behaviors exhibited by adults (see p. 237) but do not occur in the same contexts (i.e., when contesting a piece of food, or during struggles associated with changes in reproductive

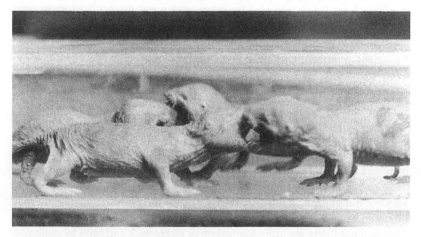

Plate 8-8 Juvenile naked mole-rats incisor fencing These juveniles are about 2-mo old Photo: D. Hammond, *Cape Town Argus* (South Africa)

status). Among the pups, incisor fencing and batting appear to occur in all contexts in which two or more pups interact.

INTERACTIVE BEHAVIORS

Naked mole-rats spend considerable amounts of time in close proximity to one another; 74% ± 5% of data points per individual collected during scan sampling (n = 65 animals in three colonies at Cornell; ≥ 125 scans per animal, ≥ 10-min scan intervals) consisted of mole-rats in physical contact with one or more colony mates. As a result, interactions among colony members are common. Animals interact when they contact one another in the nest, as well as when they are resting in or moving through the tunnels. We observed six behaviors that occurred during interactions between individuals:

1. Nose pressing
2. Nuzzling
3. Ano-genital nuzzling and sniffing
4. Sniffing (at another animal)
5. Head deflecting
6. Pawing

In *nose pressing*, two individuals face each other with their heads slightly lowered and the blunt ends of their muzzles pressed together. The nose press is brief, typically lasting only 1–2 s.

During *nuzzling*, an animal rubs the sides of its muzzle against the body of a second individual. Often one side of the muzzle is rubbed several times in succession before switching to the other side.

In *ano-genital nuzzling and sniffing*, an animal of one sex sniffs and uses its muzzle to nudge the genitalia of an animal of the other sex. One animal may mount the head of another, so that the genitalia of the animal on top are in contact with the muzzle of the animal on the bottom. More commonly, both animals lie on their sides, head-to-tail, such that the genital area of each is in constant contact with the muzzle of the other (plate 8-9).

In *head deflecting*, an individual turns its head to the side and down, such that the area of the head near one ear is closest to the muzzle of the second animal.

An animal *pawing* another reaches out with one forepaw and repeatedly drags that appendage along the body of a second individual.

A variety of responses are observed when two mole-rats encounter each other in a tunnel. Individuals may simply bump into one another and then move apart, exhibiting no apparent interactive behaviors. Alternatively, contact may be followed by sniffing, nose pressing, and head deflecting, or individuals may interact agonistically (see below). It was our impression that nose

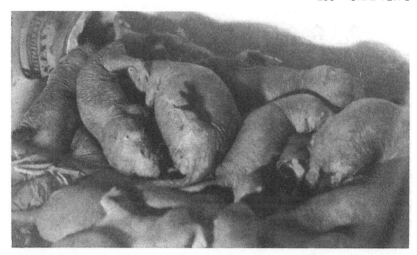

Plate 8-9 The breeding female (*right*) and breeding male (*left*) engaged in ano-genital nuzzling Photo· J U M Jarvis

pressing was most frequently exhibited by the colony's breeding female, whereas head deflections were more frequent among nonbreeders (in particular, young animals) when they encountered the breeding female or a breeding male. Pawing occurred primarily when one mole-rat tried to pass another that was blocking a tunnel, nest, or food-box entrance. Ano-genital nuzzling typically occurred between breeding mole-rats, although nonbreeding animals occasionally participated (see Jarvis, chap. 13). Nuzzling was observed during all phases of the female's reproductive cycle, particularly when the breeding female was in estrus.

AGONISTIC BEHAVIORS

Naked mole-rats exhibit agonistic behaviors in several contexts, including competition for food or other resources (e.g., access to a digging site), defense of the colony against foreign mole-rats, competition for breeding status, and stimulation of colony-maintenance activities by the breeding female. There are seven types of agonistic interactions:

1. Open-mouth gaping
2. Incisor fencing
3. Batting
4. Biting
5. Shoving
6. Tugging
7. Tetany

Fig. 8-15. A naked mole-rat giving an open-mouth gape The jaws are opened, with top and bottom incisors separated. The open-mouth gape is typically given when an animal is standing face to face with a conspecific, and it is often accompanied by a hissing vocalization (Pepper et. al , chap 9).

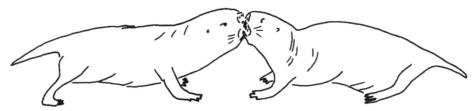

Fig. 8-16. Two naked mole-rats engaged in incisor fencing The animals push back and forth against each other with their heads at an oblique angle and their incisors locked together The incisors may be briefly released and then locked together again.

Two animals engaged in *open-mouth gaping* stand face to face, with their muzzles almost touching. Both individuals open their mouths, with the top and bottom incisors separated (fig. 8-15; plate 8-10). Air is then rapidly inhaled and exhaled through the open mouth, producing a hissing sound (Pepper et al., chap. 9).

Incisor fencing occurs when two mole-rats stand face to face, with their mouths at right angles and their incisors locked together. The incisors may be released briefly, allowing the animals to reposition their heads before locking incisors again (fig. 8-16). The animals typically shove back and forth against each other and rock their heads from side to side while the incisors are locked together. Incisor fencing in juveniles is depicted in plate 8-8.

Two animals that simultaneously swat at each other's muzzles with their forepaws are *batting*. The forefeet may also be placed on the other individual's muzzle and held there, preventing the second animal from contacting the muzzle of the first.

In *biting*, the jaws of one animal close on the body of another individual (see plate 8-10).

Shoving occurs when two mole-rats stand face to face, with their heads slightly lowered and the blunt ends of their muzzles pressed together. One

Plate 8-10. *Top*, Mole-rats threatening each other with open-mouth gapes (accompanied by hissing) The animals on the left and right are from different colonies *Bottom*, Open-mouth gapes can occur at close range and may precede biting Again the individuals are from different colonies Photos· R A Mendez.

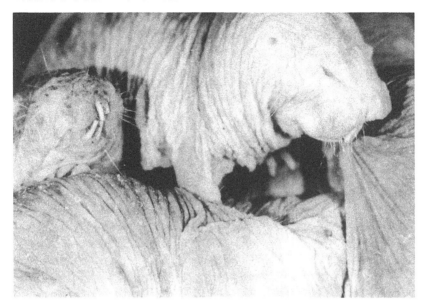

Plate 8-11. A naked mole-rat (a breeding male) tugging at a colony mate. The male is holding the loose back skin in its incisors Photo R. A Mendez.

animal moves forward, pushing the second animal backward for a distance of up to 1 m. The animal doing the shoving frequently hisses (Pepper et al., chap. 9) while performing this behavior. Participation in shoving is analyzed in chapter 11.

A *tugging* individual grasps the skin of another with its incisors and pulls backward, often while bracing against the sides of the tunnel with its feet (plate 8-11). The nape of the neck, the loose skin on the hips, and the tail are most frequently grabbed by another individual.

Tetany occurs when an individual doubles up and lies perfectly still, often with its feet in the air (plate 8-12). Tetany is a response to being shoved or gaped at by the breeding female. Individuals may remain in this contorted posture for several minutes, long after the interaction with the breeding female has ended.

Within colonies, agonistic behaviors associated with competition for resources (incisor fencing, tugging) and reproductive dominance/colony-activity stimulation (shoving) do not result in injury to either participant. In contrast, agonistic interactions associated with colony defense (biting) and competition for a breeding vacancy among females (batting, tugging, and biting) do sometimes result in injury and death. During 1981–1986, a total of 12 animals in four colonies (at Cornell) died as a result of injuries received during fights following the removal of a breeding female (Lacey and Sherman, chap. 10); deaths resulting from struggles for breeding opportunities have also been reported by Jarvis (chap. 13) and Faulkes et al. (chap. 14). Agonistic interactions

Plate 8-12 A small nonbreeding naked mole-rat (*left*) exhibiting the tetany posture after having been shoved and gaped at by the breeding female (*right*) Photo R A Mendez

associated with resources are usually accompanied by a loud chirping vocalization (see Pepper et al., chap. 9), whereas agonistic interactions associated with colony defense or reproductive status are accompanied by a hissing sound.

All 12 animals at Cornell that died as a result of agonistic interactions were ostracized before their deaths. Once injured, these animals were excluded from the colony's nest box and were attacked by other colony members, particularly the breeding female, if they attempted to enter the nest. These shunned animals subsequently crouched and shivered in the corner of an empty box or toilet; these mole-rats rarely attempted to leave the empty box and frequently failed to show any response when approached by other colony members. Death of the injured animal typically followed within 2–3 days of its being excluded from the colony's nest box. Even when removed from their colony and provided with a localized heat source and ad libitum food (see Jarvis, Appendix), these individuals usually died in a few days; it was never possible to reintroduce them to their colony without reinitiating aggressive behaviors by the breeding female and other colony members.

ALARM REACTIONS

Naked mole-rats respond strongly to disturbances such as loud noises, vibrations, and the introduction of foreign objects. Responses to such disturbances, termed alarm reactions, include:

1. Freezing
2. Scrambling

In *freezing*, an animal suddenly ceases all activity, holding itself in the same position it had been in at the time of the disturbance.

Scrambling occurs when numerous colony members simultaneously dart (see "Locomotion," p. 223) forward and backward in the tunnels or nest box, with no single coordinated direction of locomotion evident among individuals.

The alarm reactions of naked mole-rats fall into two categories: those elicited by sudden environmental disturbances (e.g., bumping the tunnel system, slamming the door), and those elicited by more minor disruptions (e.g., the introduction of food). Major disturbances elicit an immediate response by all colony members: the mole-rats freeze, and then begin scrambling, typically while trying to exit the nest and immediately adjacent tunnels. Unweaned or recently weaned pups may be carried from the nest during scrambling, or they may be trampled. In the absence of further stimuli, scrambling gradually ceases and is replaced by less frenzied forms of locomotion (e.g., walking and running). Colony members slowly return to the nest and resume activities unrelated to alarm reactions.

In contrast, reactions to introduced objects (e.g., food, snakes, or unfamiliar conspecifics) are less immediate; rather than the explosion of activity associated with scrambling, reactions to introduced objects consist of a gradual increase in colony activity. Animals encountering the stimulus typically approach it several times using the crouch advance (see "Locomotion," p. 223) and sniff briefly at the object before retreating. If the object is food, some colony members begin feeding, followed by tugging and incisor fencing as colony mates arrive in the vicinity. If the object is a snake, the animals may hiss (Pepper et al., chap. 9) or bite at it, and some recruitment of (larger) nonbreeders may occur. If the object is an unfamiliar conspecific, one of the first individuals to contact it will give a trilled vocalization (Pepper et al., chap. 9) while running back toward the nest; this results in an increase in colony activity and recruitment to the site of the developing conflict (see Lacey and Sherman, chap. 10).

Summary

A total of 72 nonvocal behaviors exhibited by naked mole-rats have been characterized. These behaviors have been grouped into 17 categories according to presumed functional similarities. As such, these categories represent hypotheses regarding the adaptive significance of individual behaviors. The behavioral repertoire of *Heterocephalus glaber* is clearly complex, and we suspect that many subtleties of the animals' behavior remain to be discovered. Elucidating the adaptive significance of individual behaviors is essential to understanding the complex, cooperative social organization of *H. glaber*.

Acknowledgments

We thank contributors to this volume for sharing their observations of behavior. B. M. OConnor helped with the identification of ectoparasites; line drawings were made by S.H.B. E.A.L. was supported by a National Science Foundation (NSF) predoctoral fellowship; J.U.M.J. was supported by the Council for Scientific and Industrial Research of South Africa and the University of Cape Town; and P.W.S. was supported by NSF grant BNS-8615842 and by Cornell University. We thank J. W. Pepper, S. F. Payne, and C. Kagarise Sherman for assistance.

9 Vocalizations of the Naked Mole-Rat

John W. Pepper, Stanton H. Braude,

Eileen A. Lacey, and Paul W. Sherman

Naked mole-rat colonies are among the most highly structured and complex societies known in nonhuman vertebrates (Jarvis 1981; this volume). Communication is essential to the coordination of any functioning group, and some of the most elaborate animal communication systems occur in eusocial species (e.g., the remarkable waggle dance of honey bees; Seeley 1985). The nature and extent of communication in the eusocial naked mole-rat is therefore of particular interest.

Various authors have noted that naked mole-rats are highly vocal animals, both in the field and the lab. In the field, Hill et al. (1957) reported hearing "squeaking noises" from naked mole-rats that were kicking soil from burrow openings (especially when they were working close together) and also from a juvenile as it was carried away from an exposed section of burrow by an adult. Jarvis and Sale (1971) found that naked mole-rats in nature squeaked softly while investigating burrow openings, and Brett (1986, chap. 5) noted both a "chirping" sound associated with digging and a "loud squeaking" accompanying feeding in the field. Based on observations of captive colonies, Hill et al. (1957, p. 468) reported that naked mole-rats produce sounds in connection with "working, panic, intimidation with squeals to gain possession of desired food, and during the assimilation of a newcomer into the colony." Jarvis and Sale (1971) noted that captive animals produced a "loud squeaking" while digging, especially when a mole-rat working at a tunnel end was replaced by another individual; likewise, Isil (1983) heard "chirping" sounds associated with digging in captive colonies.

Although vocalizations accompany most activities of naked mole-rats in captivity, many of these sounds are soft, high pitched, and of short duration, making them difficult to distinguish on the basis of casual observations. Here we present the first published attempt to catalog and describe *H. glaber* vocalizations. In addition, we discuss the contexts and behaviors associated with specific calls and develop hypotheses for their functions. Our descriptions of vocalizations, like those of behaviors in the ethogram (Lacey et al., chap. 8), introduce a common terminology that is used throughout this volume and is intended as a starting point for future studies.

Methods

STUDY ANIMALS

Information about naked mole-rat vocalizations was compiled from several independent studies of captive colonies housed at the University of Michigan and Cornell University. All of the colonies originated with animals captured around Mtito Andei in southeastern Kenya from 1977 to 1981 (see Lacey and Sherman, chap. 10; Reeve and Sherman, chap. 11). Each colony was housed in its own acrylic plastic (plexiglass) or glass tunnel system, which included one to four clear acrylic plastic boxes; every mole-rat was tattooed with a unique identifying number. Colonies were maintained similarly at Michigan and Cornell (see Jarvis, Appendix), except that at Cornell the animals were kept in virtual darkness (illuminated by dim red or yellow bulbs) whereas at Michigan they were in constant light.

CHARACTERIZATION OF VOCALIZATIONS

Four *Heterocephalus glaber* colonies with a combined population of 74 individuals were observed (by J.W.P.) from March through June 1988 in the Museum of Zoology at Michigan. An Altec 681 microphone was introduced into nest boxes through holes in their acrylic plastic lids, and vocalizations were amplified and monitored using a Nagra 4.2L tape recorder and headphones. Selected vocalizations were recorded at a tape speed of 38 cm/s, and accompanying behavioral observations were recorded either on tape or in written notes. Recordings were made when mole-rats were as close as possible to the microphone, to minimize echoes and distortions introduced by the tunnel system. To aid in classifying vocalizations, a Uniscan II quantitative-spectrum-analysis system (Unigon Industries, Inc.) was used to generate sonagrams (sound spectrograms showing frequency plotted against time) for each recorded vocalization. To create a library of vocalizations for reference, selected sonagrams were printed on a dot matrix printer (Epson FX-80). A total of 4.5 h of recordings containing more than 5,000 vocalizations were examined, and sonagrams of more than 800 vocalizations were printed. In an effort to observe as many different contexts as possible and to increase the sample size of rare vocalizations, much more time was spent observing colonies than making recordings. Because sampling was not random, characteristic duration and frequency measurements rather than means and standard deviations are reported for each vocalization. Figures showing representative sonagrams of each vocalization type were produced using a DSP Sona-Graph Model 5500 (Kay Elemetrics) and then edited to remove extraneous noises.

Vocalizations associated with defense of a colony's tunnel system against intrusions by foreign conspecifics and potential predators were documented during colony-defense trials conducted (by E.A.L. and P.W.S.) at Cornell (see chap. 10). Three colonies containing about 89 animals were studied between May 1980 and July 1985. By digging through a plug of dirt, a colony came into contact with either a second colony or a live snake. The responses of the mole-rats to these potential threats, including their vocalizations, were recorded.

To test for the presence of ultrasonic vocalizations (which are inaudible to humans), colonies at Cornell were recorded (by P.W.S.) using a Lockheed high-speed tape recorder and a mini bat detector (Q.M.C. Instruments, Inc.) capable of detecting sounds at frequencies of up to 160 kHz. Recording sessions lasted 30–40 min and were repeated five times over a 2-yr period.

The characterization of vocalizations was supplemented by ad libitum observations and recordings of captive colonies at Michigan and Cornell, discussions with J.U.M. Jarvis, and comparisons with the results of an undergraduate honors thesis done at Cornell (Kaufmann 1986). Vocalizations were classified into types based solely on their physical characteristics, as determined by ear and by examining sonagrams. The majority of vocalization types enumerated here have been noted by more than one researcher, and all were exhibited by members of more than one captive *H. glaber* colony. Each vocalization type was given a descriptive label and characterized as to (1) its physical structure, (2) the contexts in which it occurred, and (3) the behaviors that typically accompanied it. Nonvocal behaviors were labeled following the terminology and descriptions of Lacey et al. (chap. 8).

INVESTIGATION OF THE LOUD CHIRP

The loud chirp is a common and conspicuous vocalization that is apparently associated with conflict within colonies and that is often given when one mole-rat physically displaces another. Because the prevalence of this apparently agonistic vocalization in otherwise highly cooperative naked mole-rat colonies is so striking, a special investigation of the loud chirp was conducted (by S.H.B.). The relationship between loud chirping and the physical displacement of one animal by another was studied in two colonies containing, respectively, 20 and 22 mole-rats, from January through March 1983 at Michigan.

Mole-rats were observed digging or eating in an 80-cm long dead-end tunnel that was added to the existing tunnel system. The closed end of this tunnel was filled with a large carrot or sweet potato, and access to this food was blocked by a 1–3 cm thick plug of cork (fig. 9-1). The cork approximated the extreme hardness of the soil that wild naked mole-rats excavate during the dry season in Kenya. The thickness of the cork was adjusted according to its density so that about 1 h was required for the animals to dig through it and reach the food. In each of 24 replicates, a colony was observed for about an hour of digging through the cork followed by an hour of eating. The following information was recorded: (1) the identities of the vocalizing individual and the

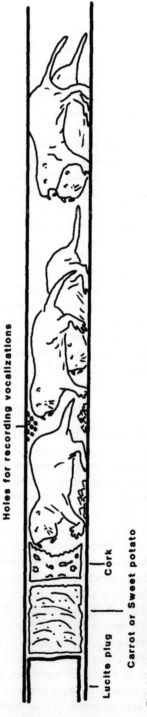

Fig. 9-1. Naked mole-rats gnawing their way through a plug of cork in the glass tunnel apparatus used to investigate the functions of the loud-chirp vocalization. A section of carrot or sweet potato is wedged behind the cork plug; a lucite plug securely blocks the end of the tunnel (drawing modified from Jarvis and Sale 1971).

individual it was facing, (2) the position of the vocalizing animal in the tunnel, (3) the number of loud chirps given, and (4) whether or not the chirping animal subsequently passed over the colony mate in front of it. Data were analyzed using nonparametric statistics in the MIDAS statistical package on the Michigan Computer Terminal System.

Results

Most activities in *Heterocephalus glaber* colonies were accompanied by frequent vocalizations, and one or more call types were audible virtually any time the mole-rats were awake and moving. Because most vocalizations were produced without opening the mouth or giving other obvious cues, the source of a call was often difficult to determine. Some of the softer vocalizations were rarely audible outside the tunnel system. Other sounds were very brief (< 200 milliseconds [ms]), and difficult to distinguish by ear without considerable practice. Although some of the sounds included ultrasonic frequencies, all had a substantial audible component; no purely ultrasonic calls were recorded.

The vast majority of vocalizations fell into 1 of 17 categories based on their physical structure. Of these types, 11 were primarily "tonal" (see figs. 9-2, 9-3, 9-5, 9-8, 9-9, 9-12 – 9-16). Tonal sounds consisted of a narrow, modulated frequency band within the range of 1-9 kHz. Most of these vocalizations also included one or more apparent harmonics: distinct frequency bands at whole-number multiples of the fundamental, or lowest, frequency. The remaining 6 vocalizations were primarily "atonal" (figs. 9-4, 9-6, 9-7, 9-10, 9-11, 9-17). Atonal vocalizations lacked distinct frequency bands, instead consisting of a single broad band of noise in the range of 0.2 to 40+ kHz. All 17 vocalization types are described below, grouped according to the contexts in which they were observed.

ROUTINE VOCALIZATIONS

This category includes sounds produced by adult naked mole-rats during everyday activities. Routine vocalizations constituted the background noise heard from colonies in the absence of juveniles, disturbances, and sexual or aggressive interactions.

SOFT CHIRP

The *soft chirp*, a quiet birdlike sound, was reminiscent of the peeping of chicks. A typical soft chirp consisted of a short upward sweep from 4 to 4.5 kHz, followed by a longer downward sweep from 4.5 to 2.5 kHz, with a total duration of about 130 ms (fig. 9-2). A single harmonic frequency was usually present. Soft chirps were produced both singly and in bouts of 1–2 chirps/s.

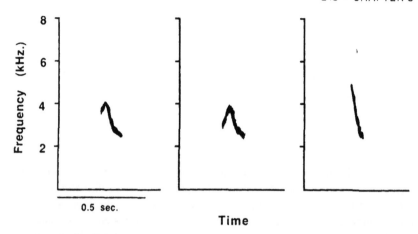

Fig. 9-2. Sonagrams of soft chirps The vertical axis represents sound frequency, the horizontal axis represents time, and the dark areas represent the relative energy of the sound *Left,* The most common form of the soft chirp; *center, right,* less common variants. Filter bandwidth, 117 Hz

The structure of this vocalization varied conspicuously in its duration, frequency range, and the extent of the upsweep portion; the upsweep ranged from nonexistent to nearly as long as the downsweep (fig. 9-2). All the soft chirps in one series, however, were similar in form.

Soft chirps were by far the most commonly heard vocalization (see fig. 9-18). In addition to accompanying routine activities, soft chirps were also given during times of arousal, such as sexual activity and human disturbance of the colony. They were often given by several animals concurrently, and a pair sometimes alternated chirps in a "duet." Soft chirps often appeared to be elicited by physical contact with a colony mate, and an investigator could usually elicit a soft chirp by touching a mole-rat. Kaufmann (1986) reported that in 17 of 20 observations of two animals bumping into one another in a tunnel, one or both gave soft chirps. He also found that 31 of 37 soft chirps were produced during physical contact with another individual.

TOILET CALL

To the unpracticed ear, the *toilet call* was not conspicuously different from the soft chirp. It consisted of whistlelike notes repeated at a rate of about 3/s, in bouts lasting 1–3 s. Each note consisted of two parts: a sharp upsweep followed by a less steep downsweep (fig. 9-3). Between the two segments, the note dropped to a low intensity or broke off entirely. A typical toilet-call note had a frequency range of 1.5 to 4 kHz and a duration of 200 ms. On some occasions a series of toilet calls graded into calls resembling soft or loud chirps (see below).

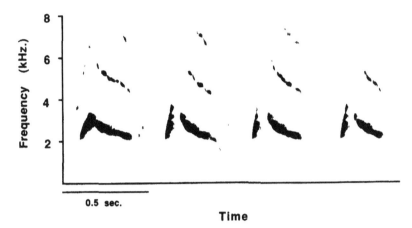

Fig 9-3. A sonagram of a series of toilet calls. These vocalizations were produced only by breeding animals when they were urinating. Filter bandwidth, 150 Hz.

The toilet call was produced by urinating mole-rats while they were in the characteristic urination posture (see Lacey et al., chap. 8). It was recorded only from breeding (i.e., reproductively active) animals of both sexes and only in the box being used by the colony as a toilet chamber. Although nonbreeders sometimes vocalized while urinating, these sounds did not have the distinctive form of the toilet call, being more similar to soft or loud chirps. Kaufmann (1986) also found that toilet calls were given only by breeding animals while urinating. Breeding mole-rats did not always give toilet calls when urinating in the toilet chamber; often they were silent or gave soft chirps.

TOOTH GRINDING

Tooth grinding is a mechanical sound rather than a true vocalization; it is included here because it was a conspicuous part of the noise produced by naked mole-rats, and because many rodents make similar sounds in connection with threat behavior (Eisenberg and Kleiman 1977). Tooth grinding is a scraping or chattering noise produced by rubbing the upper and lower incisors together. Sonagrams (fig. 9-4) reveal that both long and short strokes of the incisors produced a rapid series of discrete clicks, with most of their energy concentrated between 1 and 20 kHz. Short strokes had a typical duration of 20 ms and a repetition rate of 5/s, whereas long strokes lasted about 300 ms and were repeated twice per second. Series of long and short strokes often alternated in bouts that lasted up to several minutes.

Tooth grinding was typically exhibited by animals that were reclining in the nest or gnawing on the tunnel system, and it apparently functioned to sharpen the incisors (Lacey et al., chap. 8). Tooth grinding was also noted several times during fights both between and within colonies, but it was not clear whether

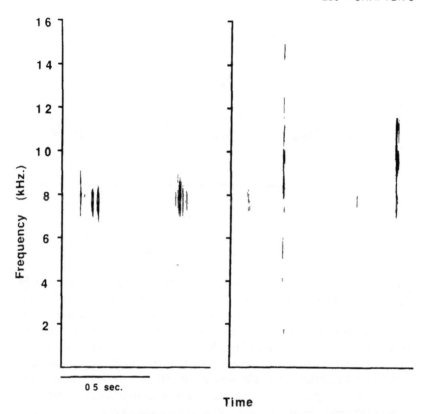

Fig. 9-4 Sonagrams of tooth grinding produced by two long (*left*) and three short (*right*) strokes of the bottom incisors against the top incisors Filter bandwidth, 600 Hz.

the sound functioned as a communication signal in this context or was merely a by-product of incisor honing.

MATING VOCALIZATIONS

Mating behavior is described by Lacey et al. (chap. 8) and by Jarvis (chap. 13). A single unique vocalization was associated with mating behavior. This vocalization, the V-*trill* consisted of a long trill with a distinctive twittering sound. It was composed of a rapid series of notes, each of which swept downward and then upward in frequency, producing a V shape on sonagrams (fig. 9-5). Each note lasted about 50 ms; it started at 2.5 to 6 kHz, dropped to 1.5 to 3.5 kHz, and then rose to near its starting frequency. Up to two harmonics were present. A typical V-trill contained four notes/s and lasted about 5 s, during which time the average pitch of the notes sometimes gradually shifted.

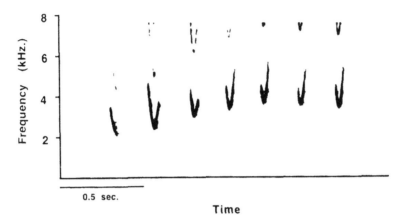

Fig 9-5 A sonagram of a portion of a V-trill It was made during sexual activity, particularly by a breeding female soliciting copulations. Filter bandwidth, 117 Hz

The V-trill was produced primarily by sexually receptive breeding females, especially while soliciting copulations. It has also been recorded while the breeding female was not nearby and may have been produced occasionally by breeding males.

DISTURBANCE OR ALARM VOCALIZATIONS

The following vocalizations were associated with alarm or distress and were typically heard after a colony had been disturbed in some way.

TAP

The *tap* was a soft, low-pitched, atonal sound somewhat like the noise made by tapping on a hard surface. Although common, the tap was easy to miss because of its low volume and its similarity to mechanical sounds made by the mole-rats as they moved through their tunnel system. Production of the tap, however, was associated with a visible movement of the caller's thorax. The tap had an abrupt onset, beginning with a very transient frequency spike to about 4 kHz (fig. 9-6). The rest of the sound, which lasted about 100 ms, fell within the range of 0.2 to 1.2 kHz. Taps were always produced in bouts at a rate of 2–4/s.

Taps were heard following mild disturbances, such as a relatively soft noise or the removal of a cover from the tunnel system, but not after more severe disturbances (e.g., banging the door). Taps were often produced simultaneously by several animals in the vicinity of the disturbance and were commonly associated with sniffing and crouch advances toward the site of the disturbance. Taps typically began 5–10 s after a disturbance and continued for sev-

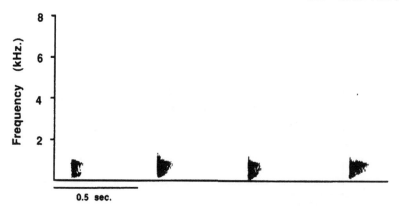

Fig. 9-6 A sonagram of a series of taps This sound was generally made after a mild disturbance of a naked mole-rat colony Filter bandwidth, 150 Hz

eral minutes thereafter. They were often associated with an unusual quietness in the colony (i.e., the absence of routine vocalizations, see above).

SNEEZE

The *sneeze* resembled a brief, explosive exhalation of air (somewhat reminiscent of a sneeze). Sonagrams of sneezes (fig. 9-7) resembled those of taps (fig. 9-6) but contained energy at higher frequencies (up to 5–16 kHz) and usually had a less abrupt onset. Unlike taps, sneezes were usually produced singly; although the two vocalizations graded into one another, they were usu-

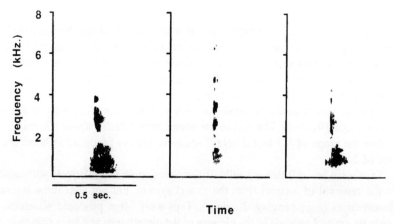

Time

Fig. 9-7. Sonagrams of several sneezes This sound was often made after a mild disturbance of a colony, or when a naked mole-rat encountered an unusual stimulus Filter bandwidth, 150 Hz

ally distinguishable by ear. A sneeze was sometimes interjected during a series of taps.

Sneezes were usually triggered by a disturbance, such as an unusual noise or an upsweep trill by a colony mate (see below). A mole-rat sometimes gave a sneeze when it had just encountered an unexpected stimulus and appeared to be startled. Nearby animals were often silent for several seconds after hearing a sneeze.

LOW-PITCHED CHIRP

Low-pitched chirps (fig. 9-8) were similar in form to soft chirps (fig. 9-2), but occurred in a lower frequency range (1–3 kHz vs. 2.5–4.5 kHz). Low-pitched chirps occasionally contained three to four frequency sweeps, rather than the single upward and downward sweep typical of soft chirps.

Low-pitched chirps were associated with strong arousal and were most often produced following a severe disturbance, such as a loud noise in the colony room or a strong vibration of the tunnel system. Low-pitched chirps were given while the mole-rats were running through their tunnels and were nearly always produced by several animals simultaneously. Breeding males and females also occasionally produced low-pitched chirps during sexual activity.

SCREAM

The *scream* varied in conspicuousness depending on its duration, which sometimes approached 1 s. Each scream consisted of two connected segments (fig. 9-9). The first segment typically swept from 2 kHz to 4 kHz and back to

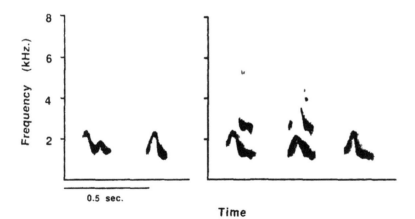

Fig 9-8. Sonagrams of low-pitched chirps. These sounds were typically heard after a severe disturbance while the mole-rats were scrambling through their tunnels. The low-pitched chirps shown here were recorded after a colony was severely disturbed by a loud noise. Filter bandwidth, 150 Hz

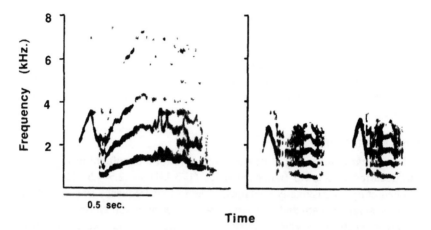

Fig 9-9. Sonagrams of screams produced by (*left*) an animal injured during fighting between two colonies and (*right*) an individual attacked and injured by its colony's breeding female. Filter bandwidth, 117 Hz

2 kHz over about 130 ms, and included up to two harmonics. The second segment contained a lower frequency band, which started at 1–1.5 kHz and rose slightly before dropping back to its starting frequency. The second segment was variable in duration, lasting up to 510 ms. It included four to seven harmonic frequencies, with atonal noise at its start and finish. This second segment usually had many small frequency modulations, giving it a tremulous quality (see fig. 9-9). Screams were repeated up to three times per second at irregular intervals.

Screams have been recorded from two mole-rats at Michigan and one at Cornell, in each case after the screamer had been attacked and severely wounded by conspecifics. In both cases at Michigan, the attacks were from colony mates, whereas the Cornell animal was injured during fighting between two colonies. Screams were heard only once from a mole-rat that was at that moment under attack; in all other cases, screams were produced after an attack had ended and the injured animal was huddled alone in a box away from the nest. Two of the mole-rats that screamed later died from their injuries, and the third recovered after being removed from the colony and housed in isolation.

AGONISTIC VOCALIZATIONS

The following vocalizations occurred in association with agonistic interactions between members of either the same or different colonies.

HISS

The *hiss* resembled a panting or huffing sound and was apparently produced by a rapid inhalation and exhalation of air. It was a broad-band, noisy vocaliza-

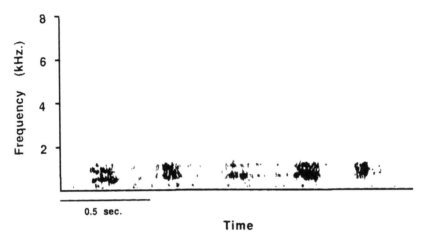

Fig 9-10. A sonagram of a series of five hisses produced by a breeding female shoving a colony mate. A hiss corresponds to one exhalation of air. Hisses were also produced in agonistic encounters with a second colony Filter bandwidth, 117 Hz.

tion that was often inaudible without the aid of a microphone located inside the tunnel system. The loud part of the hiss, apparently corresponding to an exhalation, lasted about 150 ms and contained most of its energy between 400 and 1,300 Hz (fig. 9-10). Hisses were produced in bouts of three to seven, at a rate of about three per second.

Hisses occurred in two contexts. The most common context was during shoving of colony mates by the breeding female (see Reeve and Sherman, chap. 11). While shoving a colony mate, the breeding female produced soft hisses that were often nearly inaudible from outside the tunnel system (see fig. 9-10). Although shoving and hissing were usually restricted to breeding females, a breeding male at Michigan was observed to shove and hiss at the breeding female several times while she was in estrus. A second context of hissing was observed during colony-defense trials at Cornell. Hisses were given by the mole-rats as they gaped at (threatened) snakes or unfamiliar conspecifics and also when they were presented with novel foreign objects, such as a cricket, a large beetle, and a mouse. Hisses given in these contexts were loud enough to be clearly audible, and on the basis of a single recording, appeared to include higher frequencies than the hisses that accompanied shoving. Kaufmann (1986) also reported hisses during group attacks on an ostracized colony member.

GRUNT

The *grunt* resembled a short, abrupt grunt, or if produced by several animals at the same time, a low guttural chuckling sound. It covered a frequency range from about 0.2 to 20 kHz or higher. Distinct frequency bands were present up to about 5 kHz, above which they blurred into broad-band noise

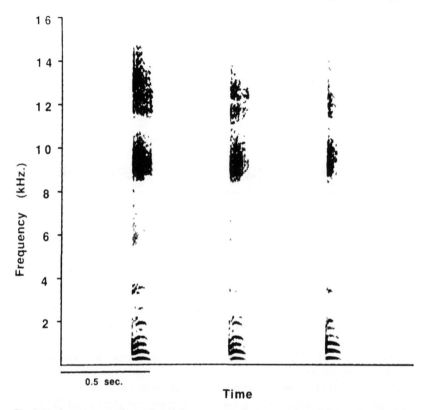

Fig. 9-11. A sonagram of a series of three grunts. Grunts were typically produced while threatening another mole-rat or other creature Filter bandwidth, 100 Hz

(fig. 9-11). Grunts had a duration of 40–100 ms, with a sharp onset and ending, and were given either singly or in bouts.

During colony-defense trials (Cornell), grunts were produced by mole-rats as they threatened a snake that had been introduced into the colony; they subsequently bit the snake. Grunts were also sometimes given by captive animals during handling and were often followed by attempts to bite. Brett (1986, chap. 4) observed in the field that when a mole-rat kicking dirt from a tunnel opening was caught by a snake, nearby colony mates grunted while sealing the tunnel off from the snake and the bitten individual. Mole-rats caught by hand while volcanoing (see Braude, chap. 6) also grunted as they were pulled from their tunnels. In the lab, Kaufmann (1986) reported hearing grunts during injurious attacks on conspecifics both within and between colonies.

UPSWEEP TRILL

The *upsweep trill* was a very conspicuous vocalization because of its high volume and rapid pitch changes. An upsweep trill consisted of a rapid series of

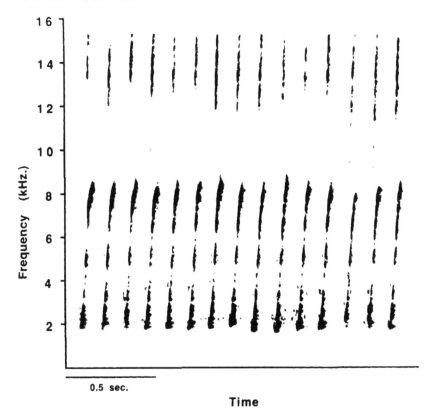

0.5 sec.

Time

Fig 9-12. A sonagram of part of an upsweep trill. Upsweep trills were produced by naked mole-rats that had just been involved in agonistic interactions with conspecifics This call was given by a breeding female just after she attacked and bit a (male) colony mate, it caused other colony members to approach the caller Filter bandwidth, 100 Hz

notes, with the fundamental frequency of each rising from about 1 to 9 kHz over a period of 80 ms (fig. 9-12). The notes were most often simple upsweeps, but they sometimes leveled off or dropped in frequency at the end. Up to 10 harmonic frequencies were present, and these sometimes extended to more than 80 kHz. A typical trill contained seven notes per second and lasted about 3 s, but some continued for as long as 6 s.

The upsweep trill has been observed almost exclusively in two contexts, both involving violently aggressive encounters with conspecifics. The first of these was during colony-defense trials in which two colonies were brought into contact, resulting in open-mouth gapes and biting between their members (see Lacey and Sherman, chap. 10). The identities of animals that gave upsweep trills (n = 76) were recorded in two different colonies, and in all cases upsweep trills were produced by nonbreeding mole-rats. Of the 76 calls, 68 (89%) were produced by individuals that actively threatened or bit members of the foreign colony during the trial, suggesting that upsweep trills are given by animals directly involved in conflicts between colonies. In 42 of these cases, the location of the caller was recorded. Of these, 37 (88%) were given either as the caller ran to the nest or immediately after it reached the nest; only 12% of the upsweep trills were given at the site of interactions between the two colonies. The response of colony mates to upsweep trills was dramatic: mole-rats rushed out of the nest, and the activity level of the colony increased markedly; many animals recruited to the site of the conflict.

In a second context, observed at Michigan, upsweep trills were given by breeding females after attacking nonbreeding colony mates. On two separate occasions, a breeding female attacked the same individual repeatedly over a period of several days. Immediately following each attack, the breeding female gave upsweep trills as she ran toward the communal nest, regardless of whether or not the bitten mole-rat attempted to fight back. One or more large males also attacked the same colony mate, but these males were never heard to give upsweep trills. In both cases, the victims of the attacks were males, and both were eventually killed. Colony mates responded to upsweep trills by markedly increasing their activity and rates of vocalizing (mostly soft chirps). They often rushed to the breeding female and pressed noses with her; at times the breeding female pressed noses with virtually every member of the colony after giving a series of upsweep trills.

Kaufmann (1986) also reported upsweep trills in both of the contexts described above, including four episodes in which male colony members were ostracized and killed. He found that during these within-colony conflicts, upsweep trills were given by breeding mole-rats of both sexes and were followed by colony members' approaching the caller and participating in group attacks on the victim. J.U.M. Jarvis (pers. comm.) has heard upsweep trills following disturbances of colonies that contained unweaned pups.

LOUD CHIRP

The *loud chirp* graded into the soft chirp, but intermediate forms were infrequent and most occurrences were readily classifiable. The loud chirp was louder and harsher sounding than the soft chirp and was always produced in rhythmic bouts. The fundamental frequency of the loud chirp (fig. 9-13) was similar to that of the soft chirp (fig. 9-2), but the former had a wider frequency range (ca. 1–6 kHz) and a longer duration (320 ms). The loud chirp contained 4 to 12 harmonic frequencies and sometimes included broad-band noise either as a short grunt at the onset or as noise overlaid on the downsweep portion. The second frequency band was the most intense in the loud chirp, unlike all other tonal vocalizations in which the lowest band had the greatest amplitude. Loud chirps were produced about twice per second in repeated bouts separated by pauses of 0.5–10 s; each bout contained an average of 17 or 18 chirps. The

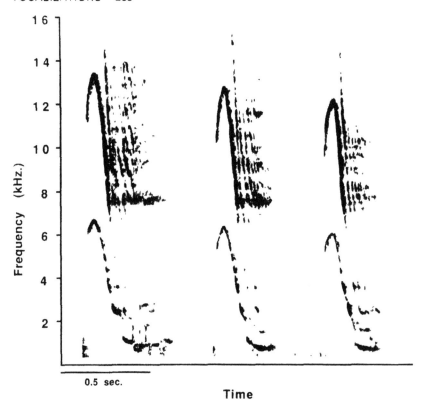

Fig. 9-13 A sonagram of three loud chirps that were part of a longer bout of loud chirping. This vocalization was produced by all colony members except the breeding female, most often during contests for access to food or digging sites. Filter bandwidth, 100 Hz.

caller's body usually jerked forward with each chirp, giving the impression that considerable effort was involved in its production. This jerking movement made the identity of the caller relatively easy to determine.

Loud chirps were produced during mild conflicts between colony mates, often involving competition for food or digging sites, and were usually directed toward a specific nearby individual. Loud chirps were most often heard when a mole-rat was touched by a colony mate while gnawing at a corner of a box, or when colony mates tried to eat from the same piece of food. Naked mole-rats also sometimes gave loud chirps when they were (1) blocked from moving through a tunnel, especially to reach food, (2) shoved or bitten by a colony mate, or (3) targets of loud chirping. Bouts of loud chirping usually involved no physical contact and ended with one animal moving away or yielding its position. If both animals continued chirping, the "contest" sometimes escalated to open-mouth gapes and (very rarely) to biting. In scan samples of three colonies at Cornell, it was noted (by P.W.S.) that of 68 open-mouthed aggressive interactions between nonbreeding individuals, 62 (91%) were preceded by loud chirps. Although loud chirps were directed at breeding as well as nonbreeding animals, no breeding female was ever observed giving a loud chirp, and breeding males seemed to give loud chirps infrequently (not quantified). Loud chirps were never heard from attackers during either injurious aggression against colony mates or conflicts between colonies.

The causes and effects of loud chirps in one context were investigated further (by S.H.B.) using the dead-end tube apparatus depicted in figure 9-1. Inside the narrow tunnel, the mole-rats generally faced toward the end containing the cork plug and the food. Forward movement was possible only if an individual could displace the colony mate in front of it; for this to occur, the mole-rat in front had to squat down and back up. Instead, however, the digging or eating animal often shifted its body in such a way as to prevent the animal behind it from passing.

Loud chirps produced in this context were given both by mole-rats that were behind another animal and apparently attempting to displace the one in front and by the animal in front. In an identical apparatus at Cornell (see fig. 9-1), it was found (by P.W.S.) that loud chirps were given by the mole-rat at the digging face 51% of the time, and by an animal behind it 49% of the time ($n = 317$ loud chirps recorded in three colonies). Another study (by S.H.B.), however, focused on loud chirping by mole-rats that were behind the foremost animal and attempting to displace it.

In the course of the latter study, a total of 3,806 bouts of loud chirping comprising 67,157 individual chirps was recorded . A single bout contained up to 120 loud chirps (mean = 17.6), and as many as 100 bouts were produced before the chirping mole-rat displaced the animal in front of it. Of all the chirping bouts observed, 32% were immediately followed by the chirping mole-rat passing over the individual ahead of it. On 21 occasions (0.55% of bouts) the digging or eating mole-rat turned to face the animal behind it and both mole-rats loud chirped with their jaws wide open; on 9 of these occasions (0.24% of bouts) the open-mouthed gapes escalated to biting (which never resulted in visible wounds). All of the threats and biting occurred between the digging or eating mole-rat and the animal directly behind it. In contrast to nonbreeding animals, which always gave loud chirps before passing over another individual, breeding females were never heard to give loud chirps even though they passed over colony mates frequently. Nonbreeders gave loud chirps at breeding females that were eating at the end of the tunnel but seldom passed over the breeder.

Regardless of whether the mole-rat at the head of the tunnel was digging or eating, other individuals gave loud chirps more times before passing over it than they did before passing over mole-rats farther back in the tunnel

(Wilcoxon rank sum analyses: for the colony of 20 mole-rats, $p < 0.01$; for the colony of 22, $p < 0.05$). Animals also gave more loud chirps before passing over colony mates that were digging than before passing over individuals that were eating (Wilcoxon rank sum analysis: for the colony of 20 mole-rats, $p < 0.01$; for the colony of 22, $p < 0.05$). Whether the animal at the head of the tunnel was digging or eating did not affect the number of loud chirps given before an individual passed over colony mates farther back in the tunnel.

The number of loud chirps given before passing over another mole-rat depended on the size of both individuals involved. Larger mole-rats chirped significantly fewer times than smaller mole-rats before passing over colony mates (Spearman's rank correlation, $r_s = -0.587$, $p < 0.01$, colonies combined). Conversely, mole-rats gave more loud chirps before passing over large colony mates than they did before passing over small individuals (Spearman's rank correlation, $r_s = 0.476$, $p < 0.01$).

JUVENILE-SPECIFIC VOCALIZATIONS

The 12 vocalizations described above were characteristic of adult naked mole-rats and were not produced by unweaned pups. The first "adult" vocalization given by juveniles was the loud chirp, which was first recorded when pups ($n=14$ pups in two litters) were 27 days old (i.e., about the time they were completely weaned). Before 27 days, pups produced five vocalizations that were apparently unique to young animals; these sounds were generally less stereotyped and more structurally variable than adult vocalizations. The five juvenile-specific vocalizations were: the juvenile downsweep, the juvenile upsweep, the juvenile chevron, the juvenile squawk, and the beg-feces chirp.

JUVENILE DOWNSWEEP

The *juvenile downsweep* was the most common vocalization produced by pups. Although sometimes flat in pitch, this vocalization typically consisted of a downsweep in frequency from 4 to 3 kHz over about 80 ms (fig. 9-14). One to two harmonics were also present. The juvenile downsweep was most commonly heard in undisturbed situations, particularly when the breeding female was present in the nest with her pups.

JUVENILE UPSWEEP

The *juvenile upsweep* rose very sharply, typically from 3 to 8 kHz in about 90 ms, and included one harmonic (fig. 9-15). The juvenile upsweep was heard as pups in the nest were pushed by adults with a characteristic rapid jerking motion (see Lacey et al., chap. 8). The sound was produced just as the adult's muzzle struck the pup's body. Although the source of the sound was difficult to confirm, it was our impression that the upsweep was produced by the pup

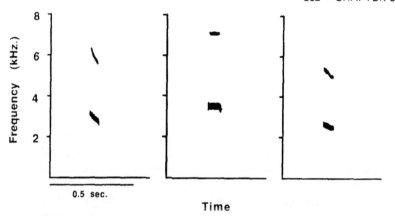

Fig. 9-14 Sonagrams of juvenile downsweeps produced by young pups when un-disturbed (e g , in the nest with their mother) Filter bandwidth, 150 Hz

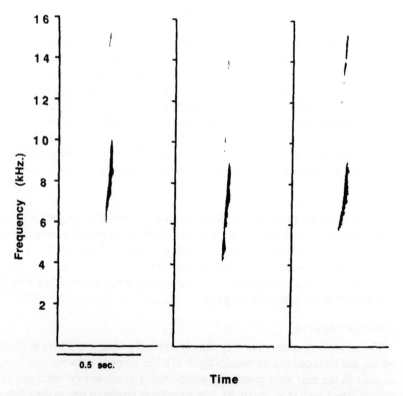

Fig 9-15. Sonagrams of juvenile upsweeps, which occurred when adults pushed pups violently with a characteristic vibration of the head; the sound coincided with the impact of the adult's muzzle on the pup's body. Filter bandwidth, 234 Hz

rather than the adult. Sounds very similar to juvenile upsweeps have also been recorded from juveniles that were removed from their colony.

JUVENILE CHEVRON

The *juvenile chevron* was symmetrical in shape, with the fundamental frequency rising from 1.2 to 1.5 kHz and dropping back to 1.2 kHz over a period of about 110 ms (fig. 9-16a). The call also contained multiple harmonic frequencies extending up to approximately 14 kHz, and sometimes also included atonal noise, particularly at the beginning and end of the call.

The juvenile chevron was produced by pups in the nest when they were being pushed by adults, but unlike the juvenile upsweep, the chevron was not given specifically at the moment of impact between the adult and the pup. The chevron was also occasionally heard when several pups were nursing together. As pups matured, the chevron increasingly came to resemble the loud chirp of adults (cf. fig. 9-16b with fig. 9-13). The jerking motion of the body associated with the loud chirp in adults was first observed when pups were 25 days old, the age when the chevron began to resemble the loud chirp. By 28 days of age, pups were producing clearly recognizable loud chirps (fig. 9-16b, right panel).

JUVENILE SQUAWK

The *juvenile squawk* was a harsh rasping sound revealed in sonagrams (fig. 9-17) to consist of multiple clicks, which covered a frequency range of 0.2–20 kHz and were too closely spaced to be discernible by ear. Although squawks were typically atonal (containing only broad-band noise), some also contained pure-tone segments that either preceded or followed the atonal portion without any break. The duration of squawks was variable, but averaged about 80 ms. Squawks were often produced by pups that were stepped on or pressed against the sides of the nest box by adult colony mates (plate 9-1) or by pups that had been removed from the tunnel system by an investigator.

BEG-FECES CHIRP

The *beg-feces chirp* was a distinctive mewing cry produced by juveniles begging for cecotrophes (nutritive feces) from adults (see Lacey et al., chap. 8; Jarvis, chap. 13). Although this call has been heard in captive colonies at Michigan, Cornell, and Cape Town (Jarvis, pers. comm.), it was largely restricted to the weaning period and was not recorded during this study. Beg-feces chirps were given by juveniles while nudging and tugging at an adult's ano-genital area before feces were produced by the adult. These vocalizations were given in bouts that lasted up to several minutes, at intervals of roughly once per second.

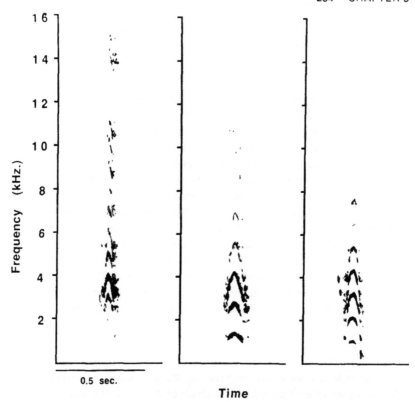

Time

Fig. 9-16a. Sonagrams of juvenile chevrons. The form of this vocalization changed with the age of the mole-rat. The chevrons above were produced by 2-day-old pups, those in fig. 16b (*opposite*) by 26-day-old pups (*left*) and 28-day-old pups (*right*) The chevrons above were given singly and with no body movement; those in fig 16b were produced in bouts and accompanied by jerking of the body. Notice the similarity between the latter call and the loud chirp (fig. 9-13) Filter bandwidth, 100 Hz

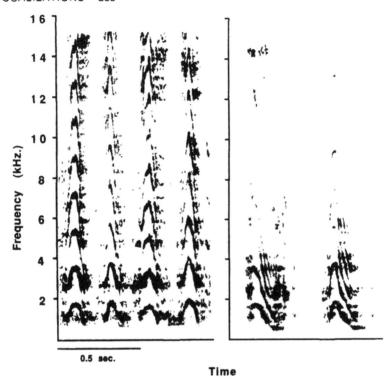

Fig. 9-16b. Chevrons made by (*left*) 26- and (*right*) 28-day-old pups See fig 9-16a.

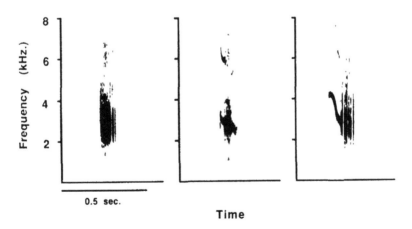

Fig. 9-17. Sonagrams of juvenile squawks, which were often produced by pups in distress. *Left*, A squawk made by a pup that had been temporarily removed from its colony; *center, right*, squawks from inside the nest box (i.e , when another mole-rat was lying on a pup; see plate 9-1). Filter bandwidth, 150 Hz

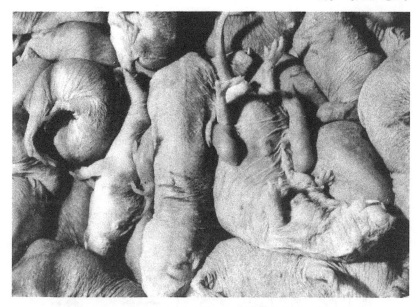

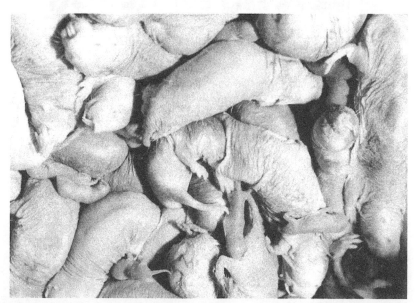

Plate 9-1. Two views of a naked mole-rat colony's communal nest, showing that pups (< 1-wk old) huddled in the midst of adults and, as a result, were often stepped on or lain upon, this often elicited squawks (see fig 9-17) from the pups Other juvenile-specific vocalizations described in this chapter also occurred in the nest Photos J U M Jarvis

Discussion

Vocalizations appear to be the primary mode of communication in naked mole-rats, although acoustic signals are certainly not their only means of communication. Visual signals are not feasible because naked mole-rats are fossorial and effectively blind (Jarvis and Bennett, chap. 3), but other communication modes may be important. Odor may play a key role in identifying colony mates (Jarvis, chap. 13, Appendix), and individuals may even possess unique odor signatures (B. Broll, pers. comm.); some evidence suggests that the reproductive status of the breeding female is also communicated via semiochemicals (Jarvis 1982, 1984, chap. 13; but see Faulkes et al., chap. 14). Sniffing and nuzzling, especially of the ano-gential region, as well as behaviors associated with urination, such as crotch dragging, scratching, and wallowing (see Lacey et al., chap. 8), are all probably involved in olfactory communication.

Tactile communication is also quite likely in naked mole-rats. There is evidence suggesting that shoving by the breeding female may regulate colony activity levels, as well as reinforce her reproductive dominance (Reeve and Sherman, chap. 11). Other behaviors such as nose pressing, sparring, nuzzling, skin tugging, head deflections, and biting also probably function in tactile communication. Vibrational communication (via foot thumping or drumming) has been observed in other species of mole-rats (Jarvis and Bennett, chap. 3), but never in *Heterocephalus glaber*.

Although olfactory and tactile communication certainly occur, the almost continuous stream of noise that emanates from active *H. glaber* colonies suggests that acoustic signals are the predominant mode of communication. Many of the 17 vocalizations described in this chapter were quite context-specific, and only 4 were commonly heard in undisturbed colonies (fig. 9-18). Given the limited range of stimuli present in the captive environment and the highly specific contexts of many vocalizations, it is possible that not every *H. glaber* vocalization has been described.

In comparison to other small rodents, however, the vocal repertoire of naked mole-rats clearly is unusually large. For example, the solitary subterranean mole-rat *Spalax ehrenbergi* reportedly has a repertoire of 6 call types (Capranica et al. 1974). In a survey of 17 species of hystricomorph rodents, Eisenberg (1974) found that the maximum acoustic repertoire size (including mechanical sounds) ranged from 11 to 14, with 9 to 10 of those being true vocalizations; sciurid rodents typically have even smaller vocal repertoires (e.g., Betts 1976; Owings and Leger 1980; Leger et al. 1984). Given the advantages of communication signals that can carry information rapidly over intervening distances and the limited utility of visual communication, it is not surprising that naked mole-rats have a large, complicated vocal repertoire. This is especially true considering that communication is often more highly developed in species

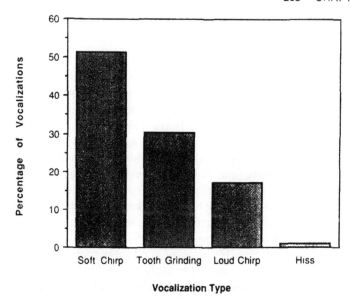

Fig 9-18 Histogram of the relative frequencies of the four vocalizations most commonly heard in undisturbed *Heterocephalus glaber* colonies Data are from a random sample of 382 sounds recorded from three Cornell colonies by Kaufmann (1986)

with complex social organizations (Otte 1974); for example, the highly social dwarf mongoose and the group-living guinea fowl have convergent, unusually complex vocal repertoires (Maier et al. 1983).

Aside from its size, another unusual feature of the vocal repertoire of *H. glaber* is the number of vocalizations that are produced only by particular subgroups of the colony. The subgroup with the largest exclusive set of calls is the juveniles, which give five vocalizations not heard from adults. Several other vocalizations are restricted to subgroups on the basis of reproductive status, either entirely or in particular contexts. Only breeding animals produce the V-trill and toilet call, and within the context of conflict within a colony, the hiss and upsweep trill were also given only by breeding animals. In contrast, the loud chirp has never been heard from a breeding female. This evidence of vocal specificity adds to the emerging picture of behavioral specialization among subgroups within naked mole-rat colonies (Jarvis 1981; Isil 1983; Lacey and Sherman, chap. 10; Jarvis et al., chap. 12; Faulkes et al., chap. 14). Further investigation into differences in vocal behavior corresponding to sex, age, body size, and frequency of performing colony maintenance and defense behaviors is clearly warranted.

A central question concerning any communication signal is that of its functional significance. Although this study does not establish the specific functions of *H. glaber* vocalizations, it does provide a basis for some initial hypotheses. Questions of cooperation and conflict are particularly interesting in naked mole-rats because of the functional integration and genetic homogeneity of their colonies (this volume; also Reeve et al. 1990). Thus our discussion of the hypothesized functions of mole-rat vocalizations is organized in terms of whether they are primarily cooperative or agonistic in nature.

One category of cooperative signals is those involved in courtship and mating. The V-trill is the only such vocalization in naked mole-rats; it is definitely given by the breeding female as a solicitation signal and may also be used by males during courtship. Interestingly, certain types of vocalizations that are typically associated with reproduction in hystricomorph (Eisenberg 1974) and sciurid rodents (Leger et al. 1984) and other mammals (Gould 1983), such as mate-attractant signals, separate male courtship and female solicitation sounds and male post-copulatory cries, are apparently absent from the naked mole-rat repertoire. One inference from this is that locating and competing for sexual partners is a less important problem for naked mole-rats than it is for other species; this is supported by the lack of overt competitive behavior surrounding estrus and copulations (see Lacey and Sherman, chap. 10; Jarvis, chap. 13).

Several different vocalizations appear to be involved in cooperative predator avoidance and colony defense. The tap and the sneeze are produced under relatively low arousal and may alert colony mates to investigate possible dangers or damage to the tunnel system. The low-pitched chirp is given by animals that are in rapid retreat from a perceived danger. This call is more easily triggered by lower-frequency sounds, and in nature, it might help to coordinate escape from digging predators, snakes, or tunnel cave-ins. These vocalizations may contribute to the efficiency of naked mole-rats at locating and blocking off tunnel sections that are disturbed by would-be collectors (Hill et al. 1957; Brett 1986, chap. 4; Jarvis, Appendix). Other agonistic vocalizations that seem to be aimed at specific targets of aggression (e.g., hisses and grunts) may also direct the attention of colony mates to potential dangers (e.g., see Brett, chap. 4). The number of different vocalizations involved in predator avoidance and colony defense suggests that coordinated responses to predators play an important role in the social biology of naked mole-rats (see Alexander, chap. 15).

Another vocalization associated with colony defense, the upsweep trill, is particularly intriguing because it was restricted to conflicts with conspecifics. Naked mole-rats that threatened or bit members of a foreign colony gave upsweep trills only after leaving the area and approaching their colony mates in the nest. Furthermore, although opponents from foreign colonies did not react noticeably to upsweep trills, previously uninvolved colony mates responded strongly, becoming active and approaching the caller. Taken together, these observations suggest that the upsweep trill may function as a recruitment call, rousing colony mates and prompting them to join in repelling conspecific intruders. In light of Lacey and Sherman's (chap. 10) finding that the first mole-rats to encounter members of a foreign colony were not always the same individuals that engaged them in conflict, it would be interesting to know if frequent colony defenders (i.e., large nonbreeders) show a different response to upsweep trills than do other size or age classes in the same colony.

Besides occurring during the defense of a colony against intruders, upsweep trills also occurred during intense conflicts within colonies, which probably revolved around reproductive status (Alexander et al., chap. 1; Lacey and Sherman, chap. 10; Jarvis, chap. 13). As in conflicts between colonies, targets of within-colony attacks showed little response to upsweep trills, whereas other colony members reacted strongly. Interestingly, the attacker (usually the breeding female) gave upsweep trills even if her target had not reacted aggressively to the attack or had fled; the victim was then sometimes attacked by other mole-rats as well. In this context, upsweep trills seemed to be involved in ostracizing a specific individual, almost as if emphatically labeling it as an enemy of the colony, and thereby recruiting others to join in attacking it. Such conflicts have lasted as long as several weeks, and in the absence of human intervention, the targeted individual was invariably killed. This type of ostracism with associated upsweep trills has been observed in captive colonies at Cornell, Michigan, and Cape Town (Jarvis, pers. comm., Appendix).

Aside from agonistic recruitment calls in several macaques (Gouzoules and Gouzoules 1989), baboons (Cheney 1977), and the dwarf mongoose (Rasa 1987), there have been few reports of vocalizations that function exclusively to recruit aid during conflicts with conspecifics (Harcourt 1988). The apparent role of the upsweep trill as a specialized recruitment signal suggests that cooperative aggression against conspecifics is unusually important in *H. glaber*. If so, this represents a striking parallel with eusocial insects such as ants, which engage in "tournaments" and "wars" between colonies of the same species (Mabelis 1979; Hölldobler 1987). It should be noted, however, that the upsweep trill has never been heard in the field, and that the role of intraspecific conflict in wild naked mole-rats, both within and between colonies, is unknown.

Although some *H. glaber* vocalizations seem clearly cooperative or competitive, others are harder to categorize. Thus, it is difficult to assign a function to the most commonly heard vocalization in undisturbed naked mole-rat colonies, the soft chirp (fig. 9-18), because it is heard during most kinds of interactions. One possibility is that it is an example of a "contact" or "cohesion" call (see Dittus 1988), which informs colony mates of the caller's presence and movements. However, the soft chirp is typically given after two animals have already come into contact. The conspicuous variation in the form of the soft chirp, and its production almost automatically upon physical contact, raises the possibility that it identifies the caller as either a specific individual or a member of a relevant class of individuals (or both). Such a vocal signature could provide an effective recognition mechanism for nearly blind animals that share a common colony odor. Recently, Dittus (1988) has suggested that the highly variable cohesion calls of certain macaques function in both group and individual recognition. Patterns of variation in the structure of soft chirps would seem well worth investigating in light of this hypothesis.

The functional significance of the toilet call is also somewhat obscure. The role that semiochemicals in the urine of the breeding female play in the integration of naked mole-rat colonies is currently under investigation in several labs (e.g., Jarvis 1981, 1982, 1984; Faulkes et al., chap. 14; B. Broll, pers. comm.). These chemicals may function to limit subordinates' reproduction or to define a unique colony odor (as do chemicals emanating from the queen in some ants; Carlin and Hölldobler 1986, 1987). In any case, it seems plausible that the toilet call, which only accompanies urination by breeding animals (especially the reproductive female), draws attention to and reinforces any chemical signal being broadcast.

Some *H. glaber* behaviors that appear overtly aggressive may contain an element of cooperation in the sense that they promote the functioning of the colony as an integrated whole. Shoving of colony mates by the breeding female, for example, is probably involved in both the maintenance of reproductive dominance and the stimulation of colony activity (Reeve and Sherman, chap. 11; Alexander, chap. 15). The hiss that normally accompanies shoves probably reinforces these messages.

Other *H. glaber* vocalizations seem to be more purely aggressive in nature. Nonbreeding mole-rats produced hisses while fighting with conspecifics and also upon encountering animals of other species within their tunnel system. Grunts were also produced in aggressive contexts and often seemed to indicate a tendency toward attack or a high level of arousal. If a caller was very excited, for example, it sometimes added a short grunt just before each note of an upsweep trill or a bout of loud chirps. Sounds much like the hiss and the grunt are also produced by various other hystricomorph rodents in aggressive contexts (Eisenberg 1974).

The most striking aggressive vocalization among naked mole-rats is the loud chirp, both because it is frequent and conspicuous and because it is apparently restricted to within-colony interactions. Soft and loud chirps seem to lie on a continuum of increasing aggression, and there is some suggestion that both may carry information about the caller. During the cork-digging trials, a mole-rat's response to loud chirping from behind depended on the size of the chirper, implying that information about the caller's size was conveyed either through qualities of the chirp that correspond to size or through individual voice recognition.

The hypothesis that the loud chirp is an aggressive vocalization was strongly supported. It was usually produced when a mole-rat was either thwarted by a colony mate from approaching a desired position or object or was approached by a colony mate while in possession of a desired object. It was one of the loudest *H. glaber* vocalizations, apparently requiring considerable effort to produce, and was repeated in long bouts, consistent with the idea that its function is to impel, not merely to inform. Loud chirpers that attempted to displace another animal were resisted, often with loud chirps, and loud chirping sometimes escalated to more direct forms of aggression. Indeed, open-mouthed gapes were virtually always preceded by and accompanied by loud chirping.

During the digging trials, loud chirps appeared to result from competition for the opportunity to dig or eat. Mole-rats had to give loud chirps more times before being allowed to displace digging or eating colony mates than before displacing animals that were simply standing in tunnels; furthermore, all of the open-mouthed gapes and biting involved the mole-rat that was digging or eating. These observations suggest that access to a resource was being contested. The size of both the vocalizing and the responding individual influenced the response to loud chirps, further supporting the idea that potential for escalated aggression is part of the message of this vocalization.

The frequency of loud chirping during the cork-digging trials (e.g., fig. 9-1) was extremely high. This supports the association of loud chirps with digging and eating that had been noted anecdotally both in captivity (e.g., Lacey and Sherman, chap. 10) and in the field (Hill et al. 1957). It is also consistent with field observations that feeding is often accompanied by loud "squeaking" vocalizations (Brett 1986, chap. 5). These field observations suggest that the competition observed in the lab was not merely an artifact of captivity. Loud chirps were produced more often during conflicts over digging sites than over food, possibly because of restricted opportunities to dig; for example, Jarvis et al. (chap. 12) found that competition over digging sites disappeared when soil was made continuously available (but see Lacey and Sherman, chap. 10).

Although it was clearly associated with aggression, the loud chirp was not heard during all types of conflict. Loud chirps were directed only toward colony mates and never at potential predators or members of foreign colonies. Moreover, although serious and often fatal conflicts occurred within colonies (see also Lacey and Sherman, chap. 10; Jarvis, chap. 13; Faulkes et al., chap. 14), they were never accompanied by loud chirps. Instead, the loud chirp appeared to be associated with a distinct class of conflict that (1) occurred only within colonies, (2) included only two individuals, and (3) involved primarily nonbreeders. This type of conflict, which did not result in injuries even on the rare occasions when it escalated to physical contact, stands in sharp contrast to fights accompanied by upsweep trills, which often involved multiple attackers and inevitably resulted in serious injuries or death. Even the limited form of conflict associated with loud chirps, however, indicates that colony mates (including nonbreeding individuals) did not share identical interests despite their intensely cooperative behavior and close genetic relatedness (see also Honeycutt et al., chap. 7; Reeve and Sherman, chap. 11; Alexander, chap. 15; Reeve et al. 1990).

The functions of juvenile vocalizations are even less clear than those of adults, both because they fall into less discrete categories, and because fewer samples were recorded. The juvenile downsweep, the squawk, and the beg-feces chirp all appear to be involved in soliciting care from adults. The young of many rodent species give a stereotyped call when displaced from the nest or otherwise distressed (Eisenberg and Kleiman 1977); the juvenile squawk probably fills this role in naked mole-rats. The beg-feces chirp involves solicitation of cecotrophes from nonbreeders (Jarvis, chap. 13), and the downsweep may solicit nursing from the breeding female.

Some juvenile vocalizations apparently represent early developmental stages of adult calls. The juvenile chevron, for example, seemed to play a role similar to that of the loud chirp among adults and later developed into the loud chirp (see fig. 9-16). The juvenile downsweep may be related to the soft chirp in the same way, although this relationship was less clear. We suspect that the large number of vocalizations specific to *H. glaber* juveniles and their structural variability are both reflections of an elaborate developmental process in this highly vocal species.

Because the present study was restricted to captive animals, some of the results may not be completely representative of wild colonies. However, the basic findings concerning the large repertoire size, the types of vocalizations produced, and the associations of call types with specific contexts probably do reflect the behavior of *H. glaber* in nature. Although the technical problems are somewhat daunting, field studies are essential both to verify the specific contexts associated with vocalization types and to clarify their adaptive significance. Lab studies, including playback investigations of the effects of various vocalizations on the behaviors of colony mates, will also help to test hypotheses about functions. Other promising avenues for future research include documenting the extent to which vocalizations identify individuals or classes of individuals, the ontogeny of vocal behavior, and the relationship among vocal, chemical, and tactile signals. The diversity of social structures found within the rodent family Bathyergidae, which includes species ranging from solitary to eusocial (see Jarvis and Bennett, chap. 3), also presents opportunities for comparative investigations of the relationship between communication modalities and complexity and social organization.

Summary

This chapter describes the vocal behavior of naked mole-rats, as observed in captive colonies at the University of Michigan and Cornell University. Vocalizations were monitored by ear or through microphones, and their associated behavioral contexts were noted. Selected vocalizations were tape recorded and analyzed using real-time or printed sonagrams and classified into call types based on their physical characteristics. Vocal behavior was studied under both "naturalistic" lab conditions and in manipulated situations in which the mole-rats were exposed to unfamiliar colonies of conspecifics, novel stimuli, or potential predators. Colonies were also provided with limited access to digging sites and food to study competitive behaviors and their associated vocalizations.

Naked mole-rats were found to be highly vocal, both in terms of their frequency of vocalization and their repertoire size. A total of 17 different sounds were characterized. Most of these were strongly associated with particular contexts and behaviors, including many contexts not associated with vocalizations in other rodent species (e.g., urination, begging for feces, and agonistic recruitment). The production of several call types was also restricted to mole-rats of particular age and reproductive classes.

Vocalizations were involved in behaviors that ranged from highly cooperative to fiercely aggressive. Vocalizations that were apparently cooperative were associated with danger, alarm, and recruitment of aid in conflicts. Other calls seemed to be involved in colony integration, both by informing colony mates of an individual's presence and identity and by reinforcing behavioral dominance and reproductive suppression. One agonistic vocalization, the loud chirp, invariably accompanied low-intensity conflicts (e.g., over food) that occurred frequently among nonbreeders within colonies. Finally, some vocalizations were associated primarily with violent, often fatal struggles that occurred both within and between colonies.

Communication is clearly a central topic in naked mole-rat biology, and this study is only a first step toward understanding the complex vocal behavior of these intriguing animals. We hope this chapter will provide a common frame of reference and stimulate further research into this fascinating subject.

Acknowledgments

We thank R. D. Alexander and J.U.M. Jarvis for loaning tape recordings of mole-rat vocalizations, and R. D. Alexander for many stimulating discussions. We are also grateful to P. Beddor for the use of the Kay sound-analysis work station. Useful comments on earlier drafts were provided by M. J. Ellis, R. D. Alexander, W. G. Holmes, J.U.M. Jarvis, W. C. Stebbins, and an anonymous reviewer. Work completed by J.W.P. was supported by a Regents Fellowship from the University of Michigan. Work completed by E.A.L. during 1984–1988 was supported by a National Science Foundation (NSF) Predoctoral Fellowship. Support for P.W.S. was provided by NSF grant BNS-8615842 and by Cornell University (U.S. Department of Agriculture Hatch Funds); P.W.S. thanks J. N. Davis and H. K. Reeve for other assistance.

10 Social Organization of Naked Mole-Rat Colonies: Evidence for Divisions of Labor

Eileen A. Lacey and Paul W. Sherman

Recent studies of naked mole-rats have emphasized the social behavior of this species. This interest in behavior is due to both pioneering behavioral and ecological studies by Jarvis (1978, 1981) and to widespread curiosity about the evolution of cooperative breeding and eusociality (e.g., Alexander 1974; Andersson 1984; Alexander et al., chap. 1). Whereas all 11 species in the 4 other genera of bathyergid rodents are solitary or live in small groups (Bennett and Jarvis 1988; Jarvis and Bennett, chap. 3), naked mole-rats are extremely colonial. Although colonies of 100 animals (Thomas 1903) and more than 295 animals (Brett 1986, chap. 4) have been unearthed, isolated individuals, pairs, or small groups (i.e., < 10 animals) are seldom seen.

Most interestingly, habitual coloniality in the naked mole-rat is associated with a remarkably complex social organization. Jarvis (1981) hypothesized that naked mole-rat colonies exhibit the three characteristics that define eusociality in insects (Wilson 1971): a reproductive division of labor, an overlap of generations, and cooperative care of the young. In support of her hypothesis, Jarvis pointed out three unique characteristics of captive *Heterocephalus glaber* colonies.

1. Mole-rats exhibit *a reproductive division of labor*, which Wilson defined as "more or less sterile individuals working on behalf of fecund individuals" (1971, p. 4). Jarvis reported that in her laboratory breeding was usually restricted to a single female per colony; the remaining females in the colony did not breed and had imperforate vaginas and quiescent ovaries (Kayanja and Jarvis 1971). Colonies captured in the wild also contained only one female with developed teats (Jarvis 1985; Brett, chap. 4), suggesting that restriction of breeding is not simply an artifact of captivity. Breeding is also restricted among males, but this restriction must be maintained by a different mechanism, since spermatogenesis occurs in the testes of over 75% of adult males in each colony (Jarvis 1981, chap. 13).

2. Mole-rats are organized into *a system of discrete castes*, which Wilson defined as "a particular morphological type, or age group, or both, that performs specialized labor in the colony" (1971, p. 462). Jarvis described two castes, "frequent workers" and "nonworkers," as well as a possible third

caste that she labeled "infrequent workers." Jarvis assigned animals to these castes based on the frequency with which they performed three tasks: nest building, digging, and transporting food and soil. Body weights also differed among castes: frequent workers were smaller animals whose body weights were thought to always remain less than those of the larger infrequent workers and nonworkers. Each caste contained both males and females, and Jarvis indicated that behavioral and morphological differentiation into all three castes could occur within a single litter of young.

3. The behavioral ontogeny of mole-rats is characterized by *age polyethism*, which means that "the same individual passes through different forms of specialization as it grows older" (Wilson 1971, p. 467). Jarvis indicated that all colony members became frequent workers soon after weaning (1–2 mo of age). Whereas slow growing animals permanently remained frequent workers, other faster growing individuals decreased their work efforts, eventually becoming nonworkers and potential replacements for breeding animals. The ontogeny of infrequent workers was not specified by Jarvis, but she suggested that members of this caste could also become breeders.

Prior to Jarvis's observations, eusociality was thought to occur only among certain insects: the termites (Isoptera), the ants, and numerous species of bees and wasps (Hymenoptera: Wilson 1971; Brockmann 1984). This suggested that something unique to insects (especially the Hymenoptera) was necessary for the evolution of eusociality and led to two decades of speculation regarding haplodiploidy as the genetic basis for this type of social system (e.g., W. D. Hamilton 1964, 1972; Trivers and Hare 1976). However, the discovery of an apparent convergence between the social system of *Heterocephalus glaber*, a diploid mammal (George 1979; Capanna and Merani 1980) and the structure of eusocial insect societies reinforces a growing conviction that hypotheses concerning the evolution of eusociality must consider common "extrinsic" (ecological) as well as "intrinsic" (genetic) selective forces (Evans 1977; Andersson 1984; Alexander et al., chap. 1).

Given the obvious importance of the naked mole-rat to hypotheses concerning the origins and maintenance of eusociality, we felt it was essential to replicate and expand Jarvis's (1981) observations. Here we present the results of our laboratory studies of three wild-caught colonies of *H. glaber*. In captivity, these colonies reproduced and, when offered appropriate stimuli, colony members displayed a variety of behaviors associated with care of the young, colony maintenance, and colony defense. For each behavior, variation among colony members was analyzed as a function of sex, body weight, reproductive status, and (whenever possible) age. These data were used to quantitatively examine the three phenomena reported by Jarvis (1981). Our results support the hypothesis that naked mole-rats are eusocial but also reveal intriguing complexities in colony organization not previously documented. These new findings allow comparisons to be drawn between the social structure of naked

mole-rat colonies and the societies of other highly social vertebrates and invertebrates.

Methods

STUDY ANIMALS

PROCUREMENT AND COLONY COMPOSITION

From May 1980 through July 1985 we studied the behavior of three captive
colonies of naked mole-rats. These colonies, designated K, A, and TT, were
captured in southeastern Kenya, in the vicinity of Mtito Andei (2° 41' S, 38°
10' E), between Tsavo East and Tsavo West National Parks. Colony A, which
contained individuals from two separate wild-caught colonies, was captured
and brought to the United States in December 1979 (by Sherman and Alexander; see the Preface). Colony K, captured in June 1980 and containing individuals from two separate wild-caught colonies, and colony TT, captured in October 1977 and containing individuals from only one field colony, were shipped
to the United States in March 1981 by Jarvis. None of the three study colonies
was complete on capture (i.e., none contained a field-caught breeding female).
The composition and laboratory demography of the study colonies is presented
in table 10-1.

Every mole-rat in our colonies was permanently marked at capture by clipping a unique combination of toes; no more than two toes per foot were
clipped. To facilitate recognition of individuals during behavioral observations, each animal was tattooed with a unique combination of letters (males) or
numbers (females) on its shoulders and back. Tattoos were done in black ink
using a hand-held, battery-powered machine; tattoos typically lasted 3–4 yr
before needing retouching. The sex of each animal was determined by examining its external genitalia. Because the genitalia of males and females are nearly
identical, sex assignments were checked by autopsy whenever an animal died
(see Jarvis, Appendix).

All mole-rats were weighed at capture on a spring balance (with an accuracy
of ± 0.5 g). In the laboratory, animals were weighed on an electronic balance
(± 0.2 g) every 3–4 mo, for a total of 15–18 weighings per individual during
this study. Weighings usually took place just before or just after a series of
behavioral experiments was conducted.

HOUSING AND MAINTENANCE

Each colony was housed in its own artificial tunnel system. Each tunnel
system comprised 20–25 m of 4.5-cm (inside diameter) clear acrylic plastic
(plexiglass) tubing, with four clear acrylic plastic (plexiglass) boxes (20 ×
20 × 12 cm) used as nesting, food, and toilet chambers (see plate 10-1). Before

Table10-1

A Summary of the Composition and Laboratory Demography of the Three *Heterocephalus glaber* Colonies Studied

Colony Capture Date	Colony Size				Births			No. Pups Surviving*		Deaths	Comments
	Initial		Final				No. in				
	F	M	F	M	Litter	Date	Litter	F	M		
Colony K 8–12 Jun 80	8	16†	10	13	A	8 Jun 81	6	3	3	4 females, 5 males, from wounds received fighting	Breeding female, 1 breeding male, 2 nonbreeders removed from colony 1 Dec 82
					B	14 Oct 81	7	2	2		
					C	5 Jan 82	10	0			
					D	28 Mar 82	7	0			
					E	20 Jun 82	11	0			
					F	16 Sep 82	8	0	0	1 male, cause unknown	1 male, very emaciated, unable to walk; excluded from data analyses
					G	8 Dec 82	13	0	0		
					H	28 Oct 83‡	5	0	0		
					I	22 Mar 84	10	2	2		
					J	20 Oct 84	10	0	0		
Colony A 5 Dec 79	15	10§	15	12	A	29 Apr 80	4	2	2	1 female, cause unknown	Breeding female removed 5 Feb 83
					B	18 Feb 81	8	0	0		
					C	9 Apr 81"	6	0	0		
					D	11 May 81	9	0	0		
					E	18 Sep 81	5	0	0		
					F	2 Dec 81	3	0	0		
					G	20 Feb 82	9	0	0		
					H	14 May 82	5	0	0		

Colony	Date					Notes
Colony TT	24 Oct 77	4	9	18	21	
A	30 Jun 81	6	0			1 breeding male removed 5 Feb 83
B	21 Nov 81	9	1	3		1 female, 1 male from wounds received fighting
C	30 Mar 82	5	0			
D	26 Oct 82	5	0			
E	29 Mar 83	12	6	5		Hind legs of 1 male paralyzed; excluded from data analyses
F	22 Sep 83	10	6	4		
G	22 Jan 84	8	1	3		
H	24 May 85	4	0			

NOTE. Each colony was captured near Mtito Andei, Kenya. Data on the initial size, sex ratio, and number of births and deaths in each colony are presented for the period from capture to August 1985.

* Number of pups surviving past weaning (age 4 wk).
† Athi Colony (2F, 5M) and Ithumba Colony (6F, 11M).
‡ First litter born to new breeding female.
§ Two females added from different colony.
" Born to female other than usual breeding female.

Plate 10-1. Acrylic plastic (plexiglass) tunnel systems used to house and observe naked mole-rat colonies at Cornell University; a nest box is in the foreground. Photo: P W Sherman

September 1, 1982, all three colonies were housed in a single room at the Langmuir Laboratory of Cornell University. Thereafter, they were housed in two separate rooms in Mudd Hall at Cornell, with colony K in one room and colonies A and TT in the other. To simulate the animals' natural environment, all rooms housing the colonies were maintained at 26°–28°C and 50%–65% relative humidity. No overhead lighting was used; all illumination came from strategically placed goose-necked lamps (three to five per colony) fitted with 25-W red bulbs (these lamps also provided localized sources of heat). A more complete description of housing and husbandry of lab colonies is provided by Jarvis (Appendix).

DATA COLLECTION

GENERAL PROCEDURES

Behavioral data were gathered primarily from January 1981 to August 1985. Most observations were made by the authors, with periodic assistance from various undergraduate students. An average of 2–3 h of observations were completed daily, generally between 8:00 A.M. and 8:00 P.M. Individual observation periods were usually 1 h but varied according to the type of data being collected. Most behavioral data were recorded on prepared check sheets; ad libitum observations of unexpected events (e.g., copulations, parturition) were recorded as narratives.

During behavioral observations, animals were identified using a pen-sized flashlight; this did not appear to disrupt colony members' behavior. However, the animals were disturbed by vibrations, sudden movements, or human voices. Thus, we did not begin data collection until at least 10–15 min after (stealthily) entering the observation room or until activity in the focal colony had returned to normal. During observations, our movements were kept to a minimum, and communication between observers consisted primarily of hand signals. If for any reason the animals were disturbed during an observation period, data collection was halted; depending on the severity of the disturbance, observations were either resumed after 10–15 min or suspended for several hours.

Throughout most of this study, colony K contained both the greatest total number of animals ($n = 34$) and the largest number of laboratory-born pups ($n=10$; see table 10-1); we initially focused our attention on this colony because of its size and regular pattern of reproduction. However, our emphasis shifted to colony TT once it began reproducing regularly in March 1983. As a result of this change in emphasis, maintenance activities (below), studied in 1981–1984, were most thoroughly documented in colony K, whereas defense behaviors, studied in 1985, were best documented in colony TT. No pups were successfully reared by colony A during the course of this study. Since we were concerned that a lack of recruitment might affect colony organization, we focused our research efforts on colonies K and TT; as a result, behavioral data are least complete for colony A.

REPRODUCTIVE STATUS

Colony members were categorized as breeders or nonbreeders on the basis of observations of copulations (males and females) and records of pregnancies and births (females). The single breeding female in colonies K and TT each mated with more than one male. We attempted to determine the paternity of pups in these colonies using starch-gel electrophoresis (e.g., Foltz and Hoogland 1981; Hanken and Sherman 1981; Sherman 1989) and DNA fingerprinting (Reeve et al. 1990), but low levels of genetic variability made paternity-exclusion analyses impossible (e.g., of 34 loci surveyed, none were polymorphic; Honeycutt et al., chap. 7). Thus, we do not know how many of the males we observed mating actually sired young.

CARE OF PUPS

We observed the care given to 35 pups in four litters: colony K's litter F, and colony TT's litters D, E, and F (table 10-1). All 13 pups in the first two of these litters died of unknown causes before weaning; data were gathered from birth to the death of the last remaining pup (ca. 4 wk after birth). In the other two litters, 21 of 22 pups survived beyond weaning; data on these litters were gathered from birth to weaning (at age 4–5 wk). Behavioral data were gathered in hour-long samples, with one to three samples completed daily. During each sample, all occurrences of grooming, nudging/handling, and pushing pups were recorded (these behaviors are described in Lacey et al., chap. 8), as were the identities of the individuals performing these behaviors. For data collection, a push consisted of one uninterrupted bout of pushing, separated from other bouts by a pause during which an individual often gently nudged or handled the pup; a "groom" consisted of one uninterrupted bout of grooming, separated from other bouts by a pause during which the grooming individual did not touch the pup.

REMOVAL EXPERIMENTS

To observe the process by which nonbreeders established themselves as reproductives, we simulated the death of several breeders by permanently removing them from their colonies. On December 1, 1982, the breeding female, a breeding male, and two small (<18 g) nonbreeding animals (one male, one female) were removed from colony K and placed in a separate burrow system. On February 5, 1983, the dominant female in colony A was removed (she had not bred during the previous 9 mo; table 10-1) and placed in a separate burrow system with a breeding male removed from colony TT. Following these removals, aggressive and reproductive behaviors in each colony were recorded.

COLONY-MAINTENANCE BEHAVIORS

During ad libitum observations, we frequently observed behaviors that were not directly associated with reproduction or pup care but that seemed to have a common colony-maintenance function. These behaviors, described in chapter 8, included carrying food, carrying and sweeping nesting material, digging at dirt, and chewing on small irregularities or protrusions in the acrylic plastic tunnel system.

Four sets of experiments were conducted to increase the frequency and predictability of these behaviors. Food carrying, nesting-material transport, digging, and clearing the tunnels of obstructions were chosen for detailed study

Plate 10-2. A naked mole-rat carrying a piece of tuber, a primary food source, from the site where it was excised toward the colony's nest Data presented in this chapter show that the majority of food carrying is done by small colony members Photo. R A Mendez

because they were readily observable, quantifiable, repeatable, and likely to resemble colony maintenance activities in nature. For example, food carrying (plate 10-2) may approximate the collection and transport to the nest of bulbs, corms, and other small food items (in the field, nest chambers often contained fresh, partially eaten food remnants; see Brett, chap. 5). Carrying and sweeping nesting material may mimic the collection of materials used to construct the colony's nest (the bark of roots and tubers). Digging involves behaviors apparently associated with the excavation of tunnels, and chewing on obstructions (plate 10-3) probably resembles the way mole-rats clear their burrows of rootlets and other protruding objects (in nature, tunnels are smooth-walled and free of debris). Each colony member's participation in these behaviors was quantified as follows.

1. Food trials. For each trial, 20–30 similarly sized (ca. 1 cm^3) pieces of food were placed in a colony's food box. During trials in colony K, commercial mixed frozen vegetables were used; these were warmed to room temperature before being introduced to the animals. During trials in colonies A and TT, a 1:1 mix of thawed vegetable pieces and comparably sized pieces of fresh produce was used. Trials began as soon as food was introduced to a colony; every time an item was carried out of the food box, the identity of the carrier was recorded. Trials continued until either all food was removed, or 30 min

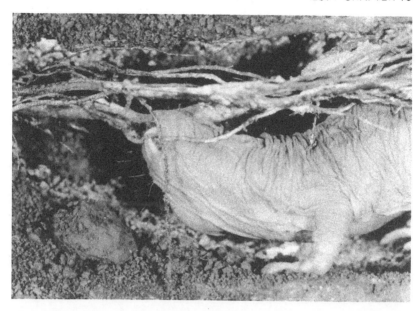

Plate 10-3 *Top,* A naked mole-rat gnawing at a rootlet that is protruding into the colony's burrow system, because of the consistent removal of such obstructions, tunnels in the field are smooth-walled and free of debris *Bottom,* A naked mole-rat carrying away a small rock that was obstructing its colony's burrow system Data presented in this chapter show that smaller individuals perform the majority of such maintenance tasks Photos R. A Mendez

had elapsed. Food trials were conducted 12–24 h after the animals had last been given fresh food; although a tuber was continuously available to each colony, no food trials were conducted if any fresh food items remained in a tunnel system.

For trials conducted in colony K, we also recorded the location to which food was carried and the fate of each item: food pieces were either eaten by the carrier or "provisioned" (i.e., dropped in the nest or less than 5 cm from it). These data were used to determine whether an item removed from the food box was primarily for an individual's own consumption or for consumption by others.

To determine if participation in food carrying differed with the effort required to obtain food, a series of experiments was conducted in colony K in which the distance between the main tunnel system and the food box was varied. Initially, 10 trials were conducted with the length of the connecting tunnel (L) equal to 0.6 m. Ten to 15 more trials were then conducted under each of the following conditions: $L = 1.2$ m, $L = 1.8$ m with the tunnel extending straight to the food box, and $L = 1.8$ m with an elevated ($20°$ angle, 10-cm high) C-shaped curve built into the connecting tunnel. The tunnel system was returned to its original configuration (i.e., $L = 0.6$ m), and a final series of 12 trials was completed. Trials conducted at $L = 0.6$ m were treated as controls, because this was also the distance between the food box and main tunnel system in colonies A and TT.

2. Nesting-material trials. For each trial, approximately 5 g of fresh pine shavings were introduced to a colony through the hole created by removing the clear acrylic plastic plate covering a selected tunnel joint (located 0.5 –1.0 m from the nearest nest box; see Jarvis, Appendix). Trials began as soon as the nesting material was introduced and continued until either all shavings had been removed from the introduction site or all carrying and sweeping had ceased (typically 10–15 min). The number of times that each animal carried pine shavings in its mouth or swept them with its hind feet was recorded (see Lacey et al., chap. 8). For data collection, a single sweep consisted of one uninterrupted bout of sweeping along a stretch of tunnel, regardless of the number of times the hind feet were kicked; simultaneous carrying and sweeping of nesting material was recorded as both a carry and a sweep. To standardize trials and to increase the accuracy of data collection, we recorded only responses to new nesting material (i.e., carrying or sweeping of shavings already present in the tunnel system was not counted).

3. Obstruction trials. For each trial, a 2–3 cm long wooden dowel (0.5 cm in diameter) was inserted through a hole 0.5 cm in diameter bored in a tunnel located 0.5 – 1.0 m from the nearest nest box. The dowel was positioned so that it partially obstructed but did not bisect the tunnel. Trials began as soon as the dowel was touched by a mole-rat; thereafter, the identities of animals chewing on the dowel were recorded every 30 s. Trials were continued until the obstruction was sheared off at the inside surface of the tunnel. For convenience, we sometimes conducted obstruction trials simultaneously with either food or nesting material trials; in these cases, a trial lasted until either the dowel was sheared off or the accompanying food or nesting material trial was completed.

4. Digging trials. For each trial, a 20–30 cm plug of fresh dirt was packed into the center of a clean tunnel (0.5 m long), which was then fitted into a convenient junction located 0.5 – 1.0 m from the nearest nest box. Data collection began when the dirt plug was first touched by a mole-rat and ended when the plug was demolished (typically 15–20 min later). The identities of all animals digging or gnawing at the dirt (see chap. 8) were recorded every 30 s, and the identities of all animals in the dirt-laden tunnel were recorded once per minute. At least 36 h were allowed between consecutive digging trials.

Digging trials were first conducted during June 1983, using a capped tunnel segment containing only dirt. A second series of digging trials was conducted during June 1985, in association with colony-defense trials (below); the protocol used was identical to that described above, except that mole-rats from separate colonies were allowed to dig toward each other from opposite sides of the dirt plug. A final series of digging trials was conducted following the completion of colony-defense trials; the protocol used during these control trials was identical to that used during the initial series of digging trials, with the addition that all occurrences of threats and bites (below) were recorded once the dirt plug had been demolished.

COLONY-DEFENSE BEHAVIORS

Quantitative data were also gathered on two behaviors associated with defense of a colony: attacking intruding mole-rats and attacking snakes. Again, these behaviors were chosen because they were readily quantifiable and repeatable, and because they probably resemble mole-rats' responses to naturally occurring threats. Encounters between colonies may simulate what occurs in the wild if one colony breaks into the tunnel network of another. Encounters with snakes were an attempt to mimic naturally occurring interactions with such predators (e.g., Brett 1986, chap. 4).

Each colony member's participation in these behaviors was quantified as follows.

1. Defense against conspecifics. At least 24 h before a defense trial, the tunnel systems of two colonies were positioned 1–2 m apart so that they could be joined by inserting a single connecting tunnel. The colonies had always been separate (i.e., they were not produced by artificially fissioning a larger colony).

Colony TT was joined six times to a colony comprising four mole-rats removed from colony K, and colony K was joined six times to a second small colony comprising two adults (from colonies A and TT) and their six offspring.

Trials were conducted as described under "Digging trials" (above), except that the tunnel containing the dirt plug was fitted into place so that mole-rats from the two test colonies could enter opposite ends of the tunnel and simultaneously dig toward each other. When the two colonies had demolished the 20–30 cm plug of earth that separated them, they invariably (12/12 trials) attacked each other. All occurrences of threats and bites (Lacey et al., chap. 8; plate 10-4) and the identities of the animals involved were recorded. Trials continued for 40 min following the initial contact between colonies. At the end of each trial, all mole-rats were returned to their colony of origin, and the connecting tunnel was removed. At least 36 h were allowed between successive trials involving a given pair of colonies. Participation in colony defense was analyzed for members of colonies K and TT only; defense behaviors by members of the two small colonies used were not analyzed.

2. Defense against predators. A 76-cm milk snake (*Lampropeltis triangulum*) was captured near Ithaca in June 1985. This snake served as the predator in all trials. To minimize the disturbance associated with introducing the snake, a 6–8 cm plug of dirt was packed into the middle of a section of tunnel 0.5 – 1.0 m long. The snake was then placed into one end of the tunnel, and that end was capped. A trial began when the open end of the snake-containing tunnel was quietly attached to either colony K ($n = 6$ trials) or TT ($n = 6$) so that the mole-rats were separated from the snake by the dirt plug. The mole-rats usually dug through the dirt plug within 15 min and, upon encountering the snake, began sniffing it. The mole-rats attacked the snake vigorously if it moved quickly or struck at them (see also Alexander, chap. 15). The identities of all individuals that threatened or bit the snake were recorded. Trials usually lasted 20 min, at which point the snake was removed; 3 of 12 trials were terminated early to prevent injury to the snake.

3. Facing-out behavior. A third behavior that we thought might be associated with colony defense was the tendency of colony members to face out of the nest box. At any given time, the majority of mole-rats in a colony were found lying in or near the communal nest. Whereas most of these animals lay with their heads and bodies entirely within the nest box, one to four individuals per colony lay with their heads and shoulders extending out of the nest and into an adjacent tunnel. These animals were not simply pressed into such positions by overcrowding in the nest; often they lay 1–3 cm from the nearest colony mate. Moreover, the head-out position was typical: individuals seldom lay with their hindquarters extending into a tunnel.

Plate 10-4 *Top,* Naked mole-rats from two different colonies threatening each other at close range after having demolished a plug of earth that separated them *Bottom,* Close-up of two colony defenders (from different colonies) biting each other Data presented in this chapter show that larger individuals are the primary colony defenders Photos R A Mendez

Because animals lying near nest entrances should be the first to encounter any approaching intruder, and because the facing-out posture appeared to be deliberately assumed, we hypothesized that individuals facing out of the nest might be acting as guards (see also Faulkes et al., chap. 14). To determine if mole-rats lying at nest entrances comprised a particular subset of individuals, we recorded the identities of these animals in colonies K and TT. Data were gathered by stealthily entering an observation room and identifying individuals lying with their heads and shoulders extending out of the nest. Such data were recorded five to six times during a 24-h period, with a minimum of 60 min between consecutive observations. As a preliminary test of the nest-guarding hypothesis, the identities of the animals most frequently seen facing out of the nest were compared with the identities of those animals most frequently observed threatening and biting either foreign conspecifics or snakes.

GROWTH DATA

Mean weight change per month was calculated for all animals during the period from capture (or birth) through September 1982 (before the experimental removal of any animals). The weights of laboratory-born animals were used to construct growth curves for males and females whose ages were known. The weights at capture of field-caught mole-rats (ages unknown) were then plotted on these curves. From these initial points, the growth curve for each wild-caught animal (while in captivity) was plotted, giving a curve for body weight as an approximate function of age. This procedure assumes that weights are similar for wild-caught and laboratory-born animals of the same age (but see Jarvis et al., chap. 12). Some mole-rats were heavier at capture than the maximum weights of our oldest laboratory-born animals, and thus growth curves for these individuals could not be constructed.

Data Analyses

Data for males and females in each colony were analyzed separately. Body weights used in these analyses were from the weighing completed closest to the time of data collection; weights used were never more than 1 mo out of date. Data from all experiments were analyzed by first tallying the number of times per trial that each individual participated in the focal behavior(s); for pup-care behaviors, each day's observation period was treated as a separate trial. Per-trial totals for food carrying, nesting material transport, and the various pup-care behaviors were summed, yielding the total number of times each animal performed each activity.

Because data from digging and obstruction trials were recorded as scan samples every 30 s and because the durations of these trials varied, the data were standardized by converting each per-trial total to a 30-min time scale. These standardized per-trial totals were then averaged across all trials, yielding a mean number of scans spent digging or chewing per 30 min for each animal. Similarly, because the duration of colony-defense trials varied (trials were sometimes terminated early to prevent injuries), per-trial totals for defense activities were standardized by converting to a 30-min time scale and then averaged for each individual. Data on facing-out behavior, recorded on an ad libitum basis, were simply summed, yielding the total number of times that each animal was seen facing out of the nest.

Preliminary scatter plots of individual effort (total or mean) versus body weight revealed that the behavior of breeders and nonbreeders differed considerably. Consequently, data points for breeders were excluded from statistical analyses (but not scatter plots) in order to focus these analyses on behavioral patterns among nonbreeders. F-tests (Sokal and Rohlf 1981, p. 219; Rohlf and Sokal 1969) were used to examine behavior differences among nonbreeders, as suggested by Brockmann and Dawkins (1979). For each behavior, the F-statistic was computed as the sum of the variances among all same-sex colony members' per-trial totals divided by the sum of the variances among each individual's per-trial totals; this is, in essence, a one-way ANOVA, with individuals used as "treatments." Before statistical analyses, per-trial totals for pup-care behaviors were standardized by converting to a 30-min time scale, and observations of facing out were divided into groups of 10, with each group considered a separate trial (such trials usually took 2–3 days to complete). For each behavior, individual totals for males and females were then compared using Mann-Whitney-U tests. Finally, the relationships between individual totals and body weight were analyzed using Spearman's rank correlation tests. All analyses were performed using the MINITAB statistical package on Cornell's IBM computer system.

To compare the results of food trials conducted at different tunnel lengths (colony K), data from each length were first normalized by converting the number of times that each colony member carried food to a proportion of the total number of food carries by all colony members of that sex. These proportions were not normally distributed; this was corrected with an arcsine transformation ($\theta = \sqrt{p}$; p = proportion of food carries at a given L; Sokal and Rohlf 1981, pp. 427–28). Linear regression analyses (transformed proportion of food carried vs. body weight) were then completed for each distance L. Regression coefficients for different tunnel lengths were compared using an F-test (Sokal and Rohlf 1981, p. 505); only coefficients for statistically significant regressions were compared.

It should be noted that because limited quantities of stimuli (e.g., food, nesting material, or dirt) were added in each of our trials, the amount of work that an individual could perform might have been limited by the behavior of colony mates. For example, if one animal carried 20 of the 30 pieces of food available during a given trial, the largest number of pieces that any other animal could carry was 10. We tried to minimize the potential effects of this problem by conducting multiple trials and by introducing copious quantities of food, nesting material, and dirt. Nonetheless, data from our trials do not strictly conform to the assumption of independence inherent in our statistical analyses.

Results

REPRODUCTIVE BEHAVIOR

BREEDING

In our laboratory, only one female per colony bred. The breeder was the only female in each colony to mate and to bear young; she was physically distinguished from other females by her large, elongate body and well-developed teats (table 10-2). Behaviorally, this female was aggressively dominant over other colony members, as indicated by the frequency with which the breeder shoved and threatened nonbreeders (see Reeve and Sherman, chap. 11; Jarvis, chap. 13). Three males in colony K, three males in colony TT, and one male in Colony A mated with the dominant female; these were among the heaviest males in their respective colonies (table 10-2). In addition to mating, breeding males were regularly observed nuzzling the breeding female's genitalia (even when she was not sexually receptive); this behavior was rarely observed among nonbreeding males or females in colonies containing a reproductive female (see also Jarvis, chap. 13; Faulkes et al., chap. 14).

Breeding females produced up to five litters per year (table 10-1). Matings usually occurred 7–10 days after the birth of a litter (mean = 7.9 ± 2.1 days; n = 8 litters); mating behavior is described in chap. 8 and 13. Breeding females exhibited behavioral estrus for 24 h or more; during estrus, females actively solicited copulations from the largest males in the colony. Upon encountering one of these breeding males, the female often nuzzled him, turned to present her ano-genital region, and assumed a lordosis-type posture; copulation usually followed immediately, and generally lasted less than 1 min (mean = 34 ± 14 s, n = 38). All mountings and apparently completed copulations that we observed occurred in the tunnels rather than in the nest. In both colonies containing more than one breeding male, copulatory frequency among males increased with increasing relative body weight.

Although the dominant female did not avoid nonbreeding males when she was sexually receptive, neither did she appear to solicit their attention. Nonbreeding males, in turn, did not exhibit any detectable interest in this female. We saw no evidence that breeding males guarded the female during her receptive period, and curiously, there was no aggression among males on the day when the female mated. Aggression among breeding males, including

Table 10-2

Comparisons of the Weights (in grams) of Breeding and Nonbreeding Animals in Colonies K and TT

Colony	Year	Breeding Female(s)	Nonbreeding Females				Breeding Males	Nonbreeding Males			
			$\bar{x}$	SD	Range	n		$\bar{x}$	SD	Range	n
K	1982	44 3	31	±13.3	14 7–56 3	14	59 8	33 5	±8 5	18 4–44 8	15
							52 7				
							45 7				
	1983	—	35 0	±8 0	28 5–50 2	11	45 8	34 2	±7 9	24 8–54 3	13
	1984	48 0	28.7	±2 9	26 4–34 7	7	49 3	30 5	±5 8	23 6–44 9	11
							42 2				
	1985	47 6	28 2	±5 2	19 0–35 1	9	52 6	28 2	±5 0	20 0–34 5	10
							45 4				
							45 0				
TT	1982	53 0	30 9	±12 3	16 2–43 7	5	46 9	31 5	±11 3	14 2–49 0	8
							52 1				
	1983	55 4	33 1	±7 7	23 4–44 4	5	47 4	35 7	±7 1	30 2–51 1	8
	1984	56 3	22 6	±3 0	13 9–40 3	18	45 3	24 5	±8 4	14 0–36 6	24
							49 6				
							39 5				
	1985	56 3	24 9	±6 3	17 4–39 8	19	46 2	26 6	±6 0	18 5–36 6	17
							46 8				
							40 3				

NOTE Data are from weighings completed each May–June from 1982 through 1985 The weights of each breeding animal are shown, data for nonbreeders are presented as the mean ± SD and range of weights for these animals (weights of lab-born young are not included)

threats and bites, was occasionally observed during the week preceding a female's receptive period.

Mean litter size was 7.7 ± 2.3 pups (n = 26 litters). The lengths of gestation for colony K's litters E, G (both litters born to female +), and H (born to female three-bars), were calculated from observations of copulations and subsequent parturition; these gestation periods were 74, 72, and 77 days, respectively. The gestation period for litter D in colony A was 75 days, and gestation for litter F in colony TT lasted 73 days (table 10-1).

Regardless of whether the breeding female was pregnant, colony members spent most of their time huddling in the nest (below); on the days immediately before parturition, this huddling appeared to be even more pronounced. During actual parturition, however, the breeding female frequently left the nest, and, as a result, pups were often born in the tunnels (30 of 42 pups in six litters, or 71%). At birth, pups were not immediately cared for by the breeding female. Instead, they were cleaned and transported to the nest by nonbreeders. Breeding males never assisted during parturition, or cleaned or carried any of the 42 pups whose births we witnessed. Depending on the size of the litter, parturition lasted 1–3 h, with pups born every 10–30 min. Further descriptions of parturition are given by Lacey et al. (chap. 8) and Jarvis (chap. 13).

CARE OF PUPS

Of the 26 litters born during this study (n = 195 pups), only 47 pups (24%) from eight litters (31%) survived to weaning; in only two of these cases did an entire litter survive (table 10-1). The care given to successful (i.e., surviving to weaning) litters differed markedly from that given to unsuccessful litters, such that it was often possible to predict within 24 h of birth if a litter would survive. For example, before weaning, pups belonging to successful litters were kept in the nest box and were quickly retrieved if they crawled or fell out of the nest and into the tunnels. In contrast, pups in unsuccessful litters were not retrieved, and, furthermore, were frequently dragged or kicked out of the nest by nonbreeders. We do not yet know why only some litters survived. Interestingly, the same dichotomy in pup-care behaviors and pup survival was observed by both Jarvis and Alexander in their laboratory colonies (see Lacey et al., chap. 8; Jarvis, chap. 13).

All mole-rat pups were nursed exclusively by the breeding female. In litters that were successfully reared, lactation lasted about 4–5 wk. Roughly 4 wk after parturition, the breeding female began preventing pups from nursing by rolling onto her belly or by moving away from the young. At the same time, pups began to eat the feces of other colony members and to nibble on solid food that had been brought into the nest. Pups actively solicited feces by chirping (Pepper et al., chap. 9) and nudging at the anal area of breeding males and nonbreeding males and females. Unfortunately, coprophagy was not observed often enough to permit an analysis of the relative frequencies with which various adults fed pups (for such an analysis, see Jarvis, chap. 13).

Breeders were the primary participants in all other pup-care behaviors in our colonies. Fifteen females were present in colony K during observations of litter F. Of these, the breeding female alone performed 58% of all pushes, 66% of all groomings, and 63% of all nudges/handles of pups (figs. 10-1–3; table 10-3). Of the 18 males in this colony, the 3 breeders (17%) together performed 80% of all pushes, 32% of all grooming bouts, and 47% of all nudges/handles by males, far more than expected if participation by males were random. Similarly, during observations of litters D, E, and F in colony TT, the breeding female (1 of 6 females in the colony) was responsible for most (31.2%–80.7%) pup-care behaviors (table 10-3). During observations of litter D in colony TT, the 2 breeding males (of 10 males in the colony) accounted for 52% of all pushes, 33% of all grooming bouts, and 49% of all nudges/handles by males. Of the 9 males in colony TT during observations of litter E, the single breeder performed 34%, 17%, and 36% of these behaviors, respectively. This male and 2 "replacement" breeding males (see below) accounted for 50% of all pushes, 39% of all grooming bouts, and 51% of all nudges/handles of pups in litter F

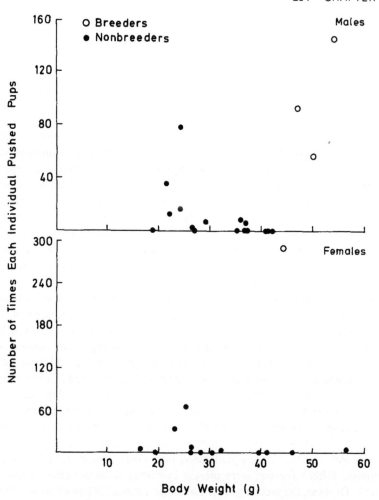

Fig 10-1 Total number of times that each mole-rat in colony K pushed pups (i e , rapidly shoved pups with its muzzle, this behavior is described in chap 8) as a function of body weight Data are from observations of a litter (F) born on September 16, 1982 (see table 10-1) A total of 867 pushes were recorded between the birth and death of all the pups in this litter (at ca 4 wk of age) Males (*top*) and females (*bottom*) are separated, and breeders are indicated Statistical analyses of these data and pushing behavior in colony TT are presented in table 10-5

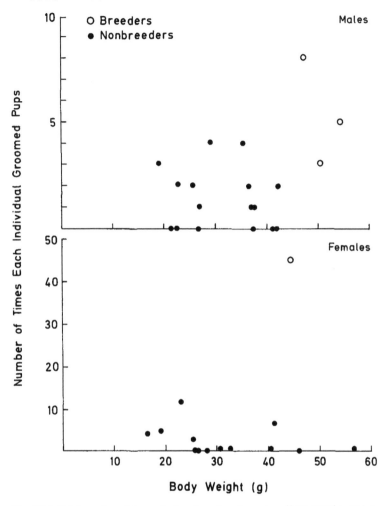

Fig. 10-2. Total number of times that each mole-rat in colony K groomed pups (i e , held pups with its forepaws and licked or nibbled at them with its incisors; described in chap. 8) as a function of body weight Data are from observations of litter F during September–October, 1982; a total of 118 grooming bouts were recorded Males (*top*) and females (*bottom*) are separated, and breeders are indicated, see table 10-5 for statistical analyses.

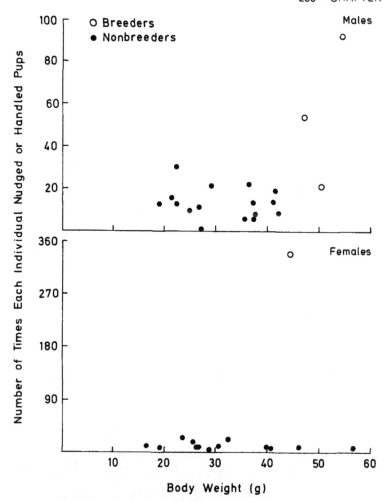

Fig. 10-3. Total number of times that each mole-rat in colony K nudged or handled pups (i e , nosed pups with its muzzle or manipulated them with its forepaws, described in chap 8) as a function of body weight Data are from observations of litter F during September–October, 1982; a total of 896 nudges and handles were recorded Males (*top*) and females (*bottom*) are separated, and breeders are indicated; see table 10-5 for statistical analyses

Table 10-3

Percentages of Pup-Care Behaviors Performed by Breeding Animals in Colonies K and TT

Colony/Litter		% Pushes		% Grooming Bouts		% Nudges/Handles	
Female	Male	Female	Males	Female	Males	Female	Males
Colony K							
F		57 9	80.2	66 2	32 0	63 4	46 6
1	3*						
15	18†						
Colony TT							
D		79.6	51 8	79.5	33 0	80.7	48.5
1	2*						
6	10†						
E		54 0	33.5	70 8	16.7	56 8	35 5
1	1*						
6	9†						
F		31.2	49 9	60 0	39.3	50 0	51 4
1	3*						
6	14†						

NOTE: See chapter 8 for a description of these behaviors Data from observations of one litter in Colony K and three litters in Colony TT are presented For each litter, the total number of times the focal behavior was performed by males and by females was determined, for males and females, the percentage of the total for each sex performed by its breeding members was then calculated The number of breeding and nonbreeding animals present in the colony at the time each litter was born is indicated

* Number of breeding females and males in the colony when each litter was born
† Total number of males and females in the colony when each litter was born

(14 males were present in the colony at that time). Among nonbreeding males and females in colony K, the number of pushes, grooming bouts, and nudges/handles by individuals were negatively correlated with body weight (i.e., the smaller mole-rats did more). In colony TT, these behaviors were also negatively correlated with weight for data from litters D and E but were positively correlated with weight for litter F (table 10-5).

REMOVAL EXPERIMENTS: COLONY K

Violent fighting occurred in colony K following the removal of the breeding female, one breeding male, and two nonbreeders on December 1, 1982. Fights were first observed on January 12, 1983, 42 days after the breeders were removed, and continued sporadically for 16 mo (until April 21, 1984). During this period, nine animals died as a result of fighting: four males and five females (sexes confirmed by autopsy). Fights occurred between pairs of mole-rats and consisted of clawing and biting at an opponent's head and shoulders. The same individuals would battle repeatedly over a period of several days to several weeks, often until one of the pair was severely injured. Death of the

Table 10-4
Summary of Aggression in Colony K Following the Removal of the
Breeding Female and One Breeding Male on December 1, 1982

Date	Individuals Fighting*		Colony of Origin[†]	Outcome
Jan 83	female 7	male K[‡]	same	male K dies
Feb 83	female 7	male C	different	male C dies
May 83	female 7	female 5	same	female 7 dies
Jul 83	female three-bars	male A	same[§]	both survive"
Jul 83	female three-bars	female 5	same[§]	female 5 dies
Nov 83	female 11	male A	same	female 11 dies
Apr 84	female three-bars	male T	same[§]	male T dies

NOTE Fights occurred between pairs of animals, for each pair, the identities of the combatants and outcome of the conflict are indicated The dates shown provide an approximate time scale for aggression in colony K
* Two males and one female also died as a result of injuries from fights Although the identities of their combatants are unknown, in each of these cases the breeding female was the only animal with blood on her muzzle
[†] Colony K contained 24 mole-rats from two wild-caught colonies (17 from one, 7 from the other)
[‡] Breeding male
[§] Males A and T and female 5 were collected from the same field colony as the mother of female three-bars
" Male A and female three-bars subsequently mated

injured animal generally followed within a few days, either as a direct result of injuries or hemorrhage or as a result of massive internal infections.

Both female-female and female-male pairs were seen fighting (table 10-4), but no male-male conflicts were observed. With one exception (male A/female 11; table 10-4), all fights involved the colony's current alpha (behaviorally dominant) female. Whereas fights between females were typically initiated by individuals attacking the alpha (23 of 28 cases = 82%), all 18 male-female conflicts observed were instigated by the dominant female. Only one of the extended conflicts that we witnessed was between colony members originally derived from different field-caught colonies (table 10-4); all other fights were between animals derived from the same field-caught colony, indicating that fights were not simply due to the mixed origin of colony K (see also Reeve and Sherman, chap. 11).

During the first 16 mo after breeders were removed, three females sequentially rose to dominance and developed enlarged teats, apparently in preparation for breeding. The first female in the sequence (7) was dominant from January to May 1983. Female 7 mated in January, but did not become pregnant; she was killed in May 1983 by a second female (5), who subsequently became dominant. No matings involving female 5 were observed, and she was killed in July 1983 by a third female (three-bars). Female three-bars was 1 of 6 pups born on June 8, 1981 (table 10-1), an offspring of the colony's original breeding female (whose removal initiated this sequence of killings). Three-bars remained dominant for the final 2 yr of this study. She mated with one male in July 1983 and subsequently bore a litter of 5 pups in October 1983; none of these young survived past weaning. In March 1984, she gave birth to a litter of 10 pups, the result of unseen matings which must have occurred in January 1984; 4 of these pups were still alive at the end of the study (table 10-1).

Male K, one of two breeding males remaining in colony K after the removals, was killed in January 1983 by female 7. The second breeding male left in the colony (male U) remained a breeder for the duration of the study; he mated with both females 7 and three-bars. In addition, male A was observed mating for the first time in late July 1983. Male A mated with female three-bars, with whom he had previously fought (table 10-4); three-bars apparently did not become pregnant as a result of these copulations. A third male (B) was observed mating with three-bars for the first time in May 1984; again, these copulations apparently did not result in pregnancy.

REMOVAL EXPERIMENTS: COLONIES TT AND A

In contrast to the upheaval associated with the removal of breeders from colony K, no aggression was observed following the removal of the behaviorally dominant female from colony A or the removal of one breeding male from colony TT. The single breeding male remaining in colony TT continued to mate, and two new males were observed copulating for the first time in July 1983 (5 mo after the removal). In colony A, the dominant female (+) was not replaced; that is, no female ever assumed aggressive dominance over other colony members or showed signs of preparation for breeding (e.g., enlarged nipples). No pups had been born into colony A for 9 mo before the removal of female +. Subsequent to her removal, however, female + mated with the male from colony TT with whom she was housed and gave birth to a litter of six pups. This indicates that the absence of breeding in colony A was not due to the sterility of this female. As of July 1987, no mating had been seen in colony A, and there was still no identifiable dominant or breeding female in this colony.

COLONY-MAINTENANCE BEHAVIORS

ACTIVITY VERSUS INACTIVITY

At any time, most mole-rats were found lying in their colony's communal nest; only a few animals were active outside the nest, either feeding or engaged in maintenance behaviors. Scan sampling (10-min intervals) revealed that the 16 nonbreeding males in colony K spent an average of $61.0\% \pm 11.6\%$ ($n \geq 125$ intervals/animal) of their time dozing or reclining in the nest (described by Lacey et al., chap. 8); for the 10 nonbreeding males in colony A and the 9 nonbreeding males in colony TT, these figures were 60.9% ± 7.9% (n = 201 intervals) and 61.3% ± 9.7% (n = 186 intervals), respectively. Nonbreeding females in colonies K (n = 9) and A (n = 15) spent a mean of 59.6% ± 13.2% and 56.2% ± 9.0% of their time, respectively, dozing or reclining in the nest; data from the four females in colony TT (table 10-1) were not analyzed because the sample size was small. The percentage of time that nonbreeding males and females spent in the nest did not differ in either colonies K or A (p > 0.05).

FOOD TRIALS

Once discovered, pieces of food were rapidly removed from the food box. Food pieces were usually carried to the nest, where they were either eaten by the carrier or dropped; occasionally food was dropped or consumed in the tunnels. The number of times that an individual (nonbreeder) carried items out of the food box varied significantly among both sexes in colonies K and A and among males in colony TT (table 10-5); the small number of females in colony TT (n=4) precluded analyses for these animals. The amount of food carried per mole-rat did not differ between males and females in either colony K or colony A (p > 0.05). All correlations between the number of food carries and body weight were negative for animals in colonies K and TT (fig. 10-4; table 10-5), although this relationship was not statistically significant for colony TT males. In colony A, however, food carrying was positively correlated with body weight for both sexes (table 10-5), and food pieces were rarely dropped in the nest (provisioned), suggesting that members of colony A simply foraged for themselves (i.e., the same individual that removed an item from the food box consumed it in the tunnels).

Data from the experiment in which we gradually increased the distance between the food box and central tunnel system in colony K (table 10-6) consistently yielded significant negative regressions between the number of times an individual carried food and body weight (p < 0.05 at L = 0.6 m, 1.2 m, 1.8 m, 1.8 m C-shaped, elevated curve, and 0.6 m). No differences were found between regression coefficients from different tunnel lengths for either sex (all p > 0.05), indicating that increases in the effort required to obtain food did not alter the pattern that smaller (lighter) individuals carried food most frequently. However, even the greatest distance between the food box and tunnel system used in our laboratory studies (1.8 m.) was small relative to distances between food sources and the nest observed in nature (see Brett 1986, chap. 4, 5).

Food provisioning was examined quantitatively in colony K. *Food provisioning* referred to the subset of food pieces that were dropped by the carrier in or around (≤ 5 cm) the nest. By this criterion, roughly 50% of all carries in colony K were classified as provisioning. Because provisioning did not include food pieces eaten by the carrier, it yielded a more sensitive measure of cooperation in foraging than did food carrying. The number of times that an

Table 10-5

Summary of F-Statistic Analyses for Significance of Interindividual Differences in Each Behavior and of Correlation Analyses Relating the Frequency of Each Behavior to Body Weight

F Statistic	Colony K				Colony TT				Colony A			
	No. of Trials	Litter	Males	Females	No of Trials	Litter	Males	Females	No of Trials	Litter	Males	Females
PUSH PUPS												
F	30	F	2 57	3 00	20	D	1 80					
df			15, 463	12, 377			5, 114					
p			< 0.01	< 0 05			< 0 05					
r			- 0.595	- 0 424			- 0 638	*				
n			16	13			6					
p			< 0 05	< 0 05			< 0 05					
F					28	E	1 88					
df							8, 243					
p							> 0 05	*				
r							- 0 075					
n							9					
p							> 0 05					
F					19	F	1 84	0 92				
df							11, 216	9, 180				
p							< 0.05	> 0.05				
r							0 777	0 714				
n							12	10				
p							< 0.01	< 0 05				
GROOM PUPS												
F	30	F	0 89	1.38	20	D	0 85					
df			15, 464	12, 377			5, 114					
p			> 0.05	> 0.05			> 0 05					
r			- 0.141	- 0 356			- 0.580	*				
n			16	13			6					
p			> 0 05	> 0 05			> 0 05					
F					28	E	3.60					
df							8, 243					
p							< 0.01	*				
r							- 0.347					
n							9					
p							> 0 05					
F					19	F	1 54	1.06				
df							11, 216	9, 180				
p							> 0.05	> 0 05				
r							- 0.070	- 0.133				
n							12	10				
p							> 0 05	> 0 05				

Table 10-5, cont.

F Sta-tistic	Colony K				Colony TT				Colony A			
	No of Trials	Litter	Males	Females	No of Trials	Litter	Males	Females	No of Trials	Litter	Males	Females
					NUDGE/HANDLE PUPS							
F	30	F	1 08	0 69	20	D	1.0					
df			15, 464	12, 377			5, 114					
p			> 0 05	> 0 05			> 0 05					
r			- 0 161	- 0 419 '			- 0 657	*				
n			16	13			6					
p			> 0 05	> 0 05			> 0 05					
F					28	E	2 40					
df							8, 243					
p							< 0 05					
r							- 0.410	*				
n							9					
p							> 0 05					
F					19	F	2.08	1 12				
df							11, 216	9, 186				
p							< 0 05	> 0 05				
r							0.404	0 259				
n							12	10				
p							> 0 05	> 0 05				
					CARRY FOOD							
F	10		7 45	3.04	18		1 69		19	2 51	3 34	
df			15, 144	12, 117			5, 102				13, 238	12, 221
p			<0 01	<0 05			>0 05				<0 05	<0.01
r			- 0 609	- 0.761			- 0 143	*			0 007	0 586
n			16	13			6				14	13
p			<0 05	<0 05			>0 05				>0 05	<0 05
					PROVISION FOOD							
F	10		8 28	2 23								
df			15, 144	12, 117								
p			<0 01	<0 05								
r			- 0.670	- 0 688								
n			16	13								
p			<0 01	<0.05								
					CARRY/SWEEP NESTING MATERIAL							
F	14		2 52	3 75	13		2 25		13		1 30	1 39
df			15, 192	12, 156			8, 108				13, 126	12, 117
p			< 0 01	< 0 01			< 0 05				> 0 05	> 0 05
r			-0 349	-0.143			-0 317	*			-0 487	-0 249
n			16	13			9				14	13
p			> 0 05	> 0 05			> 0 05				> 0 05	> 0 05

Table 10-5, cont.

F Sta- tistic	Colony K				Colony TT				Colony A			
	No. of Trials	Litter	Males	Females	No of Trials	Litter	Males	Females	No of Trials	Litter	Males	Females
					CHEW ON OBSTRUCTIONS							
F	24		2 08	2.45	10		1 50	1 44				
df			11, 275	5, 138			15, 144	15, 144				
p			< 0 05	< 0 05			> 0 05	> 0 05				
r			-0 622	0.371			-0.353	-0.189				
n			12	6			16	₌16				
p			< 0.05	> 0 05			> 0 05	> 0 05				
					DIG THROUGH DIRT (DEFENSE TRIALS)							
F	6		4 68	0.83	6		7 93	1 38				
df			10, 42	7, 31			18, 92	16, 82				
p			< 0 01	> 0 05			< 0 01	> 0.05				
r			0.372	-0 156			0 715	0.696				
n			11	8			19	17				
p			> 0 05	> 0 05			< 0 01	< 0 01				
					THREATEN/BITE FOREIGN MOLE-RATS							
F	6		6.09	1 30	6		6 32	4 25				
df			10, 52	7, 39			18, 92	16, 82				
p			< 0.01	< 0.05			< 0 01	< 0 01				
r			0 187	0 675			0 669	0.392				
n			11	8			19	17				
p			< 0 05	> 0.05			< 0.01	> 0 05				
					THREATEN/BITE SNAKE							
F	6		2.53	2 56	6		8 25	2 73				
df			10, 54	7, 40			17, 89	15, 80				
p			< 0.05	< 0 05			< 0 01	< 0 01				
r			0 515	0 611			0.533	0 577				
n			11	8			18	16				
p			< 0.05	< 0 05			< 0 05	< 0 05				
					FACE OUT OF NEST							
F	10		30 26	13.75	10		8 81	5.47				
df			10, 99	7, 72			18, 171	16, 153				
p			< 0.01	< 0 01			< 0 01	< 0 01				
r			0 550	0.012			0 588	0.432				
n			11	8			19	17				
p			> 0 05	> 0.05			< 0 01	> 0 05				

NOTE F, F-statistics; df, degrees of freedom, p, probability, r, correlation coefficient, n, sample size Data for males and females were analyzed separately For each behavior, the number of trials completed in each colony is indicated, quotation marks denote data that were grouped into "trials" in order to conduct F-tests of behavioral variation among colony members Data from breeding animals were not included in these analyses

* Data were not analyzed because the sample size was too small

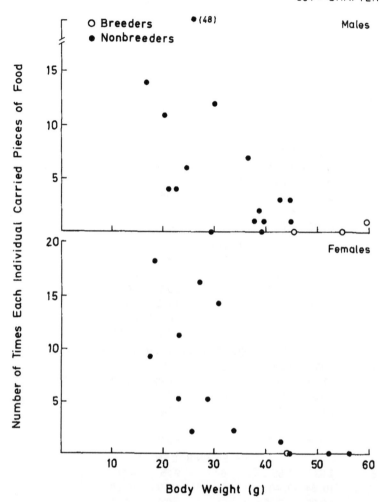

Fig 10-4 Total number of times that each mole-rat in colony K carried food out of the food box as a function of body weight Data are from 10 food trials (June–August 1982), during which 201 food pieces were removed from the food box Males (*top*) and females (*bottom*) are separated, and breeders are indicated Statistical analyses of these data and food carrying in the other colonies are presented in table 10-5

Table 10-6

Summary of Linear Regression Analyses of Colony-K Food Trials Conducted with Four Different Lengths of Tunnel Connecting the Food Box to the Main Tunnel System

Length	Males			Females		
	Y	F	p	Y	F	p
6 m (first series)	30 9– 563x*	9 95	<0.01	31.9– 616x*	32 83	<0 01
1 2 m	34 7– 640x	41 06	<0 01	30 4– 552x	28 90	<0 01
1.8 m (straight run)	32 6– 584x	22 33	<0.01	27 9–.467x	16 37	<0.01
1 8 m (with elevated curve)	23 1– 306x	4.56	<0 05	25.1– 375x	7 39	<0 025
6 m (second series)	30 2–.510x	17.56	<0 01	21 5– 274x	2 18	<0 25

NOTE: Data for males and females were analyzed separately; data points for breeders were not included in these analyses Regardless of the configuration, the smaller mole-rats carried food more often than did larger colony mates Pairwise comparisons of slopes (F-tests) were completed for each combination of tunnel lengths Only slopes from significant ($p < 0.05$) regressions were compared All F-tests were not significant ($p > 0.05$), indicating that the inverse relationship between body weight and food carrying did not differ for either sex as a function of distance traveled to obtain food

* For males, df = 1, 16, for females, df = 1, 13

individual in colony K provisioned food varied significantly among nonbreeders and was inversely correlated with body weight for both males and females (fig. 10-5; table 10-5); males and females did not differ in the amount of food provisioning per individual ($p > 0.05$).

NESTING-MATERIAL TRIALS

Wood shavings were quickly removed from the site of introduction so that the area was usually cleared within 10 min. Depending on the size of the wood chips being transported, they were either carried in the jaws, or swept with the hind feet. The largest wood chips were carried to the nest box first, leaving smaller shavings that were then swept to the nest. Individuals shifted from carrying to sweeping during the course of a trial, and sometimes simultaneously carried and swept shavings. For these reasons, data points for carrying and sweeping were combined for analyses. Although sweeping nesting material may also function to clear the tunnels of debris (rather than to build a nest), most of the pine shavings introduced during trials were transported to the nest box (rather than to the toilet box), suggesting that the primary function of the behaviors recorded was indeed nest construction. The number of times that an individual carried/swept nesting material varied significantly among nonbreeding members of colonies K and TT, but not of colony A (table 10-5). Carrying/sweeping was negatively (but not significantly) correlated with body weight in all three colonies (fig. 10-6; table 10-5). Males and females in colonies A and K did not differ significantly with respect to the number of times an individual carried/swept shavings ($p > 0.05$).

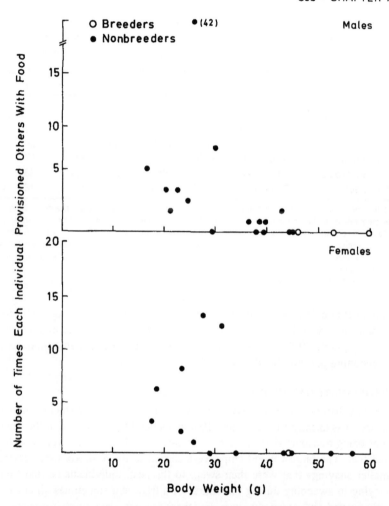

Fig 10-5 Total number of times that each mole-rat in colony K provisioned others with food (i.e., carried food out of the food box and dropped it in or near the nest) as a function of body weight. Data are from 10 food trials (see fig. 10-4), during which 119 of 201 food items were provisioned. Males (*top*) and females (*bottom*) are separated, and breeders are indicated; statistical analyses are presented in table 10-5

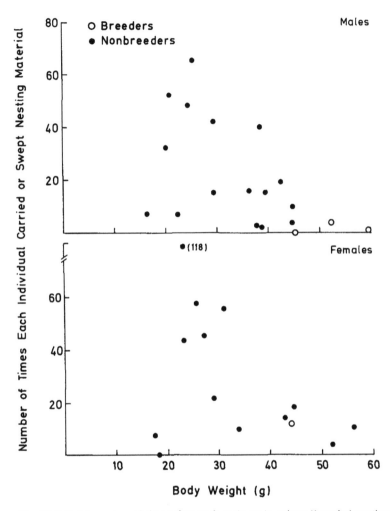

Fig. 10-6. Total number of times that each mole-rat in colony K carried nesting material (wood chips) in its mouth or swept nesting material with its hind feet as a function of body weight Data are from 14 nesting material trials (July–August 1982), during which 797 carries and sweeps were recorded. Males (*top*) and females (*bottom*) are separated, and breeders are indicated; statistical analyses are presented in table 10-5.

OBSTRUCTION TRIALS

Dowels inserted into colonies K and TT were chewed off close to the inside surface of the tunnel. The number of times that a nonbreeding individual chewed on obstructions varied significantly among members of colony K, but not colony TT (table 10-5). The frequency of chewing was negatively correlated with body weight among males in colony K (fig. 10-7) and both sexes in colony TT; the correlation for females in colony K was positive but not significant (table 10-5). The number of times an individual chewed on obstructions did not differ between males and females in either colony ($p > 0.05$).

DIGGING TRIALS

Digging began as soon as the dirt plug in the experimental tunnel was discovered. The animals worked quickly, reducing the plug to small pieces in 15–20 min; excavated dirt was swept out of the experimental tunnel and into the toilet or an unused nest box. Sweeping behavior associated with digging resembled that observed during nesting material trials; however, dirt was usually swept to the toilet (and not to the active nest box).

No more than two mole-rats could stand side-by-side and dig simultaneously. In all trials, additional animals were present in the experimental tunnel but were prevented from reaching the dirt face by the digging animals. Individuals approaching colony mates working at the dirt face (from behind) would often shudder violently and give a loud chirping vocalization (a behavior analyzed by Pepper et al., chap. 9); the digging individual typically responded by chirping as well or by turning and shoving the approaching individual, forcing it to retreat.

Digging and chewing at the dirt plug occurred in rapid alternation and were thus treated as a single behavior for analysis. For digging trials conducted during June 1983, the number of digs/chews per individual did not vary significantly among either sex in colonies K and TT ($F \leq 0.77$, all $p > 0.05$). Correlations between number of digs/chews and body weight were positive but not significant for both sexes in colony TT ($r \leq 0.82$; $p > 0.05$) and negative but not significant for both sexes in colony K ($r \leq -0.57$; $p > 0.05$). For trials conducted in association with colony-defense trials, the number of digs/chews varied significantly among nonbreeding males, but not females, in both colonies K and TT (table 10-5). Correlations between the number of digs/chews and body weight were positive and significant for both sexes in colony TT (fig. 10-8). In colony K, the relationship was positive for males but negative for females; neither of these correlations was significant (table 10-5). The number of digs/chews per individual did not differ between males and females in either colony ($p > .05$). During the final series of control trials (conducted after colony-defense trials were completed; no foreign animals present), the number of digs/chews per individual was again positively correlated with body weight for

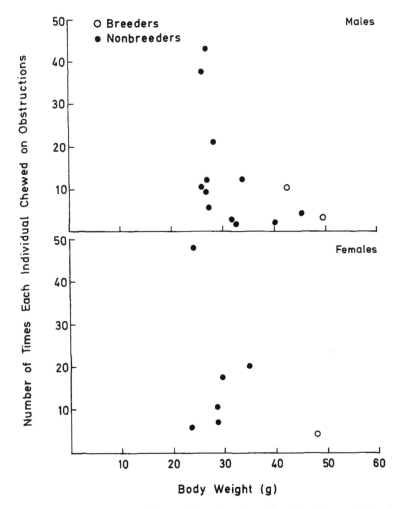

Fig. 10-7. Mean number of times that each mole-rat in colony K was observed chewing on a dowel 0.5-cm in diameter as a function of body weight. Data are from 24 obstruction trials (February–May 1984), in which a tunnel was partially blocked by the dowel. During each trial, the identity of the animal chewing on the dowel was recorded every 30 s; a total of 827 data points were obtained. For analysis, the length of each trial was standardized to 30 min. Males (*top*) and females (*bottom*) are separated, and breeders are indicated; statistical analyses are presented in table 10-5.

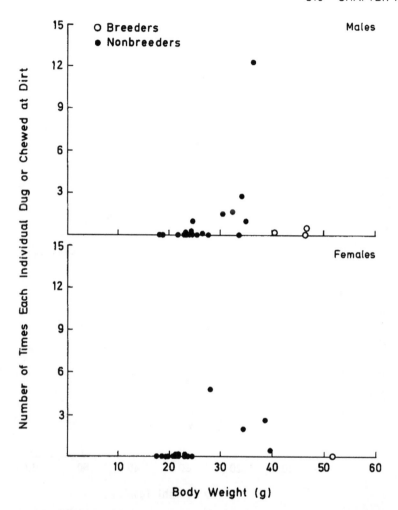

Fig. 10-8. Mean number of times that each mole-rat in colony TT dug dirt with its forepaws or chewed dirt with its incisors as a function of body weight. Data are from six digging trials conducted in association with colony-defense trials during June–July 1985. During each trial, the identity of the animal(s) digging or chewing at the dirt were recorded every 30 s; a total of 323 data points were gathered. For analysis, the length of each trial was standardized to 30 min. Males (*top*) and females (*bottom*) are separated, and breeders are indicated; statistical analyses are presented in table 10-5.

both sexes in colony TT and for males in colony K ($p < 0.01$); for females in colony K the relationship was also positive, but not significant.

COLONY-DEFENSE BEHAVIORS

DEFENSE AGAINST CONSPECIFICS

Members of colonies that were experimentally joined began threatening one another as soon as the dirt plug separating them was demolished. Typically, two colony mates would stand side-by-side and threaten, thus effectively blocking the tunnel against intruders. Occasionally three to four colony mates would fill the tunnel by climbing onto each other's back, creating a two-tiered wall of threatening mole-rats. Animals threatening each other stood face to face, jaws open wide, with their teeth and noses almost touching (see Pennisi 1986; plate 10-4). An individual was occasionally able to grab an opponent's muzzle with its incisors, and a brief tug-of-war ensued as the bitten mole-rat struggled to extricate itself. Scuffles between colonies were accompanied by considerable noise, including hissing, tooth grinding, chirping, and trilling (see Pepper et al., chap. 9). Because threatening and biting occurred in rapid succession and were performed by the same individuals, these behaviors were combined for analysis.

The number of times that an individual threatened/bit foreign mole-rats varied significantly among nonbreeding males in colonies K and TT and among females in K and TT (table 10-5). The number of threats/bites per individual was positively correlated with body weight for males and females in both colonies (fig. 10-9; table 10-5), although this correlation was significant only for males (both colonies). Both sexes defended their colony with equal vigor: the number of threats/bites per individual did not differ between the sexes in either colony ($p < 0.05$). With one exception (in TT), breeding males did not threaten or bite intruders; although the breeding female in colony TT occasionally attacked intruders, the breeding female in colony K never participated directly in this type of colony defense. Breeding females in both colonies occasionally shoved nonbreeders into the tunnel where the interaction between the colonies was occurring (see also Reeve and Sherman, chap. 11).

Although digging and defense against conspecifics were both positively correlated with body weight, individuals defending the colony were not necessarily the same ones that excavated the dirt plug separating the colonies, as indicated by the following observations.

Of the five most frequent diggers in colony K, only one was also among the five most frequent defenders of the same colony. In colony TT, two of the five most frequent diggers were also among the five most frequent defenders. In both colonies, the five most frequent defenders were significantly heavier than the five most frequent diggers ($p < 0.05$, Mann-Whitney U-tests).

In two trials in colony K and two trials in TT, the animals digging at the time of first contact between colonies were individuals that never participated in colony defense. These individuals ceased to dig and retreated from the scene of the conflict as soon as contact between colonies occurred.

The primary participants in defense against conspecifics were not simply animals that happened to be present at the time of initial contact between colonies; rather, defenders arrived at the site of the conflict several minutes later. Although we did not quantify individuals' arrival times, we did record the identities of all animals in the dirt-containing tunnel every 30 s. In every trial that we conducted, 50% or more of the mole-rats that participated directly in the conflict were not in the tunnel during the 30-s interval immediately after contact was made.

DEFENSE AGAINST SNAKES

Colony members began investigating the snake soon after demolishing the dirt plug that had isolated it. Often the mole-rats would closely approach the snake and spend several seconds sniffing at it before opening their jaws, threatening, and backing away. These behaviors were frequently accompanied by a hissing vocalization (see Pepper et al., chap. 9); the trilled vocalization heard during intercolonial interactions was never heard during snake trials. In contrast to their behavior toward foreign conspecifics, the mole-rats approached and threatened the snake singly, rather than in pairs or groups. The mole-rats bit the snake if it became agitated and struck at them or if it moved hurriedly through their tunnels.

Data on threats and bites were combined for analysis. The number of times that an individual threatened/bit the snake varied significantly among nonbreeding males and females in colonies K and TT (table 10-5); the number of attacks on the snake was significantly and positively correlated with body weight for both sexes in both colonies (fig. 10-10). The number of threats and bites per individual did not differ between males and females in either colony (table 10-5). No breeding animal in either colony threatened or bit the snake; however, the breeding female in colony TT sometimes pushed nonbreeders into the tunnel containing the snake.

We did not record the identities of the mole-rat(s) that broke through the dirt plug, and thus we do not know if these were the same individual(s) that threatened/bit the snake. However, the number of different individuals attacking the snake was greater than the number of individuals digging, suggesting that animals that threatened/bit the snake were again not simply the first mole-rats to encounter it following demolition of the dirt plug.

In one of our trials, there was a large quantity of loose sand in the snake-containing tunnel. In this case, several large nonbreeders approached the snake, turned around, and vigorously kicked sand at it. No breeding animals were seen behaving this way. Entombment may be one way that free-living mole-rats deal with snakes (Brett 1986, chap. 4), but we did not quantify this aspect of antipredator behavior.

CONTROL TRIALS

No threats or bites were ever observed during control trials (p. 308) in which mole-rats were allowed to dig through a small dirt plug but were not confronted with either the snake or other conspecifics ($n = 5$–7 trials, each 20 min long, in both colonies K and TT). After the animals finished removing the dirt plug, they walked to the blind end of the tunnel, sniffed it, and soon resumed activities unrelated to digging. These observations confirm that the snake or foreign mole-rats were the stimuli that elicited threatening and biting and that these behaviors did not result solely from the introduction of the dirt plug or from habituation to encountering foreign stimuli once the dirt was removed.

FACING-OUT BEHAVIOR

The number of times that an individual lay with its head and shoulders facing out of the nest varied significantly among nonbreeders in colonies K and TT. For both sexes in both colonies, the number of times that an individual was observed facing out was positively correlated with body weight (fig. 10-11; table 10-5); this relationship was significant only for males in colony TT. The number of times that an individual faced out of the nest did not differ between males and females in either colony ($p < 0.05$). The behavior of breeding animals, however, did differ between colonies. Whereas the number of times that breeding males in colony TT lay facing out was consistent with their large size (fig. 10-11), breeding males in colony K faced out of the nest less often than predicted on the basis of body weight. In both colonies K and TT, the breeding female lay facing out of the nest less often than predicted on the basis of her body weight (but see Jarvis, chap. 13).

Facing out of the nest may reflect individuals' efforts to position themselves for easy escape from the nest. For example, mole-rats at nest entrances should be the first to exit in response to a disturbance. However, our observations suggest that this behavior is instead related to defense of the nest. In colony K, the 5 males and 5 females most frequently seen facing out (all nonbreeders) were the same 10 individuals most frequently involved in threatening both snakes and foreign colonies. In colony TT, 3 of the 4 females and 3 of the 5 males most frequently seen facing out of the nest were the same as those most frequently involved in colony defense; the reduced correspondence between facing out and colony defense in colony TT is due to the large number of times that breeding animals in this colony were observed in the facing-out position (fig. 10-11).

The behavior of mole-rats near the nest at the start of each colony-defense trial was not recorded, and thus we do not know how animals facing out of the

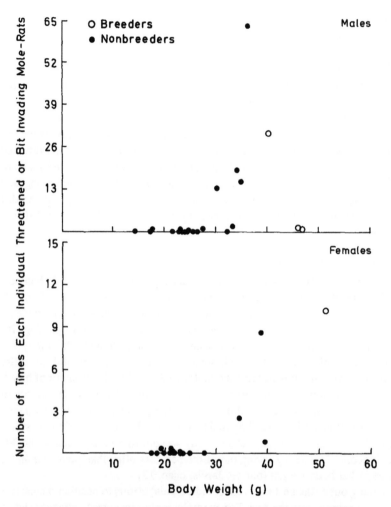

Fig. 10-9. Mean number of times that each mole-rat in colony TT threatened (i.e., gaped and then lunged; see chap 8) or bit members of an invading colony as function of body weight. Data are from six trials (June–July 1985), in which animals in colony TT and a second colony (containing no known relatives of animals in colony TT) were allowed to interact for 40 min after digging through a dirt barrier. A total of 1,342 threats and bites were recorded. For analysis, the length of each trial was standardized to 30 min. Males (*top*) and females (*bottom*) are separated, and breeders are indicated; statistical analyses are presented in table 10-5.

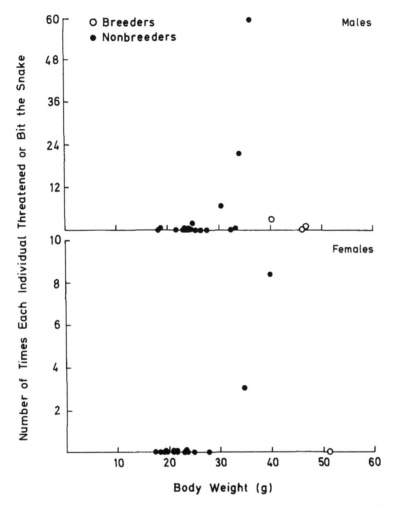

Fig. 10-10. Mean number of times that each mole-rat in colony TT threatened or bit a snake as a function of body weight. A total of 389 threats and bites were recorded during six trials (June–July 1985) in which the mole-rats and the snake were allowed to interact for 20 min or less after the mole-rats dug through a dirt barrier. For analyses, the lengths of all trials were standardized. Males (*top*) and females (*bottom*) are separated, and breeders are indicated; statistical analyses are presented in table 10-5.

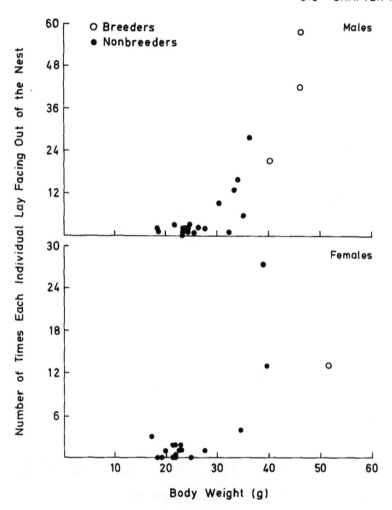

Fig 10-11. Total number of times that each mole-rat in colony TT lay with its head and shoulders extending out of the nest box as a function of body weight Identities of animals facing out of the nest were recorded at irregular intervals, with at least 60 min allowed between consecutive observations. A total of 107 scans of facing out of (guarding) the nest were recorded during June and July 1985. Males (*top*) and females (*bottom*) are separated, and breeders are indicated; statistical analyses are presented in table 10-5.

nest responded to the introduction of the snake or other mole-rats. However, in a follow-up study, a nail file was used to make a rasping noise (i.e., a nonspecific disturbance) at a tunnel junction located as far as possible from the colony's nest. In 13 of 18 trials in colony K and 9 of 12 trials in colony TT, the first two individuals to arrive at this junction were nonbreeders that had been lying facing out of the nest just before the disturbance.

It might be argued that facing-out behavior simply reflects attempts by large animals to thermoregulate (cool off) or to intercept colony mates returning to the nest with food. Although we cannot reject either possibility, both seem unlikely given that (1) the breeding female, one of the largest individuals in each colony, rarely faced out, and (2) neither the frequency with which at least one mole-rat faced out nor the total number of animals facing out varied significantly with room temperature ($25°-30°C$) or the length of time since the colony was last given fresh food ($p > 0.2$ for all tests in colonies TT and K; Spearman's rank correlation tests). Collectively, these observations suggest that facing-out behavior serves a nest-guarding function (see also Faulkes et al., chap. 14).

GROWTH DATA

Growth curves for members of colony K (fig. 10-12) reveal that among laboratory-born animals (ages known) body weight increased steadily from birth to about 15 mo, after which growth slowed. Growth curves constructed for wild-caught animals (ages unknown) also indicate an eventual leveling of individuals' body weights. Mean weight change per month was significantly negatively correlated with initial weight for the nonbreeding animals in each colony ($p < 0.05$); see also Brett, chap. 4). Mean weight changes per month for nonbreeding males and females did not differ (Mann-Whitney U-test, $p > 0.05$).

Our data indicate that the growth patterns of breeders and nonbreeders differed. In colony K, the fastest growing male for the final 20 mo of the study was A, who became a breeder in July 1983. The growth spurt by male A at about 45 mo (fig. 10-12) coincides with the 6-mo period immediately preceding his shift in status from nonbreeder to breeder. The pattern is even more pronounced among colony-K females, for which curves for breeding and nonbreeding animals are clearly divergent. The three breeding females whose growth rates were plotted are the colony's first breeding female (+), and females 5 and three-bars. As described previously (see table 10-4), the latter two animals were involved in the succession of dominants that occurred following the removal of the colony's original breeding female, one breeding male, and two nonbreeders. Data for the third female involved in this succession (7) are not presented in figure 10-12 because the initial capture weight of this female exceeded the range of weights recorded for laboratory-born animals. For fe-

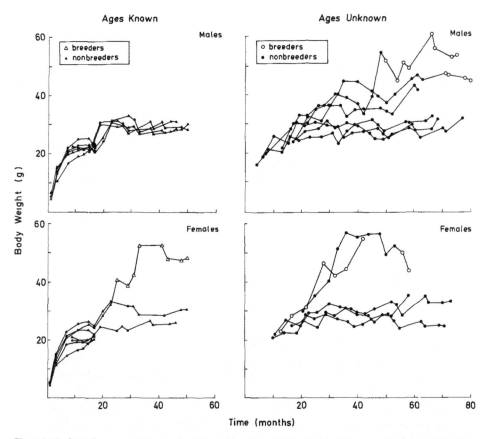

Fig. 10-12. Growth curves for naked mole-rats in colony K. Curves for laboratory-born animals (*left*) show body weight as a function of exact age. Curves for wild-caught animals (*right*), whose ages were unknown, were constructed by plotting the capture weights of these animals on the curves for laboratory-born mole-rats. From these initial points, the growth of each wild-caught animal while in captivity (June 1980–August 1985) was plotted, yielding curves for body weight as function of approximate age Curves for 13 males and 12 females are shown. Capture weights of 12 other colony members were too great to be fitted to curves for laboratory-born mole-rats Males (*top*) and females (*bottom*) are separated, and breeders are indicated.

males 5 and three-bars, the growth spurts depicted (at ca. 30 mo) again coincide with the 6-mo period immediately preceding each animal's shift from nonbreeder to breeder. Calculations of mean weight changes per month through September 1982 reveal that the colony's first and third breeding females (+ and 5) grew more than twice as fast as nonbreeding animals with comparable initial weights. Similar growth spurts did not accompany the changes in breeding status of two males in colony TT. No changes in breeding status ever occurred in colony A, and no growth-rate differences among colony members were observed.

Summary

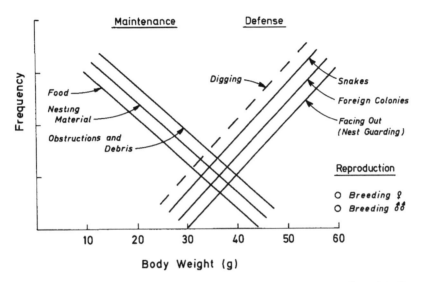

Fig. 10-13. Graphical summary of the key behavioral results of this study (figs 10-1–10-11; table 10-5). Breeding individuals (one female, one to three males) were heavily involved in directly caring for pups but seldom defended or maintained the colony; this reproductive division of labor is depicted by the placement of the breeders at the lower right corner of the graph (although breeders are not always the largest individuals of their sex in a colony). Nonbreeders of both sexes exhibited a size-based division of labor. Small males and females performed the majority of colony-maintenance behaviors, whereas large males and females defended the colony and excavated dirt. Males and females of similar size participated equally in these activities. Note that the relative horizontal placement of lines on this figure is for visual clarity; that is, participation in food carrying, nesting material transport, and obstruction and debris removal all decreased with increasing body weight, but food carrying did not occur less frequently at any given weight than debris removal. Dirt digging is presented as a dashed line because (1) digging is not obviously a defensive behavior, and (2) although our data indicate a positive relationship with body weight, the results of Jarvis et al. (chap. 12) do not (see the text).

Discussion

Naked mole-rats in our study colonies did not participate equally in reproduction, colony maintenance, or colony defense. Within each colony, some animals performed these behaviors far more often and others considerably less often than expected if participation were random (table 10-5). Behavioral differences among colony members were defined primarily by reproductive status and, among nonbreeders, were correlated with size. Behavioral trends in our data are summarized in figure 10-13.

EVIDENCE FOR A REPRODUCTIVE DIVISION OF LABOR

The primary behavioral asymmetries in our colonies were associated with differences in reproductive status. By definition, breeding animals were the only colony members that mated; they were also the only animals to exhibit anogenital nuzzling or other sexual behaviors on a regular basis. A single breeding female per colony gave birth and nursed pups. Breeding females and the males that mated with them were also the primary participants in pup-care activities subsequent to parturition (figs. 10-1 – 3, table 10-3; see also Jarvis, chap. 13). Although small nonbreeding animals carried newborn pups to the nest and sometimes retrieved pups if they fell out of the nest, on a per capita basis nonbreeders handled, groomed, or pushed pups less frequently than did breeders. Whereas the breeding female's participation in pup care resulted in part from her role as the colony's sole lactating female, similar constraints were not placed on the colony's breeding males. However, breeding males frequently handled, groomed, and pushed pups, indicating that direct participation in pup care was influenced by whether or not a male had mated.

Conversely, breeding animals rarely participated in colony maintenance or defense activities (figs. 10-4 – 11); these behaviors were performed primarily by nonbreeding colony members. This asymmetry between breeders and nonbreeders was a qualitative difference, and not simply a quantitative effect resulting from the large body sizes of breeders (table 10-2). In colony K, digging (during defense and control trials), defense against conspecifics and snakes, and facing out of the nest all increased with body weight among nonbreeders, but breeding animals, among the heaviest individuals in the colony, rarely performed these behaviors. In colony TT, this same pattern was evident for data on digging and defense against snakes but not for defense against conspecifics and facing out of the nest. The data from colony TT are confounded by the deaths of two large nonbreeding colony members as a result of injuries received during fights with foreign mole-rats. Following the loss of these animals, participation by breeders (in particular, the breeding female) in colony defense increased greatly. Before these deaths, data on defense against conspecifics and facing out of the nest in colony TT indicated a sharp behavioral transition between nonbreeders and breeders similar to that observed in colony K. Thus, we concur with Jarvis (1981), Brett (chap. 4), Reeve and Sherman (chap. 11), and Jarvis (chap. 13) that breeding and nonbreeding mole-rats represent behaviorally distinct subsets of colony members.

Changes in growth rate were associated with the transition from nonbreeding to breeding status among members of colony K, as well as among females in Jarvis's colonies (1981, fig. 1; see also chap. 12). Colony K's original breeding female (+) was one of three large, fast-growing females (fig. 10-12). The two other fast-growing females depicted (5 and three-bars) both participated in the succession of breeders that occurred following the removal of the original breeding female (table 10-4). Thus, growth spurts were exhibited by both the colony's original and current breeding females, as well as a third individual killed while trying to establish herself as a breeder. Similarly, the colony's fastest growing male (A) became a breeder during this study (fig. 10-12, table 10-4).

Comparable growth-rate differences between breeders and nonbreeders were not observed in colonies A or TT. No succession of breeders occurred in colony A, and no individuals exhibited growth spurts. The situation in colony TT is more problematic, because two males did begin to breed during our study. The initial weights of these two males were greater than the initial weights of fast-growing animals in colony K (36 and 61 g at capture vs. 22, 23, and 30 g for the wild-caught, fast growing breeders in colony K), suggesting that the two colony-TT males may have already undergone a period of rapid growth before the beginning of our study. We do not know why these animals attained such large sizes but apparently did not begin to breed until July 1983. This is particularly puzzling given that rapid growth and the attainment of large body weights were so clearly associated with the onset of breeding in colony K (however, see Jarvis et al., chap. 12; Jarvis, chap. 13).

We found behavioral and morphological differences between breeding and nonbreeding *Heterocephalus glaber* to be flexible, at least among young animals. As evidenced by data from colony K, nonbreeding females up to at least 2 yr of age (i.e., the age at which female three-bars became reproductive) can attain reproductive status after the loss of the previous breeder; among males, this figure can be extended to at least 3 yr, the length of time that male A had been in captivity when he became a breeder (table 10-1). Furthermore, breeders were not simply the oldest animals in a colony; colony K's current breeding female (three-bars) was born in June 1981 and was thus at least 1 yr younger than all the wild-caught animals in her colony (captured in June 1980). Jarvis (1981) suggested that female naked mole-rats retain the ability to reproduce for only a limited portion of their lives. Our data are consistent with this menopause hypothesis, but are inconclusive. The observations presented by Jarvis (chap. 13) suggest, however, that both females and males can resume breeding after long periods of reproductive quiescence, apparently casting doubt on this hypothesis.

EVIDENCE FOR A DIVISION OF LABOR AMONG NONREPRODUCTIVES

Behavioral variation among nonbreeders was nonrandom and was correlated with body weight (fig. 10-13). As in previous studies by Jarvis (1981), Payne (1982), and Isil (1983), smaller nonbreeders were the most frequent participants in food carrying, nesting-material transport, tunnel clearing, and obstruction removal; similar results are presented by Jarvis et al. (chap. 12) and

Faulkes et al (chap. 14). In addition, by challenging the animals in ways that previous researchers had not, we were able to document colony members' participation in defense against foreign conspecifics and snakes; our data indicate that larger nonbreeders were the primary participants in these activities. This result is particularly interesting in light of Brett's report (chap. 4) that, in the field, large, nonbreeding individuals were usually the first colony members to be captured (i.e., the first animals to visit tunnel breaks), and Braude's (chap. 6) observations suggesting that larger individuals are the primary participants in volcanoing.

Temporal stability of these behavior patterns (fig. 10-13) is suggested by two observations: (1) regression coefficients for food carrying versus body weight in colony K did not differ despite changes, albeit small, in the distance traveled to food, and (2) similar negative correlations for participation versus weight were obtained in colonies K and TT for food and nesting-material trials conducted more than 1 yr after the original series of trials was completed.

Surprisingly, food carrying was positively correlated with body weight in colony A. Each colony member appeared to forage for itself, and food items were rarely provisioned. Larger individuals, presumably those with greater energy requirements, carried and consumed food pieces most frequently. Behavioral differences between members of colony A and colonies K and TT may have been related to the lack of reproduction in colony A during this study. Because there were no births, no small individuals (the primary food carriers in other colonies) were present. Furthermore, performance of all maintenance activities may have been reduced in the absence of a breeding female (Reeve and Sherman, chap. 11). For these reasons, we believe that the patterns of maintenance and defense observed in colonies K and TT, rather in than colony A, best reflect the social organization of *H. glaber* colonies.

Considerable scatter was evident in plots of maintenance and defense activities versus body weight for all colonies. For behaviors for which scatter was greatest among the smallest colony members, this variation may reflect immaturity. The smallest animals in both colonies K and TT were laboratory-born pups; observations (unpubl. data) of the behavioral ontogenies of these animals indicate that they began performing different behaviors at different ages. Hence, for any given behavior (e.g., provisioning, fig. 10-5), it is possible that pups had not yet fully begun participating in that activity at the time of data collection. In addition, individuals may specialize or "major" (Heinrich 1976, 1979) on a particular activity or set of activities, and these specializations may be reflected in the interindividual differences seen in our data. Finally, these behavioral variations may result from as of yet undocumented differences between individuals, such as differences in reproductive potential (see below and chap. 12).

One apparent discrepancy between our results and those of Jarvis (1981; Jarvis et al., chap. 12) merits discussion. Our laboratory observations suggested that (in the context of colony defense) larger nonbreeding mole-rats were the primary participants in digging (fig. 10-8; pp. 308–11). For digging trials not associated with defense trials, correlations between number of digs/ chews and body weight were positive for members of colony TT but negative for members of colony K (table 10-5). Data from other laboratory colonies are also contradictory; Braude (1983) found a positive correlation between body weight and frequency of digging during experiments in which he provided colonies (at Michigan) with a 1–3 cm thick chunk of cork (with a large piece of food located behind it), which the mole-rats readily dug through. In contrast, Jarvis et al. (chap. 12) reported that smaller nonbreeders were the most frequent diggers during trials in which animals were continuously exposed to loose dirt.

Data from free-living colonies also provide potentially contradictory results. Brett (1986, chap. 4) found that in one field-caught colony, the percentage of animals with soft, wet dirt caked between their incisors was inversely correlated with body weight; on the basis of this observation and the "small" size of foraging burrows, he suggested that smaller individuals were the primary participants in tunnel excavation. However, Brett (1986, p. 157) also reported that during a digging experiment performed in the field "In general, . . . large, dominant animals tended to occupy the excavating position more by virtue of being more aggressive in contests over this position." Furthermore, Braude (chap. 6) found that in two of three field colonies, individuals that (in 1-h sampling periods) most frequently kicked loose dirt out of burrow systems to form molehills were significantly heavier than non-volcanoers.

Thus, at present, it is unclear which colony members are the primary participants in digging. Although the amount of dirt caked on an individual's incisors may reflect participation in tunnel excavation, these data are confounded by the order in which colony members were captured. Brett (1986) reported that the last animals caught from a colony (those with dirt-caked incisors) were relatively small. These animals may have had additional time to escape Brett's disturbance by tunneling, and thus the incisors of these individuals may have been more caked with dirt at the time of capture than the incisors of (previously captured) larger colony members. Alternatively, the incisors of larger animals may be less likely to become caked with dirt for reasons unrelated to digging (e.g., larger animals may clean their incisors more frequently or effectively than smaller animals).

Data from laboratory studies of digging may have been influenced by the artificial environment in which these trials were conducted. Whereas members of our study colonies were continuously exposed to the stimuli used in most other colony-maintenance trials (e.g., food, nesting material), the animals were exposed to fresh, hard-packed dirt only during digging trials. As a result, our digging trials may not reflect participation in tunnel excavation, but rather the mole-rat's responses to a novel stimulus or attempts to escape confinement.

Jarvis et al.'s trials (chap. 12) involved continuous exposure to soil but always presented the animals with soft, workable earth that could not be tunneled through. Thus, the trend toward small animals digging reported by Jarvis et al. may actually reflect the mole-rats' responses to a plugged or debris-filled tunnel (i.e., a tunnel requiring cleaning), rather than participation in tunnel excavation per se.

This discrepancy notwithstanding, our results are generally consistent with those presented elsewhere in this volume and clearly indicate that behavioral differences among nonbreeding mole-rats are associated with differences in body weight. Two mechanisms have been suggested to explain variation in body weight (and thus variation in behavior) among these nonbreeders.

THE PHYSICAL-CASTE HYPOTHESIS

Jarvis (1981) originally proposed that the division of labor in *H. glaber* colonies is based on a system of two or three morphologically distinct castes. Behavioral differences between castes were thought to be associated with substantial variations in body weight that resulted from consistent differences between the growth rates of colony members. According to this hypothesis, data points for nonbreeders should have fallen into two or three clouds when participation was plotted against body weight (e.g., figs. 10-1–10-11). That is, a group of small mole-rats with data points high on all ordinates (frequent workers) should have been clearly distinguishable from a second group of larger animals with data points low on the same axes (nonworkers); infrequent workers, if they exist, might be expected to appear as a third and intermediate cloud of points.

Our data on the behavior of nonbreeders do not support this hypothesis. Although behavioral variation among nonbreeders is evident, data points do not fall into clearly separable clouds. Instead, differences between frequent and infrequent participation in colony-maintenance and colony-defense behaviors appear continuous (a similar conclusion was reached by Payne [1982], Isil [1983], and Faulkes et al. [chap. 14]). Moreover, the body weights of individuals are continuously distributed along the abcissas of figures 10-1 through 10-11; no differentiation into large and small nonbreeders is apparent. With the exception of breeding animals, laboratory-born individuals in each of our colonies exhibited similar growth patterns (fig. 10-12), suggesting that behavioral differences among colony members were not caused or obscured by growth-rate differences. Thus, we find no evidence of discrete behavioral or morphological castes among nonbreeding colony members.

THE TEMPORAL-CASTE HYPOTHESIS

Jarvis (1981) also suggested that some naked mole-rats exhibit age polyethism. She proposed that (1) all animals enter the frequent-worker caste at the age of 1–2 mo, (2) slow-growing animals permanently remain in this caste, and (3) faster-growing individuals may become infrequent workers or non-workers and potential reproductives.

Because most of our animals were field-caught (and their ages were unknown), we could not determine directly how age affected participation in colony activities. However, we attempted to test Jarvis's (1981) age-polyethism hypothesis indirectly, using body weight as an indicator of age. Two lines of evidence suggest that body weight is correlated with age. First, growth curves for known-age individuals indicate that body weight increases with age. Nonbreeding mole-rats in our colonies followed similar growth patterns (fig. 10-12). In particular, growth trajectories for all laboratory-born mole-rats (except potential or actual breeders) were similar over the entire 0–4 yr age span studied ($n = 37$ individuals in 5 litters), suggesting that, in general, larger animals were indeed older. Second, there was an inverse relationship between age and growth rate: laboratory-born mole-rats, known to be the youngest members of each colony, were also the fastest growing. Among our field-caught animals, mean weight change per month decreased with increasing weight at capture; similar growth curves were presented by Jarvis (1981), and by Brett (chap. 4). This suggests that small, fast-growing individuals were indeed young animals whose growth would slow as they approached adult size (however, see Jarvis et al., chap. 12).

If the body weights of colony members do co-vary with age, then behavioral variation among nonbreeders may reflect differences in age, as well as in size. Whereas the age polyethism described by Jarvis (1981) would occur as a secondary consequence of morphological differentiation into castes, we suggest that size may co-vary with age such that both affect behavioral differences among nonbreeders. This age/size polyethism is not limited to a particular subset of colony members (e.g., Jarvis' "fast-growing" mole-rats) but may occur among all nonbreeding individuals. We emphasize that this polyethism is not a rigid temporal (as opposed to morphological) caste system but rather represents a flexible (albeit predictable) series of behavioral changes that occur as an individual ages and grows. Our hypothesis does not require all individuals to shift from colony maintenance to colony defense at a predetermined age; indeed, as suggested by Jeanne et al. (1988) for social wasps, the timing of this behavioral transition is probably affected by ecological or demographic conditions such that the age at which the transition occurs varies among individuals. As a result, behavioral differences between large and small (old vs. young) colony members tend to be gradual and continuous, rather than abrupt.

The relationship between age and body weight suggested here has been challenged by Jarvis et al. (chap. 12), who argue that body weight is not a reasonable predictor of relative age. Known-age animals in three of Jarvis's lab colonies (including littermates) exhibited considerable variation in growth rate, such that the body weights of animals of the same age were sometimes vastly different and, conversely, the body weights of animals of different ages were occasionally similar. The reasons for this discrepancy between our results and those of Jarvis et al. are unknown. Although we attempted to house and maintain our laboratory colonies under conditions similar to those in Jarvis's lab (Appendix), variations in laboratory conditions may have affected the growth patterns of colony members. Alternatively, the growth rates of individuals may have been affected by demographic or social factors such as the degree of crowding or differences between the reproductive potentials of colony members. Regarding crowding, Jarvis et al. (chap. 12) reported that the first two litters born to two colonies started from pairs (note that we did not start any study colonies from pairs) grew considerably faster than subsequent litters; it seems possible that in these cases subsequent litters grew more slowly because of increased competition and space limitations in Jarvis's colonies or because of decreased chances of becoming a breeder in a numerically large colony. Regarding differences in reproductive potential, growth curves for laboratory-born animals in our colony K (fig. 10-12) indicated that all nonbreeding females ($n = 3$) exhibited a sudden increase in growth rate shortly after the removal of that colony's original breeding female. However, the extent of this growth spurt varied among females. As a result, the body weights of same-age females showed considerable variation (25–42 g vs. 24–28 g) following this sudden increase in growth rate. Similar growth spurts were associated with changes in breeding status among females in Jarvis's colonies (1981; chap. 12), suggesting that variations in the body weights of our animals and hers could represent present or past differences in colony members' attempts to exploit reproductive opportunities.

Clearly, additional data, especially from wild colonies, are needed on the factors affecting growth patterns of individual mole-rats. To examine directly the relationship of age, body weight, and behavior, individuals must be aged accurately (e.g., using some nonfatal technique such as tail-collagen strength; Sherman et al. 1985), then observed over the course of their lifetimes. The latter project is currently underway in our labs but, given the long life span of *H. glaber* in captivity ($\geq$ 16 years; Jarvis and Bennett, chap. 3; Jarvis, chap. 13), neither our data nor Jarvis's are yet sufficient to address this question.

INDIVIDUAL-LEVEL VERSUS COLONY-LEVEL SELECTION

Although the demography of free-living colonies of *H. glaber* is under investigation by Braude (pers. comm.), we do not yet know how the division of labor described here affects the fitness of individual colony members. However, both observations of captive colonies and ecological data (e.g., Brett 1986, chap. 4, 5) suggest that this division of labor can be interpreted using an ecological-constraints approach (e.g., Emlen 1984; J. L. Brown 1987). According to this model, individuals should remain in their natal colony when the costs of dispersal and independent breeding are prohibitively high. Once philopatry is enforced by ecological circumstances, individuals may be able to reproduce only by mating with relatives and helping to rear the offspring of close kin (see Reeve et al. 1990; Alexander et al., chap. 1).

The mechanisms by which naked mole-rats disperse are not known; possibilities include colony fissioning (i.e., a few individuals are isolated by a flood or other natural disaster or actively sequester themselves in a portion of the tunnel system) or coordinated movement aboveground to locations in which new burrow systems can be established (Brett, chap. 4). Both options are likely to be costly for three reasons: (1) the difficulty of tunneling through the rock-hard soil in which the animals live (see, e.g., Jarvis and Sale 1971); (2) the difficulty of finding patchily distributed food resources, particularly for small numbers of animals (Brett, chap. 5); and (3) the likelihood of predation, especially for animals dispersing aboveground. Hence, most individuals may be forced to remain and breed in their natal colonies and may increase their inclusive fitness by providing assistance to pups born to that colony (Alexander et al., chap. 1).

Within a colony, behavioral differences among nonbreeders may reflect the different reproductive options available to specific animals (see West-Eberhard 1981). For example, the potential for reproduction may decrease as an individual ages because of the decreasing probability that the current breeder(s) will cease to reproduce during the individual's lifetime or the increasing number of reproductive competitors as additional animals (with longer potential reproductive lifetimes) are born into the colony. Consequently, older animals with low reproductive value and little chance of direct reproduction may gain most by protecting the colony's current breeders, including participating in such dangerous activities as molehill production (Braude, chap. 6) and defense against predators or foreign conspecifics. In contrast, younger animals with a comparatively higher probability of becoming breeders might be expected to protect their reproductive interests by refraining from such activities, and instead performing apparently safer tasks such as gathering food, transporting nesting materials, and cleaning tunnels (see fig. 10-13); youngsters will also be sibs of new litters.

Alternatively, the division of labor in naked mole-rat colonies can be examined from the perspective of colony-level selection (e.g., Oster and Wilson 1978; Wilson 1985). Colonies with a more efficient division of labor (i.e., a more precise match between individuals' abilities and the tasks they perform) should be better able to survive and reproduce than those exhibiting a less effective organization, and thus the inclusive fitness of both breeders and nonbreeders in efficient colonies should be greater than that of members of less efficient colonies. Among naked mole-rats, the behavior of individuals appears to match their presumed abilities in that smaller, younger (i.e., weaker, less experienced) animals perform seemingly safer activities such as gathering food and nesting material, whereas larger, older (i.e., stronger, more experienced) animals participate in apparently more risky activities such as volcanoing, fighting off external threats, and (probably) digging. The efficacy of this system is enhanced in mole-rats because the animals can change in size and morphology with age, unlike holometabolous insects whose exoskeleton size is fixed at metamorphosis. Thus, the same mole-rat might be an especially rapid and effective forager when small and a particularly powerful colony defender when older and larger. The ability of individuals to alter both size and defensive capabilities (e.g., incisor size) with age is an important attribute of vertebrate age/size polyethisms not found among those of insects.

Thus, the infrastructure of naked mole-rat colonies contains elements that can be interpreted as the product of either colony-level selection (i.e., a "superorganism"; Wheeler 1928; Wilson 1985) or individual- and gene-level selection (Dawkins 1976; West-Eberhard 1981). Clearly, however, understanding how selection has shaped the social structure of naked mole-rat colonies will require an understanding of the demography of these animals and the dispersal and reproductive options available to colony members of different sizes and ages. Gathering such data represents an important challenge for the future.

THE EUSOCIALITY HYPOTHESIS

Wilson (1971; p. 4) characterized eusocial species as those exhibiting a reproductive division of labor, an overlap of at least two generations, and cooperative care of the young. By these criteria, naked mole-rats in our study colonies are indeed eusocial, thus confirming Jarvis's (1981) original suggestion. In our laboratory, a single female and one to three males per colony bred; breeding females were fed, defended, and assisted in rearing offspring by young from one or more previous litters. Data on wild-caught animals are also consistent with the eusociality hypothesis; as summarized by Brett (chap. 4), no wild-caught colony ever contained more than one female with developed teats. Furthermore, several colonies contained groups of similar-sized small individuals that were clearly juveniles and probably littermates. This suggests that *H. glaber* colonies in nature exhibit both a reproductive division of labor and an overlap of generations resulting from the retention of young in the colony. At present, no observational data exist regarding individual participation in pup care, colony maintenance, or colony defense in the field. However, Brett's radio-tracking data (1986, chap. 5) suggested that smaller, lighter mole-rats frequently shuttled between food sources and the presumed nest site, whereas larger, heavier animals were the first to be caught and were most frequently recaptured; Braude's (chap. 6) data indicated that larger individuals were primarily involved in volcanoing. These results are consistent with our laboratory observations.

COMPARISONS WITH EUSOCIAL INSECTS

We compared the social organization of *H. glaber*, as documented here, to the social organizations of three species of eusocial insects (table 10-7): the paper wasp *Polistes fuscatus*, the honey bee *Apis mellifera*, and the wood termite *Kalotermes flavicollis*. Like naked mole-rats, these species are highly social, and reproduction is typically limited to one female per colony. Both naked mole-rats and wood termites feed primarily on plant material, and their guts contain flagellate endosymbionts that facilitate the break down of cellulose (Porter 1957). These microorganisms are continually being passed among colony members via allocoprophagy. However, neither honey bees nor paper wasps feed exclusively on cellulose-containing plant parts, and allocoprophagy is not commonly observed; thus, sociality cannot be exclusively related to diet or endosymbiont transmission.

The breeding female's control of worker activity and reproduction in *P. fuscatus* colonies is mediated behaviorally (Reeve and Gamboa 1983, 1987); this also seems to be the case in *H. glaber* (Reeve and Sherman, chap. 11; Alexander, chap. 15). Active maintenance of breeding status through behavioral dominance also occurs in a variety of other primitively eusocial bees and wasps (e.g., *Lasioglossum zephyrum*, Michener and Brothers 1974; Breed and Gamboa 1977), all of which inhabit relatively small colonies (< 100; see Alexander et al., chap. 1). In contrast, reproductive inhibition of workers in the vastly larger colonies of *A. mellifera* and *K. flavicollis* (> 10,000 workers) is more passive and is primarily chemically mediated. The chemical in question is actively sought by worker honey bees and may serve to assure them of their queen's continued survival, vigor, and reproductive competence (Seeley 1979; 1985, pp. 22–31).

Conflict between breeders and nonbreeders appears to be more continuous and prolonged in species that live in small colonies (e.g., *P. fuscatus*), as opposed to those that live in large colonies (e.g., *A. mellifera*). This difference may be caused by the greater probability that members of small groups will have an opportunity to breed when the current breeder dies or is deposed and by the conflict between breeders and workers over the optimum activity levels of nonbreeders (Alexander et al., chap. 1; Reeve and Sherman, chap. 11). The former consideration may explain why a larger proportion of individuals retain the physiological ability to reproduce in species with small colonies, as opposed to species with very large colonies. The constant threat of reproduction by subordinates (i.e., nest usurpation from within) may in turn have favored aggressive behavioral dominance by breeding females in *P. fuscatus*, *L. zephyrum*, and *H. glaber* (see Alexander, chap. 15).

The ability of workers to reproduce directly also varies among the species considered here. Unlike reproductively neuter *A. mellifera* and *P. fuscatus*

Table 10-7

Comparisons of the Social Organization of *Heterocephalus glaber* and the Social Systems of Other Eusocial Vertebrates and Invertebrates

Species	Similarities to *Naked Mole-Rats*	Differences from *Naked Mole-Rats*	Source
Paper wasp (*Polistes fuscatus*)	single breeding female per colony aggressive domination of other colony members by reproductive female no permanent sterility in subordinate foundresses size—based subdivision of nonbreeding caste similar colony sizes (20–100 workers) slightly larger size in queens than in workers	all female workers haplodiploid genetics new nests each spring outcrossing promoted by dispersal of reproductives from nest site carnivore	West-Eberhard 1969; Reeve & Gamboa 1983, 1987; Noonan 1981
Honey bee (*Apis mellifera*)	single breeding female per colony colony reproduction by fissioning (swarming) age polyethism among nonreproductives slightly larger size in queens than in workers	all female workers haplodiploid genetics primarily chemical, not behavioral, reproductive suppression of workers permanently "sterile" nonreproductive females vastly larger colonies in honey bees outcrossing promoted by aerial mating aggregations nectarivore	Seeley 1982, 1985; Winston 1987
Wood termite (*Kalotermes flavicollis*)	single breeding female per colony diploid genetics male and female workers delayed cast fixation (?) division of labor among nonbreeders extremely long life of reproductives diet of plant material, breakdown of cellulose by gut endosymbionts allo- and autocoprophagy opportunity for workers to become reproductives if they escape the breeder's influence working behavior begins before adulthood	chemical, not behavioral, reproductive suppression vastly larger colonies in termites queens many times larger than workers outcrossing promoted by aerial mating aggregations	Wilson 1971; Alexander et al., chap. 1

Species			References
Florida scrub jay (*Aphelocoma coerulescens*)	alloparents of both sexes alloparents are previous offspring no simultaneous breeding by alloparents and parents diploid genetics	dominant breeding pairs small groups (2–10 birds) granivore-insectivore similarity in size of alloparents and breeders outcrossing breeding male as primary colony defender	Woolfenden 1975; Woolfenden & Fitzpatrick 1978, 1984; Francis et al. 1989
White-fronted bee-eater (*Merops bullockoides*)	alloparents of both sexes no simultaneous breeding by alloparents and parents alloparents are previous offspring diploid genetics	dominant breeding pairs small groups (2–7 birds) within colonies of 50–500 possibility of alloparents' having been or becoming breeders insectivore similarity in size of alloparents and breeders	Emlen 1981, 1984; Emlen & Wrege 1986, 1988
Silverbacked jackal (*Canis mesomelas*)	alloparents of both sexes mammals; diploid genetics alloparents are previous offspring division of labor in allofeeding young	dominant breeding pairs small groups (2–5 animals) alloparents usually become breeders after several years carnivore-scavenger similarity in size of alloparents and breeders	Moehlman 1983, 1986
Dwarf mongoose (*Helogale parvula*)	dominant female as primary breeder; multiple mating alloparents of both sexes mammals; diploid genetics divisions of labor in defense and allofeeding of young; larger nonbreeders as primary group defenders alloparents are previous offspring	mating by alloparents; pregancy possible for subordinate females occasional lactation by nonpregnant females small groups (10–30 animals) carnivore-insectivore similarity in size of alloparents and breeders	Rood 1978, 1980, 1983, 1986; Rasa 1977, 1987
Wild dog (*Lycaon pictus*)	dominant breeding female plus several male consorts aggressive suppression of subordinate females by the dominant food located by foraging in groups mammals; diploid genetics alloparents are previous offspring	small groups (8–12 animals) estrus in female alloparents; possibility of their mating occasional killing of subordinate's pups by dominant females usually male alloparents carnivore similarity in size of alloparents and breeders	Frame et al. 1979; Malcolm & Marten 1982; Lawick 1973

workers, which at best can reproduce only (sons) parthogenetically, at least some nonbreeding *H. glaber* females retain the ability to reproduce sexually. In *K. flavicollis*, reproductive castes are initially flexible (e.g., both sexes can become reproductives) but later become fixed in response to stimuli within the colony. At present, the limits of reproductive flexibility among male and female naked mole-rats are uncertain (see Jarvis, chap. 13).

Whereas the nonreproductive castes of *P. fuscatus* and *A. mellifera* comprise only females, *H. glaber* and *K. flavicollis* are similar in two important ways. First, workers in both species may be either males or females, and working behavior begins early in life. Second, whereas the two hymenopteran species are haplodiploid, mole-rats and termites are diploid. Naked mole-rats and termites are also similar in that the tasks performed by nonbreeders can be roughly divided into maintenance and defense behaviors, although these distinctions are less rigid in *H. glaber* than in *K. flavicollis*. Whereas behavioral distinctions between worker and soldier termites are fixed, we found that *H. glaber* individuals exhibit a size-based and perhaps age-based polyethism (fig. 10-13). Temporal castes also occur in honey bees (Seeley 1982; cf. Seeley 1986 and Kolmes 1986) and, like *A. mellifera* individuals, individual mole-rats seem to progress from frequent to infrequent participation in maintenance behaviors and infrequent to frequent participation in defense behaviors as they grow (age). Thus, the oldest individuals in honey bee and (perhaps) naked mole-rat colonies are involved in the most dangerous tasks, in particular colony defense.

Clearly intriguing parallels exist between the social organization of *H. glaber* and the colony structures of several eusocial insect species (see also Alexander et al., chap. 1). Can similar comparisons be drawn between naked mole-rats and other cooperatively breeding vertebrates?

COMPARISONS WITH "EUSOCIAL" VERTEBRATES

The social organization of two species of cooperatively breeding birds and three species of cooperatively breeding mammals are compared with that of naked mole-rats in table 10-7. The social units of Florida scrub jays (*Aphelocoma coerulescens*), white-fronted bee-eaters (*Merops bullockoides*), silver-backed jackals (*Canis mesomelas*), and dwarf mongooses (*Helogale parvula*) can all be characterized as a dominant, nearly monogamous pair assisted by a variable number of subordinate, nonbreeding individuals known as *alloparents*. Somewhat in contrast, *Heterocephalus glaber* colonies and wild dog packs (*Lycaon pictus*) are dominated by a breeding female, with the breeding male or males behaving as subordinates to her. Although each of the six species considered here is highly social, the sizes of groups differ between taxa, with naked mole-rat colonies being largest by an order of magnitude. With the exception of naked mole-rats, all species are either carnivorous or insectivorous. Among the carnivores/insectivores, social foraging only occurs in wild dogs; naked mole-rat colonies also cooperate to locate and obtain their patchy, hard-to-find food (subterranean tubers; Brett, chap. 5).

The degree of reproductive suppression imposed on alloparents varies among the species under consideration. Whereas alloparental Florida scrub jays and silver-backed jackals do not breed, subordinate female dwarf mongooses and wild dogs may mate and occasionally even give birth; however, the offspring of subordinates survive less often than do the offspring of dominants. The situation in white-fronted bee-eaters is even more complex since alloparents may be either nonbreeders or breeders whose nests have failed, and individuals (especially males) may switch between attempting to breed in one group and serving as an alloparent in their natal group. In contrast to these species, subordinate female mole-rats usually do not become sexually receptive (except under peculiar circumstances; see Jarvis, chap. 13). Furthermore, whereas scrub jay and jackal alloparents frequently become breeders after several years of helping, it seems likely that many subordinate naked mole-rats never breed.

Divisions of labor in group vigilance and defense have been studied in dwarf mongooses (Rasa 1977, 1987), scrub jays (Francis et al. 1989), and naked mole-rats. In the scrub jays, male breeders mobbed snakes most intensely, and female breeders and helpers participated significantly less often but about equally. In the mongooses, the largest nonbreeders were the first to attack intruders from other packs and to investigate frightening stimuli; breeding females participated infrequently in these behaviors. The division of labor in dwarf mongooses rather closely approximates our results for naked mole-rats (see fig. 10-13).

All of the cooperatively breeding vertebrates considered here are similar in that only a subset of the group reproduces, both males and females serve as alloparents, and offspring from previous years are retained as helpers. In other words, all of these animals are eusocial according to Wilson's classic (1971) definition. Some readers may object to this terminology because there is as yet no evidence of physiological sterility among vertebrate alloparents; however, physiological sterility is not a necessary condition for eusociality. Moreover, few data are available regarding the frequency with which alloparents in any species of cooperatively breeding bird or mammal live out their lives without ever directly reproducing (i.e., the rate of sociological sterility). We suspect that socially imposed lifetime sterility is more widespread than is currently realized.

We hypothesize that naked mole-rats (and perhaps dwarf mongooses) represent extreme, complex, or "advanced" vertebrate eusociality, scrub jays and silver-backed jackals somewhat less elaborate ("primitive") vertebrate eusociality, and white-fronted bee-eaters an in-between stage. Regardless of the classificatory terminology eventually adopted, it is obvious that social systems which entomologists (e.g., Wilson 1971; Eickwort 1981) would term "primitive eusociality" occur in birds and mammals; eusociality can thus no longer be considered phylogenetically unique or taxonomically limited to insects. We join Andersson (1984) and Alexander et al. (chap. 1) in suggesting that the extrinsic and intrinsic factors (Evans 1977) favoring the evolution of eusociality can best be understood by including vertebrate social systems in comparative studies of this topic. Probably the same ecological pressures and constraints (e.g., predation, food distribution) that favored solitary life in some species of bees and wasps and sociality in others (Brockmann 1984; Michener 1985) also favored the same spectrum of sociality in birds and mammals. Apparently nepotism (assisting kin) has played the same pivotal role in enhancing the complexity of all these social systems. Further studies of naked mole-rats, as well as additional comparative analyses of the social systems of other cooperatively breeding vertebrates and invertebrates are clearly warranted.

Summary

Three colonies of individually marked naked mole-rats were studied at Cornell University from 1980 to 1985. Each colony or its ancestors had originally been captured in Kenya, East Africa, during the period 1977 to 1979. In the laboratory, the animals were housed in artificial burrow systems constructed of clear acrylic plastic (plexiglass) tubes. All three colonies reproduced in captivity such that each contained both wild-caught and laboratory-born animals during this study.

Long-term, unobtrusive observations were combined with simple behavioral experiments designed to increase the frequency of specific behaviors. Both colony-maintenance behaviors (e.g., digging, transporting food and nesting material, and clearing the tunnels of debris or protruding objects), and colony-defense behaviors (e.g., threatening and biting either snakes or mole-rats from different colonies and guarding the nest) were studied. The data indicated that, within colonies, individuals exhibited behavioral variations associated with differences in reproductive status, body weight, and perhaps age. These results are summarized in figure 10-13 and below.

1. Reproductive status. Only one female and one to three males per colony bred; these were the only colony members to exhibit sexual behavior. Breeding animals also performed the majority of direct care given to pups. In contrast, breeding animals seldom participated in colony-maintenance or defense activities. If the breeding female died or was removed, several other females grew rapidly and competed aggressively for reproductive dominance.

2. Body weight. Colony-maintenance and defense activities were performed by nonbreeding animals; both sexes participated equally. The specific behaviors exhibited by individuals varied with body weight. Smaller nonbreeders were the primary participants in all colony-maintenance activities; digging and colony-defense activities were performed primarily by larger nonbreeders.

3. Age. Growth curves for nonbreeding mole-rats followed similar trajectories, providing no evidence for discrete morphological castes among these animals. Instead, growth curves for nonbreeders suggested that body weight increased as a function of age. Thus, behavioral differences among nonbreeders may reflect an age/size polyethism in which smaller (younger) animals gradually switched from performing colony-maintenance activities to performing colony-defense activities as they grew larger (and older).

The data indicated that naked mole-rats exhibit a reproductive division of labor, an overlap of two or more generations of colony members, and cooperative care of young. These three characteristics qualify *Heterocephalus glaber* as eusocial. However, naked mole-rats are diploid mammals, and thus the social system of *H. glaber* is probably best viewed as an extreme form of vertebrate cooperative breeding in which (1) the probability of an individual's becoming a breeder is low but not insignificant, (2) reproductive inhibition is behaviorally enforced (and perhaps chemically mediated), (3) nonbreeding offspring are retained in the colony on a long-term basis, and (4) individuals of different sizes and ages specialize in either collecting the resources necessary for survival or in defending the group.

To place these findings in perspective, the social organization of *H. glaber* was compared to the social systems of three species of eusocial insects and five species of cooperatively breeding vertebrates (two birds and three mammals). The remarkable convergence between the social organization of these mammals and birds and the social systems of certain eusocial insects suggests that the evolutionary origins of eusociality may best be understood by examining both extrinsic (ecological) and intrinsic (genetic) selective factors and that haplodiploidy alone is neither necessary nor sufficient to account for the evolution of eusociality.

Acknowledgments

We thank J.U.M. Jarvis for her pioneering studies of naked mole-rats, and R. D. Alexander for ideas and encouragement. The 1979 expedition during which colony A was captured included R. D. Alexander, L. A. Alexander, J.U.M. Jarvis, and C. Kagarise Sherman and was accompanied by A.S.N. Chondo; we thank these people for diligent efforts in the field, and the National Geographic Society for financial support. In the laboratory, our efforts were aided by V. Demas, M. A. Fleming, A. F. Greene, M. Morero, N. Myers, G. F. Streidter, S. F. Payne, C. Vispo, C. Kagarise Sherman, and J. R. Wieczorek. Helpful reviews of the manuscript were provided by R. D. Alexander, S. T. Emlen, J.U.M. Jarvis, T. D. Seeley, C. Kagarise Sherman, G. Borgia, and especially H. K. Reeve. Support was provided by a National Science Foundation (NSF) predoctoral fellowship (to E.A.L., 1985-1986) and by a John Simon Guggenheim Memorial Foundation Fellowship (to P.W.S., 1984-1985). Continuing research support has been provided to P.W.S. by NSF grant BNS-8615842 and by the College of Agriculture and Life Sciences at Cornell (Hatch Grants). The first draft of this paper was made available to other authors in this volume in 1985.

11 Intracolonial Aggression and Nepotism by the Breeding Female Naked Mole-Rat

Hudson K. Reeve and Paul W. Sherman

Naked mole-rat colonies are not perfectly harmonious societies. Colony operations are frequently punctuated by "mild" agonistic interactions such as tugging (Braude 1983), tooth fencing, batting, shoving (Lacey et al., chap. 9), and occasionally by serious aggression such as gaping and biting (Lacey and Sherman, chap. 10; Jarvis, chap. 13); agonistic vocalizations, especially hissing and chirping, accompany many such behaviors (Pepper et al., chap. 8). This chapter presents a detailed analysis of one agonistic behavior, the shove. Shoves are vigorous, prolonged pushes, usually involving nose-to-nose contact, that can cause a mole-rat to be displaced up to 1 m in the tunnel system (see Lacey et al., chap. 8). Shoving is particularly interesting to us because it occurs more often than other forms of mild aggression and appears to be initiated frequently by the breeding female (see plate 11-1).

The appearance of agonism by the breeding female mole-rat is intriguing in view of evidence that queens of many social Hymenoptera that inhabit small-sized colonies are the most active and aggressive colony members (e.g., in *Polistes* paper wasps; Reeve and Gamboa 1983, 1987; in halictine bees, Michener 1988; see also Alexander et al., chap. 1). To understand why the breeding female naked mole-rat is aggressive, we initiated a descriptive and functional analysis of shoving behavior in six lab *Heterocephalus glaber* colonies, three at Cornell University and three at the University of Michigan.

We address four questions: (1) How does the rate of shoving by the breeding female compare with that by other colony members, and how is the breeding female's rate of shoving influenced by the (2) genetic relatedness, (3) body size (weight), and (4) sex of the potential recipient? To address these questions, we analyzed the characteristics of initiators and recipients of 633 shoves recorded in 150 h of detailed, nonintrusive observations of the six colonies. The results indicated that breeding females were the primary shovers and that their behavior varied with the genetic relatedness and body size of potential recipients. In light of the data, the possible adaptive significance of shoving is discussed and comparisons with queen aggression in social insect colonies are drawn.

Plate 11-1 An open-mouth threat by the breeding female toward a small colony mate; such threats often precede or accompany shoves by the breeding female. Photo: R. A Mendez

Methods

COMPOSITIONS OF STUDY COLONIES

We observed three laboratory colonies (K, A, and TT) between 1985 and 1988 at Cornell University. The histories of these colonies, their compositions, and the tunnel layouts are described in the preceding chapter. In addition, we observed three laboratory colonies, A(M), B(M), C(M), in the Museum of Zoology of the University of Michigan from July 25 to August 10, 1989; these were the same colonies studied by Braude (1983) and Pepper et al. (chap. 9). Each of the six colonies contained a single breeding female. Breeding individuals in each colony were inferred by direct observations of parturition (the breeding female) or copulations with the breeding female (the breeding male(s)). In two of the Michigan colonies, A(M) and B(M), no males were observed copulating with the breeding female. For these two colonies, a breeding male was identified by its reciprocal nose-to-anus contacts with the breeding female; this interaction occurs primarily between the breeding female and copulating males (Lacey et al., chap. 8; Jarvis, chap. 13; Reeve, unpubl. data).

COLONY MAINTENANCE

The three Cornell colonies were maintained at 26°–28°C and 50%–65% relative humidity, under low levels of red or yellow lighting (see Lacey and

Sherman, chap. 10; Jarvis, Appendix). Colonies were fed ad libitum except during periods of data gathering. Before and during the 2–4 mo long periods of data gathering, each colony was fed about 5.2 g/individual/day of a diet consisting of 50% sweet potato, 15% lettuce, 20% apple, and 15% banana; this is approximately the amount of food a colony consumes in a 24-h period. These diets were supplemented every 1–2 wk with lentils and a vitamin solution mixed in a bolus of baby food. The three Michigan colonies were maintained under constant (dim) light and fed ad libitum amounts of fresh vegetables, even during the 2-wk data-gathering period in 1989. All colony members were weighed every 3–4 mo (Cornell) or 4–6 mo (Michigan). For statistical analyses, weights taken closest to the midpoint of the period of observation were used.

INDIVIDUAL COLONIES

COLONY K

Colony K (Cornell) contained 21 individuals during the study and exhibited the greatest range in kinship among the study colonies. Colony K had been formed by joining two separate field-caught colonies shortly after collection (in 1980). The breeding female during this study ("three-bars") was born on June 8, 1981; she was conceived in a laboratory mating between two field-caught members of the same colony. Three-bars fought her way to reproductive status, as detailed in the previous chapter, and bore her first litter on March 22, 1984. Thus, three-bars was the female breeder in a colony composed of (1) individuals derived from a different wild-caught colony ($n = 5$), (2) individuals derived from her own colony but no more closely related to her than aunts and uncles ($n = 5$), (3) her littermate siblings ($n = 6$), and (4) her own offspring ($n = 4$).

The average genetic relatedness among field-caught colony members (relative to individuals from the different colonies) has been estimated from DNA fingerprint data as 0.81 (Reeve et al. 1990). This value, derived using the regression technique of Pamilo and Crozier (1982), estimates the probability that an allele in one individual is identical by descent to that in a randomly chosen colony mate, relative to the probability that alleles in two randomly chosen gametes of the first individual are identical by descent. Thus, assuming that field-caught animals from the same colony were predominantly full siblings (as is reasonable if there is an asymmetry in the fertilization success among the one to three breeding males per colony), the relatednesses between the breeding female and her four different classes of colony mates (above) are, respectively, (1) $r = 0$, (2) $r = 0.72$, (3) $r = 0.81$, and (4) $r = 0.81$. For convenience, we term the members of these four relatedness classes as nonrelatives, aunts/uncles, siblings, and offspring, respectively. Thus, colony K provided us an opportunity to study the behavior of the breeding female toward various kin and non-kin.

COLONY A

This Cornell colony was composed of 28 individuals during the study. Of these, 22 were wild-caught from one source colony (in 1979), 4 were surviving offspring conceived in lab matings between field-caught males from the source colony and 4 different field-caught females, and 2 were adults captured from a second source colony (again in 1979). Unfortunately, we cannot assign either maternity or paternity to the 4 offspring: Since 4 breeding females bore young simultaneously in this, the first breeding event in the lab, and since their pups could not be tagged, the surviving offspring could have been the progeny of 1, 2, 3 or all 4 of the females.

During our study, the breeding female (4) was a field-caught animal from the larger source colony. She reproduced for the first time on April 29, 1980, and then entered a long period of reproductive quiescence (during which no other female bred); on January 17, 1988, she bore another litter. We observed female 4 both in 1985, while she was reproductively quiescent, and in 1988, after she had resumed reproduction. Unless otherwise noted, our analyses refer to the period in which female 4 was reproducing regularly. In the 1988 observations, female 4 was observed both before and after she bore a litter.

COLONY TT

Cornell colony TT consisted of 32 individuals during the study. Of these, 27 were offspring of the breeding female (+), and 5, including the breeding female herself, were collected from a single wild colony by Jarvis in 1977 (see Lacey and Sherman, chap. 10). Female +, like the breeding females in all the other Cornell colonies, became established as a breeder after capture; she was not the original breeding female in the wild colony. The offspring, which were conceived in lab matings between female + and field-caught male(s) from the same colony, were born on or after July 12, 1981.

COLONY A(M)

Michigan colony A(M) consisted of 26 animals during the study. The breeding female (61) and 7 other adults (all siblings of female 61) were descended from a breeding female (deceased at the time of the study) that had been field-caught in 1979, along with 3 other adult males in the colony. Thus, at least 2 of these males must have been no more closely related than uncles to female 61 (1 male may have been her father). In addition to 2 uncles and a father (or 3 uncles) and 7 siblings, female 61 had 15 offspring in the colony, all born in the lab beginning on May 12, 1988.

COLONY B(M)

Michigan colony B(M) contained 28 animals during the study. The breeding female (33) had been field-caught in 1979 (as a young nonbreeder) along with 10 other animals from the same source colony. Six other adults in this colony had been captured from a second source colony. The 11 remaining animals were offspring of female 33, born in the lab beginning on February 11, 1985.

COLONY C(M)

Michigan colony C(M) comprised 20 animals during the study. The breeding female (49) had been field-caught in 1979 (as a young nonbreeder), along with 7 other animals from the same source colony. Eight other adults in this colony had been captured from a second source colony. The other 4 animals were offspring of female 49, born in the lab beginning on January 11, 1988.

Observation Protocol and Schedule

Observations were made daily in 30–60 min blocks, beginning at various times between 10:00 A.M. and 4:00 P.M. Shoving was readily detected, even if the animals involved were not being observed closely at the time, because of the audible hissing and substrate scratching that accompanies shoving (see Pepper et al., chap. 9). We defined a bout of shoving as a continuous push, during which contact between the participants was not broken. For every shoving bout that occurred, we recorded (1) the identity of the aggressor, (2) the identity of the recipient, and (3) the location (nest box or tunnel system) of the shove. For observations of colonies A, TT, A(M), B(M), and C(M), the duration (in seconds) of the shove was also recorded. All observations were blind with respect to relatedness; that is, observers were unaware of the colony genealogies until after the data had been tabulated.

Colony K was observed between August 28, 1985 and December 29, 1985 for a total of 31.7 h (55 observation periods) and between February 26, 1986 and May 2, 1986 by a different observer for a total of 32 h (32 observation periods). Only shoves initiated by the breeding female were recorded in the 1986 observations. Colony A was observed between October 3, 1985 and November 21, 1985 (8 h, 15 observation periods) when the breeding female (like all other females in the colony) was reproductively quiescent and from March 14, 1988 to June 28, 1988 (26.5 h, 47 observation periods) when the breeding female was reproducing regularly. Colony TT was observed from February 13, 1988 to June 22, 1988 (23 h, 44 observation periods). Colonies A(M), B(M), and C(M) were observed from July 25, 1989 to August 10, 1989 (9.65, 9.18, and 10.37 h, respectively; a mean of 11 observation periods/colony).

Because the breeding female's activity level is low for several hours following feeding (Reeve, unpubl. obs.), observations were begun 6–48 h after the colony had been fed. The mean time between feeding and the beginning of observations was 22.4 h (SD = 9.7 h) for colony-K observations, 23.3 h (SD = 6.5 h) for colony-TT observations, and 23.3 h (SD = 12.7 h) and

21.09 h (SD = 8.3 h) for colony-A observations when the breeding female was reproductively quiescent and reproducing regularly, respectively. The latency between feeding and observation ranged from 15 to 23 h for the Michigan colonies, but, as a result of the ad libitum diet, food was almost always present in the food hampers of these colonies (this was not generally true for the Cornell colonies).

STATISTICAL ANALYSES

Unless otherwise noted, all correlations are Pearson product-moment correlations, and all statistical tests are two-tailed. Partial-correlation analyses of the influences of recipient relatedness and body size on the breeding female's shove rate in colony K (the most extensively studied colony) were performed in three ways: Pearson product-moment partial correlations were calculated on the assumption that nonrelatives, aunts/uncles, siblings, and offspring were related to the breeding female by 0, 0.72, 0.81, and 0.81, respectively (see above). Spearman partial correlations, in which only the rank order, not the absolute values, of the relatednesses are important, were also calculated. In one analysis, field-caught colony mates were assumed to be predominantly full siblings (i.e., they had the same mother and father; relatedness rank order was 1, 2, 3, 3, for the 4 relatedness classes, respectively). In a second analysis, we assumed that colony mates were predominantly half siblings (i.e., they had the same mother but different fathers; relatedness rank order was 1, 2, 3, 4, respectively).

Pearson partial correlations between the breeding female's shove rate and recipient relatedness or body weight were also calculated for the Michigan colonies, which contained either a mixture of nonrelatives, siblings, and offspring (colonies B(M) and C(M)), or a mixture of uncles, siblings, and offspring (colony A(M)). The three putative uncles in colony A(M) were assigned relatednesses of 0.75 to take into account the possibility that one of these males was the breeding female's father, but this adjustment did not affect the results of the statistical analysis. Partial-correlation analyses in which each of the three males in turn was assumed to be the father ($r = 0.81$) of the breeding female also did not affect the overall trends. The small number of nonrelatives in colony A, and uncertainty over the maternity of offspring in this colony, prevented calculation of the partial correlation between the breeding female's shove rate and recipient relatedness for this colony. Colony TT consisted only of the breeding female's presumed siblings and her offspring, ruling out calculation of the latter partial correlation in this colony as well.

Statistical comparisons of shove rates of a given breeding female between two time periods were carried out with Wilcoxon matched-pairs tests; the rates of shoves initiated toward specific recipients in the first period were compared to the corresponding rates in the second period. When possible, tests of the significance of correlations between two variables were conducted with one-sample, two-tailed t-tests across all six colonies, with the correlation within each colony serving as a single, independent data point. Likewise, tests of differences between the sexes or among individuals of different breeding status were carried out across all six colonies with paired comparisons t-tests or, when appropriate, a Friedman two-way ANOVA. For statistical comparisons between rates of shoving by the breeding female and other colony members in colony K, we used only the 1985 data, because only shoves initiated by the breeding female were recorded in 1986.

Results

SHOVE-INITIATION RATES

In all six study colonies (Cornell and Michigan), the breeding female initiated more shoves per unit of time than did any other colony member (fig. 11-1). The rate at which the average breeding female shoved (3.8 shoves/h) was 12 times the rate for the average breeding male (0.32 shoves/h) and nearly 400 times that for the average nonbreeder in the respective colonies (0.1 shoves/h); these differences were highly statistically significant ($p < 0.005$, Friedman two-way ANOVA and multiple-comparisons tests). In fact, the breeding female alone accounted for 74%, 67%, 78%, 84%, 69%, and 97% of all recorded shoves in colonies K, A, TT, A(M), B(M), and C(M), respectively.

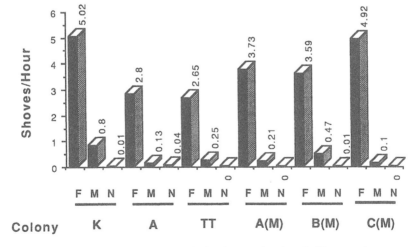

Fig. 11-1 The mean shove rate (shoves/h) for the breeding female (F) versus the mean shove rate for breeding (copulating) males (M) and nonbreeders (N; sexes averaged) in all six study colonies of naked mole-rats (three from Cornell, three from Michigan).

Breeding males and large nonbreeders initiated most of the remaining shoves. On the average, a breeding male initiated significantly more shoves per unit of time than did a nonbreeder in a colony ($p < 0.005$; Friedman two-way ANOVA and multiple-comparisons tests). All of these trends and significance patterns held whether nonbreeding males and females were considered together or separately. In colony A, where an appreciable number of shoves (22% of the total) were initiated by nonbreeders, there was a significant, positive correlation between the nonbreeder's body weight and its rate of shove initiation ($r = 0.56$; $p < 0.01$). Among these nonbreeders, the shove rates for males (0.03 shoves/h) and females (0.05 shoves/h) did not significantly differ ($p > 0.05$, Mann-Whitney U-test). Similarly, in colony B(M), where 5% of the shoves were initiated by nonbreeders, the correlation between the nonbreeder's body weight and its rate of shove initiation was positive and significant ($r = 0.44$; $p < 0.05$), and the mean shove rate for nonbreeding males (0.02 shoves/h) and nonbreeding females (0.03 shoves/h) did not significantly differ ($p > 0.10$; Mann-Whitney U-test). In sum, the breeding female initiated the majority of shoves, and the remainder were initiated primarily by breeding males (which generally are large animals) and large nonbreeders of both sexes.

The high shove rate of breeding females was associated with their reproductive activity. Breeding-female 4 in colony A shoved significantly more frequently after she resumed a regular cycle of reproduction (2.8 shoves/h) than she did during her period of reproductive quiescence (0 shoves/h; $p < 0.0001$; Wilcoxon matched-pairs test). The mean shove rate for female 4 during the latter 30 days of her pregnancy (3.7 shoves/h; 9.4 h obs.) was nearly twice her postpartum mean shove rate (2.2 shoves/h; 17.1 h obs.), but this difference was not statistically significant ($p = 0.14$; Wilcoxon matched-pairs test). Nevertheless, a time plot of the daily shove rate for female 4 revealed a temporary (ca. 1 wk) depression in her shoving behavior in the week following the birth of her litter (fig. 11-2); this period corresponds to the time during which her pups were alive in the colony (all the pups in this litter died in less than 7 days). A similar depression in the shove rate for female 61 is evident when her newborn offspring were still present in colony A(M) (fig. 11-2); her pups all died within 2 days.

DURATION AND LOCATION OF SHOVES BY THE BREEDING FEMALE

The durations of the breeding female's shoves were quite variable, ranging from 0.5 s to 10.0 s. The mean shove duration across all colonies was 2.1 s (mean SD = 1.3 s). Of the shoves initiated by the breeding female, 87%, 53%, 87%, 50%, 78%, and 87% occurred in the nest box of colonies K, A, TT, A(M), B(M), and C(M), respectively; this probably reflects the relatively high proportion of time spent by the breeding female in the nest box (see plate 11-2). The mean durations of shoves of the same individuals in the nest box (2.2 s) and the tunnel system (2.5 s) did not significantly differ across colonies ($p > 0.05$; Wilcoxon matched-pairs test).

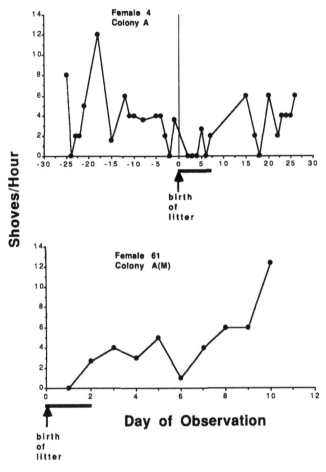

Fig. 11-2. Hourly shove rates (*top*) for breeding female 4 (colony A) 1 mo before and 1 mo after the birth of her litter (on April 9, 1988), and (*bottom*) for female 61 (colony A(M)) for 11 days following the birth of her litter (on July 27, 1989). *Horizontal bars*, Periods during which live pups were present in each colony

EFFECT OF RECIPIENT'S RELATEDNESS

The shove rate of female three-bars, the breeding female in colony K, was strongly influenced by the relatedness of the recipient: less-related individuals were shoved more. The Pearson partial correlation between rate of shoving and the recipient's relatedness (controlling for recipient weight) was significantly less than zero for both the 1985 observations ($r = -0.55$) and for the 1986 observations by a different observer ($r = -0.55$; $p < 0.02$ for both). For the combined observation periods, this correlation was -0.65 ($p < 0.01$). The partial correlation was $r = -0.59$ when breeding males were excluded from the analysis ($p < 0.02$). The partial Spearman (rank-order) correlations for the

Plate 11-2. A breeding female lies atop a pile of colony mates in the nest chamber; because of her high levels of activity and frequent exits from and entrances to the nest, the breeding female is usually found on top of the pile of huddling colony members Photo: R. A Mendez

combined observations were also significantly less than zero ($p < 0.01$), both when field-caught animals were assumed to be full siblings ($r_s = -0.63$) and when they were assumed to be half sibs ($r_s = -0.67$). The Pearson partial correlation between the breeding female's shove rate and the relatedness for male recipients ($r = -0.79; n = 12$) did not differ significantly ($p > 0.05$) from the corresponding partial correlation for female recipients ($r = -0.40, n = 8$).

A further analysis of the distribution of the shoves by female three-bars across kin classes revealed that (1) uncles/aunts were shoved as frequently as nonrelatives, (2) both were shoved significantly more often than either siblings or offspring, and (3) siblings and offspring were shoved at about the same (low) rate (fig. 11-3). Removal of the breeding males from the analysis did not affect any of these trends or the significance patterns (the mean rate at which aunts/uncles were shoved increased to 0.34/h, which is almost identical to that for nonrelatives, 0.33/h).

The pattern of shove rates for the different relatedness classes cannot be attributed simply to differences in mean weight among these classes. When breeding males were excluded from the analysis, the only significant differences in mean weight were between offspring (21.6 g) and the other relatedness classes, which tended to include larger individuals (nonrelatives, 31.4 g; aunts/uncles, 33.5 g; siblings, 28.9 g; $p < 0.05$ only for comparisons involving offspring, one-way ANOVA and Scheffé's multiple-comparisons tests).

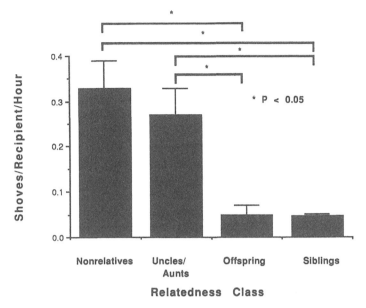

Fig 11-3. Hourly shove rates by breeding female three-bars in Cornell colony K as a function of the recipient's relatedness class. The relatedness classes differed significantly in the number of shoves received ($p < 0.005$, Kruskal-Wallis test) Uncles/aunts ($n = 5$) and nonrelatives ($n = 5$) received significantly more shoves from three-bars than did siblings ($n = 6$) and offspring ($n = 4$) ($p < 0.05$, multiple-comparisons tests). Siblings and offspring were not shoved at significantly different rates by the breeding female ($p > 0.05$).

Thus, the breeding female behaved as if she had an all-or-none relatedness threshold for shoving: Individuals of very high relatedness (ca. 0.81) were seldom shoved, whereas slightly less related (ca. 0.72) colony mates and colony mates that had originated from a different field colony both were frequently shoved. In further support of this threshold hypothesis, breeding female 4 in colony A and breeding female + in colony TT, both of which were exposed mostly to offspring or presumed siblings (field-caught colony mates), displayed about the same low rates of shoving (0.10 shoves/recipient/h and 0.09 shoves/recipient/h, respectively) exhibited by female three-bars in colony K toward her siblings and offspring (0.05 shoves/recipient/h).

For the Michigan colonies, the partial correlation between breeding-female shove rate and recipient relatedness was significantly negative in colony A(M) ($r = -0.64; p < 0.01$) and negative, though not significantly so, in colonies B(M) and C(M) ($r = -0.21$ and -0.30, respectively; $p > 0.05$). The mean partial correlation, over all four colonies for which this correlation could be calculated, was significantly negative ($r = -0.45; p = 0.03$; one-sample t-test). Thus, overall, the breeding female shoved less-related individuals more often, independently of their body weight. The Michigan colonies also provided evidence of the threshold effect of relatedness on shoving: breeding female 49's mean rate of shoving nonrelatives in colony C(M) (0.49 shoves/recipient/h) was similar to both the rates at which female three-bars in colony K shoved nonrelatives and aunts and uncles (above) and to the rate at which female 61 in colony A(M) shoved presumed uncles (0.73 shoves/recipient/h). In addition, the mean rates at which the breeding female shoved siblings and offspring in the Michigan colonies (0.06, 0.12, and 0.06 shoves/recipient/h for colonies A(M), B(M), and C(M), respectively) resembled the low rates at which the breeding females of the Cornell colonies shoved their siblings and offspring.

Surprisingly, female 4 in colony A did not discriminate between mole-rats from her own source colony and the two individuals from a different source colony: The mean number of shoves/recipient/h toward the latter (0.04) was actually lower than, but not significantly different from, the mean number of shoves/recipient/h toward members of female 4's own source colony (0.11; $p = 0.22$; Mann-Whitney U-test).

EFFECT OF THE RECIPIENT'S BODY WEIGHT

Across all six colonies, the rate at which the breeding female shoved was significantly positively correlated with the body size of the recipient (figs. 11-4, 11-5; table 11-1), and this correlation did not significantly differ between male and female recipients ($p > 0.10$, Wilcoxon matched-pairs test; table 11-1). The correlations between rate of shoving by the breeding female and the recipients' body weight were also significantly greater than zero if breeding males were excluded from the analysis.

The mean duration of shoves by the breeding female was positively, but not quite significantly, correlated with recipients' body weight across all colonies ($r = 0.32$, $p = 0.11$, one-sample t-test). However, when relatedness was controlled for, this correlation was significantly positive within both colony A(M) and colony B(M) ($r = 0.78$ and 0.50, respectively; $p < 0.01$ for both), indicating a tendency in at least some colonies for larger recipients to be shoved longer.

EFFECT OF THE RECIPIENT'S SEX

Breeding females shoved without regard to the sex of the recipient. Across all colonies, the mean rate at which the breeding female shoved a male (0.30 shoves/recipient/h) was higher than, but not significantly different from, the corresponding mean rate at which females were shoved (0.15 shoves/recipient/h, $p > 0.05$, Wilcoxon matched-pairs test). The tendency for males to be shoved more often was probably an artifact of the tendency of males to be heavier than females in our study colonies; the mean weight of males (36.9 g) was significantly greater than the corresponding mean weight of females (33.3 g) across colonies ($p = 0.045$, matched pairs t-test). Indeed, the mean

Table 11-1

Correlations between Rate of Shoving by the
Breeding Female and Recipients' Body Weight
for the Six Study Colonies

Colony	Males	Females	Both Sexes
Cornell			
K	+0 36	+0.20	+0.26
A	+0 66	+0 53	+0.59
TT	+0.48	+0.72	+0.55
Michigan			
A(M)	+0 42	+0.27	+0.54
B(M)	+0 34	+0 67	+0 66
C(M)	–0.15	+0 07	+0.11
Means	0.35**	0 41*	0.45*

NOTE· Correlations are given for all colony members and for males and females separately (for those animals whose sex was known) In colonies with variability in kinship, partial correlations (effects of recipient relatedness removed) are given Mean correlation coefficients are significantly greater than zero (one-sample t-tests) Males and females did not differ in the correlation between breeding-female shove rate and recipient's body weight ($p > 0$ 05; Wilcoxon matched-pairs test)

* $p < 0$ 01
** $p < 0$ 05

regression slope of breeding-female shove rate on male-recipient body weight (0.17 shoves/h/g) was close to and not significantly different from the corresponding mean regression slope for females (0.14 shoves/h/g, $p > 0.10$, Wilcoxon matched-pairs test).

As discussed above, male and female recipients did not differ significantly in the correlations between the rate at which individuals were shoved and their relatednesses or body weights. In addition, we found no evidence that the breeding female shoved male breeders and nonbreeders at different rates when recipient's body weight was controlled: 10 breeding males were shoved at rates higher, and 9 at rates lower, than those predicted by linear regressions of breeding-female shove rate on recipient's body weight for their respective colonies (the mean standardized residual across all colonies was only 0.24, which was *not* significantly different from 0; one-sample t-test).

We found a slight, but significant, difference between the overall mean durations of shoves directed at males and females across colonies (2.4 s vs. 1.8 s, respectively; $p < 0.05$; Wilcoxon matched-pairs signed ranks test), but this probably resulted from (1) the tendency of larger individuals to receive longer shoves from the breeding female (above) and (2) the greater average

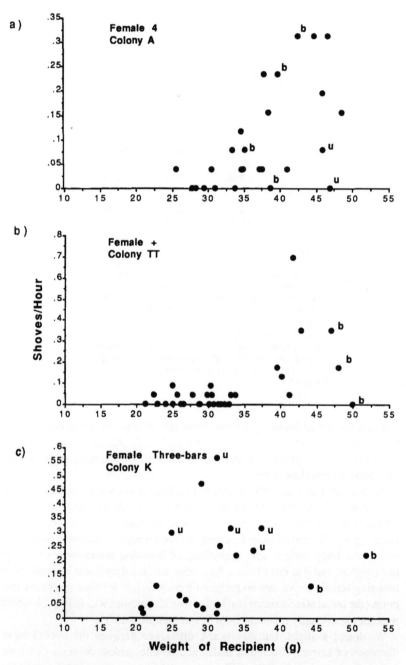

Fig. 11-4 a, Rate of shoving by breeding female 4 in Cornell colony A versus recipients' body weights (when female 4 was reproducing regularly); n = 27, u, nonrelatives; b, males that mated with the breeding female; and n, the number of (potential) recipients of shoving in the colony. b, Rate of shoving by breeding female + in Cornell colony TT versus recipients' body weight; n = 31. c, Rate of shoving by breeding female three-bars in colony K versus recipients' body weight (combined 1985 and

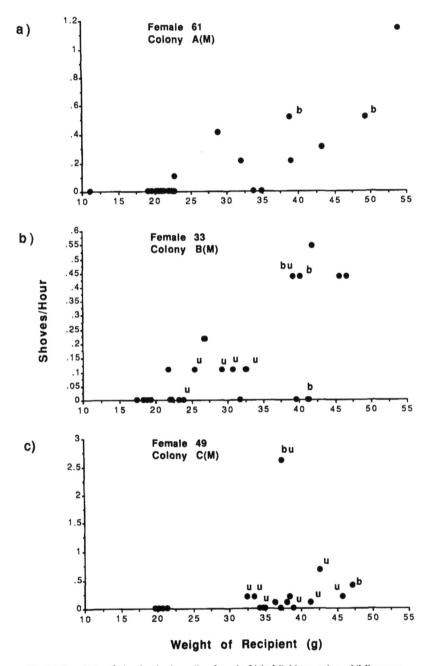

Fig 11-5. a, Rate of shoving by breeding female 61 in Michigan colony A(M) versus recipients' body weight; $n = 25$; b, males that mated with the breeding female; n, the number of (potential) recipients of shoving in the colony. b, Rate of shoving by breeding female 33 in Michigan colony B(M) versus recipients' body weight; $n = 27$; u, nonrelatives. c, Rate of shoving by breeding female 49 in Michigan colony C(M) versus recipients' body weight; $n = 19$.

size of males versus females within colonies (above). In sum, we were unable to detect any consistent effect of the recipient's sex on the probability of being shoved by the breeding female or on the duration of the shove received.

Discussion

Results presented in this chapter indicate that breeding-female naked mole-rats initiated more shoves per unit of time than did any other colony members. Moreover, this shoving behavior was most pronounced when the breeding female was engaged in regular cycles of reproduction, suggesting that shoving may be functionally related to the maintenance of reproductive dominance and/or the stimulation of colony activity. Our data further show that the breeding female was more likely to shove less related, larger individuals, and that her rate of shoving did not vary with the sex of the potential recipient.

There are at least two alternative (but not mutually exclusive) hypotheses for the function of shoving by the breeding female, both of which postulate evolutionary conflict between the breeding female and her workers. The alternatives are both examples of Type II conflicts described by Emlen (1982) in which a breeder gains but a worker loses personal fitness if the worker remains in the natal group. We label these alternatives the *threat-reduction hypothesis* and the *activity-incitation hypothesis*. The first hypothesis is that the breeding female shoves colony mates to reduce the threat that they will challenge her for breeding rights. Shoving might accomplish this by informing a recipient of the breeding female's willingness or ability to fight, of her high probability of breeding successfully (Alexander et al., chap. 1), and/or by actually reducing the recipient's fighting ability (e.g., by forcing the recipient to engage in risky or energetically costly colony tasks).

Under this first hypothesis, less related individuals may be shoved more often because their ascension to breeding status would entail the most severe inclusive-fitness losses for the reproductive female. Relatively large individuals would be shoved more and longer because they are the most immediate threats to the breeding female's reproductive dominance (unless some large individuals are reproductively senescent). A connection between body size and fighting ability in naked mole-rats is implied by the observation that breeding vacancies are typically filled by winners of fights among individuals that exhibit growth spurts after and sometimes before the breeding vacancy arises (Lacey and Sherman, chap. 10; Jarvis, chap. 13). It might be argued that, under this hypothesis, the lack of shoving by reproductively quiescent "breeding" females is anomalous, since, when colony reproduction ceases, others should increase their attempts to take over the breeding role. However, the breeding female's reproductive quiescence may simply indicate her assessment that any attempted reproduction is likely to be unsuccessful under current conditions;

this assessment may be shared by the other females in the colony (thus causing them not to challenge the breeder at that time).

The alternative activity-incitation hypothesis is that shoving serves primarily to incite activity in "lazy" workers. Under this hypothesis, breeding females shove less related recipients more often because less related recipients tend to work less. This hypothesis assumes that a worker's activity level varies inversely with its relatedness to the breeding female, which is reasonable if its activity level represents an optimal balance between the kin-selective benefits of colony-maintenance activities, and the personal fitness costs of such efforts. Thus, less related workers should be in greater conflict with the breeder over how active they ought to be. Larger individuals might also tend to work less because of their greater likelihood of replacing a breeder, leading to increased rate and intensity of shoving by the breeding female. Larger individuals are apparently the colony's defenders and infrequently perform colony-maintenance activities (Lacey and Sherman, chap. 10; Faulkes et al., chap. 14); thus, the breeding female might benefit by inciting larger animals both to maintain and defend the colony. Alternatively, perhaps smaller individuals can be incited into action by fewer or less intense shoves.

At present, we have not critically tested between these two hypotheses, and indeed many of our data seem to support both alternatives. For example, both hypotheses can explain why a breeding female might shove less after she gives birth to a litter (fig. 11-2). At that time it might be unfavorable both for workers to challenge the breeding female (threat-reduction hypothesis) and for them to be lazy (activity-incitation hypothesis), given the enhanced inclusive fitness payoffs for cooperation when related, needy young are present.

However, the threat-reduction hypothesis predicts that females, the likely candidates to overthrow the breeding female, should be the most frequent targets of shoving by the breeding female, unless males and females rise to reproductive dominance as pairs (see Jarvis, chap. 13). However, even if a male and female become breeders as a pair, the optimal level of aggression by the breeding female toward the challenging female should be higher than that toward the challenging male. The challenging male would be selected to yield to the breeding female at a lower level of aggression because, unlike the challenging female, he may still be able to reproduce in the presence of the breeding female (by later mating with her). Hence, the complete lack of sex bias in shoving by the breeding female in all our colonies makes the activity-incitation hypothesis the more parsimonious explanation at the present time. In addition, the activity-incitation hypothesis more easily explains why the breeding female shoves the breeding males at rates similar to those for nonbreeders of comparable body size. Breeding males are not obvious threats to the breeding female's reproductive dominance, but they might be in conflict with the breeding female over how frequently to perform maintenance or defense activities. For example, when the breeding female mates with more than one male, multiple paternity may result; this, in turn, lowers the average relatedness between male breeders and the pups, but not that between the female breeder and her offspring. Although our results are not conclusive, they suggest that the activity-incitation hypothesis is the better of the two explanations for the function of shoving by breeding females (but see Alexander, chap. 15).

Interestingly, the queens of many social insects that inhabit small colonies (i.e., fewer than 100 workers) also are the principal initiators of intracolonial interactions, many of which are aggressive in nature (e.g., paper wasps, *Polistes* sp.: Dew 1983; Reeve and Gamboa 1983, 1987; Strassmann and Meyer 1983; and sweat bees, *Dialictus* sp.: Breed and Gamboa 1977, Buckle 1982; review by Michener 1988). In paper wasps, it is well documented that the queen is more aggressive toward higher ranking workers (e.g., Downing and Jeanne 1985). Moreover, the queen's aggression toward a dominant worker is most likely to occur when the worker is inactive, and this aggression is often followed by increased activity on the nest or by the worker's departure on a foraging trip (Reeve and Gamboa, 1983, 1987). Reeve and Gamboa (1983) suggested that such aggression may reflect conflict between the queen and a dominant worker over how active the worker should be. Thus, a central role for the queen in colony activation may be a convergent feature of the social organizations of naked mole-rats and at least some small-colony eusocial Hymenoptera (see also Alexander et al., chap. 1).

The observed relationship between the breeding female's shoving and the recipient's relatedness (fig. 11-3) is evidence of nepotism. The pattern of nepotism is intriguing in view of the high intracolonial relatedness generated by extensive inbreeding in *Heterocephalus glaber* (Reeve et al. 1990; see also Honeycutt et al., chap. 7). Uncles and aunts, which may be related to the breeding female by as much as $r = 0.72$, were treated as aggressively as individuals from a different colony. Siblings and offspring, which are probably related to the breeder by only a slightly greater amount ($r = 0.81$), were shoved markedly and significantly less often than uncles and aunts.

One evolutionary explanation for this pattern may be that intracolonial aggression reflects conflicts of genetic interests during breeder succession; that is, when an asymmetry develops (e.g., due to the death of a breeder) between an individual's relatedness to the current breeder's offspring and its relatedness to its own offspring. As the asymmetry increases, the conflict between a breeding female and a worker is expected to increase until a point is reached at which elevated aggression by the breeding female is favored. Further increases in the asymmetry might not result in increased queen aggression because (1) such extreme asymmetries occur so rarely that selection has not consistently favored enhanced aggression in response to them, or (2) the form of the personal-fitness cost function for queen aggression or risks of misdirected aggression result in selection against further increases in aggression.

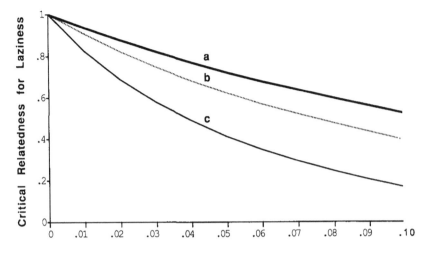

Worker's Proportional Contribution
to Breeder's Output (k)

Fig 11-6 The critical values of relatedness between a worker and breeder below which worker "laziness" is favored, as a function of the worker's proportional contribution to maximum breeder output (see the text). a, $P = 0.25$; $P' = 0\ 40$; b, $P = 0.25$; $P' = 0.35$; and c, $P = 0\ 25$; $P' = 0\ 30$ P and P' are the worker's probabilities of becoming a breeder if a task is, or is not performed, respectively. These graphs illustrate that laziness is favored for a higher relatedness as the worker's proportional contribution decreases and as the positive effect of laziness on an individual's breeding prospects increases

The question remains, Why is the resulting aggression threshold set at such a high level of genetic relatedness? To try to get at this, consider a simple activity-incitation model (fig. 11-6) in which a worker that is in line to become a replacement breeder is confronted with a certain colony-maintenance task (e.g., food-carrying). Let P and P' be the probabilities of becoming a breeder if the task is, or is not, performed, respectively, let Q and $Q - c$ be the breeder's reproductive output if the task is, or is not, performed by the worker, respectively, and let r be the genetic relatedness between the worker and the current breeder's offspring, divided by the genetic relatedness between the worker and its own offspring. By Hamilton's rule (W. D. Hamilton 1964; Grafen 1984), not performing the task (laziness) is favored, and breeder-worker conflict and breeder aggression result, when

$$r < (P' - P - kP')/[P' - P + k(1 - P')]$$

where k is the ratio c/Q. Now, as the number of individuals aiding the breeder increases, and as the breeder's maximum output Q increases, k will approach zero, because the worker's proportional contribution to the breeder's output declines. Decreasing k raises the value of r below which laziness will be favored (for $k = 0$, laziness is favored for any $r < 1$; see fig. 11-6). Thus, even if being lazy only slightly increases the chance of becoming a breeder, laziness is favored even for high r if the colony is large enough and if the fitness payoff for becoming a breeder is great enough, as it is likely to be in *H. glaber* (for a different but complementary idea about the evolutionary reasons for aggression among closely related colony mates, see Alexander, chap. 15).

From this simple model, we might expect no conflict between a breeder and its own offspring, because $r = 1$ when a breeding female mates with only one male. However, our data (fig. 11-3) indicate that the breeding female does shove her own offspring, albeit at a relatively low rate. What factors might generate at least some evolutionary conflict between a breeder and her nonbreeding (worker) offspring in naked mole-rats? There are at least four possibilities. (1) If the breeding female is at least sometimes fertilized by multiple males, a nonbreeder would be, on the average, more closely related to its own offspring than to the breeder's offspring. (2) Conflict may develop when the breeder's fecundity or fertility relative to that of the nonbreeder is reduced through disease or senescence (although this probably would lead to strong sex bias in aggression by the breeding female, unless both members of the breeding pair are simultaneously affected). (3) A nonbreeder may be favored to work less hard than is optimal from the breeder's point of view because a lazy worker might be more likely than a (hypothetical) harder working sibling to replace the breeder upon the latter's death (which is beneficial to the nonbreeder because it is more related to its own offspring than to that of a sibling). (4) A nonbreeder would be more closely related to its own offspring than to even a singly mated female breeder's offspring if the nonbreeder could displace its same-sex parent and (in)breed with its other parent. This seems plausible given that naked mole-rats apparently consistently inbreed (Reeve et al. 1990), which has likely led to a reduction in inbreeding depression, hence a lowered cost of consanguineous mating.

In sum, our results illustrate that high average genetic relatedness within social groups, which may have existed in *H. glaber* over long periods of evolutionary time, does not necessarily preclude the appearance of vigorous forms of kin-correlated aggression within these groups (see also Alexander, chap. 15). Slightly different, although high, values of relatedness may make the difference between tolerance and frequent aggression, especially if, as occurs in naked mole-rats, there is a high reproductive skew (in the sense of Vehrencamp 1983) in a large group. When the reproductive skew in a large group is high, the fitness benefits for becoming the dominant reproductive are so large relative to an individual worker's effect on colony reproduction that strong evolutionary conflicts of interest are generated between even closely related colony members.

Summary

We analyzed shoving behavior in six laboratory colonies of naked mole-rats, three at Cornell University and three at the University of Michigan. In all colonies, the breeding female initiated more shoves per unit of time than did any other colony member. This shoving behavior was pronounced only when the breeding female was engaged in regular cycles of reproduction. The breeding female was more likely to shove less related, larger individuals, and her probability of shoving did not depend on the sex of the potential recipient.

Shoving may be functionally related to the maintenance of reproductive dominance and/or the stimulation of colony activity. Thus, shoving by breeding-female naked mole-rats may be analogous to queen aggression in many small-colony social insects. Our data support the hypothesis that breeding females shove to incite nonbreeders to become more active; the alternative hypothesis, that shoving deters female nonbreeders from reproductively challenging the breeding female, was not ruled out but was somewhat less well supported. A high level of average relatedness within naked mole-rat colonies evidently has not prevented the evolution of frequent intracolonial aggression and nepotism in this species. We suggest that the potential for conflict within a large group consisting of even closely related individuals may be high when, as in naked mole-rats, there is a high reproductive skew and thus the payoffs for becoming a breeder are large relative to a single helper's effect on colony reproduction.

Acknowledgments

We thank J. Carey, A. Freedman, J. Shellman-Reeve, and C. Saldern for assistance with the observations. For comments on the manuscript, we thank R. D. Alexander, G. C. Eickwort, G. J. Gamboa, J.U.M. Jarvis, T. D. Seeley, and an anonymous reviewer. R. D. Alexander kindly allowed H.K.R. to observe the Michigan colonies and provided genealogical data. Financial support was provided to H.K.R. by a National Institutes of Mental Health Training Grant in Integrative Neurobiology and Behavior. Research support was provided by National Science Foundation grant BNS-8615842 (to P.W.S. and C. F. Aquadro) and by U.S. Department of Agriculture Hatch Grant NYC-191412 administered by Cornell University.

12 Growth and Factors Affecting Body Size in Naked Mole-Rats

Jennifer U.M. Jarvis, M. Justin O'Riain, and Elizabeth McDaid

A feature common to many bathyergids is variation in adult body size within a population (De Graaff 1964a; Taylor et al. 1985; Jarvis 1981, 1985; Bennett and Jarvis, 1988b). In some instances, such as in the solitary genera (e.g., *Georychus*), this variation may be attributable to the generally slow growth rates of the Bathyergidae; in many species, individuals take a year or more to attain adult body mass (Jarvis 1981; Taylor et al. 1985; Jarvis and Bennett, chap. 3). However, in three species of social mole-rats, *Heterocephalus glaber*, *Cryptomys damarensis*, and *Cryptomys hottentotus hottentotus*, the body mass and shape of an individual may remain nearly constant for many years and then change dramatically if its reproductive status changes (e.g., if the reproductive female is removed; Jarvis 1981; Bennett 1988; Bennett and Jarvis 1988b; Lacey and Sherman, chap. 10). Furthermore, the number of individuals in a colony apparently affects the rate at which the pups in subsequent litters grow (Bennett 1988; Bennett and Jarvis 1988b). Ecological factors such as food availability and soil hardness also appear to affect body size in the Bathyergidae (Jarvis 1985; Jarvis and Bennett 1990; Brett, chap. 4) and in other subterranean rodents such as the Geomyidae (Patton and Brylski 1987).

Because naked mole-rats live in colonies showing a degree of division of labor not found in other mammals, several attempts have been made to determine whether they have physical castes and/or exhibit age polyethism, a process defined by Wilson (1971, p. 461) as the "regular change of labor roles by colony members as they age." These previous age-related behavioral studies on *H. glaber* (Jarvis 1981; Isil 1983; Lacey and Sherman, chap. 10) were done on wild-captured colonies in which the exact ages of many of the mole-rats were unknown and in which growth curves were only available for the captive-born animals. Understandably, the assumption was then often made that body mass co-varied with age. Because we now have entire colonies of known age, and detailed knowledge of growth patterns from several hundred mole-rats, we were able to repeat and enlarge on results presented by Jarvis (1981) and Lacey and Sherman (chap. 10).

We studied three captive colonies at the University of Cape Town, among which the ages of all the nonreproductive individuals were known, to examine the age polyethism and physical-caste hypotheses. We also examined factors affecting body size and growth of captive animals, some of which were more than 10 yr old, and followed the growth patterns of multiple wild-caught and laboratory-initiated colonies and of specific individuals within those colonies.

Methods

STUDY ANIMALS

Colonies of mole-rats used in this study, and from which new colonies were founded in the laboratory, were collected from 1974 to 1980 in two localities in Kenya, at Lerata water hole and at Mtito Andei; these localities are shown in figure 7-1 (p. 197). The principal field-caught colonies we studied (from capture until June 1989) were Lerata 4 (collected in 1980), and colony 80 collected at Mtito Andei (in 1977; see chap. 13). Eleven lab-reared colonies, each established from one pair of mole-rats and consisting entirely of known-age animals (except, sometimes, the breeding pair) were also studied.

The mole-rat colonies were housed in similar artificial burrow systems and were fed ad libitum on a similar diet (see Jarvis, Appendix). At capture, the mole-rats were weighed (to the nearest gram) and were individually marked by toe-clipping. Mole-rats born in captivity were either weighed and toe-clipped at birth (occasionally) or at weaning (usually). All captive animals were weighed at irregular intervals (to ± 0.1 g), the interval between weighings depending on the type of study being conducted. For example, when monitoring changes in the masses of pups, the animals were weighed every 2–5 wk (except when newborn pups were present in the colony); during long-term studies, however, the mole-rats were weighed every 3–12 mo.

GROWTH CURVES AND AGE-RELATED COMPARISONS OF BODY MASSES

GROWTH CURVES

The change in masses of pups born to three colonies were used in this analysis. The first colony, Lerata 4, contained 71 wild-captured animals. The change in mass of a single litter of 22 pups born to Lerata 4 on June 19, 1981, was monitored from weaning until December 22, 1984. Two colonies begun from pairs were also studied. One colony (700) was being subjected to "predation" (see chap. 13 and below), and changes in the masses of pups in the first five litters (*n* = 18 pups) born to the colony were monitored from the birth of the first litter on July 10, 1984 to May 1, 1986 when the fifth litter was 1-yr old. The second colony (7000) was founded from a pair and had no pups removed over the study period. We analyzed the changes in masses of individuals in the first three litters ($n = 21$ pups) from the birth of the first litter on March 20, 1982 to June 21, 1985 when the last litter was 2 yr old.

All the individuals in a colony were weighed on the same day (to reduce disturbance). As a consequence, there was no uniformity in the ages at which different litters in a colony were weighed. Direct comparisons between the body masses of pups from different litters at the same age were therefore not always possible. To try to obviate this problem, mathematical models for the approximation of postnatal growth were constructed using a differential of the Von Bertalanffy equation, a nonlinear regression. Using two parameters, the growth-rate constant and asymptotic mass, we employed the likelihood-ratio test (Draper and Smith 1966) to see if statistically valid mean growth rates could be derived from the modeled data.

BODY MASS AT DIFFERENT AGES

The relationship between actual (not estimated) body mass and the age of each nonbreeding mole-rat was also examined by graphically comparing all the litters from 19 captive colonies that had been weighed at four different ages:

1. 30–35 days: 184 pups from 21 litters in 11 colonies; 4 of these litters (23 pups) were first or second litters of colonies begun from pairs

2. 60–70 days: 116 pups from 16 litters in 11 colonies; (none of these were first or second litters of colonies founded from pairs

3. 119–129 days: 80 pups from 11 litters in 7 colonies; 1 was the second litter (7 pups) from a colony begun from a pair

4. 198–210 days: 120 pups from 19 litters in 12 colonies; 8 litters (45 pups) were first or second litters of colonies founded from pairs; in this age group, the mean body masses of pups in the first and second litters were compared with the mean body masses of the rest of the pups (using a two-tailed t-test)

Three of these ages (1, 3, 4) were chosen to allow us to examine the range and the median body masses of newly weaned pups and those of pups that were 80–90 days apart in age (the approximate time between successive litters). The other age (60–70 days) allowed us to study the rate at which the spread in body masses occurs after weaning. Using the pups weighed at 60–70 and 119–129 days, the masses of pups in each of the litters and colonies represented in this sample were analyzed with the Kruskal-Wallis test to see if it was statistically valid to calculate a mean for the whole population of pups at each age.

AGE-RELATED BEHAVIORAL STUDIES

In 1988, three captive-born colonies of known age (table 12-1) were observed for 3 mo. The relationships between the age of each mole-rat, its body mass, and its participation in work activities associated with excavating burrows were investigated (e.g., digging, sweeping, gnawing, plugging, and carrying small stones; see below). Two of the colonies (300 and 700, both established from one pair) were breeding and rearing pups; the third colony (70, established from three mole-rats) had recruited no pups for 5 yr, but before that had experienced a good history of recruitment (table 12-1). The former two colonies had a history of "predation" in that on several occasions before our study, about equal numbers of male and female subadult animals had been removed, thereby slowing the rate at which the colonies grew numerically (Jarvis, chap. 13).

Each of the above three colonies was provided with a digging system (2.4-m long), which allowed the mole-rats access to a continuous flow of sand (fig. 12-1). The sand was placed in a bucket (A) with a hole at its base (B). The mole-rats dug at this soil and swept it along a length of transparent acrylic plastic (plexiglass) tunnel to a chamber (F) positioned directly above the bucket and floored with a metal grid. The sand fell through this grid and re-entered the bucket (A) thus completing the cycle. At the link between the digging system and the rest of the burrow system, two structures, a 1-cm high barrier across the entrance and a section of metal mesh flooring, prevented the animals from moving sand out of the digging areas. The continuous supply of soft sand could be interchanged with solid cores of soil. These cores were introduced to the digging system by replacing a section of tunnel (C), with cylindrical tubing packed tightly with 30 cm of hardened dry sand to which a small quantity of clay had been added. It took the mole-rats at least 60 min to dig to the end of such a core.

Colony behavior was observed for 1-h periods at irregular times of the day (but not between 1:00 A.M. and 9:00 A.M.), not during or immediately after feeding or cleaning the burrow systems. Observations were terminated immediately preceding the birth of a litter and recommenced after the litter had been weaned or had died. Each 60-min observation period began when the first animal arrived at the soil face. Data were collected by the scan-sampling method (Altmann 1974), in which the activity of each mole-rat in the colony was recorded every 2 min, yielding a maximum of 30 samples per hour.

Five digging behaviors—digging, sweeping, gnawing, plugging, and carrying stones—were recorded during each trial period; these behaviors are described in greater detail by Lacey et al. (chap. 8).

1. *Digging* is either a rapid scrabbling movement with the forelimbs (at the loose soil face) or biting at the earth face with the incisors (hard-soil trials).

Table 12-1
Details of the Known-Age Animals in the
Three Colonies Used in the Behavioral Studies
Commencing on February 1, 1988

Colony Name	Litter No	Birth Date	No. of Surviving Mole-Rats	
			M	F
70	1	6 Oct 80	1	2
	2	16 Mar 81	2	1
	3	1 Dec 81	0	1
	4	22 Mar 82	4	2
	5	22 May 82	4	2
	6	10 Aug 82	3	1
	7	18 Jan 83	2	1
	8	9 Apr 83	2	2
	9	1 Jul 83	6	4
Total			24	16
700	1	23 Oct 84	5	1
	2	8 Jan 85	2	1
	3	12 Jun 85	0	1
	4	30 Aug 85	1	1
	5	2 Feb 86	2	0
	6	3 Jun 86	4	3
	7	21 Aug 86	2	0
	8	15 Nov 86	2	2
	9	22 Sep 87	1	1
	10*	6 Dec 87	3	1
Total			22	11
300	1	30 Oct 84	0	1
	2	15 Feb 85	1	1
	3	7 Oct 85	1	0
	4	25 Dec 85	1	0
	5	13 Mar 86	4	1
	6	31 May 86	1	4
	7	19 Aug 86	2	4
	8	8 Nov 86	1	0
	9	20 May 87	1	0
	10*	8 Jan 88	0	3
Total			12	14

* Animals less than 20 g, which were excluded from the analyses

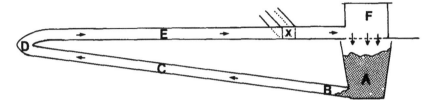

Fig. 12-1 Plan of the apparatus used in digging trials A, sand- filled, soil-recycling bucket with a tunnel at its base; B, sand face where soil is excavated; C, D, E, path along which sand is swept to F, a chamber with a mesh floor; X, link with the main burrow system.

2. *Sweeping* mole-rats move backward along the burrow kicking either soil (in the digging system) or wood shavings (in the main burrow system) with a simultaneous backward movement of the hind limbs.

3. *Gnawing* involves biting or scraping the incisors along the walls of the tunnels and at junctions or corners in the burrow system.

4. *Plugging* is a behavior in which a mole-rat uses backward sweeps of the hind feet to send sand or dirt into the blind end of the tunnel (B) while anchoring its body with its forelegs. This only happened in "continuous exposure" sand trials (below).

5. *Carrying stones* is the act of carrying small stones in the mouth, often as the animal continues to sweep sand. The stone is eventually dropped in some other part of the burrow system. This behavior is shown in plate 10-3.

These five behaviors were recorded under two conditions.

1. The mole-rats were continuously exposed to soft sand ($n = 6–9$ trials per colony). In these circumstances, the mole-rats were habituated to the presence of soil, could dig and move soil at any time, and the trials presented them with no novel experience.

2. The mole-rats were freshly exposed to sand (5–6 trials per colony). Under these "novel-experience" circumstances, the mole-rats either were presented with soft sand after a break in digging of at least 1 wk (the animals were denied access to the apparatus by a gate) or were given a hard core of sand after a similar period of deprivation.

By varying the methodology associated with the sand trials, we hoped to see what influence different conditions may have on a mole-rat's inclination to dig. Mole-rats in their natural habitat have seemingly unlimited opportunities to dig and move earth, and in the continuous-exposure trials we attempted to mimic this aspect of field conditions. Previous studies (Braude 1983; Brett 1986; Lacey and Sherman, chap. 10) provided limited digging opportunities and behavioral observations consisted of a series of one-off trials, most like trial 2 above.

All statistical analyses were performed with standard software on an IBM-compatible personal computer. The relationships between work, age, and body weight were investigated using scatter plots and Spearman's rank-correlation coefficients (Zar 1984). Data for breeding animals were excluded to allow for "statistical focus" on the behavioral patterns of nonbreeders as suggested by Lacey and Sherman (chap. 10). The body masses of the mole-rats at the beginning of the 3-mo study were used in this analysis. The Mann-Whitney U-test was used to determine whether digging performance between the sexes differed with body size. Data for most trials did not conform strictly to the assumptions of normality inherent in the statistical analyses; however, the effects of this problem were minimized by conducting multiple trials and by providing the mole-rats with large quantities of sand and wood shavings. Many aspects of this study repeated and enlarged on that of Lacey and Sherman (chap. 10); the major differences are that the ages of all the colony members were known and that the mole-rats were sometimes continuously supplied with sand.

BODY-MASS CHANGES OF THE BREEDING MALE(S)

Spearman's rank-correlation coefficients were used to assess sequential patterns of change in body masses of breeding and nonbreeding adult males in five colonies. Breeding males were identified by observing them mate with the breeding female and engage in frequent ano-genital nuzzling and sniffing with her. Two of the colonies (300 and 700) were established from pairs, the third (70) from 3 animals, and the fourth (80) from 5 animals; the fifth (Lerata 4) contained 71 wild-caught animals. Nonbreeding males studied were those closest in age or body mass to the breeding male. Thus, in the colonies established from a pair or from fewer than 5 animals, the oldest males were used in the comparison, whereas in Lerata 4, wild-caught males of comparable masses at capture were used. In addition, the body-mass changes of six males used in founding colonies from pairs (from colonies 300, 700, 70, and three other newly founded colonies) were assessed before and after they became breeders. One other male (467 from Lerata 4) was weighed before and after the death (from an intestinal infection) of a second breeding male (473) in the colony.

BODY-MASS CHANGES ON THE REMOVAL/DEATH
OF THE BREEDING FEMALE

In two colonies, the body masses of the adult members of the colony (excluding the new breeding female) were assessed for about the same length of time before and after the removal ($n = 4$) or death ($n = 2$) of the current breeding female. The Wilcoxon matched-pairs test (two-tailed, 95% confidence limits) was used to determine whether significant changes in body mass had occurred before or after the loss of the breeding female. The assumption was made that growth just before the removal of the breeding female (the time tested) was similar to the pattern of growth preceding this time.

Changes in Body Shape of Females on Becoming Breeders

Five breeding and four nonbreeding females of about the same body masses were anesthetized with Halothane and X-rayed. Using prints of the X-rays, the length of each lumbar vertebra (occurring from the last rib to the level of the pelvis) was measured, then the combined mean vertebral lengths of the breeding females were tested against those of the nonbreeding females, using Wilcoxon's nonparametric t-tests.

Results and Discussion

Weight at Birth

The mean birth weight of pups from 14 litters (five different females) was 1.86 $\pm$ 0.33 g ($n=155$ pups). Male and female pups did not differ in mass at birth ($p > 0.05$). Although mass at birth may vary by as much as 1 g, no significant correlation could be found between birth weight and mass at age 1 yr ($n = 20$ animals).

Growth Curves, Age, and Body Mass

The likelihood-ratio test, using the growth-rate constants and asymptotic body masses of individuals, showed that, in a litter of 22 mole-rats (Lerata 4) and the two known-age colonies tested (eight litters), individual variation in the growth of pups was so great that mean growth curves could not reasonably be constructed for a litter. It was therefore not possible to compare statistically the mean growth rates of litters born to a colony. This variation in body mass within the litter of 22 is illustrated in figure 12-2; when the juveniles were 11-mo old, the spread in their body masses was 17.7 g (range, 25.7–43.4 g).

Considering pups weighed at the same age (fig. 12-3), it is apparent that with increasing age the spread of their body masses increased. Thus, the range and median (M) body masses of newly weaned pups (30–35 days old) was 2.5–8.9 g ($M = 5.7$), that of the preceding litter was 13.2–26.5 g ($M = 19.51$), and that of the next oldest litter was 13.7–44.3 g ($M = 24.95$). The variation in body masses within and between litters in each of the two age groups tested (60–70 and 119–129 days old) was highly significant ($p < 0.001$). It was therefore not possible to calculate statistically valid mean body masses for pups at these ages ($H > 50.4$, df = 10; $H > 32.89$, df = 10, respectively; Kruskal-Wallis tests, corrected for ties). It is also clear (fig. 12-3) that the youngest pups in the colony could only be reliably distinguished (based on body mass) from those in the preceding litter up to an age of about 60–70 days. Therefore, given only

(a) WILD-CAUGHT   **(b) LITTER OF 22**

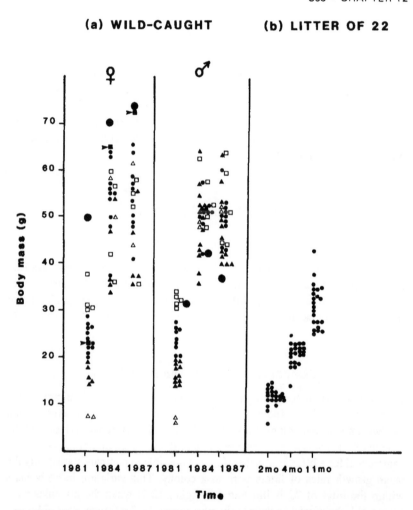

Fig. 12-2 Variation in growth of individual naked mole-rats in Lerata-4 colony. a, Change in body masses of the same individuals over 6 yr. b, Change in body masses of a litter of 22 captive-born animals *Open triangles*, Mole-rats smaller than 10 g on capture (August 1981); *filled triangles*, 15–19 g on capture, *small filled circles*, 20–29 g on capture; *open squares*, 30–38 g on capture; *large filled circles*, breeding females or males; *arrow*, mole-rat that became a second breeding female in 1983

its body mass, we were unable to predict the exact age of a pup over 3-mo old in our study colonies.

In our laboratory, extremely rapid increases in body mass occurred in pups born to young colonies that had been founded from a pair of mole-rats. Most (but not all) of the mole-rats in the first and second litters also tended to become the heaviest animals in the colony. Subsequent litters showed a wide

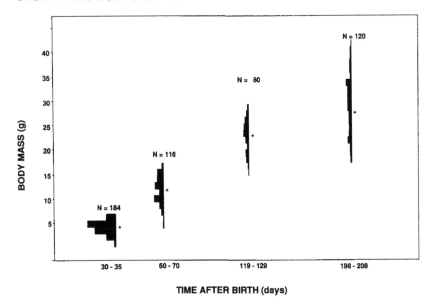

Fig. 12-3 The range in body masses of captive-born *Heterocephalus glaber* pups of the same age. Three of the ages (i e , excluding 60–70 days) approximate the usual gap between three successive litters in nature. (*N*, sample size, asterisk, median body mass)

range of body masses that overlapped with those of litters born before or after them (fig. 12-4). This difference in mass was apparent in the sample of 198–210-day-old pups (fig. 12-3): the 45 pups born in litters 1 and 2 were significantly heavier (28.6 ± 5.8 g; $p < 0.001$, two-tailed $t = 5.46$, df = 118) than the 75 pups born in later litters and to larger colonies (23.2 ± 4.41 g). This wide scatter in body masses between successive litters was maintained for many years. The first litter of colony 70 (in which all the animals were 5 yr or older) and the first two litters of colonies 300 and 700 remained larger than all the others (fig. 12-4). The rest of the litters could not be distinguished from each other on the basis of body mass.

Because of the variation in growth rates, mole-rat pups may take 4–24 mo to attain the mean adult body mass of wild captured colonies (ca. 32 g; Jarvis 1985; Brett 1986, chap. 4). A few mole-rats may never attain this mass. For example, in colony 70 (fig. 12-4a), 12 animals more than 5-yr old weighed less than 32 g.

If the body masses at capture (August 1981) of mole-rats from Lerata 4 (fig. 12-2) are compared with their masses in 1984 and 1987, it can be seen that the lightest (fig. 12-2, *open triangle*) and heaviest (*open square*) individuals at capture did not retain these relative positions in the colony. The mole-rats in a colony are therefore not all growing at the same rate. Furthermore, captive individuals seldom attain the maximum body mass of approximately 75 g (in

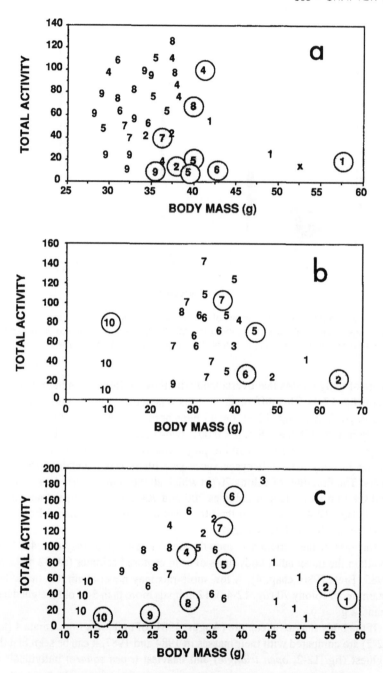

Fig 12-4. The relationships between the age of a mole-rat, its body mass, and the total amount of digging activity (excavating, sweeping, gnawing, plugging, carrying) performed when continuously exposed to soft soil. a, colony 70; b, colony 300; c, colony 700 Numbers indicate the litter in which the individual was born (see table 12-1); *circled numbers* indicate the heaviest individual from each litter; numbers in shaded circles represent the heaviest individuals

the field and in captivity); indeed, few exceed 60 g despite having access to ad libitum quantities of preferred foods. Of 39 mole-rats in our colonies aged 11–15 yr, the 3 heaviest individuals had masses of 60–65 g and the 4 lightest were 30–40 g, again demonstrating that a comparatively small body mass can occur in old animals.

The data on known-age animals (figs. 12-3, 12-4), the histories of weight changes in our colonies (fig. 12-2), and the statistical tests on the growth rates of litters, suggest that it may be impossible to determine reliably the age of captive *Heterocephalus glaber* from either body mass or growth trajectory. The growth patterns of marked individuals in wild colonies also show considerable variation (Brett 1986; Braude, pers. comm.), and it may be impossible to examine a newly captured colony and pick out successive litter cohorts within it. The only exceptions are the most recently weaned litter (ca. 30–70 days old) and possibly the preceding one (fig. 12-3; see also Brett, chap. 4).

The possibility of using morphological information to determine age was investigated by Hagen (1985), using 23 cranial and body measurements taken on a captive colony ($n = 71$). This colony contained 29 known-age animals and the amount of work performed by each mole-rat had been determined (by B. Broll, unpubl. data) just before their accidental death when the room overheated. Using univariate and multivariate analyses, Hagen found a significant positive correlation between the height of the coronoid process of the lower jaw and age. After stepwise reduction of variables, Hagen found that the thickness of the lower incisor could also be used to estimate age but that the correlation was not as good as with coronoid height. Morphometric studies have limited value in aging live animals of any species, however, and the use of other methods (e.g., tail-collagen strength; Sherman et al. 1985) has yet to be investigated in *H. glaber*. One possible, but very loose, behavioral indication of age is the amount of sparring and tooth fencing (see Lacey et al., chap. 8) a mole-rat does (Jarvis, chap. 13); these behaviors occur frequently in young animals but are seldom observed in mole-rats 2 yr or older.

BEHAVIORAL STUDIES

It was shown by Isil (1983), Payne (1983), and Lacey and Sherman (chap. 10) that males and females did not perform significantly different amounts of maintenance work. We confirmed this result for our three study colonies (70, 300, and 700), using the total work performed as our behavioral measure ($n = 38$ trials, $p > 0.05$, Mann-Whitney U-tests for the three colonies). Therefore, both sexes were considered together for subsequent analyses. In the analyses that follow, pups with body masses smaller than 20 g were excluded because of insufficient data, and body mass and work are discussed separately from age and work.

Body Mass and Work

TOTAL BURROWING ACTIVITY AND BODY MASS

The sum of all the burrow-excavating activities (e.g., digging, sweeping, gnawing, plugging, and carrying stones) was analyzed with respect to body mass. When presented with a novel exposure to soft soil or hard-soil cores, a greater amount of work was performed by the larger-sized mole-rats in colonies 300 and 700. Thus, body mass and total burrowing activity in these colonies were positively correlated, and the correlation was significant ($p < 0.01$) for colony 300 (table 12-2). In contrast, the correlation for colony 70 was significantly negative ($p < 0.01$; table 12-2).

In trials where the mole-rats had been "continuously exposed" to soft soil, total activity was negatively (but not significantly) correlated with body mass in all three study colonies. It is evident from the scatter plots (fig. 12-4) that although animals with body masses smaller than 45 g tended to perfom more work than those with body masses over 45 g, there were always some individuals smaller than 45 g that did little. The negative relationship between body mass and "work" in these trials is in accord with the results of Lacey and Sherman (chap. 10) and Faulkes et al. (chap. 14).

DIGGING SAND AND BODY MASS

In every novel-exposure trial, the mole-rats actively engaged in digging at the soil face prevented the animals behind them from digging. Brett (1986) and Pepper et al. (chap. 9) reported similar behavior in freshly captured mole-rats and laboratory colonies, respectively. We observed that the nest often emptied soon after the onset of the trial, and every mole-rat in the colony became active but was not necessarily working. When this happened, the tunnels containing sand were often packed with excited mole-rats climbing over one another, pulling each other's tails, and sweeping sand. A consequence of this heightened activity and overcrowding of the digging system was that the smaller individuals were competitively excluded from the sand face. Indeed very young animals with body masses of less than 20 g were frequently pushed and pulled entirely out of the digging area by larger mole-rats.

In these novel-exposure trials there was a significant positive correlation (table 12-2) between digging activity and body mass for colonies 300 and 700. By contrast, colony 70 showed a negative, but not significant, correlation between digging activity and body mass.

When mole-rats were continuously exposed to sand, there were always fewer animals present in both the digging and main burrow systems than in the novel-exposure trials (i.e., mean number present = 15.8 ± 2.95 vs. 24.3 ± 5.38; colony 700). Continuous-exposure trials were therefore much less competitive in nature, and digging was negatively correlated, although not significantly, with body mass in all three colonies (table 12-2).

Table 12-2

The Correlation between Digging Activities (in the digging system) and
Body Masses for the Three Known-Age Colonies

Behavior	Colony 70		Colony 700		Colony 300	
	N	C	N	C	N	C
Digging						
r	-0 13	-0 18	0 41	-0.09	0 85	-0.13
F	0.70	1 27	6 30	0.25	8.97	2 46
p	0.41	0.21	0.02	0 61	0 00	0.13
Sweeping						
r	-0 38	-0.14	0 23	-0.30	-0 20	-0.28
F	6.43	0 85	1.75	2 56	1 02	2.24
p	0 01	0.12	0 19	0 12	0 32	0.15
Gnawing						
r	*	0.00	*	0.16	*	0.38
F		0 01		0 77		4 26
p		0.97		0.38		0.49
Combined activities[†]						
r	-0.45	-0 33	0 19	-0 31	0 85	-0 28
F	9.70	4 59	0 98	2.67	9 64	0 26
p	0 01	0.03	0 32	0.11	0 01	0 61

NOTE Breeding females and pups less than 20 g are excluded C, constant exposure
to soil, N, novel exposure to soil, r, correlation coefficient, p, probability level All
values were rounded off to two decimal places

* Insufficient data

† Including digging, sweeping, gnawing, plugging, carrying stones in digging system,
and sweeping in tunnels

The different responses of the mole-rats in colonies 300 and 700 to continuous and novel exposure to soil suggests the importance of these conditions
for the behavioral outcome. Indeed, it appears that the behavior of an individual can differ depending on whether it is or is not excited by a novel stimulus.
We suggest that mole-rats in a wild colony would rarely experience the excitement associated with a novel exposure to sand since, in most circumstances, all
individuals would have unrestricted access to soil and unlimited opportunities
to dig. Two circumstances that may excite the mole-rats and elicit responses
similar to our novel-exposure trials and also to those of Lacey and Sherman
(chap. 10) would be predation and the first heavy rainfall after a dry spell. Brett
(1986, chap. 5) reported heightened digging after the onset of the rainy season
when the soil was soft and easily worked. In these circumstances, the cost of
digging would be lower (Vleck 1979), the constraints on body size less severe
(Jarvis and Bennett, chap. 3), and considerable benefits would accrue to a
colony that could mobilize a large work force to extend the burrows rapidly.
Brett (chap. 4) also observed that predation on naked mole-rats by snakes

elicited an alarm response by the colony in which they vocalized and speedily sealed off the area. In these emergency circumstances, many animals may be recruited to the working force, and large animals may also act directly as defenders (Lacey and Sherman, chap. 10).

AGE AND WORK

From figure 12-4, it is evident that individuals born in the same litter do not appear as discrete behavioral or morphological groups, and that mole-rats of a similar age can have very different body masses. In all three study colonies, the heaviest individuals were born in the first and second litters; thereafter, no clear trend is discernible between age and body mass. It is also apparent that, with the possible exception of the first and second litters, the age of an individual is not necessarily related to the amount of digging work it performs (Spearman's rank-correlation coefficient, $p > 0.05$, for all three colonies).

Examination of the body masses and amount of work performed by mole-rats in the 23 litters of two pups or more (in colonies 300, 700, 70) revealed that the heaviest one to two mole-rats in 15 of these litters performed the least digging work. All the mole-rats in one other litter (litter 1, fig. 12-4c) were heavy and performed little work. A mole-rat that did little digging work would thus appear to be the one within a given litter that attained the greatest body mass relative to its littermates (fig. 12-4a, b). This is most clearly seen in colony 70 where there has been no "predation" or recruitment for 5 yr, and in which individuals have shown little or no change in body mass over time. In seven of eight litters comprising two or more pups, the heaviest mole-rat(s) in each litter did the least digging work (fig. 12-4a). Within the same absolute weight range of the mole-rat that was heaviest in the litter are the lighter individuals from other litters; the rate of digging work by these latter animals may be more or less than that of the mole-rat in question. Thus, for example in colony 70, of the mole-rats weighing 35–36 g (fig. 12-4a), two were the heaviest in their litters (litters 9 and 7) and performed the least amount of work in their litters. Some of the older mole-rats in the same weight grouping (litters 2 and 5) were actually the lighter mole-rats from their litters and performed more work than their heavier littermates (e.g., heaviest in litter 3, 39 g; in litter 6, 38.5 g and 39.5 g). Thus, the largest individual in a litter may differ markedly from its littermates in both the amount and type of work performed.

Therefore, it may be best not to group all the individuals in a colony with respect to their ages or their body masses and the amount of work performed. Instead, each litter must be examined independently. When this is done, a pattern begins to emerge in which the heaviest one or two mole-rats (irrespective of their actual mass or work rating within the entire colony) tend to do less digging work than their lighter littermates. Additionally, it appears that most (but not all) individuals born to the first and second litters (especially in colonies started from pairs) become the heaviest individuals in the colony and perform little maintenance work. The reasons for these patterns are unknown.

The division of labor seen in our three study colonies appears to differ from that observed in certain insect societies having age polyethism (e.g., the honey bee *Apis mellifera*, Seeley 1982; the ant *Pheidole dentata*, Wilson 1976) and/or morphologically distinct worker castes (ants and certain termites; see Alexander et al., chap. 1). From Wilson's (1971) definition of age polyethism, we would expect individuals of the same age to perform tasks within the colonies' work schedule that are quantitatively and qualitatively similar. However, it appears that individuals from the same litter may be both morphologically (in terms of body mass) and behaviorally disjunct from one another (fig. 12-4). Application of the term "age polyethism" to our study colonies is thus an oversimplification of the ontogeny of each mole-rat's behavior. Perhaps the only individuals that exhibit a functionally important temporal polyethism are the firstborn litters (1 and 2) to a colony. They are older and larger than the rest of the colony, seldom work, and would appear to be primarily concerned with colony defense (Lacey and Sherman, chap. 10).

Interestingly, polyethism (i.e., variation in size and role) is apparent within litters and our data support the suggestion (Jarvis 1981) that slow-growing animals may remain as workers and faster-growing individuals may change roles and work less frequently (here we use "work" as originally defined by Jarvis [1981], namely, colony-maintenance activities). Recent recruitments into two of our study colonies (300 and 700) and also experimental "predation" on individuals in both these colonies may have slightly obscured the picture that was clearly apparent in colony 70; nevertheless, many of the trends are still apparent in these two colonies. It should be noted that Lacey and Sherman (chap. 10) present data and arguments about size and age polyethism in *H. glaber* that contrast somewhat with our interpretations. The question of age polyethism offers intriguing possibilities for future research.

Hagen (1985) found that of the 71 mole-rats from one colony that she examined, four morphological dimensions (weight, total length, tail length, and lower incisor length) could be used to separate the colony members into frequent workers, infrequent workers, and nonworkers (as defined by Jarvis [1981]). All these dimensions were negatively correlated with total work. We could discern no trend in incisor length in the three colonies studied here; however, this may have been due to difficulties in accurately measuring the incisors of live animals. Because of problems of accuracy, we did not measure body or tail length but, as indicated above, neither our study nor Hagen's (1985) found any visual way that mole-rats could be unambiguously divided into different work categories based on morphological attributes. The largest individuals in the colony tend to do little work, but the converse is not always true. This clouds the picture and makes it impossible, from body mass alone, to distinguish between small-sized workers and small-sized nonworkers. Furthermore, even among the larger mole-rats, some work far more frequently than others. Data showing similar scatter were presented by Lacey and Sherman (chap. 10), who suggested that interindividual variation might represent further role specialization.

Young pups (< 20 g) born to established colonies such as the three examined here participated in sweeping activity but rarely dug or gnawed (not quantified). We are currently examining the ontogeny of pups with respect to their involvement in work and will also be comparing the ontogeny of first litters and later litters in a colony. We do not yet know if the fast-growing pups from the first and second litters of new colonies become involved in digging activities at an earlier age than do subsequent litters. However, we hypothesize that benefits would accrue to a newly founded colony if the first recruits could speedily join the work force and participate in activities that would locate new food supplies for the colony. Once a colony is larger, the converse may be true, and greater advantages may be gained by having a reserve work force composed of small, slow-growing animals imposing low energy demands on the colony. The young animals also participate more than the older colony members in other activities such as assisting the breeding individuals in caring for new recruits to the colony (Jarvis, chap. 13; Lacey and Sherman, chap. 10).

CHANGES IN BODY MASS AFTER THE REMOVAL/DEATH OF THE BREEDING FEMALE

In two of our colonies that have had a succession of breeding females, the mass of adults of both sexes (> 2 yr old) increased significantly ($p < 0.05$) following the removal ($n = 4$) or death ($n = 2$) of the breeding female (table 12-3). Before the removal or death of the breeding female, the body masses of the same mole-rats assessed over comparable periods either showed no significant change ($n = 5$) or a significant decrease ($n = 1$; table 12-3). Thus, the data suggest that the increase in body mass was a response to the removal of the breeding female.

We have corroborative (but not exactly comparable) data to those of Lacey and Sherman (chap. 10), who reported an increase in the body masses of females for several months after the death or removal of the breeder, during which the growing females aggressively competed for reproductive dominance. We observed that when the first breeding female was removed from a colony (see table 12-3, colony 2,A) the two oldest laboratory-born females (sisters) both showed a marked rise in body mass but no aggression (e.g., shoving, incisor fencing, or biting). They also did not become the reproduc-

Table 12-3
Analysis, Using Wilcoxon Matched-Pairs Test (two-tailed), of the Body Masses of
Adult Mole-Rats (> 2 yr old) in Two Colonies Before and After the
Removal/Death of the Breeding Female

Colony and Treatment	Before Removal				After Removal			
	Time (days)	n	T	Critical Values of T	Time (days)	n	T	Critical Values of T
Colony 1								
A. First (removal)	70	13	35	17	66	13	15.5	17*
B. Second (removal)	120	13	24 5	17	110	11	0	11*
C. Third (removal)	177	9	8 5	6	165	9	1	6*
Colony 2								
A. First (removal)	87	8	7 5	2	51	15	3.5	25*
B. Second (death)	81	17	7	35†	100	15	17 5	25*
C. Third (death)	107	17	38	35	167	17	0	35*

NOTE. The new breeding females are not included in these analyses n, number of mole-rats in the sample (sexes combined), the time interval is between weighings used in matched pairs
* Significant increase in mass (p < 0 05)
† Significant decrease in mass (p < 0 05)

tives until 16 mo later, when a second breeding female in the colony died (table 12-3, colony 2,B). Both sisters then mated and no signs of aggression were seen until after the sisters each had one litter; they then fought, and one was killed. This killing, 148 days after the death of their predecessor, was not preceded by prolonged aggression between the two females and occurred overnight. Interestingly, the female that had gained the least weight was the survivor. During the reign of the surviving sister, the next female to become the breeder (table 12-3, colony 2,C) also showed an increase in mass, some 14 mo before her predecessor died. In the instances in table 12-3, the increases in mass of the successor to a breeding female occurred a year or more before the predecessor died and were not accompanied by (observable) prolonged aggression. This contrasts with the prolonged aggression observed after the removal/death of the breeding female in several other of our colonies (see Jarvis, chap. 13). Results from colony 2 may indicate that in an otherwise undisturbed colony (with respect to removals of individuals), the successor to the reigning breeding female is determined long before she actually assumes this role.

THE SHAPE OF THE BREEDING FEMALE'S BODY

The breeding female has a noticeably elongated and often a dorso-ventrally deep body. Her body shape begins to change during the period of accelerated growth before attaining reproductive status in the colony (Jarvis 1981). The breeding female's shape becomes distinctive after she has had several litters; therefore, the change results not simply from gaining weight but also from a marked change in her body length. The breeding female's body shape is so distinctive that she can usually be visually distinguished from the rest of her colony.

Measurements from X rays showed a highly significant difference between the vertebral length of breeding females (5.09 ± 0.58 mm/vertebra, $n = 5$) and (large) nonbreeding females (3.56 ± 0.31 mm/vertebra, $n = 4$; $p < 0.01$, Wilcoxon matched-pairs t-test). Thus, the elongation of the breeding female's body is due to lengthening of the vertebrae and not to a widening of the intervertebral spaces (R. Buffenstein and Jarvis, in prep.). Preliminary data (Buffenstein, pers. comm.) on changes in the length of the vertebrae of the same individual before and after becoming the breeding female also show that the vertebrae elongate.

A consequence of the large litter sizes of captive *H. glaber* (Jarvis 1984, chap. 13) is that, just before the pups are born, the breeding female may have increased her mass by as much as 87% (Jarvis, chap. 13). The long body of the breeding female may enable her to accommodate large numbers of fetuses without expanding too much in girth. This would reduce the danger of the gravid breeding female wedging in the narrow burrows and would enable her to continue visiting the toilet area and patroling parts of her burrow system; it would also facilitate her escape from predators (e.g., snakes).

THE SIZE OF THE BREEDING MALE

In five colonies where males were weighed consistently, all eight breeding males lost body mass over time, whereas nonbreeding males gained weight (table 12-4). All the alpha breeding males (i.e., the ones that mated most often and did the most ano-genital nuzzling) in these five colonies steadily lost weight before leveling off at body masses of 9.4–16.2 g, 17%–30% below their greatest mass. Eventually the breeding male is clearly identifiable in a colony as an emaciated animal. The marked decrease in the masses of breeding males occurred most consistently in colonies exhibiting a high recruitment of pups.

There were 88 nonbreeding males over 2-yr old in the above five colonies. If the difference between the greatest mass and the mass at last weighing is examined, only two males , both of which were over 8-yr old, showed excessive weight losses (9–10 g); 82 nonbreeding males either gained weight, remained static, or lost no more than 2 g, and 6 lost 3–4 g. Few mammals exhibit

Table 12-4
The Sequential Changes in the Body Masses of Breeding and
Nonbreeding Males in Five Naked Mole-Rat Colonies

	Breeding Males			Nonbreeding Males		
Colony	r_s	No. of Weighings	Time Span (yr)	r_s	No. of Weighings	Time Span (yr)
Lerata 4	-0.28	17	8.0	+0.92	18	8.0
	-0.70	7	4 0	+0 91	20	8.0
	-0.97	6	3.0	+0 77	18	8.0
				+0.88	19	8 0
				+0 96	16	8.0
				+0 86	18	8.0
				+0 92	14	8.0
				+0.89	18	8.0
80	-0.60	30	8.5	+0 96	21	6 0
				+0 95	22	7 0
				+0 96	25	7.0
				+0 88	24	7.0
				+0.96	23	7.0
				+0 97	15	5 5
70	-0.62	10	7 5	+0.94	15	7 5
	-0.03	12	7 5	+0 98	11	7 0
				+0.98	11	7.0
				+0.98	11	7 0
				+0.97	11	7 0
				+0.98	11	6 0
				+0.96	11	6 0
700	-0 67	16	4 0	+0 89	15	4.0
				+0 92	17	4 0
				+0.94	16	4.0
				+0.96	16	4.0
				+0 89	16	4.0
				+0.88	16	3.5
				+0.71	15	3.5
300	-0.90	16	4.0	+0.90	16	3 5
				+0.94	9	2.5
				+0.92	15	3.0
				+0.90	15	3.0
				+0.61	12	2.5
				+0.93	12	2 5

NOTE· Spearman's rank-correlation coefficients (r_s) are presented for body weights of individual males Where there was more than one breeding male in a colony, the original or alpha breeder is listed first Time span is the time covered by the analysis, for breeding males, the total time (or the time in captivity, if wild-caught) that the animal had been the breeding male

Table 12-5
Pattern of Change in Body Masses with Time of Male Naked Mole-Rats
Before and After Becoming Breeding Males

Colony	Before Breeding			After Breeding			
	r_s	Weighings	Time Span (yr)	r_s	Weighings	Time Span (yr)	Mass Loss* (%)
Lerata 4[†]	+0 96	6	1 0	-0 96	11	7	30 7
Lerata 4	+0 99	11	3 0	-0 70	7	4	5.9
Lerata 4	+0 97	12	4 0	-0 94	6	3	8 6
70	+0 97	10	3 0	-0 62	10	7 5	17.4
700	+1 00	6	1.0	-0 67	16	4	20 6
300	+0.90	14	3.5	-0 90	16	4	28 7
7112	+1 00	8	4 0	-0 89	9	1.7	18 7
7119	+1 00	8	4 0	-0 92	5	1	16 2
7101	+0 98	11	4 5	-1.00	7	1 7	34 4

NOTE Spearman's rank-correlation coefficients (r_s) are presented for body weights of individual males in different colonies
* Percentage loss in mass is from the time males became breeders until the last weighing
† Male 467, showing the change in his growth pattern before and after becoming the alpha male

a static body mass, and we suggest that (with the two exceptions above) these fluctuations fall within the normal variations in a colony.

The growth patterns of eight breeding males were analyzed before and after they attained breeding status. Positive correlations were obtained for all eight before they became reproductive males, and negative correlations were obtained after they attained breeding status (table 12-5). The before-and-after growth pattern also held true for two subordinate breeding males in Lerata 4, although their percentage drop in mass was small (5.9% and 8.6%; table 12-5). There were two breeding males when Lerata 4 was captured. One male (473) died a year later; the change in mass of male 467 (the other breeder) was positive until 473 died and strongly negative after his death. All but the two subordinate Lerata-4 males became reproductives when they were removed from their natal colony and paired with a female. The "before" changes in mass therefore occurred while they were nonreproductive animals and part of a large colony, and each showed positive correlations with respect to mass while nonreproductive (table 12-5).

MEAN COLONY MASS, BODY SIZE, AND FOOD AVAILABILITY

Brett (chap. 4, table 4-1) gives details on the sizes and body masses of colonies (or parts of colonies) captured in Kenya. Four of these colonies were captured at Lerata in northern Kenya, and four in southern Kenya near Mtito Andei (all were captured by Jarvis). A striking difference between these two populations was the very low mean mass of mole-rats (19.5 ± 5.05 g) found in Lerata 1, 2, and 3. The mole-rats in these colonies were so small that until those of Lerata 4 were trapped (mean body mass 27.9 ± 8.9 g), we believed that the northern Kenya *H. glaber* belonged to a smaller subspecies. Even the breeding females were small (22 g and 28 g) compared to the breeding female from Lerata 4 (52 g). Of 124 mole-rats caught in Lerata 1, 2, and 3, only 2 weighed more than 30 g. In contrast, the weights of 26 of the 82 mole-rats in Lerata 4 exceeded 30 g. The mean body mass of mole-rats from the four southern Kenya colonies (31.5 g) and the large standard deviation from the mean (± 10.9 g) demonstrate that Mtito Andei animals were heavier and had a much larger variation in body mass. It is improbable that colonies as numerically large as Lerata 1 (60 mole-rats) and Lerata 2 (40 mole-rats) could be composed entirely of young animals, and other reasons for this size discrepancy must be considered.

Although we were unable to estimate the biomass of food available to the mole-rats (the prevailing conditions were too dry and many plants with subterranean storage organs were dormant and difficult to locate), it appeared that Lerata 1, 2, and 3 were located in an area with a paucity of large tubers; indeed, no large tubers were found during our excavations of these three burrow systems. In contrast, the Lerata-4 burrow system was located close to the Lerata water hole, where the vegetation was noticably more lush. An *Asparagus* sp. with a large rhizome as well as a plant with a large tuber (> 4 kg) grew in this area. In southern Kenya, the colonies were collected in areas where large *Pyrenacantha kaurabassana* tubers (often weighing > 20 kg) were common (Jarvis 1985; Brett 1986, chap. 5). Jarvis (1985) hypothesized that a difference in food availability may account for the different mean body masses of northern and southern colonies. If this hypothesis is correct, it would imply that when food is limiting, body size rather than the number of mole-rats in a colony is reduced, providing support for the arguments of others (Lovegrove and Wissel 1988; Jarvis and Bennett, chap. 3) that a colony must contain a certain minimum number of animals to reduce the risks of not finding a dispersed food resource. Interestingly, most of the individuals in Lerata 1, 2, and 3 increased in mass in captivity, and within a year they were no different in size from those from Lerata 4 or from the Mtito Andei region.

Conclusions

We have shown that the body shape and mass of naked mole-rats is extremely plastic and not necessarily related linearly to age. The possible exceptions to this are the first and second litters born to newly established colonies and very young animals. In colonies established from pairs, most individuals in the first two litters became the largest individuals in the colony, and subsequent litters showed no distinct pattern with respect to body size and the age of the individual. Even in captive conditions with ad libitum food, many colony members did not attain the maximum body mass possible for *Heterocephalus glaber* (ca. 75–80 g in captivity and the field); this even holds true for 15 yr-old mole-rats. We also found that (in agreement with Lacey and Sherman, chap. 10), upon removal of a breeding female, most adults in the colony (both sexes) exhibited a significant increase in body mass before leveling off again at a higher body mass. Our field data indicate that whole colonies of *H. glaber* may consist of small animals and that they will grow to standard body sizes in captivity where there is ad libitum food.

These results indicate that the breeding female, colony numbers, and probably also the food resources available to a colony, influence the growth of colony members such that few individuals ever become very large. All apparently retain the ability to increase in mass should certain restraints be removed. We suggest that this dynamic situation is adaptive in that the colony will not carry an excess of large-sized individuals (as could happen if all the colony exhibited a strictly related age polyethism in size). Furthermore, gaps in the body-size distribution of a colony, created by predation or fighting/defense, could be rapidly filled by a change in body mass of individuals already present in the colony. If, as our data and those of Brett (1986), Hagen (1985), Faulkes et al. (chap. 14), and Lacey and Sherman (chap. 10) suggest, there is a link between the role fulfilled by an individual and its body mass, the sudden loss of individuals in one size range could affect the efficient functioning of the colony, especially if the loss was of large-sized defenders or of the breeding female. To replace either of these latter two categories from the ranks of newly weaned mole-rats could take several years, but an average-sized individual could increase in mass and fill the gap in a few months (Jarvis 1981). It is perhaps this phenomenon, that of behavioral flexibility among individuals of the same age, that is the biologically significant aspect of naked mole-rat behavioral ontogeny.

Plasticity in body mass and disparate growth within colonies is not unique to naked mole-rats, having now been documented in two other social bathyergids: *Cryptomys damarensis* and *C. hottentotus hottentotus* (Bennett 1988). In these species, individuals change in mass if the colony structure is perturbed by removal of key animals, such as the breeders. Furthermore, some individuals remain small throughout their lives. Bennett (1988) has limited data suggesting that in *Cryptomys*, pups born in the first and second litters of new colonies grew 25% faster (in grams per day, *n* = 2 colonies) than those in subsequent litters, so that they reached average adult mass at an earlier age. Even the solitary *Bathyergus janetta* shows some plasticity in mass, such that individuals living in harsh habitats (with respect to rainfall, food availability, etc.) are substantially smaller (ca. 30%) than those occurring in more equitable areas (Jarvis, unpubl. data).

Our results on digging activity (in continuous soil trials) in three known-age colonies indicate that, apart from the mole-rats in the first and second litters and the youngest pups in the colony, individuals of any age can be involved in much or little work. We found no evidence that the whole colony underwent age polyethism. There was still division of labor in colony 70, which had had no recruitment for 5 yr, suggesting that individuals do not necessarily show a concomitant change in their size and roles with age. Our data provide further evidence (Jarvis 1981) that there are fast- and slow-growing individuals among nonbreeders and that those mole-rats in a litter that become the heaviest perform less digging work than their smaller-sized littermates. These trends are only apparent within litters, and the largest individual from one litter could, in fact, be smaller than a slow-growing animal from another litter. It has been suggested (Jarvis 1981) that fast-growing individuals are perhaps potential nonworkers and reproductives. We plan to test this by removing animals from known-age colonies and observing which animals succeed them.

Lacey and Sherman (chap. 10) present data suggesting that large-sized mole-rats perform more digging work than the smaller ones. Their experimental procedure was similar to our novel-exposure trials, and our results show considerable agreement. We contend, however, that the converse results that we obtained in the continuous-exposure trials more truly reflect digging behavior in wild mole-rats that are surrounded with soil and have unlimited opportunities to dig. Field studies to determine which animals dig have yielded ambiguous results. Braude (chap. 6) found a positive correlation between body size and volcanoing in two of three colonies he studied. Brett (1986, chap. 5), however, found that the superficial (foraging) burrows are small in diameter (2.5–3.0 cm), suggesting that they were dug by small animals. Brett also found a significant positive correlation between the percentage of animals with their incisors clean of soil (presumably, therefore nondiggers) and body mass in a colony that he excavated after heavy rains. Energetic constraints (Vleck 1979, 1981; Lovegrove 1987a) would tend to preclude the larger animals from being the chief extenders of the burrows. Because small size does not necessarily imply that the animal is young and inexperienced, the digging force could contain experienced mole-rats well able to dig efficiently and effectively. The reader is referred to Lacey and Sherman (chap. 10) for further discussion of the issues raised here.

To our knowledge, no other family of mammals exhibits the plasticity in body mass exhibited by the Bathyergidae and more particularly by *H. glaber*. This plasticity is such that even adult animals that have ceased to grow apparently retain the potential to resume growth. We suggest that this allows the colony to be dynamic and rapidly responsive to changes that may occur in its social structure.

Summary

Naked mole-rats live together in large colonies showing evidence of a division of labor and having a complex relationship between body mass and the role of an individual within the colony. We examined the growth patterns of a large sample of captive naked mole-rats, many of known age. We also examined factors contributing to the disparity and plasticity of body size and shape shown by naked mole-rats and attempted to determine whether colonies have a system of physical and/or temporal castes and whether the division of labor is a product of age polyethism. We reached five main conclusions.

1. Except for very young animals and the mole-rats in the first two litters born to a newly founded colony, body mass did not usually co-vary with age. The firstborn mole-rats in a colony (begun from a pair) attained the greatest body mass, whereas many individuals from subsequent litters remained small throughout their lives.

2. There was considerable scatter in body masses and in the amount of work performed by individuals within each litter born to a colony. The heaviest one or two individuals in each litter tended to do the least amount of maintenance work, but their body masses could be greater or less than the frequent workers from other litters. We were therefore unable to use body mass to delineate physical and temporal castes unambiguously in naked mole-rats. Because age and work did not co-vary statistically, we found little evidence supporting age polyethism as a process operative in the division of labor.

3. The body mass and shape of an adult in the colony changed if it became one of the reproductive animals. The breeding female was one of the largest mole-rats in the colony and had a distinctly elongated body resulting from the lengthening of her vertebrae; the reproductive male(s) often began as large animals but many then showed a marked loss in mass with time.

4. When the breeding female died or was experimentally removed from the colony, all the individuals in the colony increased in mass and then stabilized once a new reproductive was established.

5. As with other subterranean rodents, food availability as well as the social situation in the colony apparently affected body size in wild as well as laboratory colonies of naked mole-rats.

Acknowledgments

We thank N. C. Bennett, T. Crowe, and A. Punt for advice and help in the statistical analyses of our data, C. Barnard for building the digging apparatus, and J. Booysen for helping with animal maintenance. R. Buffenstein, K. Davies, and M. Griffin helped collect the mole-rat colonies in Kenya. Funding for collecting and maintaining the animals was provided by the National Geographic Society, the Council for Scientific and Industrial Research (South Africa), and the University of Cape Town. We also thank H. Bally and P. Bally for their warm hospitality, and the Ministry of Environment and Natural Resources and the Office of the President of Kenya for permission to conduct research and to collect animals. E. A. Lacey and P. W. Sherman made a preliminary version of their chapter 10 available to us in 1985. Helpful criticisms of the manuscript were made by N.C. Bennett, P. W. Sherman, and an anonymous reviewer.

13 Reproduction of Naked Mole-Rats

Jennifer U. M. Jarvis

At birth, most mammals are highly dependent on adult care. This care may be provided by the mother alone, by both parents or, in relatively few instances, by parents assisted by a group of helpers. These helpers may be nonbreeding and only contribute to the rearing of the young, such as in the red fox (Macdonald 1979; Macdonald and Moehlman 1983) and dwarf mongoose (Rasa 1973; Rood 1980), or the helpers may form part of a communal breeding system, such as in lions (Schaller 1972; Bertram 1975), in which parentage is likely to be shared by the individuals involved in helping. Grades of helping between these two extremes also occur. Lacey and Sherman (chap. 10) and Emlen (1984) compare the various mammalian social systems that have alloparents, and those comparisons will not be reiterated here.

The reasons why some mammals live and reproduce in social groups are varied (Alexander 1974; Wrangham and Rubenstein 1986). Direct benefits may be gained by each individual in the group as a result of increased awareness of predators (Hoogland and Sherman 1976; Hoogland 1979; Rasa 1986) or reduced risks involved in finding and harvesting food resources that are difficult to locate (Jarvis 1978; Lovegrove and Wissel 1988). Emlen (1984) reviewed evidence that some social mammals are forced to remain in groups because of the high risks involved in dispersing and breeding on their own (see also Alexander et al., chap. 1). These latter groups of social mammals tend to be closely related because their dispersal from natal groups is limited. Whatever the reasons for living in a group, the helpers may contribute to the survival of both the offspring and the breeders by reducing the risk of predation and by better provisioning of food. The provisioning of food not only directly benefits the young but also reduces the energetic stress on the breeders.

In most mammal species that breed cooperatively, group size is small (< 15) and the role of helper is transient. That is, many of the helpers will eventually have an opportunity to breed on their own. The naked mole-rat is exceptional in that group size averages 70–80 and may be as great as 300 (Brett 1986; chap. 4). Moreover, laboratory data indicate that many of the individuals will remain as nonbreeders for their entire life spans. Many generations of young naked mole-rats apparently remain in the natal burrow system and directly as well as indirectly contribute to the care of the litters born to a single breeding female. This female may remain the sole reproductive female in a colony for more than a decade. *Heterocephalus glaber* therefore apparently exhibits a degree of cooperative breeding and a reproductive division of labor that has no parallels among other rodents, or indeed among most, if not all other cooperatively breeding mammals (Lacey and Sherman, chap. 10).

Previous descriptions of reproduction in captive naked mole-rats (Jarvis 1978, 1981) were based on data from wild-captured colonies in which the breeding female (among others) was not captured, and in which the colonies had been maintained in captivity for only 3–4 yr. Subsequent to this, complete field-captured colonies (including breeding females) have been maintained in captivity for as long as 9 yr. Colonies have also been established from pairs and from small groups. Consequently, many new data on reproduction in *H. glaber* are now available. These data come from colonies maintained under similar conditions (see the Appendix) in four different laboratories (University of Cape Town, Cornell University, University of Michigan, and Institute of Zoology, London) and also from field work in Kenya (Jarvis 1985; Brett 1986). I examine mainly proximate (immediate) rather than ultimate (evolutionary) aspects of reproduction. I show the mean as well as the variance found in the reproductive biology of this animal. Some hormonal aspects of reproduction are covered by Faulkes et al. in the next chapter.

Methods

Study Animals

In this chapter, reference will be made to several captive colonies housed at the University of Cape Town under conditions described in the Appendix. Most of the data were obtained from one colony, known as Lerata 4. This colony contained 82 animals at capture in August 1980, at Lerata Water Hole, in northern Kenya (Jarvis 1985; see fig. 7-1, p. 197). At capture the mean mass of colony members was 28 g (excluding 8 newly weaned pups, which were less than 8 g each); 29 of the mole-rats were considered juvenile and subadult (5–20 g). Of the 82, 11 mole-rats were accidentally killed during capture or died within the first 3 mo in captivity. At capture, the colony contained 1 breeding female (400, 52 g), and 2 breeding males (467, 35 g; 473, 33 g) were subsequently identified (i.e., they mated with female 400). Because the breeding female was captured on the day we left the study area, we were unable to confirm (by checking the burrows for several days) that the entire colony had been captured. However, few burrows had been blocked on the previous day, indicating that by then most of the colony had been captured (see Brett, chap. 4).

In early September 1980, Lerata 4, now comprising 71 animals, was housed in a burrow system constructed from 10 m of clear acrylic plastic (plexiglass) tubes (Jarvis, Appendix) linking nest, toilet, and food chambers. During the study period (November 1980 – January 1982), the colony was using two nests and two toilet areas. The toilet areas were cleaned daily, and the colony was fed daily. Behavioral studies were done on the 35 male and 36 female wild-caught mole-rats that comprised the colony (46 adults and 25 juveniles; the juveniles weighed less than 20 g at capture). All individuals were weighed and marked by toe-clipping at capture. The body masses of the animals at capture and at the end of the 14-mo study were used to determine their percentage of growth.

At least 1 day before behavioral observations, the toe identification number was written on each animal's back with a felt-tipped pen. The behaviors (below) of the mole-rats and their positions in the burrow system were recorded over periods of 1–12 h. Often the majority of the colony was resting and a 5-min scan-sampling method (Altmann 1974), similar to that employed by Lacey and Sherman (chap. 10) and Faulkes et al. (chap. 14), was used to record the behaviors of the active colony members. The scan began at one end of the burrow system, and the behavior and position of each animal encountered was recorded. If there was a period of heightened activity, all interactions involving the reproductive animals were preferentially noted. Once each observation hour and 26 times outside formal observation periods, the position in the burrow system (including the nests) of every colony member was recorded. All behaviors that could be attributed to disturbance by an outside source (e.g., a loud noise) were discounted, and data collection was halted until colony behavior had returned to "normal." The behaviors of pups born during the study period were also recorded.

During the 14-mo study period, female 400 was the only breeding female, and there were only two breeding males (467 and 473) in Lerata 4; five litters were born, and 35 pups were weaned (a 28% survival rate). Behavior was monitored over 114 h of formal observations and covered all phases in the reproductive cycle of the breeding female. My behavioral study was conducted before the completion of the naked mole-rat ethogram presented in this volume (Lacey et al., chap. 8). Consequently, some of the behaviors listed as separate in the ethogram were grouped together in my study. Although many of my observations closely parallel those presented elsewhere in this volume, they were done independently and on a very large colony with an established (field-captured) breeding female. My results therefore add to (and usually confirm) those presented by Lacey et al. (chap. 8), Lacey and Sherman (chap. 10), Reeve and Sherman (chap. 11), and Faulkes (chap. 14).

The behaviors that I studied included: (1) the number of times that each mole-rat was in the two nests or at resting sites in the tunnels (hourly scans); (2) interactions with the breeding female and males, especially ano-genital nuzzling and sniffing, allocoprophagy, and aggression, such as shoving encounters and biting; (3) work or colony-maintenance behaviors: carrying food and nesting material to the nest box, sweeping, backshoveling, and gnawing at corners of the tunnels; (4) agonistic behaviors between nonreproductives: shoving, open-mouth threats, tooth-fencing, sparring, and tugging (all were considered together) and biting; (5) interactions with and between the pups: attendance at births, carrying, nudging, grooming, pushing pups, allocoprophagy with pups, lying with pups in the tunnels, and sparring. Lengthy continuous behaviors, such as sweeping, gnawing, ano-gential nuzzling, and sniffing between the breeding animals were scored as one event every 5 min. Likewise, the identities of animals engaged in long sparring sessions with and between juveniles were recorded once every 5 min.

I combined data on sweeping soil or other materials and carrying food and nesting material for regression analyses after performing simple parametric correlations to test the relationship between the frequency with which a mole-rat swept soil and carried nesting material and food. These work behaviors showed a highly significant positive correlation ($r = 0.60$, $p < 0.001$, $n = 71$); this was also found by Lacey and Sherman (chap. 10) and Faulkes et al. (chap. 14). One further work activity, gnawing, was analyzed both independently and when incorporated with the sweeping and carrying activities. Specific behaviors related to the number of visits and duration of stay in the toilet areas were monitored over an additional 38 h in continuous observation sessions of 2, 2.5, 3 ($n = 2$), 3.5 ($n = 3$), 5, and 12 h.

When the breeding female was near the end of her pregnancy (i.e., within a week of the birth date, ca. 77 days after the last birth) the teat size of each individual in the colony was visually assessed and scored on a scale of 1 (absent) to 8 (very large). The assessment was done without reference to previous pregnancies and always by the same person. I analyzed the teat-size score for each mole-rat for the first and last pregnancies in the study period. I also noted whether the vagina of each female was perforate or imperforate. In a few individuals, the vaginal closure membrane had a very small aperture; these animals were termed "almost perforate."

The number of times that each wild-captured individual in the colony was involved in particular behaviors was tallied for the entire 14-mo period. Using these tallies, product-moment correlation coefficients (multiple linear regressions, BMDP 1R; Dixon 1983) were computed to identify possible correlations among behaviors 1–5 (above) and among the various behaviors and nest usage, teat size, body masses, and the percentage of growth of the wild-captured animals.

The percentage success in weaning pups was examined in Lerata 4, as well as in colony 1977 and four colonies begun from pairs. Two of these latter colonies were exposed to experimental "predation" of colony members, and two were not (table 13-1). Pups were counted within 24 h of birth and at weaning. In large colonies (> 40 adults) it was often difficult to count every pup in the crowded nest huddle; consequently, the figures used in estimating

Table 13-1
Details of the Simulated Predation Events in Two Naked Mole-Rat Laboratory
Colonies Begun from Pairs

Colony Name	Predation Date	Birth Date	No. in Litter		Colony Size	
			Removed	Left	Before	After
300	3 Jul 85	15 Feb 85	2	5		
		4 May 85	10	3	27	15
	2 Nov 85*	30 Oct 84	4	1	17	13
	23 Jan 87	23 Mar 86	5	6		
		31 May 86	5	8		
		19 Aug 86	5	6	49	34
	14 Oct 87†	4 May 85	3	0		
		25 Dec 85	2	1		
		23 Mar 86	1	5		
		31 May 86	2	6	37	39
700	7 Jun 85	10 Jul 84	2	2		
		23 Oct 84	2	7		
		8 Jan 85	6	4	25	15
	23 Dec 86	12 Jun 85	1	2		
		3 Jun 86	1	7		
		21 Aug 86	6	3	43	35
	7 Aug 87	8 Jan 85	1	3		
		28 Mar 85	1	0		
		12 Jun 85	1	1		
		30 Aug 85	1	2	33	29

NOTE A predation event was the removal or one of more young adults by the investigator
* These individuals were killed when they invaded another colony
† One of these was a second breeding female who was removed, together with the others, to establish a new colony

pup rearing success are for the minimum number of pups born. For 22% of the births ($n = 152$), no counts of pups were possible; in these cases, the mean of the number of pups born to that female in all her other litters was used. In only 6.6% of the litters whose sizes I estimated were pups weaned; thus, the error from these estimations was small.

The sizes of litters born in 8 colonies (54 litters) of mixed-colony parentage were compared with those in 10 colonies (63 litters) that were either captured in the field ($n = 4$) or established with animals from the same field colony. The former 8 colonies were taken as examples of cases in which outbreeding had occurred (5 were formed by pairing an animal from Lerata with a mate from Mtito Andei, and 3 were formed by pairing individuals from neighboring colonies), and the latter 10 colonies were taken to represent cases in which outbreeding had not occurred (see, e.g., Reeve et al. 1990).

Results

SEXUAL DIMORPHISM AND THE
ANATOMY OF THE REPRODUCTIVE TRACT

No obvious sexual dimorphism exists among nonbreeding naked mole-rats (Hagen 1985). However, the body size of both sexes varies with age and/or status within the colony (Jarvis 1981; Lacey and Sherman, chap. 10; Jarvis et al., chap. 12). Most breeding females have a distinctively elongated body and can be visually identified by their shape alone; they are also among the largest (heaviest) females in the colony. Apart from a faint dark-red horizontal line between the clitoris and anus in females, the external genitalia of males and nonbreeding females are also very similar. Reproductive females and nonbreeding females that are close to breeding (see below) differ from nonbreeding females in having either a perforate vagina or a very thin vaginal closure membrane; in the former, the position of the vagina is clearly apparent as a darker pink line. The actively breeding female always has prominent nipples. However, the temporary presence of visible nipples is not by itself diagnositic because, under certain conditions, nipples may be visible in colony members of both sexes (below).

The number of mammae present is variable and differs on the two sides in 65% ($n = 71$) of the animals. Nonbreeding adult male and female mole-rats from Lerata 4 ($n = 52$) had a mean of 6.4 pairs of teats (range 5–8). Of 19 breeding females in my other colonies, the mean number of pairs of mammae was 5.8 (range 4–7).

In males, the testes are undescended and lie in the abdominal cavity on either side of the bladder; the epididymis and the seminal vesicles are small (Hill et al. 1957). Dempsey et al. (1974) described the ultrastructure of the spermatozoa, and Fawcett (1973) showed that the testicular interstitial tissue (of a male of unknown reproductive status) was very well developed with many Leydig cells. These may well be the source of the high levels of androgens found in some males (Faulkes, pers. comm.). The gross anatomy of the reproductive tract of two breeding males that I examined was not obviously different from that of nonbreeding males in the colony. Autopsies of 84 wild-captured *Heterocephalus glaber* males (with body masses greater than 20 g) showed that 76% had some spermatozoa present in their vasa deferentia. Of these males, 41 weighed over 30 g, and 51% of these were producing large quantities of spermatozoa (i.e., smears examined microsopically showed the vasa deferentia of these males to be packed with spermatozoa). Since the majority of males in a colony are undergoing spermatogenesis but only 1–3 males mate with the breeding female (below; also Lacey and Sherman, chap. 10), it follows that there must be criteria, other than the presence of spermatozoa, that determine which males mate. These criteria are presently unknown.

Hill et al. (1957) commented on the embryonic appearance of the genital organs of *H. glaber* females. They were probably describing the tract of non-reproductive females, in which the uteri are thin-walled and narrow (ca. 1 mm wide) and the ovaries are small, thin, flat structures (ca. 2 mm long) lying caudal or caudo-lateral to the kidneys. The uterine horns of the breeding female are wider and thicker walled, and their dimensions vary during her reproductive cycle; the ovaries are two to three times larger and thicker than those of nonbreeding females (Kayanja and Jarvis 1971). The zona parenchymatosa of the ovary of a breeding female mole-rat contains clusters of many primordial follicles and various stages of developing follicles surrounded by large accumulations of interstitial gland tissue. The ovaries of nonbreeding females contain mainly primordial and primary follicles and little interstitial gland tissue.

Tam (1974) found that the interstitial gland tissue in the ovaries of hystricomorph rodents secretes progesterone. Perhaps the large aggregations of this tissue in the ovaries of *H. glaber* breeding females are the source of the high concentrations of progesterone in the blood and urine (Faulkes et al., chap. 14). Postmortem examination of 117 *H. glaber* females with body weights between 20 and 69 g (including 21 from one colony) showed that 115 (99%) had very small ovaries and thin uterine horns with no evidence of placental scars or of the thickening and widening of the uteri typical of other parous rodents. The remaining 2 animals appeared to be in estrus, but their uteri bore no signs of placental scars. The 115 reproductively inactive females in this sample were therefore not just temporarily sexually inactive but had never bred (see also Faulkes et al., chap. 14).

The nonreproductive members of a naked mole-rat colony are not sterile. Most can rapidly become reproductively active and breed if removed from the colony and housed with a partner of the opposite sex. In an established colony, however, they may remain nonbreeding and apparently anovulatory for many years. Indeed, in three colonies at Cape Town, six females and seven males have lived for 15 yr without reproducing. Reproductives can also continue to breed for long periods. In two different colonies, the same individuals have been the sole breeding females for 13 yr; another colony has had the same breeding female for 10 yr, and four other colonies have had the same breeding female for 6 yr (all dates are as of August 1989).

REPRODUCTIVE ANIMALS AND THEIR INTERACTIONS WITH THE COLONY

A strong behavioral bonding appears to exist between the reproductive female and male(s), as evidenced by both the number of ano-genital nuzzling and sniffing interactions that occur between them and by the large amount of time

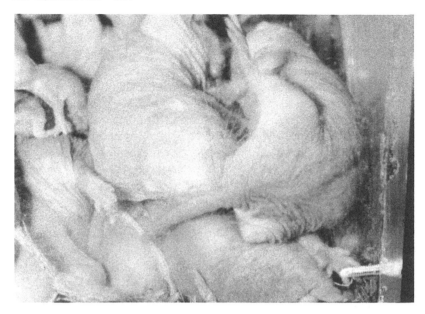

Plate 13-1. Mutual ano-genital nuzzling occurs frequently between the breeding female and breeding male in a colony Photo: C G. Faulkes

they spend together in the nest (plate 13-1). During the study period, the breeding female of Lerata 4 would recline in either of two nests during a 24-h period. Of 76 hourly observations in which the breeding female was in a nest, the dominant breeding male (467) was in the same nest as the breeding female 89% of the time, and the subordinate breeding male (473) 55% of the time. On 82% of these occasions, 467 and the breeding female were either lying in body contact or within 1 cm of each other. When the breeding female moved to the other nest, she was usually followed by at least one of the breeding males within 10 min or less. By contrast, none of the nonbreeding animals were in the same nest with the breeding female more than 55% of the time; indeed, 70% of colony members were in the same nest as the breeding female less than 31% of the time.

When a breeding animal enters the nest area, it approaches or is approached by its partner and they engage in mutual ano-genital nuzzling and sniffing. While the breeders are resting, they often lie side-by-side in the naso-anal position, rousing briefly to smell or rub their partners' genital area (plate 13-1; see also Lacey et al., chap. 8). Unless there is colony strife (below), ano-genital nuzzling and sniffing occurs only occasionally between the breeders and a few other colony members (of both sexes) and among these other colony members (both sexes). During my study of Lerata 4, 46.5% of the involvement in ano-genital nuzzling and sniffing ($n = 314$ individuals, 157 interactions) was exclusively between the breeders; considering both partners in ano-genital interactions, 289 acts (92.1%) involved at least one breeder, and 282 (89.8%) involved the breeding female. Moreover, the 47% figure is certainly an underestimate because repeated ano-genital interactions occurring almost without a break were recorded as a single datum. Of these ano-genital interactions between the breeders, male 467 was involved 57.1% of the time and male 473, 42.9%. The frequency with which 28 nonbreeding colony members engaged in ano-genital interactions with the breeding female was strongly positively correlated with the frequency of ano-genital interactions with the two breeding males by these same animals ($r = 0.715$, $p < 0.001$). There was a weaker positive correlation ($r = 0.440$, $n = 68$, $p < 0.01$) between the number of ano-genital interactions occurring exclusively between nonbreeders and their participation in a suite of interactions with juveniles and subadults (e.g., tooth fencing, sparring, wrestling, and dragging).

As was also noted by Lacey and Sherman (chap. 10), the percentage of occupancy of the nest increases close to the birth of a litter, such that about 24 h before parturition nearly the entire colony can be found resting with the breeding female. However, apart from this increased occupancy, no obvious preparation of the nest occurs before parturition.

In the nest, the breeders and some other large animals, especially those involved in ano-genital interactions, usually lie on top of the huddle of resting mole-rats, often with their heads close to the exit from the chamber. They are the first to vacate the nest if the colony is alarmed. For this reason, and also because large animals were the primary defenders of the colony, Lacey and Sherman (chap. 10) suggested that this facing-out behavior was associated with colony defense. However, breeding females also lie close to nest exits and are usually (95%, $n = 35$) the first animals out of the nest when the colony is disturbed by a noise or vibration. Since breeding females are not usually involved in defense (Lacey and Sherman, chap. 10), it is clear that they (and perhaps some other colony members) lie close to the exits for other reasons.

In colonies where there is strife between individuals, such as when females are fighting for the breeding position (below), the number of ano-genital interactions not involving the breeding animals increases. Furthermore, if one of these hopeful reproductives is reclining in the nest area, it often senses the approach of a rival before that animal actually enters the nest and rouses and violently shoves the approaching animal backward along the burrow and away from the nest. Rivals do not appear to announce their presence vocally, and in large colonies, vibrations made by an approaching rival are masked by the traffic of animals in and out of the nest; thus, I assume that odors are sensed by the rival animals. The shoving encounters may begin as violent tooth fencing and end in mutual ano-genital nuzzling and sniffing between the combatants.

When resting between interactive behaviors, the rival animals also frequently (not quantified) lie flat on their backs on top of the huddle in the nest. Descriptions of the fights associated with achieving reproductive dominance are given by Lacey and Sherman (chap. 10) and below.

MATING

Mating occurs 8–11 days postpartum (see also Lacey and Sherman, chap. 10) while the breeding female is still lactating. The breeding female is very restless during estrus, both within the nest and in the tunnels of the burrow system (Lacey et al., chap. 8). On encountering a breeding male, she emits a trilling sound (Pepper et al., chap. 9), crouches in front of the male, and adopts a lordosis posture (i.e., head raised, back stongly concave, rump slightly raised, and tail held to one side). The female may also back up and thrust her rear end into the face of the male, and her anus and clitoris are very prominent at this time. The male may then smell the female's genitalia, whereupon the female vocalizes and moves forward a short distance, followed by the male. The sequence is then repeated and may also be interspersed with mutual ano-genital nuzzling and sniffing and autogrooming of the genitalia. As courtship and soliciting by the female progresses, the male attempts to mount the female after smelling her genitalia. Throughout this sequence, courtship appears to be a process initiated primarily by the female.

Mounting appears clumsy. The male climbs onto the back of the female, grips her flanks with his forefeet, and often kneads her back or sides with his hind feet. His back is strongly arched, and he tucks the posterior part of his body to one side and partly under the female as intromission is attempted. The male does not bite the female's neck, and he often falls off the female while attempting to mate. It seems likely that mating in the wild occurs in the confines of a narrow, rough-surfaced burrow where the burrow walls would help keep the male atop the female. After the male dismounts or falls off, the female runs forward and, if followed by the male, stops, crouches, and the sequence is repeated. If the male does not follow, the breeding female often turns and goes back to him, sometimes mounting his head before turning again and repeating the soliciting behavior. These bouts of soliciting and mounting are interspersed with rests in the nest or with the breeding female patroling the burrow system. Other colony members are not solicited by the female when she is in estrus and show no apparent reproductive interest in her when they meet in the burrow or nest. Mating occurs primarily in the tunnels and only occasionally in the nest box.

If there is more than one breeding male in a colony, they do not fight over the estrous female, and she seems to solicit whichever of these males she encounters. The reproductive males show little observable excitement over the estrous female, and the mating attempts and the actual copulation appear to be under the control of the breeding female. Lacey and Sherman (chap. 10) observed some increase in male-male aggression a few days before parturition. With the exception of the fighting following the death of breeding female 5017 (below), only 7 of the 28 mole-rats killed or very seriously injured in intracolonial fighting at Cape Town were males. Nearly all these males were killed by the breeding female; there was only one confirmed instance of male-male killing. These findings are similar to those of Lacey and Sherman (chap. 10).

Behavioral estrus lasts 2–24 h. During estrus, repeated matings (the number of which I have not quantified) take place, and the 8–11-day-old sucking pups of the last litter have little chance to feed. During estrus, the vaginal opening of the breeding female is perforate, there is a reddish mucus discharge from the vagina, and the whole urogenital area is distended and colored a deep purplish red. At other times in the reproductive cycle, the female's vaginal opening is perforate or very lightly sealed (and easily broken with gentle rubbing). A characteristic feature of actively breeding females is a copious vaginal discharge of mucus when the genitalia are firmly rubbed with a finger. This is not produced by other females except when several are fighting for dominance and are also perforate. The significance of this mucus is not known; perhaps it contains semiochemicals that are transferred to the noses (and thence to the vomeronasal organs) of the males during ano-genital interactions. During a period of intense female-female strife in Lerata 4 (below), mutual ano-genital nuzzling and sniffing between the rival females was sometimes (22.7% of 286 ano-genital interactions) followed by one or both of the participants violently wiping and rubbing their noses against the walls of the nest chamber or against the bodies of animals lying in the nest. The action was so intense that it appeared as if the animals were attempting to wipe an irritant off their faces.

GESTATION

The mean interval between successive litters is 79.6 days with a range of 77 to 84 days ($n = 121$ litters, 16 breeding females). Assuming a postpartum estrus of 8–11 days, the gestation length is 66–76 days. Lacey and Sherman (chap. 10) independently calculated a mean gestation time of 72–77 days. The pregnant female only becomes obviously pregnant about 40 days after mating. Thereafter, body mass increases sharply until, just before parturition, it has increased by a mean of 49% ($n = 11$ animals, 22 pregnancies) and a maximum of 84% of her most stable mass (a month after parturition).

When close to parturition (plate 13-2), the breeding female walks with difficulty and is far less mobile than at other times. The frequency with which she patrols her burrow system decreases (Reeve, unpubl. data), but the number of her visits to the toilet area increases. Her resting posture also changes and she now usually lies on her ventral surface, occasionally on her side, and almost

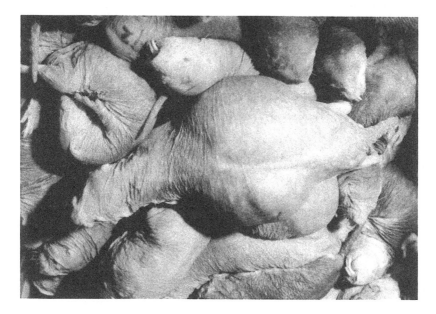

Plate 13-2. The breeding female of Lerata-4 colony, 1 day before she gave birth to 27 pups; at this time her body mass had increased by 84% of her nonpregnant weight. *Top,* side view; *Bottom,* dorsal view. Photos· J.U.M. Jarvis.

never on her back. She has difficulty in doubling up to reach her anus for autocoprophagy and she begs feces from other adults and subadults engaged in autocoprophagy in the nest area. In Lerata 4, 21 instances of allocoprophagy by the breeding female were seen. There was a strong positive correlation between nonreproductive females ($n = 8$) that fed the breeding female feces and those that performed ano-genital nuzzling and sniffing with her and the two breeding males ($r = 0.837$, $p < 0.001$). Among the males ($n = 5$), the subordinate breeding male (473) most frequently fed the breeding female feces (23.8% of 21 feedings); the alpha breeding male (467) was not observed giving feces to the breeding female.

The nonbreeding individuals at Cape Town (> 25 colonies) also showed morphological and physiological changes when the breeding female was pregnant. The nipples of all colony members (both sexes, all age groups, including the approximately 2-mo-old pups of the last litter) began to enlarge in the second half of pregnancy, peaking in size just before the birth of the pups. In addition, the line indicating the position of the sealed vagina became more obvious in all females in the colony, and a few females even became perforate.

The phenomenon of nipple development in nonreproductive colony members of both sexes is intriguing. It does not result in milk production or supplementary feeding of the pups by other mole-rats. The size of the nipples (and in the females, the state of the vaginal closure membrane) is not uniform throughout the colony but is consistent for an individual. In Lerata 4, there was a strong positive correlation between the nipple size of an individual at the beginning and end of the 14-mo study (for males, $r = 0.824$, $n = 35$; for females, $r = 0.767$, $n = 35$; for both, $p < 0.001$). In the nonbreeding individuals, there was a positive correlation between the original (capture) mass of a mole-rat and its teat size (males, $r = 0.556$, $n = 33$, $p < 0.001$; females, $r = 0.418$, $n = 35$, $p < 0.05$). The cyclical development of nipples indicates that the physiology of the entire colony is strongly affected by the reproductive state of the breeding female, so much so that even males develop pronounced nipples (plate 13-3). At Cape Town, nipple development has not been observed in colonies with a history of no survival of pups.

Nipple development among females and changes in the vaginal closure membrane suggest that, as the breeding female approaches parturition, her control over other females in the colony weakens (i.e., other individuals become more sexually active). Indeed, Reeve and Sherman (chap. 11) found that the breeding female's rate of shoving other colony members decreased during pregnancy. It is interesting that even though some females in the colony became perforate every time the breeding female approached parturition (4–5 times a year), and the colony contained many males undergoing spermatogenesis, pregnancies among secondary females rarely occurred.

Evolutionarily, the increased sexuality of the nonbreeding females as the breeding female approaches parturition may ensure that should the breeding

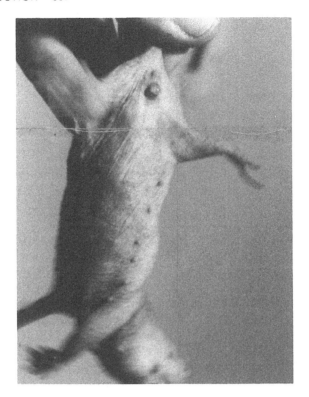

Plate 13-3 A male from Lerata-4 colony showing pronounced nipple development just before the birth of a litter. Photo: J.U.M Jarvis.

female die during parturition, she is rapidly replaced by another reproductive female. Indeed, in two instances (below), a breeding female died just before parturition. On the first occasion, two females (sisters from the same litter) had their first litters 111 and 114 days later, and on the second occasion, the re-placement female came into estrus and mated within 10 days of the death of the breeding female. A corollary to the above suggestion is that in circum-stances in which the breeding female cannot regularly assert her dominance over other females (e.g., by shoving [Reeve and Sherman, chap. 11] or other means), some of the others begin to try to breed. One circumstance in which the breeding female is less mobile and therefore less able to shove is when she is pregnant; another would be if she were ill.

Although I have no field data to substantiate this, I suggest that new colonies are most likely to be founded after several years of unusually good rainfall and a concomitant period of high recruitment to the colony (see also Brett, chap. 4, 5). On the ultimate level, the occurrence of exceptionally favorable environ-mental conditions in which to breed and, on the proximate level, a large influx of juveniles into the parent colony together with a relaxation of the breeding female's control at the end of a pregnancy could stimulate a perforate female and a small group of colony members to bud off and form a new colony.

PARTURITION

Before and during parturition, the breeding female is restless and repeatedly doubles up to lick her urogenital region (see Lacey et al., chap. 8). She repeatedly turns around in the burrow and often walks backward before straining and doubling up again. She makes frequent visits to the toilet area where again she doubles up and pulls at and licks her urogenital region. Occasionally colony members smell her vagina and attempt to mount her head or other parts of her body. The pups are born in the nest and also along the tunnels. In general, the breeding female briefly grooms pups born at the beginning of parturition and when there is a longish (ca. 5 min) gap between births. If, however, the time gap between births is short, the mother will usually drop the pup and move away. Both headfirst and breech births are common. The majority of colony members take no part in attending the breeding female, in grooming the newborn pups, or in eating the afterbirth. The breeding female may give birth on top of the huddle in the nest, with no apparent interruption to the normal colony behavior in the nest. At other times, colony members nearest to the newborn pup groom it and eat any adhering membranes and the placenta; they also lick the bodies of other mole-rats in the huddle that have been wetted with the amniotic fluids and blood.

Thus, the breeding female is not surrounded by a retinue of workers or mates at parturition. She is not assisted by any clearly identifiable individual or group of individuals. However, on a per capita basis, small mole-rats (< 20 g) show a greater involvement in the birth of pups than do larger ones (> 20 g). Two of the births to Lerata 4 (in September and December 1981) were monitored during the 14-mo study period. Of the 71 wild-captured colony members, 15 participated in the birth process by eating the placenta and/or grooming the wet pups (n = 28 times). The sex ratio of these attendants was 7 males to 8 females. Thirty-two percent of the 25 animals smaller than 20 g at capture and 16.3% of the 43 animals larger than 20 g at capture participated in the birthing process. Also participating in the December birth were 6 male juveniles that were 6-mo old (they were from a litter of 14 males and 8 females). These juveniles were involved on 13 occasions.

Not infrequently, parturition triggers heightened sweeping activity by the smaller mole-rats (not quantified). Any pups encountered by the sweeping mole-rats are kicked along the tunnel, the worker behaving as if the object it is kicking is dirt or nesting material. This sweeping activity is often interspersed with the mole-rat turning around and smelling and nudging the pup before carrying it a short distance, dropping it, and resuming sweeping. During partu-

Table 13-2
The Sex Ratios of Wild-Captured Naked Mole-Rats and of Animals Born in Captivity

Number		Sex Ratio		
Males	Females	Males:Females	Comments	Source
176	138	1.28 : 1	7 complete and other almost complete	1, 2
377	338	1.12 : 1	wild-captured colonies*	3
146	104	1.40 : 1	fragments of wild-caught colonies	2
246	173	1.42 : 1	pups born and weaned in captivity	2
164	190	0.86 : 1	newborn pups (died within 4 days of birth in captivity)	2
31	15	2.07 : 1	juveniles < 16 g at capture[†]	1, 2, 3
13	13	1.00 : 1		3

NOTE See also Brett, chap 4
* In all of the samples except juveniles under 16 g at capture, the sex ratio is significantly different from 1 1 ($p < 0.005$, χ^2 tests).
† From colonies in first two rows of this table
SOURCE 1, Jarvis 1985, 2, Jarvis, unpubl data, 3, Brett 1986

rition, larger nonreproductive mole-rats may be seen baby-sitting one or more pups in the tubes. Here the mole-rat lies on its side with the pup(s) sheltered within the curve of its body; the pups are gently nudged and may also be groomed by the baby-sitter. Usually, within 6–10 h of birth, all the pups have been gathered into the nest. However, in some captive colonies with a poor record of rearing young, the sweeping and tumbling of pups continues until they eventually die, probably through starvation and trauma. Lacey and Sherman (chap. 10) note, and I concur, that the treatment accorded to the pups soon after birth often makes it possible to predict whether or not a litter will be successfully weaned.

LITTER SIZE

Few data are available on litter size in wild colonies, but from size cohorts in newly captured colonies (Brett 1986, chap. 4) and from 5 pups found in each of two nests (Sherman, Jarvis, and Alexander, unpubl. data), it would appear that in wild colonies 5–15 pups are born per litter. Brett (chap. 4) gave the mean number of young surviving to weaning (and therefore to capture) as 9.71. More than 12 pups in a litter are often born in captivity ($\bar{x} = 12.3 \pm 5.7$, $n = 84$ litters, 16 females, Jarvis's laboratories; 7.7 ± 2.3, Lacey and Sherman, chap. 10) with a minimum of 1 and a maximum of 27 (Jarvis 1984). The range in litter size is large over all ages and body sizes of breeding females. The body mass of newborn pups is also quite variable ($1.0 - 2.4$ g, $\bar{x} = 1.86$ g), even for the same female and within one litter.

There is no evidence that "outbreeding" alters litter size. The mean litter size of the 54 litters born to 8 colonies with outbreeding was 9.94 ± 3.79 pups and that of the 63 litters born to 10 colonies with no outbreeding was 10.4 ± 5.7 pups. This difference is not significant ($p > 0.5$, two-tailed t-test). Probably, as Reeve et al. (1990) suggested, naked mole-rats have inbred for so long that there is today no observable inbreeding depression.

SEASONALITY IN BREEDING

Because I captured no breeding females during a year of sampling wild animals, I originally suggested that *H. glaber* was an opportunistic breeder whose reproduction was linked to the rather unpredictable rainfall of their habitat (Jarvis 1969, 1978). Laboratory and field data now show that *H. glaber* breeds every 76 – 84 days throughout the year and that the paucity of breeding females in field samples can be attributed to the difficulty of catching them (Jarvis 1985; Jarvis and Bennett, chap. 3; Brett, chap. 4). In the field, newly born pups (< 3 g) were caught in May (Jarvis 1969; Hill et al. 1957), August (Jarvis 1985), and December (Sherman, Jarvis, and Alexander, unpubl. data), and juveniles of various sizes (ages) were captured in most colonies. Brett (1986, chap. 4) checked carefully for seasonality of births in wild *H. glaber* and found no evidence of it.

SEX RATIOS

If the data from the Cape Town laboratory and captures of wild animals are pooled, there is a small but consistent and significant ($p < 0.005$) bias toward males (table 13-2). As pointed out by Brett (chap. 4), a male bias is also seen in complete and almost complete wild-captured colonies, and in fragments of colonies; my data show that the male bias also occurs among juveniles weaned in captivity. Brett (1986, chap. 4) found that the sex ratio of wild-caught juveniles smaller than 16 g was parity and that the bias toward males developed with age. I, however, found more males among my wild-captured juveniles. In captivity the sex ratio of pups at birth was parity (sexing done by autopsy at 1–4 d). One mechanism leading to sex ratio differences in my captive colonies was that significantly fewer males than females died before weaning (table 13-2). In other words, pup mortality was female-biased.

CARE OF THE PUPS

In the nest, the pups form part of the general huddle (plate 13-4). Other adult colony mates lie with them and on them, groom them, and perform a packing or pushing behavior whose function is unknown (see Lacey et al., chap. 8; Lacey and Sherman, chap. 10). All but 2 of the 71 colony members of Lerata

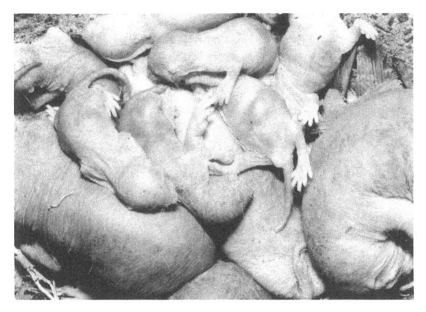

Plate 13-4. Month-old pups form part of the general huddle in a naked mole-rat colony's nest; note the pups' pot-bellied appearance Photo: R. Borland.

4 participated in the 991 recorded instances of pup care (table 13-3). The 3 reproductive animals (4% of the colony) were involved 195 times (19.7%), and the 25 animals smaller than 20 g at capture (35% of the colony) were involved in 46.2% of the remaining 796 instances. Breeders and small individuals showed a significantly higher per capita involvement than did the 43 nonbreeding colony members weighing more than 20 g at capture (table 13-3). There was also a slight negative correlation between the initial body mass of the nonbreeding colony members and the number of interactions with pups (all pup care activities combined: females, $r = -0.427$, $n = 35$, $p > 0.01$; males, $r = -0.3406$, $n = 33$, $p > 0.05$).

A foreign adult male and female were accidentally introduced into Lerata 4 in February 1981; they were apparently fully accepted by the colony. They were involved in pup care and were responsible for 24.8% ($n = 113$) of the carryings of pups born in the two litters following their introduction. This carrying was almost exclusively moving the pups from a nest box vacated by the breeding female to a second nest in use at that time. The foreigners also groomed ($n = 5$) and pushed ($n = 7$) the pups, and the foreign female was twice seen grooming the breeding female during parturition. Seven weeks after the birth of the second litter (in August 1981) the breeding female began to attack the introduced female whenever they met, and the newcomer was eventually so badly injured that I removed her from the colony. Two weeks later, the breeding female began to attack the introduced male, who eventually also was

Table 13-3
Summary of Pup Care by the Seventy-One Wild-Captured Naked Mole-Rats
in Lerata-4 Colony

Behavior	No. of Animals Participating		% Involvement of the Animals in Each Group	Frequency		
	Total	Female		Total	(Female)	%
Carry pups						
Breeding animals	3	1	100	68	6	30.9
< 20 g	20	8	80	75	26	34 1
> 20 g	24	9	55.8	77	25	35
TOTAL	47	18		220	57	
Push pups						
Breeding animals	3	1	100	88	37	17 8
< 20 g	23	10	92	188	81	38
> 20 g	38	23	88 4	219	112	44
TOTAL	64	34		495	230	
Nudge pups						
Breeding animals	3	1	100	26	15	31
< 20 g	21	9	84	32	15	38 1
> 20 g	20	11	46 5	26	14	31
TOTAL	44	21		84	44	
Groom pups						
Breeding animals	3	1	100	12	5	17 4
< 20 g	12	6	48	24	13	34 8
> 20 g	14	9	32 6	33	25	47 8
TOTAL	29	16		69	43	
Allocoprophagy involving pups						
Breeding animals	1	0	33 3	1	0	1 1
< 20 g	12	5	48	37	20	39
> 20 g	17	8	39.5	57	30	60
TOTAL	30	13		95	50	
Care at birth*						
Breeding males	0			0		
< 20 g	8	3	32	12	5	42 9
> 20 g	7	5	16 3	16	13	57.1
TOTAL	15	8		28	18	
Total pup care						
Breeding animals	13	4		195	63	19.7
< 20 g	96	41		368	160	36 8
> 20 g	120	65		428	219	43 2
TOTAL	229	110		991	442	
TOTAL (excluding breeding animals)	216	106		796	379	

NOTE The colony comprised 1 breeding female, 2 breeding males, 25 animals weighing less than 20 g at capture (10 were females), and 43 animals weighing more than 20 g at capture (27 were females) Data were obtained during 114 h of observation, covering the birth of five litters
* Groom pups or breeding female, eat placenta

removed. Neither of the introduced mole-rats was ever seen to fight back. Increased aggression (shoving) toward distantly related mole-rats by the breeding female is discussed by Reeve and Sherman in chapter 11.

Over the 14-mo study, the three reproductive animals were involved in 30.9% ($n = 220$) of the pup carryings, 17.8% ($n = 495$) of pup pushings, 17.4% ($n = 69$) of pup groomings, and 31% ($n = 84$) of the nudgings of pups. The alpha breeding male carried the pups more than any other animal (52 times), and the breeding female and beta male pushed the pups the most (both 37 times). Lacey and Sherman (chap. 10) also show that the breeders were significantly more often involved in pup care per capita than were the other colony members. In further agreement with my observations, Lacey and Sherman found negative correlations between the body mass of nonbreeding colony members and the number of pushes, grooming bouts, and nudges they gave to the pups.

Members of a litter of 22 pups born in June 1981 were involved in the care of two litters born after them. Because they were not present for the whole of the study period, their involvement in pup care was not included in the main regression analysis. If, however, their participation in pup care and that of the 71 wild-caught colony members is combined and compared (for the same two litters), it is apparent that these 22 newly recruited colony members were heavily involved. For example, they carried pups 43.7% ($n = 203$) of the time; the three breeders did so 25.8% of the time, and the remaining 68 colony members only 30.5% of the time. There was no difference (on a mean per capita basis) in the amount of pup carrying done by the 14 males or 8 females of this litter. The 22 young animals were also active in grooming (55.8%, $n = 77$) and nudging (34.2%, $n = 76$) the pups. Lacey and Sherman's (chap. 10) study colonies did not contain any large litters of recently weaned pups; therefore, the primary burdens of pup care apparently fell on the breeding animals.

On entering the nest, the breeding female does not seek out her pups but lies down and the pups converge on her, perhaps using her odors as a cue. A band of milk-laden mammary tissue, with a whitish lumpy appearance (through the skin), extends along the sides of the lactating breeding female, from her axilla to inguinal region. To nurse, the breeding female will often lie flat on her back with her legs splayed out, thus fully exposing her nipples to the pups. At times the female is almost completely covered by nursing pups (plate 13-5). Only by lying flat on her stomach (or leaving the nest entirely) can she prevent the pups from nursing. Suckling is frequently disrupted by other colony members' lying against or on top of the breeding female and also by animals' pushing the pups so violently that they lose their hold on a nipple. Because I have been unable to mark individual pups, I cannot say whether certain pups are pushed more than others, or whether each pup has a favorite nipple. The same pup has however been seen to suck from more than one nipple. In Lerata 4, the breeding female (with 11 nipples) reared 22 of a litter of 27 pups; in this instance,

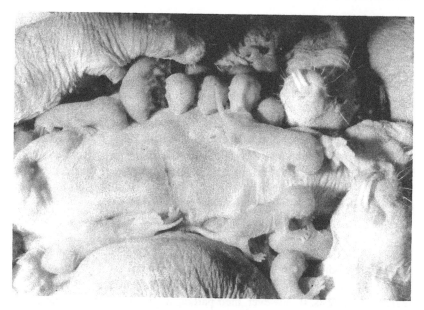

Plate 13-5. A breeding female naked mole-rat lies on her back to suckle her pups. Her milk-ladened mammary glands can be seen through her skin The pups in the photo are 2–3-days old Photo· J U.M Jarvis

several pups obviously had to share a nipple. Despite the development of prominent nipples in other colony members, none of them has ever been seen to suckle.

Lactation is demanding on the breeding female, and she loses up to 16% of her mass while lactating. During the latter half of lactation, additional demands are being made on the breeding female because she is pregnant with her next litter.

On five occasions I experimentally introduced foreign pups into colonies with pups of the same age. Before the introduction, the pups were washed in water and then rolled in soiled wood shavings from the toilet area of the foster colony (see Jarvis, Appendix). All the pups were accepted by the foster colony and carried to the nest. However, only the newborn ($n = 2$ and 3) and day-old pups ($n = 2$) were able to locate the foster breeding female. The 2–3-day-old pups ($n = 3$ and 3) apparently had imprinted on their own mother's odor and were unable to home in on another breeding female. Some of the breeding females' own pups were successfully reared, but none of the 2–3-day-old foster pups.

When newborn young are present, colony members (especially the younger animals) show a heightened alarm response to noise or vibrations, sometimes even to familiar laboratory noises associated with feeding and cleaning. This occasionally elicits an upsweep trill or "panic cry" (Pepper et al., chap. 9) and often results in the pups being carried out of the nest. When carried, the pups are held between the incisors (in much the same way as food is carried) by any part of the body, including the head (see Lacey et al., chap. 8). Most frequently, the carrier encircles the pup's chest or abdomen in its incisors. The pups lie passively while being carried. The mole-rat carrying the pup may walk forward or backward and frequently has to push past colony mates, thereby potentially exposing the pups to considerable physical trauma. It is indeed surprising how resilient the pups are to this treatment, to the trauma of being swept along burrows, and to being lain on in the huddles. It suggests a long evolutionary history of rearing pups in large colonies under close quarters. However, if the colony is disturbed too frequently the pups are rarely contained within the nest huddle; then they are often swept or kicked about in the tunnels and usually die of trauma and starvation. This heightened sensitivity to disturbance means that it is difficult to handle, mark, and weigh pups or to obtain behavioral data on individual pups until they are weaned (ca. 3-wk old).

When the pups are groomed, they are held between the forefeet of the attendant mole-rat, who licks them gently and picks at the pups with its incisors (see Lacey et al., chap. 8). Grooming the urogenital area of very young pups induces them to urinate and defecate, and the excreta are eaten by the groomer. At times, the grooming mole-rat gently shakes the pup or holds it in its incisors and brushes the pup with its forefeet, using a similar behavioral repertoire to that used when feeding. A dead pup may sometimes be licked until its skin breaks, whereupon feeding on the carcass occurs; the head is rarely eaten. The majority of dead pups are abandoned in the toilet area and, less frequently, in the nest or a tunnel. A few colonies eat all their pups within 2–3 days of birth, some within 24 h of birth while the pups are still alive. Eating live pups is not common; it has been seen in the colonies at Cape Town but not at Cornell (Sherman, pers. comm.).

In many numerically large colonies (i.e., > 40), certain mole-rats habitually rest in the tunnels away from the nest area. Among the males but not the females of Lerata 4, body mass correlated positively with the proportion of scans in which individuals rested in the tunnels away from the nests (for males, $r = 0.548$, $n = 35$, $p < 0.001$; for females, $r = 0.115$, $n = 36$, $p < 0.1$). When newborn pups are present, some individuals in these satellite clusters lie on their sides with pups in the curve of their bodies and gently nose them. The breeding males do not participate in this behavior. The breeding female never stays for long at these clusters, and unless returned to the nest, the pups starve to death.

DEVELOPMENT OF THE PUPS

At birth naked mole-rat pups weigh 1.86 ± 0.33 g ($n = 155$). They are a shiny bright pink-red color and the deeper layers of the skin are almost gelatinous.

The skin is hairless (except for vibrissae on the face), smooth, and semitransparent (plate 13-6). The position of organs such as the eyes, diaphragm, liver, and the milk-filled stomach and gut are clearly seen through the skin of newborns. Sutures and the fontanel in the skull and much of the skeleton are also visible. The digits of the forefeet and hind feet are short but distinct and tipped with short claws. The eyelids are fused and the incisors have not erupted. If placed on their backs, the pups can right themselves if they can find something to push against. Within a few hours of birth, the pups are able to walk and to crawl over adults in the nest huddle. Within a day, their skin dries out and tightens and the pups look smaller than when newborn. Their color changes to a duller and paler pink, and the tips of minute incisors begin to break through the gums. Body mass varies considerably within a single litter, and my limited data on growth of pups toe-clipped at birth ($n = 20$) show no relationship between mass at birth and growth rate or adult mass (see Jarvis et al., chap. 12).

Morphological changes in the pups over the next 2 wk are subtle. During weeks 1–2, the skin thickens, becoming more opaque and a lighter pink-brown color (see plate 13-4). Pups are slightly darker dorsally than ventrally, and isolated hairs erupt over their body surface. Their incisors grow and widen. The agility of the pups increases daily. When they are about 2-wk old, pups first attempt to eat solid food and begin to beg feces from adult and subadult colony members. The pups also begin to wander out of the nest but are usually carried back. As they walk, the pups' legs are splayed out laterally (see Lacey et al., chap. 8), and they run in a stilted manner with the whole body bouncing up and down. Month-old pups have a sparse distribution of hair similar to that of an adult (plate 13-6). The eyes open on about day 30. Like adults, pups generally keep their eyes shut unless alarmed.

By the age of 3–4 wk, nursing has become infrequent, and the pups obtain most of their nutrients from solid foods and from coprophagy (plate 13-7). The pups will occasionally also eat feces found in the toilet area. During allocoprophagy the pups approach colony members engaged in autocoprophagy (perhaps, as in rats, using odors as a cue; Moltz and Lee 1981) and pull at the skin around the anus with their incisors while making a distinctive begging cry (see Pepper et al., chap. 9). While producing soft, wet feces (cecotrophs), the donor mole-rat remains in the typical sitting posture adopted during autocoprophagy (plate 13-7), yawning frequently; its hind feet may also shake quite violently. As many as four pups have been seen clustering around and begging from a single adult. In 96 instances of coprophagous feeding of pups by the wild-captured members of Lerata 4, 33 different animals fed the pups (16 males and 17 females, of which 2 males and 6 females weighed less than 15 g at capture). The breeding female and alpha male were not involved and the beta breeding male fed pups only once.

The cecotrophs and the caudal region of the cecum of *H. glaber* contain many large white ciliate protozoa, bacteria, and fungi (Porter 1957; Jarvis and

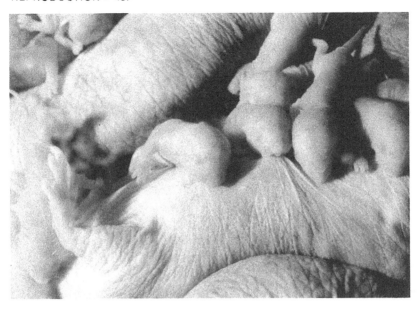

Plate 13-6. *Top*, Newborn (2–3-day old) naked mole-rat pups sucking; notice the smooth, semitransparent facial skin and the skull sutures *Bottom*, Three-wk-old pups just before weaning; notice the development of vibrissae. Photos· J U.M. Jarvis

Plate 13-7. Naked mole-rat pups begging cecotrophes from an adult. As they beg, the pups vocalize and pull at the skin next to the donor's anus. These pups are about 4-wk old Photo: J.U.M. Jarvis.

Bennett, chap. 3; S. Yahev and R. Buffenstein, unpubl. MS). It is probable that in addition to nutrients, coprophagy provides the weaning pups with the gut fauna needed to digest the cellulose in their food. Perhaps a lack of these microbial organisms is the reason pups are noticeably pot-bellied at the onset of weaning (i.e., their guts appear distended with undigested food, and their abdominal skin is tight and shiny). Within a few weeks, their bodies slim down again, and the pups are fully weaned. Begging by pups for feces becomes infrequent about 2 mo after weaning. Adults other than the breeding female rarely beg feces.

Most young pups urinate and defecate in the tunnels rather than in the toilet area. Sometimes they lie on their backs and squirt urine onto the roof of the burrow system. As the pups get older, the time spent in the toilet area increases, and juveniles seem to spend more time sweeping this area than do the older animals. In Lerata 4, percentage of increase in body weight among wild-caught females was positively correlated with the time they spent in the toilet ($r = 0.501$, $n = 36$, $p < 0.01$); however, a negative but nonsignificant correlation was found for the males ($r = -0.017$, $n = 35$, $p < 0.1$).

Upon weaning, pups begin to perform tasks in the colony. They carry small items of food and nesting material and even attempt to sweep with their hind feet. In Lerata 4, the mass of an individual at the time of its capture was

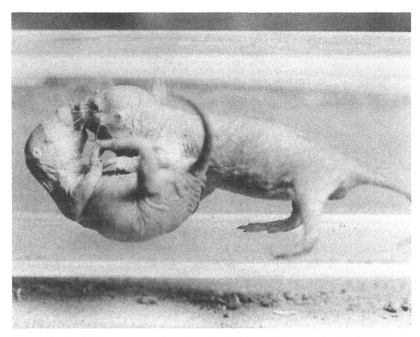

Plate 13-8. Play fighting between juvenile naked mole-rats. The pups lock incisors, wrestle, and engage in tugs-of-war, often pushing each other over. Photo: D. Hammond, *Cape Town Argus* (South Africa).

strongly negatively correlated with its total amount of maintenance work (digging, sweeping, gnawing, carrying nesting material and food; for females, $r = -0.776, n = 36$; for males, $r = -0.566, n = 35$; for both, $p < 0.001$). The percentage of growth over the study period was strongly positively correlated with the amount of maintenance work done by the females ($r = 0.836, n = 36$, $p < 0.001$) and more weakly correlated for the males ($r = 0.477, n = 35, p < 0.01$). Faulkes et al. (chap. 14) and Isil (1983) also found significant negative correlations between the total frequency of maintenance behaviors (digging, sweeping, gnawing, and foraging) and body mass in their captive colonies. Similarly, Lacey and Sherman (chap. 10) found negative correlations between body mass and tunnel sweeping, food carrying, and obstruction and debris removal; however, Lacey and Sherman (chap. 10) found that larger animals dug more frequently than smaller ones (but see Jarvis et al., chap. 12).

Between 3 wk and 2 yr of age, the pups and young adults engage in play fighting in which the animals lock incisors or wrestle, tug at each other's skin, and engage in tugs-of-war (plate 13-8). Sometimes two animals are involved, but at other times as many as four may team up against similar numbers; the animals are aligned side by side and on top of each other as they face mole-rats on the opposing side. Partners are changed frequently. This play fighting rarely occurs in older mole-rats, and when it does, it is usually in response to a younger mole-rat tugging at its skin or locking incisors with it. This suite of juvenile-specific behaviors (see also Lacey et al., chap. 8) may affect the position of young mole-rats in the colony hierarchy. Initially, the interactions between juveniles occur in the nest, but as the pups grow stronger and more agile, their bouts of tooth fencing, sparring, wrestling, and dragging occur in the burrows and particularly in the toilet area just after it has been cleaned.

In Lerata 4, juvenile-specific behaviors were analyzed together. The number of times that wild-captured individuals (of unknown age) engaged in play fighting was strongly correlated with their percentage of increase in body mass over the 14-mo study period (for females, $r = 0.898$, $n = 36$; for males, $r = 0.703$, $n = 35$; for both, $p < 0.001$). The frequency of play fighting was also strongly correlated with the frequency with which maintenance work was done (sweeping, digging, carrying, and gnawing combined), especially by female colony members (for females, $r = 0.775$, $n = 36$; for males, $r = 0.548$, $n = 35$; for both, $p < 0.001$). Subsequent to this study period, when the younger animals captured with the colony were over 2-yr old, their sparring encounters became rare; this was also true for the litter of 22 pups born to Lerata 4 when they reached about 2 yr of age. Juvenile-specific behaviors associated with play fighting thus seem to be age-linked, and they may be used as a rough indicator of the ages of small mole-rats in newly captured colonies.

REPRODUCTIVE MATURITY AND THE LENGTH OF REPRODUCTIVE VIABILITY

Individuals born into a colony may remain reproductively quiescent for many years, perhaps for their whole lives. This and the variable growth rate of pups (Jarvis et al., chap. 12) make it difficult to determine when pups become sexually mature. In my captive colonies, the youngest female to become reproductively active was 7.5-mo old when she conceived. Her seven pups did not survive but her second litter did. The youngest male to mate and sire young was 12-mo old.

The age at which sexual activity ceases has not yet been determined. Two females are still breeding (but not as regularly as before) after being in captivity for 15 yr, and their current breeding males were captured with them (observational data are not available on whether the same animals have been the breeding males throughout). Contrary to my original suggestion (Jarvis 1981), the ability to become a reproductive animal appears to be retained in sexually quiescent colony members of both sexes for many years. Thus, four males, captured as adults, had not bred for 6, 7, 8 and 8 yr, respectively, when they were removed from their colonies and placed with a female. All became reproductively active and, as of September 1989, were still siring litters; one of these males is 13 yr or older as of this writing. Two wild-captured females (weighing 30 and 38 g at capture) were 8 yr or older when they were removed from Lerata 4; they subsequently became breeding females.

LONGEVITY

Longevity is discussed earlier (Jarvis and Bennett, chap. 3). Eighteen animals captured in July 1974 were still alive in January 1990. These animals weighed more than 30 g at capture and were therefore probably 1 yr or older (Jarvis et al., chap. 12); thus, they were 15–16 yr or older in January 1990. Longevities of 15–16 yr are unknown for any other rodent of comparable body mass.

SUCCESSION OF REPRODUCTIVES

One colony, 1977, contained 40 mole-rats, but not the breeding female, when it was captured in Kenya during October 1977; a second colony, Lerata 4, contained 82 animals including 1 breeding female and 2 breeding males, when captured in August 1980. Figure 13-1 shows the distribution of their body masses at capture. At varying times subsequent to capture, both colonies have been split. In one derivative of colony 1977, known as colony 80, there has been a succession of 5 breeding females. The succession process has sometimes been peaceful and at other times violent. In both colonies there have been times (on one occasion for 5 yr) when more than 1 breeding female was present.

The histories of these two colonies illustrate the diversity of responses that can occur when a reproductive female dies or is removed. The events preceding and following the emergence of rival reproductive females are presented below. The reader is referred to Lacey and Sherman (chap. 10) for parallel observations on three Cornell colonies.

SUCCESSION IN COLONY 80

Breeding female 80 had given birth to eight litters (26 pups weaned, a 40% weaning success rate) before I removed her in September 1981. Thereafter, the colony had a succession of five reproductive females. The first succession was peaceful, with no evidence of aggression preceding or following it. Female 75, the largest of two wild-captured females in the colony, became the breeder, and another female (5002) was perforate but did not breed. Female 75 bred for 15 mo and had four litters (10 weaned pups, a 27% success rate). She died of natural causes toward the end of her fifth pregnancy (i.e., there was no evidence that she was killed).

The second succession was initially peaceful, and two sisters (5001 and 5002, both offspring of 80 and the only females in their litter) became preg-

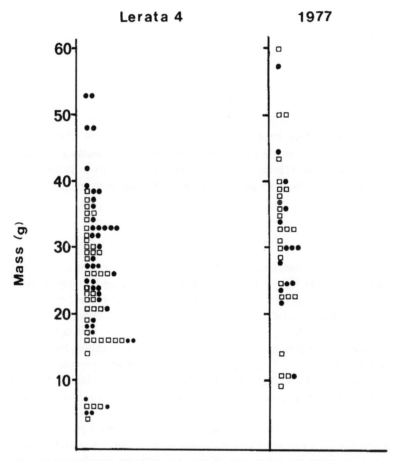

Fig 13-1. The body masses (grams) of naked mole-rats from Lerata 4 and colony 1977 at capture. *Open squares*, males; *filled circles*, females.

nant. They gave birth 114 and 111 days, respectively, after the death of 75. Then 5002 was killed by 5001. The surviving sister was the breeder for 30 mo (10 litters, 22 pups weaned, a 21.6% success rate) and then died of natural causes in July 1985, toward the end of pregnancy. The sisters were the oldest females born in captivity in this colony; however, there was one older (wild-caught) female in the colony.

I have no explanation for these natural mortalities of pregnant females. This colony was not subjected to quantitative behavioral studies, but it was being carefully watched for signs of aggression (shoving, threatening, and biting); although shoving was observed, no threats or bites were ever seen. In addition to the 2 breeding females, 11 other animals (4 of them subadults weighing less than 20 g) were found dead or dying between September 1983 and August

1987. These animals were behaving normally and eating well until 1–2 days before their deaths. Autopsy and tissue cultures by a local veterinary laboratory were unable to establish the cause of death, and I could determine no changes in the behavior of the colony to account for these deaths. Other colonies housed in the same room showed no sudden deaths.

The third succession, in which female 7004 took over the breeding role, was again peaceful; she had her first litter 81 days after the death of 5001. In this succession, unlike the second one, the only female littermate (5017) of the new breeding female showed no sign of reproductive activity. Apparently not all females of any one age group respond similarly to the loss of the breeding female (however, see below). Female 7004 bred for 20 mo (6 litters, 22 pups weaned, a 27.9% success rate). The same breeding male copulated with all of these first three breeding females.

In March 1987, when breeding female 7004 was nearing the parturition of her seventh litter, there was a breakout of a neighboring colony (Lerata 4), and two males got into colony 80. The two foreign males were killed, and eight animals from colony 80 were injured; of these, two died from their wounds. The breeding female aborted her litter. A month later, all the colonies were moved to a new building, and in the process two foreign males were again accidentally introduced into the colony. In the fighting that followed, the foreign males were severely injured (removed), the breeding male in colony 80 died of injuries, and 7004 (the breeder) was also severely injured. Female 7004 recovered from her wounds, but her littermate (5017) also became pregnant and gave birth 82 days after the April fighting. Both females then bred twice but reared no pups. On December 25, 1987, both females were found badly injured, and no other animals were hurt. The implication is that the sisters had fought. Female 7004 died from her wounds.

The remaining breeding female (5017) continued to breed until March 1988, when fierce fighting broke out. Both males and females were involved and 5017 was found dead (but uninjured); at the time of her death, 5017 was approaching parturition. At this time, the colony contained 33 males and only 10 females. Since April 1988, 12 males and 4 females have been killed; the new breeding female had her first litter in January 1989, 9 mo after the death of her predecessor. She and a female littermate (both born in October 1980) are the oldest remaining captive-born females ($n = 6$) in the colony.

SUCCESSION IN LERATA 4

In this colony there has been no succession of breeding females, but the original breeding female (400) "tolerated" the emergence of a second reproductive female (460). This second female was wild-caught with 400, and she fought for her position for 3–4 mo (October – December 1982), not with the breeding female (400), but with seven other females in the colony. Two of these other rivals died of their wounds, and four were badly injured and removed from the colony. During the fighting, the breeding female (400) did not become pregnant, and she and the breeding male (467) tended to avoid areas in the burrow system where fighting was most intense (i.e., the nest box and the burrow leading to it). The breeding pair spent much of their time reclining with satellite clusters of large mole-rats in the tunnels. Immediately before this period of fighting, one of the two breeding males (473) died from a gastric (*Escherichia coli*) infection. By the time the aggression between 460 and the rival females had ceased, a new reproductive male (412) had emerged (was seen mating), and two other males were frequently engaged in ano-genital interactions with females 400 and 460. The new breeding male appeared to be dominant over 467 (the original surviving breeding male) in mating and in the number of ano-genital interactions with the breeding female. It is noteworthy that during the 14-mo behavioral study in 1981 (ca. 10 mo before this period of aggression and before male 412 became a breeder), the number of ano-genital interactions between 400, the sole breeding female, and male 412 was only surpassed by those between her and her two breeding males (table 13-4).

During the 3–4–mo period of fighting between rival females, the number of shoves occurring in the colony increased from 0.8 shoves/h ($n = 89$ shoves in 114 h of observation; table 13-4) to 27.9 shoves/h ($n = 517$ in 18.5 h of observation). Likewise, there was a marked increase in the number of open-mouth threats, tooth fencing, and bites; these three behaviors frequently occurred in rapid succession and were recorded as one. They increased from 0.04 occurrences/h (when 400 was the sole breeding female) to 9.8 occurrences/h (during the period of fighting; table 13-4). In both these categories, the breeding female (400) showed the greatest involvement before the fighting but participated little during the fighting; the breeding males were rarely involved at any time. Before this period of heightened aggression between females, the breeding female participated in 43% of all ano-genital interactions ($n = 314$); during the fighting most ano-genital interactions occurred between rival females (table 13-4).

Breeding female 400 had never been very aggressive, as indicated by the small number of shoves and bites initiated by her and by her tolerance of the presence of a second breeding female. Encounters between breeding female 400 and would-be reproductive 460 were never more violent than a strong shove, whereas those between 460 and the seven other rival females resulted in crippled forelimbs, swollen faces, scabs on the body and (especially) around the mouth. Occasionally, deep wounds were inflicted on the face and body. In these aggressive encounters, 460 also received wounds, whereas breeding female 400 was never injured. No males were injured, nor were they more than peripherally involved in the pushing matches (table 13-4).

The weights at capture of the seven females involved in this period of fighting ranged from 18 g to 48 g, and there were no clear size-related trends (table 13-4). Twenty-four other females in the colony with capture weights falling

Table 13-4
Details of Some of the Interactive Behaviors in Lerata-4 Colony

Mole-Rat I.D.	Body Mass (g)		Shoves (%)		Other Agonistic (%)		Ano-Genital Nuzzling (%)		Fate
	Aug 1980	Dec 1982	A	B	A	B	A	B	
Reproductives (n = 3)									
400 (F)	52 0	61 1	53.9	3.5	100	0 0	43 0	10.8	remained a breeding female
467 (M)	31.5	52.7	2.3	0.0	0 0	0 0	14.0	1.4	remained a breeding male
473 (M)	30.6	30.4	5 6	—	0 0	—	10 5	—	died Oct 1982
Fighting Females (n = 7)									
460	24.0	48.0	1 1	22.6	0.0	25.3	0.6	19.9	became breeding female #2, 1983
403	33 0	48 0	0 0	5.0	0.0	5 5	0.0	1 8	badly injured; removed Nov 1982
443	16.3	49 0	0.0	9.1	0.0	9 9	0.3	9 1	badly injured; removed Nov 1982
449	20 3	45.0	1.1	20.5	0.0	17.0	0 3	13.6	killed, Dec 1982
409	44.4	47.9	4.5	15.5	0 0	21 4	0 6	12.9	badly injured; removed Jan 1983
482	32.7	45 0	0.0	4.5	0.0	3.8	0.3	7.3	killed, May 1983
471	22 8	44 5	0 0	0 0	0.0	0.0	0.0	0.0	badly injured, with no history of aggression; removed Nov 1982
Future Breeding Males (n = 3)									
412	31.4	44.4	1.1	0.6	0.0	0.0	3.5	1.0	became breeding male #2 in 1983
406	30 9	46.0	0 0	3.1	0.0	0 0	0.0	5.6	became a breeding male in 1988
477	23.6	44.6	0.0	1.1	0.0	0 0	1 0	0.0	became a breeding male in 1988
Rest of the Colony (n = 60)									
Mean (± SD)	27.9 (8 9)	39 9 (6.6)	29 2	15.9	0.0	17.0	20.3	16.4	
Sample Sizes									
No. of interactions			89 0	517 0	5.0	182.0	314.0*	286.0*	
Observation hours			114.0	18.5	114.0	18 5	114 0	18 5	
Interactions/h			0 8	27.9	0.04	9 8	2.8	15.5	

NOTE Behaviors occurred at (A) a time during the 14-mo study period (Nov 1980 – Jan 1982), when there was one breeding female and two breeding males and good recruitment of young, and at (B) a time during Oct–Dec 1982, when there was fighting among six females, the breeding female was anestrous, and breeding male 473 had just died of an *Escherichia coli* infection Descriptions of the behaviors are given in Lacey et al. (chap 8)
* Both individuals in each interactive behavior counted

within this range remained sexually quiescent during this period of aggression, and six others became perforate or almost perforate (i.e., the line of the vagina was clearly marked by a depression) but were only occasionally shoved or otherwise attacked.

The colony had two breeding females for 5 yr; during this time female 400 had 15 litters, and 460 had 16 litters. Parturition was synchronous on six occasions. Between them, the females reared 16 of an estimated 540 pups born, a success rate of approximately 3% (the number of pups could not be accurately counted when parturition was synchronous; a mean litter size of 21.2 for 400 and 12.8 for 460 was then used). In September 1987, 460 was severely injured in fighting with three other females. All four females were removed from the colony, and 400 has remained the sole breeding female up to January 1990. Since the removal of female 460, 400 has reared 33 pups (8 litters, $n = 182$ pups, an 18.1% success rate).

Before 460 became the second breeding female, 400 had had a 27% pup success rate (8 litters, 179 pups). Her success was therefore greatly reduced during the time in which she and 460 were both breeding. After the removal of 460, breeding male 412 rarely engaged in ano-genital interactions with 400; as of June 1989, there were three breeding males in the colony, the original one (467) and two new ones (477, 406). Unlike 412, these latter two males gave no evidence of being latent reproductives during the 14-mo behavioral studies in 1980–1982 (see above; table 13-4).

MULTIPLE BREEDING FEMALES IN COLONY 500

In December 1984, I attempted to alleviate the aggression in Lerata 4 (above) by reducing the total number of animals in the colony. One wild-captured male (434) and all the surviving mole-rats born to female 400 in captivity ($n = 37$) were removed to form a new colony (500). Before removal, one captive-born female (508) clearly stood out as the potential breeding female (i.e., her vagina was perforate, and she had ano-genital interactions with both breeding females, the breeding males, and male 434). Among the maternal siblings forming the new colony, one female was 4-yr old, female 508 and her eight littermate sisters were 3.5-yr old, and two females were 3.25-yr old.

Within 21 days of removal from the parent colony, 9 of the 11 female siblings over 3-yr old began to fight; of these, 3 showed sudden increases in body mass and were perforate. By April 1985, 6 of the females had been killed or severely injured (and removed). Two females over 3-yr old and 4 younger females in the colony did not show large increases in body mass or become perforate.

Fourteen months after the colony had been divided, 508 had her first litter. A second female (513, a littermate of 508) had her first litter 6 mo later. Thereafter, shoving encounters between these two females were frequent (but not quantified), and only 11 pups were reared. In October 1988, 3 yr and 10 mo after colony 500 was formed, a third female had pups. This female was under

3-yr old when the colony was formed and was not involved in the fighting at that time. There were two older nonbreeding females in the colony when the third female became pregnant. All three breeding females then bred (asynchronously), but no pups were reared.

OTHER INSTANCES OF MORE THAN ONE BREEDING FEMALE
IN A COLONY

In the Cape Town colonies (19 of which are more than 3-yr old), there have been four cases (in addition to those cited above) in which a second breeding female has emerged in a colony and has been tolerated by the original breeding female long enough to have one or more litters (although no pups survived). In one of these colonies, the second female had three litters and then killed the original female. In another case, the second female was removed during her second pregnancy because fighting broke out between the two breeding females, and five other colony members were killed. In the last colony, the second female had one litter, and a month later killed the original breeding female (which was nearing parturition). In four of the six colonies, the emergence of the second breeding female was preceded by a period of good recruitment of pups to the colony. These may be instances in which, if dispersal had been possible, rival females would have budded off from a thriving and rapidly growing parent colony and founded new colonies.

THE FATE OF REPRODUCTIVE MALES

The fate of reproductive males is more difficult to determine, largely because it is impossible to be certain which male(s) are the reproductives unless mating is actually witnessed. The same male may sire litters for as long as 10 yr (e.g., colony 80), sometimes retaining the breeding role with successive females. Alternatively, new males may emerge but not completely replace the original breeding male (e.g., 467 in Lerata 4). At no time in the 9 yr that male 467 has been a breeder has he been the sole breeding male in the colony. Alternatively, in four colonies begun from pairs, the founding breeding male has been the only breeding male for 5 yr. Data show that many breeding males in colonies with sustained recruitment may initially be quite large animals but then show a marked drop in body mass (Jarvis et al., chap. 12).

THE FATE OF REPRODUCTIVES IN COLONIES WITH
NO RECRUITMENT

One unexpected feature of captive *H. glaber* colonies is that the dominance of a breeding female can be retained even when she is no longer breeding regularly or when pups are born but are not reared to weaning. Four "breeding" females in the Cape Town laboratories and two at Cornell now no longer breed or breed very irregularly, and five others have not successfully reared pups for at least 2 yr. In each of these colonies, there is a breeding male, and the pair engage in mutual ano-genital nuzzling and sniffing; they also mate. In none of these colonies has the breeding pair been replaced, although in three of the six nonbreeding colonies, younger and thus potentially more virile mole-rats are present.

COLONIES OF MIXED ORIGIN

Nine colonies of mole-rats of mixed origin have been established in captivity (five at Cape Town, two at Cornell, two at Michigan). All were initially established without a breeding female. In at least four of these colonies (three Cape Town, one Cornell), fierce fighting occurred, leading to the deaths of a number of mole-rats (see also Lacey and Sherman, chap. 10). Two of the mixed colonies in the Cape Town lab (1700 and 2700) had similar compositions. Both contained an adult female and young animals from another colony. In colony 1700, one adult (3-yr-old) female and nine siblings (four of which were females) from another colony (ages 5 wk to 11 mo) formed the nucleus of the new colony. The adult female became the breeding female and killed three of the younger females 5, 9, and 10 wk after the birth of her first litter. These three females were the oldest of the siblings (i.e., they were 18-, 14-, and 12-mo old when killed). Between these deaths and the birth of female 1700's fourth litter, two of the founding males and the fourth female were also killed by the breeding female. In colony 2700, the adult female was a sister of the one in 1700, and the nine siblings (including four females) were from the same colony and the same litters as those that made up colony 1700. In colony 2700, however, no animals were killed by the breeding female, who successfully reared four litters of pups.

SURVIVAL OF LITTERS

Field-captured colonies usually contain a number of size (age) cohorts of young animals (Jarvis 1985; Brett, chap. 4), clearly indicating that they are successfully recruiting young. In contrast, a feature of many of the Cape Town captive colonies is that an initial phase of high recruitment is followed by few to no pups being weaned. This has also been seen in the Cornell and Michigan colonies. Cessation of successful reproduction has occurred in five of the Cape Town colonies for which there are long sequences of litters (e.g., > 18 litters). In most instances, an important source of mortality of later litters appeared to be repeated moving of newborn pups, usually by the subadult mole-rats. As shown in table 13-3, with the exception of the breeders, the younger animals in a colony are more involved in pup care than are the older animals. Thus, perhaps under captive conditions without predation, a successfully breeding colony grows numerically large too fast and contains a disproportionately large number of young mole-rats.

In wild colonies, even low levels of mortality, such as one to two animals a month, would have a profound effect on the numerical growth of a colony, whereas a loss of four animals a month would probably halt growth completely (Brett 1986, chap. 4; Jarvis and Bennett, chap. 3). To test the hypothesis that too rapid an increase in colony numbers contributes to high pup mortality, I simulated predation by removing some of the subadult animals in two colonies begun from pairs (table 13-1). Two other colonies begun from pairs but not subjected to predation were used as controls. All colonies were housed under similar conditions. The colonies subjected to predation both had 19 litters, with pup survival of 38.2% (n = 208 pups born; 22 animals removed in three predation events) and 52.4% (n = 207; 37 animals removed in four predation events). By comparison, the control colonies each had 20 litters and pup survival of 14.6% (n = 199) and 30% (n = 147). In each colony, the same breeding female gave birth to all the litters, and there was only one breeding male per colony. These data are too limited to draw firm conclusions, but they indicate that the two colonies from which subadult mole-rats were removed had higher pup-rearing success than two other colonies from which no animals were removed; moreover, the colony from which the most mole-rats were removed had the highest success of all (52.4%). In this latter colony, weaning success continued high through the last litters born; that is, pup survival in litters 16–19 was 76% (n = 46 pups) in this colony, whereas for the other three colonies, the pup-rearing rates were 0% (with predation), 0%, and 12% (without predation).

Despite the possibility that rapid colony growth contributes to pup mortality, other as yet unknown factors must also play a role in the successful rearing of pups in captivity. In all the laboratories, some colonies with no history of high recruitment rear few pups. Colony members care for the pups and the breeding female has milk, but the pups die within a few days of birth. Sometimes the breeding female does not permit her pups to suckle. Poor recruitment also appears to occur sometimes in wild colonies (S. Braude, pers. comm.).

Intercolonial Fighting

In addition to the fighting in colony 80 (above), on five occasions, two Cape Town colonies have escaped from their tunnel systems and broken into each other's tunnel system. In all cases, the animals have fought. In each instance the "invasions" occurred during the night, and injured animals were found when the colonies were cleaned the following day. No fights were actually observed, and the following data were obtained by examining all the individuals in both colonies for bite wounds.

In three invasions, the breeding males and females were completely uninjured; in one, both reproductives had slight wounds, and in the last case, the breeding male was badly bitten, and the breeding female was less severely injured. This last colony contained no adult animals other than the breeding pair; the remainder of the colony consisted of nine, 5-mo-old pups. No clear trends can be seen with regard to which mole-rats in a colony were slightly injured, badly bitten, or killed during invasions. However, in one colony begun from a pair, the four animals that invaded the neighboring colony were all from the first litter born to that colony and were therefore the oldest nonbreeders. When the mean mass of all the injured animals (42.8 ± 10.6 g, $n = 58$) is compared with the mean mass of all mole-rats (excluding breeders) in the colonies in which fighting occurred (40.4 ± 11.3 g, $n = 202$), the body masses of the injured mole-rats were slightly but not significantly greater ($p > 0.05$, t-test). These results suggest that, on the average, about 25% of the mole-rats in a colony are involved in fighting, corroborating the intercolonial aggression data of Lacey and Sherman (chap. 10).

Discussion

In this chapter I have discussed many aspects of reproduction in the naked mole-rat. Because there is considerable variability among colonies, I have had difficulty in making concise, totally unambiguous general statements. At this stage in our understanding of reproduction in *Heterocephalus glaber*, the best that can be done is to present data from captive colonies showing both the mean and the variance.

Some general statements can be made, however. Captive *H. glaber* colonies normally contain one breeding female, and she can retain this position for many years. Among the other females in the colony, some appear to remain sexually quiescent for many years. A few, although not parous, show some evidence of reproductive activity, especially just before the birth of a litter. It is not possible to predict (from growth or behavior) which females are latent reproductives, nor which one of these will assume the reproductive role should the breeding female die. On many occasions, a small-sized (presumably young) female becomes the reproductive; she grows rapidly, changes her body shape, and eventually becomes one of the largest and longest females in the colony (see also chap. 10, 12). In colony 80, a captive-born female from the oldest litter in the colony was involved in all but the first succession. In colony 500, however, the oldest female sibling in the colony did not breed.

The males and females in a colony are sensitive to the presence and reproductive state of the breeding female. Thus, succession usually occurs rapidly if the breeding female dies (see also Lacey and Sherman, chap. 10; Faulkes et al., chap. 14). The reproductive physiology of the colony cycles with that of the breeding female (e.g., males and females develop teats, and some females become perforate), and consequently the colony appears primed for the arrival of her litters. Furthermore, the entire colony (especially the females), frequently responds to the removal of the breeding female by a sharp increase in body mass (Jarvis et al., chap. 12; Lacey and Sherman, chap. 10). Indications are that succession of the breeding male is also rapid, but because most colonies have more than one breeding male, a colony is rarely bereft of all reproductive males.

My data and those of Lacey and Sherman (chap. 10) illustrate the variability that exists in the levels of aggression shown by breeding females and by potential reproductives. I document instances in which there has been more than one breeding female in a colony for up to 5 yr. The response of the colony to the removal or death of the breeding female is also varied. If only the breeding female is removed and the colony structure is otherwise undisturbed, succession may occur with little to no overt aggression between females, perhaps indicating an established dominance hierarchy among the females. However, two or more females may compete viciously for the vacant position. When the history of these competing females is known, they are often littermates, and their competition may or may not be immediate. In two instances, the sisters both had one or more litters before they fought and one was killed. If the colony structure has been severely disrupted (e.g., by the removal of the breeding pair and some other large mole-rats, or by a rapid succession of reproductive females), then fierce fighting almost always occurs, leading to the death of some or even all of the fighting animals. In this situation, even the emerging reproductive female may receive severe injuries. Although both Lacey and Sherman (chap. 10) and I documented severe fighting between females, fights are never seen in many colonies. In 20 colonies that have been established for 2 yr or longer at Cape Town (1, 13 yr; 3, 12 yr; 2, 8 yr; 3, 6 yr; 11, 2–4 yr), there has been the equivalent of 110 colony-years with no aggression (other than shoving). In each of six of these colonies, I have had no deaths or injuries from fighting in 8 yr or more.

Most naked mole-rat colonies in the lab have several breeding males and the succession of breeding males is apparently achieved without intense male-male fighting. Brett (1986; chap. 4) suggested that because the tenure of breeding status by males may be shorter than tenure by females, the probability of an individual male's breeding during its lifetime is higher than that of a female. Therefore, males may have less incentive than females to compete strongly for breeding status. It is also possible that the breeding female may suppress male-male aggression. Males that are killed in intracolonial fights are usually killed by the breeding female (see also Lacey and Sherman, chap. 10). Why they are killed is not known. Perhaps they are showing increased sexuality but are not closely bonded to the breeding female. If so, they could be regarded as potential mates for other females in the colony and might constitute a threat to the reigning breeding female by mating with other perforate females when she is approaching parturition.

Once a breeding female has established herself, the levels of fighting within her colony are generally low. Should she single out a colony member and repeatedly attack it, there is little or no fighting back. Indeed, the breeding female may even kill another mole-rat without receiving injuries herself. Once singled out by the breeding female, many individuals apparently lose the will to live and can be found sitting alone, often in the toilet area where they eventually die (see also Lacey et al., chap. 8). However, if there is fighting, other colony members do not participate in these fights until a rival has been seriously wounded, whereupon it is attacked by many members of the colony. The reasons for this ostracism are unknown.

Part of the intercolonial variability found in captive colonies can perhaps be attributed to inherent differences in the aggressiveness of the breeding females. Thus, in Lerata 4, female 400 was placid whereas 460 was far more aggressive and devoted much time to shoving large colony mates (usually the same few individuals). Breeding female 400 even tolerated the presence of a second breeding female (460) in her colony for 5 yr. At Cape Town, colonies in which the breeding female was involved in fewer shoving encounters appeared to be better at rearing young than the others; this is partly because the breeding female spent more time lying quietly in the nest with her nursing pups than did more aggressive females. These less aggressive females may also have better control over their colonies, whereas the aggressive females are more threatened and neglect their pups to devote their energies to maintaining reproductive control. Reeve and Sherman (chap. 11) showed that the breeding female shoved less closely related larger colony members more than she shoved her siblings or offspring, and they put forward several hypotheses regarding the function of shoving by the breeding female.

The potentially long reign of the breeding female, the sometimes shorter reign of the reproductive male, and the possibility of more than one breeding male in the colony mean that many well established colonies are probably composed of siblings with the same mother but not necessarily the same father. Several independent studies (Honeycutt et al., chap. 7; Faulkes et al. 1990a; Reeve et al. 1990) have shown that, within a local population of H. glaber, intercolonial as well as intracolonial genetic variation is extremely low. Inbreeding among H. glaber colonies probably occurs commonly (Reeve et al. 1990), and the wide spacing of colonies in some areas (Jarvis 1985; Brett 1986, chap. 4), the lack of genetic variation, and the fierce fighting that occurs when colonies mix (Lacey and Sherman, chap. 10; this chapter) suggest that outbreeding occurs infrequently. In the few instances in which singletons or pairs of foreign adult mole-rats have been introduced into a captive colony, they may be initially accepted by the colony and then attacked months later.

In captivity, a pair of mole-rats can be used to start a new colony (Jarvis has done this 19 times, Sherman twice). Brett (chap. 4, 5), however, argues that the odds are strongly weighted against one to two dispersers' being able to dig enough tunnels to locate sufficient food sources before they starve to death. The chances of successful food location by one naked mole-rat versus several have been calculated by Lovegrove and Wissel (1988). They developed a model to show that where geophyte densities are low (e.g., *Pyrenacantha kaurabassana*; see chap. 5) the risk of one burrowing mole-rat's not encountering a geophyte after digging 3 m is 41 times greater than when 10 animals are sharing the search. Furthermore, groups of more than 4 huddling naked mole-rats can control body temperature over the ambient temperature range normally encountered in their habitat, whereas pairs of the animals cannot (S. Yahav and R. Buffenstein, pers. comm.). The implication is that new colonies should be founded by groups of mole-rats budding off from the parent colony. Nothing is known of the proximate factors inducing a colony to split. However, as Brett (1986, p. 363) pointed out, after many litters have been reared to adulthood, there must come a time when further addition to the work force of a colony makes no difference to the likelihood of successfully rearing juveniles. Moreover, a female's opportunities for breeding diminish as the colony grows. She would gain by leaving, therefore, when the work force at home is full (i.e., when her leaving does not jeopardize the survival of future pups).

I suggested (above) that dispersal is most likely to occur after several years of good rainfall. These favorable environmental conditions would reduce the risks inherent for a small group dispersing in an arid environment where the rainfall is unpredictable and where the food is widely dispersed and energetically costly to obtain. Several years of good rainfall would also probably have increased pup survival in the natal colony and led to a high influx of juvenile animals. Reduced dispersal risks and high recruitment could trigger colony fissioning when the control of the breeding female is weakest, namely, in late pregnancy or immediately postpartum. Very small colonies are found only rarely in the wild (Brett, chap. 4, pers. obs.), and successful dispersals appear to be infrequent.

Many colony members are superficially involved in care of the pups in that they sit with them in the nest and tunnels and occasionally carry or feed them. However, the indirect contribution to pup care of these individuals is considerable. They forage for food, build and maintain the protective burrow system, and defend the colony against predators (Lacey and Sherman, chap. 10). The reproductives and younger mole-rats in the colony invest the most effort in the direct care of the pups, possibly because these animals are most closely related to the newborn pups. The breeding males would have fathered the pups, and the younger colony members are more likely than older ones to have the same parents as the pups (see Alexander et al., chap. 1). This may also be one of the reasons why the new breeding female in a colony often rises from the ranks of the younger colony members: younger mole-rats are more likely than the older animals to be full sisters and brothers to the new breeder. However, the long reproductive life of the breeding female and the probability that both she and the reproductive male(s) were born in the colony ensure that even the older colony members are kin.

Because individuals in *H. glaber* colonies work together as a unit, the breeding female is shielded from the energetic stresses and dangers of digging and procuring food. She is able to concentrate her efforts on reproduction. She leaves general colony maintenance, defense, finding of food, much of pup care, and all the weaning of pups to other mole-rats and is consequently able to bear large numbers of young throughout the year. However, the energetic cost of bearing two to three times her body mass of young each year and then of suckling them for a total of about 15 wk in a year must be considerable. Indeed, the cost would probably be unsupportable if the pups had a fast postnatal growth rate, or if the breeding female alone had to feed her pups cecotrophs and to find them solid foods after weaning. This is probably one proximate reason why the solitary Bathyergidae are seasonal breeders with small litters (Jarvis 1969; Taylor et al. 1985; Bennett and Jarvis 1988a). *Heterocephalus glaber* pups change little in mass while nursing and only really begin to grow after weaning, when the rest of the colony begins to contribute directly to meeting their energy demands. It would therefore appear that this pattern of reproduction must have evolved in *H. glaber* after group living was fixed.

Summary

Central to the organization of colonies of naked mole-rats are the reproductive animals, particularly the single breeding female, whose offspring and close relatives constitute the rest of the colony. The breeding female has a litter of up to 27 pups approximately every 80 days and captive females can continue to breed for more than 13 yr; while the breeding female is reproductively active, the remaining females in the colony are anovulatory. If the breeding female is removed from the colony, another individual assumes the reproductive role, sometimes within as little as 10 days. This succession of reproductive females may occur with or without bloodshed. The one to three reproductive males in the colony maintain close physical contact with the breeding female. All mate with her and, together with the breeding female, are the individuals most involved per capita in caring for pups. Of the remainder of the colony, juvenile mole-rats invest the most time in pup care.

A feature of captive colonies is the variability in the amount of aggression shown by the reproductive female and in the response of the colony to the loss of the reproductive female. In this chapter, I illustrate the mean as well as the variance in reproductive behaviors found within captive colonies, describe normal reproductive behavior and care of the pups, and then use case histories of colonies (some of which have been in captivity more than 10 yr) to illustrate the range of responses and behaviors exhibited within and between naked mole-rat colonies.

Acknowledgments

Numerous people have helped me collect mole-rats in Kenya. Particular thanks go to P. Bally, H. Bally, R. Bally, R. Buffenstein, K. Davies, M. Griffin, M. Jackson, A. Lerewan, and A. Ndalinga Chondo. J. Booysen is thanked for helping maintain the colonies in captivity. I gratefully acknowledge financial assistance from the University of Cape Town, the Council for Scientific and Industrial Research of South Africa, and the National Geographic Society. I also thank the Kenya Ministry of Natural Resources and the Office of the President of Kenya for permission to collect mole-rats, and N. C. Bennett, B. W. Broll, T. M. Crowe, and E. McDaid for help with statistical analysis of the data. My grateful thanks go to N. C. Bennett, R. Buffenstein, O. J. Reichman, an anonymous reviewer, and especially to P. W. Sherman for helpful criticism of the manuscript.

14 Hormonal and Behavioral Aspects of Reproductive Suppression in Female Naked Mole-Rats

Christopher G. Faulkes, David H. Abbott,
Caroline E. Liddell, Lynne M. George,
and Jennifer U.M. Jarvis

One of the most intriguing features of the biology of the naked mole-rat is the monopoly of reproduction. It is clear that breeding is usually restricted to 1 female and 1–3 males in colonies of all sizes, even those with a population of 295 or more (Jarvis 1981, chap. 13; Brett, chap. 4). The remaining animals are somehow rendered infertile by social, chemical, and/or physiological factors. This chapter examines the hormonal correlates and behavioral aspects of reproductive suppression in *Heterocephalus glaber*.

Fertility in many mammals is determined by social and environmental cues. These serve to regulate breeding and ensure that reproduction occurs when conditions for both parents and offspring are most favorable, thus increasing the chances of survival of young. Food availability (Dubey et al. 1986) and photoperiod (Lincoln and Short 1980; Karsch et al. 1984) are well-studied examples of such cues. Recently it has become apparent that the social environment of an animal can also have a profound effect on reproductive performance. In some social species, especially those living in extended families, behaviorally subordinate individuals may be reproductively suppressed and produce fewer offspring (and sometimes none at all) when compared with dominant members of the group. The extent of suppression is usually greater among females than males (Wasser and Barash 1983; Abbott 1987).

Socially induced infertility in female mammals has now been documented in species ranging from primates to rodents, including common marmoset monkeys, *Callithrix jacchus* (Abbott 1984); tamarin monkeys, *Saguinus oedipus* (French et al. 1984); yellow baboons, *Papio cynocephalus* (Wasser and Barash 1983); silver-backed jackals, *Canis mesomelas* (Moehlman 1983); dwarf mongooses, *Helogale parvula* (Rood 1980); and prairie voles, *Microtus ochrogaster* (Carter et al. 1986). In some species, suppressed females are sexually active, but conception rates and birth rates are lower than those of dominant females (e.g., gelada baboons, *Theropithecus gelada*; Dunbar 1980). Suppression can be extreme when a breeding female requires nonreproductive helpers to assist in rearing her offspring (Wasser and Barash 1983), such as in marmosets and tamarins. In these species, nonbreeding females are prevented from ovulating and are therefore completely infertile when they remain in their (natal) social group (Abbott et al. 1988).

On a proximate or mechanistic level, two external factors lead to the altered physiological state that inhibits reproduction. In primates, behaviorally induced stress has been implicated (e.g., socially subordinate talapoin monkeys, *Miopithecus talapoin*; Keverne et al. 1984), whereas in rodents, pheromones released by conspecifics mediate reproductive suppression (Vandenbergh 1986). However, these two factors may be interrelated; for example, in male lesser mouse lemurs (*Microcebus murinus*), volatile urinary odors from dominant males suppress circulating testosterone levels in male subordinates (Schilling and Perret 1987).

Dunbar (1980) noted that, among wild gelada baboons, the frequency of attack from other females increased as social rank decreased; this harassment could lead to increased stress and its physiological consequences. A similar but more extreme example is seen in the common marmoset. Breeding is monopolized by a dominant female in both wild (Stevenson 1978) and captive (Abbott and Hearn 1978) colonies. Marmosets housed in peer groups establish a social hierarchy characterized by aggressive and submissive behavioral interactions. Infertility in socially subordinate females is due to a lack of normal ovarian cycles and ovulation (Abbott and Hearn 1978) caused by insufficient gonadotrophic support from the anterior pituitary, as indicated by low plasma luteinizing hormone (LH) levels (Abbott et al. 1988). The suppression of LH secretion and lack of ovulation in nonbreeding female common marmosets is reversible and apparently results entirely from their subordinate social status. If subordinates are removed from their peer groups and housed singly, plasma LH levels rise to preovulatory values, and ovarian cyclicity commences. If they are returned to their original groups and subordinate status, LH secretion soon ceases and ovarian cycles stop (Abbott 1987).

Many rodents also exhibit varying degrees of reproductive suppression. In the species studied thus far, reproductive suppression is mediated by urinary pheromones. The so-called Lee-Boot effect was first described 35 years ago (Lee and Boot 1955), to explain the delay in sexual maturation and increase in cycle length seen in group-caged female house mice, *Mus musculus*. Exposure to urine from group-housed females delays the first estrus in singly caged mice by 5–7 days (Cowley and Wise 1972; Vandenbergh et al. 1972; McIntosh and Drickamer 1977). In large groups (e.g., 30 animals) complete suppression may occur in all the females, resulting in a state of anestrus (Whitten 1959). The chemical factors causing the reproductive suppression have also been reported from the urine of wild house mice living in high-density populations (Massey and Vandenbergh 1980; Coppola and Vandenbergh 1987), thus apparently providing a mechanism by which individuals avoid reproducing under conditions of overpopulation and resource scarcity.

Density-related suppression of reproduction has also been documented in several other species of rodents. In crowded laboratory populations of prairie deer mice (*Peromyscus maniculatus bairdii*), only about 10% of females reproduce even in conditions of excess food and water. The nonbreeding (suppressed) individuals have reduced body weights and underdeveloped reproductive tracts (Kirkland and Bradley 1986). However, when these nonbreeding animals are removed from the group, they become sexually active and 70% reproduce within 50 days (Terman 1973). Male deer mice are also inhibited in their sexual maturation: if juveniles are exposed to soiled bedding (containing urinary chemosignals) from the cages of adult males, growth of testes and seminal vesicles is retarded (Lawton and Whitsett 1979). The urine of adult females has no effect on males, implying that a specific male pheromone is involved in male suppression. However, the exact nature of the involvement of pheromones in the inhibition of sexual maturation in male and female deer mice remains unclear, because other experiments have indicated that behavioral cues may be of major importance (Terman 1987).

In the related white-footed mouse (*Peromyscus leucopus*), a pheromonally induced block to implantation has been reported in females (Haigh et al. 1988). The chemosignal appears to be a volatile component of adult female urine because airborne contact with soiled bedding was sufficient to produce the effect, and exposure for as little as 3 h per day caused suppression (Haigh et al. 1985). However, as with the prairie deer mouse, other workers have been unable to prove that pheromones are the major suppressing influence (Terman 1984). Field studies have shown that in natural populations of *Peromyscus leucopus*, reproduction is restricted to a few of the adult females (Metzger 1971; Trudeau et al. 1980).

Reproductive suppression is also well documented in the microtine rodents. Delayed puberty in females has been reported in high-density populations of bank voles (*Clethrionomys glareolus*; Kruczek and Marchlewska-Koj 1986) and pine voles (*Microtus pinetorum*; Lepri 1986; Lepri and Vandenbergh 1986). Littermates of both California voles (*Microtus californicus*) and prairie voles suppress each other's growth and reproductive maturation if they are housed in groups. This effect continues if the voles are isolated except for air contact, again pointing to the involvement of a volatile pheromone. Release from suppression requires a physical stimulus and only occurs if individuals are paired with members of the opposite sex (Batzli et al. 1977).

None of the species studied so far shows such extreme socially induced infertility as the naked mole-rat. These highly social rodents live in large subterranean colonies, and with rare exceptions (Jarvis, chap. 13), reproduction is monopolized by a single female and one to three large males in both wild

(Brett, chap. 4) and captive colonies (Jarvis 1978, 1981; Lacey and Sherman, chap. 10). The remaining mole-rats in a colony usually show no signs of sexual behavior and can be classed as nonreproductive. The external genitalia of nonbreeding animals are almost monomorphic (Jarvis 1969), and the vaginas of females are not perforate (however, see Jarvis, chap. 13). The smaller nonbreeding animals maintain the underground tunnel system and forage for food (underground roots and tubers; Jarvis 1981; Brett, chap. 5; Lacey and Sherman, chap. 10; Jarvis et al., chap. 12).

Some evidence suggests that nonbreeding female mole-rats are reproductively quiescent. Postmortem examinations of reproductive tracts of nonbreeding females revealed poorly developed uteri and ovaries compared with breeding females, and an absence of placental scars (Kayanja and Jarvis 1971; Faulkes 1990; Jarvis, chap. 13). However, this situation is changed if the breeder either dies or is removed from the colony. One of the previous nonbreeding females will come into estrus (Jarvis 1981), usually within 6 wk and often after prolonged female-female strife (Lacey and Sherman, chap. 10; Jarvis, chap. 13). In males, the block to reproduction is less clear-cut than that in females, since spermatogenesis occurs in the testes of both breeding and nonbreeding individuals (Jarvis 1978; chap. 13). However, breeders and nonbreeders are physiologically different, the latter having significantly lower urinary testosterone and plasma LH concentrations (Faulkes et al. 1989b). The implications of these endocrine deficits on fertility are unknown.

To further our understanding of the physiology and extent of the reproductive block in nonbreeding female naked mole-rats, we collected urine and plasma samples from captive colonies housed at the Institute of Zoology in London and at the University of Cape Town, South Africa. We present here preliminary results from hormonal determinations of these samples together with results of behavioral observations of two of the London colonies.

Methods

ANIMALS

A total of 37 captive colonies of naked mole-rats (19 in London, 18 in Cape Town) were used in the hormonal studies, and these ranged in size from male-female pairs up to 72 individuals. Colonies were maintained at both the Institute of Zoology and the University of Cape Town in artificial burrow systems (see Brett 1985; Jarvis, Appendix) in rooms heated to 28°–30°C with a relative humidity of 40%–60%. The burrow systems consisted of a series of interconnecting clear acrylic plastic (plexiglass) tubes with a nest chamber, a toilet chamber, and a chamber where food was introduced. The nest chambers in London were heated to 32°–34°C with a thermostatically controlled lamp.

Cape Town colonies were provided with 40-W lamps positioned at several different points. The total tunnel length varied from 2 m to 15 m, according to the number of animals in the colony. Fresh food was given daily ad libitum and included sweet corn, sweet potatoes, carrots, potatoes, apples, and bananas. A cereal supplement containing vitamins and minerals was also provided once per week (London colonies) or daily (Cape Town colonies).

To facilitate identification during behavioral observations and urine sampling, each mole-rat was numbered and marked in three ways: the toes were clipped according to a numerical pattern, a number was tattooed on the animal's back, and the tattooed number was reinforced weekly with a permanent marker pen.

Two London colonies, A and B, were used for behavioral observations. Colony A contained 25 animals, originally caught by R. A. Brett near Mtito Andei, Kenya in 1983. The original breeding female, while remaining in the colony, was replaced by another female without fatality. Successful breeding by the new female raised numbers to 48 within 2 yr. Twenty mole-rats were removed after this 2-yr period to form colony B (below). At the start of the study (October 1985), colony A contained 33 mole-rats (18 females, 15 males), only 7 of which had been wild-caught.

Colony B was formed with the 20 mole-rats removed from colony A in April 1985. Initially three females developed perforate vaginas, including the former breeding female from colony A. During fights for breeding status, this female and another of the females with a perforate vagina were killed, and female 18 became the breeder. At the start of the study (October 1985), colony B contained 24 mole-rats (14 females, 10 males), 13 of which were from the original wild-caught colony.

BEHAVIORAL OBSERVATIONS

It has been reported that a behavioral division of labor exists within captive colonies of naked mole-rats, with smaller animals performing significantly more colony maintenance or worker behaviors compared with larger individuals, many of whom have never been seen engaging in these activities (Jarvis 1981; Lacey and Sherman, chap. 10). This is consistent with behavioral observations in wild colonies (Brett 1986, chap. 4; Braude, chap. 6). To confirm the division of labor seen by others and to ensure that our animals had a social structure similar to those previously studied, observations were undertaken over a 4-mo period from October 1985 to January 1986 in colonies A and B (London).

The behavioral technique employed was based on a scan-sampling method similar to that used by Lacey and Sherman (chap. 10). Each observation consisted of starting at one end of the colony and moving around to the other end within 5 min, recording the behavior and location of each mole-rat when it was first encountered. Animals that were inactive within the confines of the nest box were not recorded. Working animals were defined as those involved in tunnel digging and foraging, including digging, chewing, sweeping nest material, and carrying either food or nest material (for descriptions see Lacey et al., chap. 8; Lacey and Sherman, chap. 10). Another type of observed behaviors are what we termed *colony-monitoring* behaviors: (1) *patroling*, moving around the colony but not carrying out maintenance behaviors or any other obvious activity before returning to the nest; and (2) *sentry*, lying in an entrance to the nesting chamber facing outward. These two behaviors were possible passive defense activities when no element of challenge or novelty was present. These and other more active defense behaviors have been investigated and described more fully by Lacey and Sherman (chap. 10).

Our behavioral observations were recorded either at half-hourly or hourly intervals between 9 A.M. and 5 P.M., and at hourly intervals between 5 P.M. and 9 A.M. (overnight). Observations were started 10 min after an observer had entered the room, and 102 scans (17 h total) were completed for each colony. In addition, casual observations of mating were recorded during the periods of scan sampling and urine collection.

URINE SAMPLING

To investigate ovarian activity in both breeding and nonbreeding females, urine samples were collected ad libitum over the period from January 1986 to May 1989. Urine was chosen for routine progesterone determination in preference to blood because collecting urine is not an invasive technique and permits minimal disturbance of the animals. Urine was therefore the most practical option for regular sampling. It is also widely used as a medium for hormonal analysis in other species, and analytical methods have been validated (e.g., Lasley 1985; Hodges 1986).

Urine sampling involved the removal of all the shavings from the toilet chamber in each colony and wiping the chamber clean with tissue paper. Immediately following an individual's urination, a sample was collected in a glass pipette; thereafter, the toilet chamber was again wiped clean with tissue paper. Samples were placed in a freezer within 1 h of collection and stored at $-20°C$ until hormone determination. Sampling was usually carried out between 9:00 A.M. and 1:00 P.M. to standardize collection time, although urinary hormone levels are much less susceptible to changes with sampling time than are plasma hormone levels, because urine summates what has been excreted by the body since the previous urination.

BLOOD SAMPLING

Animals were hand-held, and blood was removed by cutting the tip of the tail with a sterile scalpel blade; 300–400 µl of blood were collected by capillary action using heparinized micro-hematocrit tubes. The wound was treated with an antibiotic powder (Aureomycin), and the animal was returned to its colony. After collection, the samples were stored on ice for a maximum of 2 h before being centrifuged, and the plasma was stored at $-20°C$ until hormone determination.

HORMONE DETERMINATIONS

PROGESTERONE

Samples of urine from pregnant female mole-rats were categorized arbitrarily by dividing the average gestational period of about 70–73 days (Jarvis, chap. 13) into four time intervals as follows: early (days 1–20), mid (days 21–40), and late (days 41–60) pregnancy, and just before parturition (days 61–71).

Progesterone concentrations were determined in petroleum ether-extracted urine samples (50–100 μl) and plasma samples (25 μl) by radioimmunoassay without chromatography (Hodges et al. 1983). The sensitivity of the assay (determined as 90% binding) was 10 pg/tube. Expressed as the coefficient of variation for repeated determinations of a quality control, intra-assay precision was 6.0% ($n = 11$), and inter-assay precision was 7.1% ($n = 19$). This progesterone assay had previously been validated for use on naked mole-rats (see Faulkes et al. 1990a).

Before assaying for progesterone, creatinine concentrations were determined in all urine samples, as described by Bonney et al. (1982). Creatinine is excreted at a constant rate as a result of the breakdown of tissue proteins and therefore provides an index by which urinary hormone concentrations can be measured. This avoids the problem of differing dilutions of urine being produced by animals according to the time of day or their fluid intake. All urinary progesterone concentrations were expressed per milligram of creatinine.

As part of a preliminary confirmation that high urinary progesterone reflected high circulating progesterone, four plasma samples were taken from two females during times of high and low urinary progesterone concentrations. The two plasma samples taken during the high period yielded values of 35.4 and 25.2 ng/ml of plasma, respectively, and were notably higher than the two samples taken during the low period, which were below the limit of detection of the assay (< 0.5 ng/ml). These results indicate that urinary progesterone does reflect the amount of circulating progesterone.

LUTEINIZING HORMONE BIOASSAY

To investigate whether reduced pituitary luteinizing hormone (LH) secretion was involved in the suppression of reproduction in nonbreeding female naked mole-rats, LH was measured using an in vitro bioassay based on the production of testosterone by dispersed mouse Leydig cells (Van Damme et al.

1974). Details of the method have been described previously (Harlow et al. 1983; Hodges et al. 1987; Abbott et al. 1988). Only plasma samples were used, and these were assayed in duplicate at two dilutions (1:10 and 1:20) and compared with a rat LH standard (the rat LH antigen preparation was rLH-I-7) over the range 2.0 – 0.0625 mIU/ml (milli International Units per milliliter). The testosterone produced was measured by radioimmunoassay, as described by Hodges et al. (1987).

The assay was validated for use on *Heterocephalus glaber* by looking for parallelism. Dilutions of a naked mole-rat pituitary homogenate containing high levels of LH were parallel to the reference preparation. The sensitivity of the assay (determined at 90% binding) was 1.0 mIU/ml. Expressed as the coefficient of variation for repeated determinations of a quality control, intra-assay precision was 10% ($n = 15$), and inter-assay precision was 16% ($n = 9$).

Luteinizing Hormone Responses to Administration of Exogenous Gonadotrophin Releasing Hormone

To investigate possible differential pituitary secretion of LH in breeding and nonbreeding female naked mole-rats, the LH responses to administration of single and multiple injections of gonadotrophin releasing hormone (GnRH) were measured. Blood samples were taken before and then 20 min after (1) a single subcutaneous injection of 0.1 μg of GnRH in 200 μl saline (breeding females, $n = 6$; nonbreeding females, $n = 5$), or (2) four or eight consecutive hourly subcutaneous injections of 0.1 μg of GnRH in saline or saline alone (nonbreeding females only, $n = 4$ and 3, respectively).

Statistical Analyses

Behavioral data were analyzed by least-mean-squares regression analyses (Helwig and Council 1979). Logarithmic transformations of the data were carried out before the regression analyses as a standard procedure to reduce heterogeneity of variances and to counteract skew (Sokal and Rohlf 1981).

Young animals (< 7 mo) were not included in the regression analysis because they participated little in colony activities (see Lacey and Sherman, chap. 10; Jarvis et al., chap. 12). However, once they reached approximately 7 mo of age (≈ 20 g body weight), they spent increasing amounts of time in colony-maintenance activities (i.e., they began working).

Hormonal data were analyzed using either Student's t-tests or by two-way ANOVA's for repeated measures following log transformation. Plasma LH concentrations following GnRH administration were analyzed by a one-way ANOVA for repeated measures. Comparisons of individual transformed means were made post hoc using Duncan's multiple-range test (Helwig and Council 1979).

Results

BEHAVIOR

Each of the two London colonies had only one breeding female. The breeding female was physically distinct from all the other animals, in that she was the largest female and had an elongate body with prominent nipples and genitalia and a perforate vagina. From observations of mating, two animals were identified as breeding males in both colony A and colony B. Mutual ano-genital nuzzling frequently occurred between the breeding female and the breeding males whether or not mating was involved (see Lacey et al., chap. 8). This nuzzling occurred both in the acrylic plastic tubes and in the nest chambers. In the latter location, it was common to find the breeding female apparently asleep with one of the breeding males in a mutual naso-anal posture. In colonies A and B, this ano-genital nuzzling was not displayed by any nonbreeders.

There was a clear division between reproductive and nonreproductive animals in our laboratory, as was shown by Brett (chap. 4) in the field. Our observations of colonies A and B confirmed (in close agreement with Lacey and Sherman, chap. 10; Jarvis et al., chap. 12) that the mole-rats in each colony did not participate equally in colony-maintenance or "monitoring" activities. The breeding females were the only animals not observed working. The remaining mole-rats in both colonies performed all the maintenance tasks such as digging, sweeping tunnels, and foraging. Regression analyses revealed a clear relationship between behavioral activity and body weight. The results (fig. 14-1) show a significant negative correlation in the total frequency of maintenance behavior with body weight for both colony A ($F = 14.9$; df = 1, 24; $p < 0.0007$) and colony B ($F = 5.5$; df = 1, 16; $p < 0.032$). Conversely, the total frequency of colony-monitoring behavior was found to be positively correlated with body weight for both our colonies (colony A: $F = 16.2$; df = 1, 24; $p < 0.0005$; colony B: $F = 8.67$; df = 1, 16; $p < 0.0095$; see fig. 14-2).

A summary of the individual behavioral categories, together with their respective statistical values, is given in the Appendix to this chapter. A nonworker group of mole-rats (previously described by Jarvis 1981; see also Jarvis et al., chap. 12) could not be distinguished in either of the study colonies.

HORMONES

PROGESTERONE

Urine samples from both breeding and nonbreeding females were analyzed. In 237 samples from 83 nonbreeding females, urinary progesterone levels were undetectable (< 0.50 ng/mg creatinine [Cr]). Results obtained from the urine of breeding females were combined to give a range of values over the gestation period (fig. 14-3). Urinary progesterone concentrations rose significantly from 12.39 ± 3.1 ng/mg Cr (mean ± SE: 5 samples from 3 breeding females) during early pregnancy (days 1–20), to reach maximum levels of 57.02 ± 10.37 ng/mg Cr (19 samples from 4 females) during the following 20 days. Progesterone levels decreased to 39.27 ± 10.37 ng/mg Cr during days 41–60 (7 samples from 4 females) and dropped significantly just before parturition (days 61–70) to 13.8 ± 2.2 ng/mg Cr (5 samples from 3 females).

Figure 14-4a shows urinary progesterone concentrations for five nonbreeding females over periods of up to 120 days. As with the less-frequent sampling regime of urine taken from other nonbreeders, progesterone concentrations from these serially sampled nonbreeding females were undetectable, indicating that they were not ovulating. Figure 14-4b illustrates the urinary progesterone concentrations for nonbreeding female 21, an individual that was removed from her parent colony and paired with a male to form a new colony (colony K). Reproductive activation occurred rapidly on removal from the suppressing influences of her original colony, with the vagina becoming perforate for the first time 7 days after removal. Urinary progesterone was elevated after 19 days. Female 21 went on to undergo four cyclical peaks of urinary progesterone; these may correspond to luteal phases of the ovarian cycle. The durations of these elevated progesterone levels ranged from 27 to 38 days (accurate to ± 3 days). Mating was observed on days 88 and 127, corresponding to times when urinary progesterone concentrations fell below the sensitivity limit of the assay. These short time periods of low urinary progesterone probably represent the follicular phase of the cycle, and the periods of mating during these times may reflect behavioral estrus and approaching ovulation (fig. 14-4b). This female did not become pregnant during this study.

Figure 14-4c shows urinary progesterone concentrations for female 38, another nonpregnant but reproductive female from a colony composed of a male-female pair (colony G). High concentrations of urinary progesterone (up to 98 ng/mg Cr) were recorded over a 160-day period, and during this time there were two clearly defined cyclical elevations, one from day 80 to day 115 and a second from day 115 to day 150, similar in length to those seen in female 21 (fig. 14-4b). Mating was observed on four occasions, two of which coincided with periods of very low or undetectable urinary progesterone. As with female 21, these periods of low progesterone may reflect the follicular phase of the cycle. Female 38 also did not become pregnant during the study.

Figure 14-4d shows the urinary progesterone profile for breeding female 100 in colony N, which contained seven individuals. Before day 80, this female appeared to undergo two ovarian cycles, reflected by the cyclical elevation of progesterone. This female gave birth to seven pups 72 days after the end of the second cycle (at day 152).

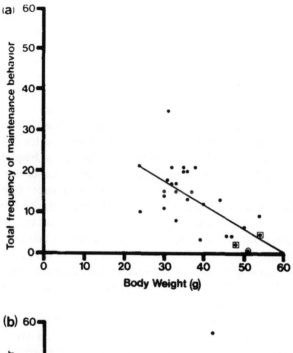

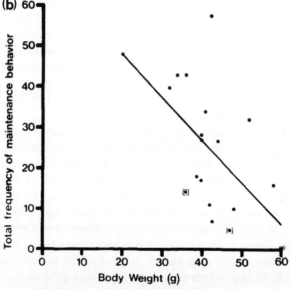

Fig 14-1. a, Correlation between maintenance behaviors (see the chapter appendix) and body weight (colony A); $y = 0.60x + 35.61$, $r = -0.727$. *Filled circle*, nonbreeding animal; *filled circle within a circle*, breeding female, *filled circle within a square*, breeding male. b, Correlation between maintenance behaviors and body weight (colony B); $y = 1.04x + 68.6$; $r = -0.635$; symbols as above. The individual values shown are the untransformed data; they were log-transformed for statistical analyses.

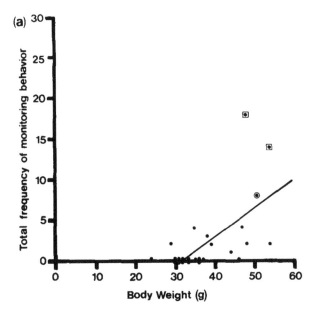

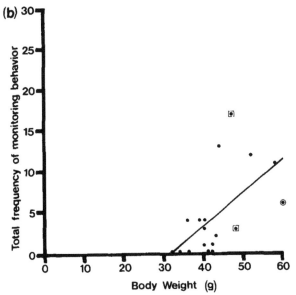

Fig. 14-2 a, Correlation between colony-monitoring behaviors
(see the chapter appendix) and body weight for colony A; $y =$
$0.34x - 10.86$; $r = 0.739$; symbols as in figure 14-1. b, Correlation between colony-monitoring behaviors and body weight for
colony B; $y = 0.41x - 13.03$; $r = 0.706$; symbols as in figure 14-1.
The individual values shown are the untransformed data; they
were log-transformed for statistical analyses.

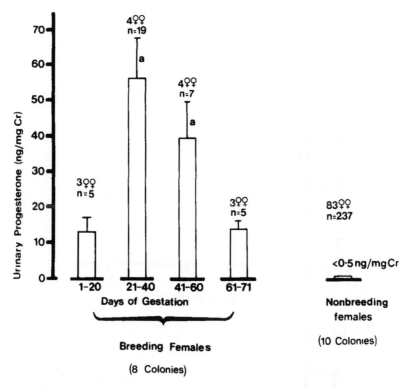

Fig 14-3 Mean ± standard error urinary progesterone concentrations in 8 breeding and 83 nonbreeding female naked mole-rats. The numbers of samples (n) are given for each mean value. Six breeding females were only sampled during one of the stages of gestation, whereas three breeding females were sampled during two or more periods a, A significant difference (p < 0.05) from the values for days 1–20 and days 61–71 (Duncan's multiple-range test following two-way ANOVA for repeated measures) Republished with permission from Faulkes et al (1989a).

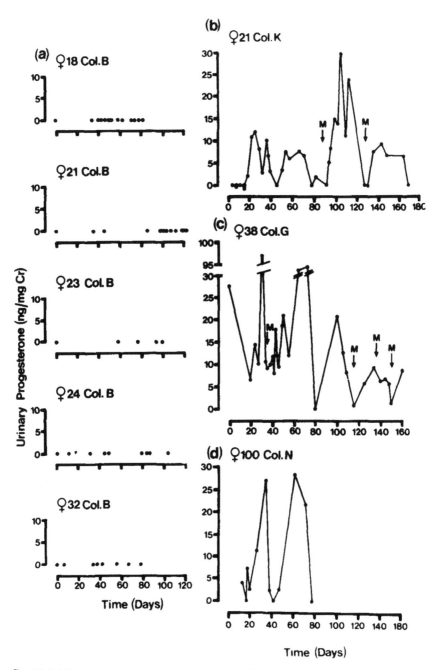

Fig 14-4. Urinary progesterone concentrations in (a) five nonbreeding females, and (b, c, and d) three reproductively active female mole-rats; M, observations of mating. In b, Day 0 is the day the female was removed from her parent colony and paired with a male.

LUTEINIZING HORMONE

Determination of plasma LH concentrations in both breeding and nonbreeding females revealed clear and consistent differences between these two groups. Breeding females had significantly higher plasma LH concentrations (3.0 ± 0.2 mIU/ml, 73 samples from 24 females), than nonbreeding females (1.6 ± 0.1 mIU/ml, 57 samples from 44 females; $p < 0.001$; df = 128, $t = 5.42$). This suggests that the failure of ovulation in nonbreeding females may reflect low pituitary LH secretion or a more rapid clearance of LH from the circulation.

LUTEINIZING HORMONE RESPONSES TO
EXOGENOUS GONADOTROPHIN RELEASING HORMONE

There was a significant difference in the LH response to a single injection of 0.1 μg of gonadotrophin releasing hormone (GnRH) between breeding and nonbreeding females ($F = 21.4$; df = 1,9; $p < 0.01$), with the latter producing only a small but significant ($p < 0.01$) increase in plasma LH from 1.3 ± 0.2 to 2.9 ± 0.5 mIU/ml (fig. 14-5). However, multiple administration of 0.1 μg of

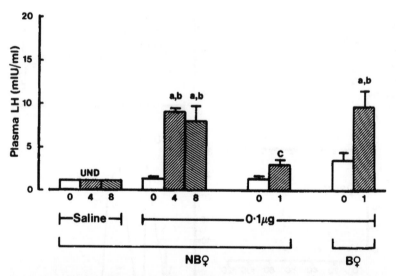

Fig 14-5. Mean ± standard error plasma LH concentrations in breeding (B ♀) and nonbreeding (NB ♀) female naked mole-rats before (0), and 20 min after a single 0 1-μg subcutaneous injection (1), or 20 min after four or eight subcutaneous injections of 0 1-μg GnRH given at hourly intervals, or a saline control a, $p < 0.01$ versus time 0; b, $p < 0.01$ versus nonbreeding females after a single 0 1-μg dose; c, $p = 0.01$ versus time 0 UND, Values below the sensitivity limit of the assay (= 1 0 mIU/ml plasma). All data were examined with Duncan's multiple-range test following ANOVA's for repeated measures.

GnRH resulted in an increased LH response in nonbreeding females. Plasma LH concentrations were significantly elevated in nonbreeding females after four (1.2 ± 0.2 to 9.0 ± 0.2 mIU/ml, $n = 4$) or eight (1.2 ± 0.2 to 7.9 ± 1.7 mIU/ml, $n = 4$) injections of GnRH, comparable with the elevations obtained after a single 0.1-μg injection of GnRH in breeding females (3.4 ± 0.8 to 9.6 ± 2.0 mIU/ml, $n = 6$; $F = 10.36$; df = 13, 4; $p < 0.001$). The magnitude of the LH responses did not differ among these three groups of females, and they were all significantly greater than that obtained with nonbreeding females given a single 0.1-μg injection of GnRH. There was no LH response to saline injections.

Discussion

Initial behavioral observations indicated that in the two London colonies we studied, there was a clear reproductive division of labor; only one female and two males in each colony bred. In addition, detailed observations confirmed that colony-maintenance activities and monitoring behaviors changed in frequency as individuals aged and grew. Juvenile mole-rats, on reaching approximately 7 mo of age, became frequent workers. As animals increased in body weight, they worked less frequently and appeared to adopt a different role within the colony, possibly related to defense (see Lacey and Sherman, chap. 10). This transition was gradual with no distinct cutoff at any body weight between animals carrying out maintenance versus monitoring behaviors. Our captive London colonies therefore appeared to behave similarly to other captive naked mole-rat colonies observed previously by other workers (Jarvis 1981; Lacey and Sherman, chap. 10; Jarvis et al., chap. 12).

Our endocrine results in females were consistent with the observation that only one female was fertile and reproductively active in each colony. Negligible levels of progesterone, poorly developed external genitalia, and no reproductive behavior in the nonbreeding females suggested that their gonadal function was inhibited. The lack of measurable levels of progesterone implies that the suppression of reproduction in nonbreeding females was due to a failure of corpus luteum development (probably resulting from a failure of ovulation), rather than a failure of copulation, conception, or of embryo implantation, all of which have been implicated in reproductive suppression in other species (Keverne 1983; Wasser and Barash 1983; Haigh et al. 1988).

The changes seen in urinary progesterone output over the gestation period may reflect differences either in secretion or excretion of the hormone. Concentrations of progesterone during gestation differ greatly between species. In rats, plasma progesterone levels rise to a maximum during the second half of pregnancy, then drop dramatically before parturition (Heap and Flint 1984). This is similar to what we observed in *Heterocephalus glaber* (fig. 14-3). A completely different picture is seen in guinea pigs, however, where a large increase in plasma progesterone over that in the nonpregnant state reaches a maximum at mid-term. This then declines and rises again prior to parturition (Heap and Flint 1984). In naked mole-rats, prolonged high levels of urinary progesterone (in excess of 40 days, e.g., fig. 14-3) normally distinguish the pregnant female from the regular oscillations found in nonpregnant, cycling females (figs. 14-4b, 14-4d; fig. 14-4c, from day 80 onward), and the constantly undetectable levels found in nonbreeding females (fig. 14-4a).

However, in a nonpregnant reproductively active female naked mole-rat (fig. 14-4c), urinary progesterone concentrations may reach comparatively high levels over a long period of time (days 0–80). It is not clear what this prolonged period of high urinary progesterone represented because no birth occurred and the animal in question, female 38, did not undergo the normal increase in body weight observed in pregnancy. The cyclical peaks of progesterone seen in figures 14-4b and 14-4d and in figure 14-4c from day 80 onward correspond to the luteal phases of the ovarian cycle, and their duration (ca. 28 days; Faulkes et al. 1990a) is similar to that in guinea pigs, in which the corpus luteum of unmated, nonpregnant females actively secretes progesterone for 15–17 days (Weir and Rowlands 1974). Total ovarian cycle lengths are relatively long in hystricomorph rodents (e.g., cuis, 22 days; chinchilla, 40 days; acouchi, 30–55 days; Tam 1974), in contrast to the brief 4-day ovarian cycles of laboratory rats and mice (suborder Myomorpha) (Short 1984).

The failure of nonbreeding female naked mole-rats to ovulate may be due to insufficient gonadotrophic stimulation of the ovary. Breeding females had significantly higher plasma luteinizing hormone (LH) concentrations compared with nonbreeders, whose LH levels were consistently low, often below the sensitivity limit of the assay. In addition, nonbreeding females were less responsive to a single injection of 0.1 µg of gonadotrophin releasing hormone (GnRH), suggesting that the pituitaries of these females were less sensitive to this dose of GnRH (Faulkes 1990; Faulkes et al. 1990b). This lack of sensitivity to exogenous GnRH in nonbreeding females may be the result of a reduction in pituitary GnRH receptors caused by a lack of endogenous GnRH priming of the pituitary in these females (for a review, see Clayton and Catt 1981). The lack of sensitivity to GnRH in nonbreeding females was reversed by four or eight consecutive hourly injections of 0.1 µg of GnRH, which produced a rise in LH comparable to that of breeding females given a single 0.1 µg GnRH dose.

Our data therefore suggest that the socially induced block to ovulation in nonbreeding female naked mole-rats may be due to reduced plasma LH concentrations resulting from impaired endogenous hypothalamic GnRH secretion. The central role of hypothalamic GnRH secretion in integrating environmental cues with reproductive function has been well documented in other examples of natural suppression of fertility, for example, the seasonally anestrous ewe (Legan and Karsch 1979), lactational amenorrhea in humans (Glasier et al. 1986), and the socially induced suppression of reproduction in female common marmosets (Abbott et al. 1988).

The reversible nature of the lack of LH responsiveness to exogenous GnRH (0.1 μg) in nonbreeding female naked mole-rats reflects the rapidity with which the block to ovulation can be reversed in nonbreeding females when their social environment is changed. For example, if a nonbreeding female is removed from her parent colony and housed singly or paired with a male, sustained elevations of urinary progesterone (indicative of the luteal phase of an ovarian cycle) occur for the first time 8.0 ± 1.9 days after separation (Faulkes et al. 1990a; Jarvis, chap. 13). This is exemplified by female 21 in figure 14-4b. Since the follicular phase of the cycle is 6.0 ± 0.6 days, the hypothalamic-pituitary axis must activate within the first 2 days after separation (Faulkes et al. 1990a).

Both behaviorally induced stress (Abbott 1987) and pheromones released by conspecifics (Vandenbergh 1986) have been shown to mediate reproductive suppression in other species. In the case of the naked mole-rat, preliminary results suggest that behavior plays the predominant role in suppression (Reeve and Sherman, chap. 11; B. Broll, unpubl. data). Indeed, daily transfer of bedding and litter from the nest, food, and toilet chambers between separated male-female pairs and their parent colonies neither prevented nor delayed the onset of ovarian cyclicity, compared with controls (Faulkes 1990). If a urinary primer pheromone were involved in suppression of reproduction, then this daily transfer of soiled material would have been expected to maintain suppression in the separated animals. The naked mole-rat therefore provides an excellent opportunity to investigate both the physiological and behavioral mechanisms that inhibit female reproduction.

Summary

Behavioral observations of two captive colonies of naked mole-rats have shown a clear distinction between the breeding female, the breeding males, and nonbreeding animals in each colony. Among the nonbreeding animals, colony-maintenance behaviors (tunnel digging and foraging) were significantly negatively correlated with body weight. Conversely, colony-monitoring behaviors (patroling and guarding the nest chamber) were significantly positively correlated with body weight. Urinary progesterone levels were found to be elevated in pregnant (2.31 – 148.44 ng/mg creatinine [Cr]) and reproductively active (< 0.5 – 97.8 ng/mg Cr) females, but they were undetectable (< 0.5 ng/mg Cr) in nonbreeding females. Breeding females had significantly higher concentrations of plasma luteinizing hormone (LH) (mean, 3.0 ± 0.2 mIU/ml), compared with nonbreeding females (mean, 1.6 ± 0.1 mIU/ml), which, in turn, were less responsive to a single administration of 0.1 μg of exogenous gonadotrophin releasing hormone (GnRH). However, this lack of sensitivity to GnRH in nonbreeding females was reversed by four or eight consecutive hourly injections of 0.1 μg of GnRH, which produced a rise in LH comparable to that of breeding females given a single 0.1 μg dose. These endocrine findings indicate that the suppression of reproduction among nonbreeding female naked mole-rats is due to a block to ovulation, possibly as a result of inadequate gonadotrophin secretion from the anterior pituitary arising from an impairment of endogenous hypothalmic GnRH secretion.

Acknowledgments

We thank the National Institutes of Diabetes and Digestive and Kidney Diseases, Baltimore, Maryland and the National Hormone and Pituitary Program at the University of Maryland School of Medicine for the rat LH preparation (rLH-I-7) and GnRH; A.P.F. Flint, J. P. Hearn, P. M. Summers and three anonymous reviewers for criticism of our manuscript; T. Noble, M. Llovett, and the laboratory animal staff for care and maintenance of the animals; T. Dennett and M. J. Walton for preparation of the figures; and G. Ray and P. Wallace for building the colonies and accessory equipment. We are indebted to J. Shasha for collecting many urine samples in Cape Town. This work was supported by an MRC/AFRC Programme Grant, by Science and Engineering Research Council (U.K.) Research Studentships to C. E. Liddell and C. G. Faulkes, and by grants from the Council for Scientific and Industrial Research (South Africa) and the University of Cape Town to J.U.M. Jarvis. We thank E. A. Lacey and P. W. Sherman for making preliminary versions of their chapter available to us in 1985.

Appendix

Summary of Behavior/Body Weight Correlations for Colonies A and B

Behavior	Data	Colony A				Colony B			
		p	F	df	r	p	F	df	r
Sweep	untransformed	0 0028	11.07	1,24	−0.681	0.0335	5.41	1,16	−0.632
	log-transformed	0.0010	14.12	1,24	−0.718	0 0217	6.47	1,16	−0 660
Chew	untransformed	0 0540	4.10	1,24	−0 527	NS	0.66	1,16	—
	log-transformed	0 0449	4.48	1,24	−0 540	NS	2.66	1,16	—
Dig	untransformed	0.0048	9 66	1,24	−0.660	0.0033	11.88	1,16	−0 752
	log-transformed	0.0007	15 12	1,24	−0 728	0.0014	14.86	1,16	−0 784
Carry food/	untransformed	NS	0 41	1,24	—	NS	0.17	1,16	—
nest material	log-transformed	NS	1 02	1,24	—	NS	0 09	1,16	—
Total	untransformed	0.0007	14 99	1,24	−0.727	0 0322	5 50	1,16	−0.635
	log-transformed	0.0001	21 20	1,24	−0 777	0.0054	10 35	1,16	−0.732
Patrol	untransformed	0.0001	24.58	1,24	0 797	0.0028	12.43	1,16	0 759
	log-transformed	0 0001	30.97	1,24	0 826	0 0023	13 15	1,16	0.767
Sentry	untransformed	0 0113	7 53	1,24	0.620	NS	1.89	1,16	—
	log-transformed	0 0012	13 54	1,24	0 712	NS	2 46	1,16	—
Total	untransformed	0.0005	16.20	1,24	0 739	0 0095	8.67	1,16	0.706
	log-transformed	0.0001	31.49	1,24	0 828	0 0033	11 88	1,16	0.752

NOTE p, probability; F, F-statistic; df, degrees of freedom, r, correlation coefficient, NS, not significant

15 Some Unanswered Questions about Naked Mole-Rats

Richard D. Alexander

Naked mole-rats are unusual for reasons beyond their eusociality. They share with only a few other species the combination of being almost completely subterranean, virtually blind, and depending for food mainly on large subterranean tubers that are located and approached from underground (Jarvis and Bennett, chap. 3; Brett, chap. 5). They are unique among rodents in being relatively hairless and ectothermic, and they may be the only social mammal that does not allogroom (Lacey et al., chap. 8). Although they are exceptionally uniform genetically (see Honeycutt et al., chap. 7; Reeve et al. 1990) and obviously cooperate extensively in their eusocial existence, surprisingly, they also continue to show frequent evidence of aggression (Reeve and Sherman, chap. 11) and even have special vocalizations associated with aggression (Pepper et al., chap. 9); sometimes colony mates fight to the death (this volume).

Some naked mole-rat characteristics that may not seem dramatic may nevertheless be of great importance in explaining why they have become so different, in particular ways, from their closest relatives. For example, they are smaller in body size than other mole-rats (Honeycutt et al., chap. 2; Jarvis and Bennett, chap. 3), and their burrows, which are typically excavated in heavy clay soil, are correspondingly smaller in diameter. These features seem to be most closely approached by *Cryptomys hottentotus hottentotus*, which lives in colonies of 2–14 animals in arid regions of South Africa, where the soil is hard, the food is not concentrated in large tubers, and, presumably, the predators differ from those in other areas. *Cryptomys damarensis*, which is endothermic and haired, lives in softer soils, has a larger body size (up to 220 g) and larger burrows, often feeds on large tubers, lives in colonies of 25 or more individuals, and has reproductive and work-related divisions of labor and overlapping generations (Jarvis and Bennett, chap. 3). The unique naked mole-rat combination of such features may actually have been crucial in allowing evolution of the more remarkable aspects of *Heterocephalus glaber* appearance and existence, because they greatly restrict the accessibility of naked mole-rat burrows to many predators. Presumably, this relative invulnerability, especially to homeothermic predators, surface predators that dig, and larger snakes, was instrumental in allowing naked mole-rats to become permanent inhabitants of burrows locatable to predators (but not accessible to them), which in turn allowed them to become more or less blind and to discard the expensive machinery of homeothermy. *Lowered mortality from predation is almost surely a correlate of still another naked mole-rat feature, their unusually long lives*; lowered mortality tends to cause an incidental retardation of the onset of senescence (Williams 1957; Hamilton 1966; Alexander 1987; Alexander et al., chap. 1).

All in all, naked mole-rats may be said to have evolved to be dramatically different from other rodents in the combined features of their morphology, physiology, and behavior. It seems appropriate to view this change as having been facilitated by their having moved into an ecological niche rather dramatically different from that inhabited by any other mammal; that is, a niche involving almost completely subterranean life in relatively invulnerable burrow systems, with access to food items sometimes big enough to feed numerous individuals for days or weeks, a niche evidently accessible as well to certain kinds of predatory snakes (Jarvis and Bennett, chap. 3; Brett, chap. 4; Braude, chap. 6). Presumably, sociality on a scale of that shown by naked mole-rats would not be possible in regions outside the cyclically wet and dry tropics where it is perpetually warm, yet numerous plants have evolved the tendency to produce large tubers that enable them better to survive droughts.

The various special features of naked mole-rat life are not likely to be understood except as a result of understanding the *general* selective environment in which naked mole-rats evolved, in other words, through considering how all (or most) of their traits, collectively, may have evolved. It is also unlikely that reasonable explanations for the unusual attributes of naked mole-rats will be developed without greater understanding of their life-style and its evolutionary history. *This volume is intended not only to answer questions but also to bring to the fore questions about special attributes as yet poorly understood* and, if possible, to cast them in a light that may assist in their eventual analysis. The present essay is written in a hypothesis-generating spirit to bring a selectionist view to bear on the distinctive attributes of *H. glaber* mole-rats. It is attempted as a parallel to Darwin's search for a plausible explanation for sterility in helpers among eusocial forms (see Alexander et al., chap. 1) and under the supposition that it is sometimes easier to criticize or test an already generated idea than to dream up an idea and develop it. I presume that ideas about how events might possibly have happened represent necessary first steps in the scientific process and that there is often value in discussing hypotheses, especially sets of hypotheses generated around an unusual situation such as this one, even if one cannot yet achieve the stage of satisfying testability in each and every case.

Here I engage two general questions: (1) What can be said about the selective forces that have caused naked mole-rats to diverge so far from their closest relatives in the particular ways they have? (2) Why did naked mole-rats apparently fail to speciate during their rather dramatic divergence from their closest relatives? To accomplish this, I first examine some special features individually, in a comparative way, and then I attempt to discuss how all of the special features of naked mole-rats may fit together; that is, how considering them as a set may contribute to understanding their selective background.

Why Are Naked Mole-Rats Virtually Hairless?

VARIATIONS IN MAMMALIAN HAIRLESSNESS

Mammals are the organisms that have hair and produce milk. There are analogues for both traits in other organisms (pigeons produce a milklike food for their young, and many organisms have hairlike structures) but no homologues.

The amount and kind of hair varies extensively among different mammals. Relative hairlessness occurs in a variety of mammals, rarely for reasons that are entirely obvious (Lyne and Short 1965; W. J. Hamilton III 1973). For some aquatic mammals, both marine and freshwater (e.g., cetaceans, elephant seals, sirenids), a layer of fat beneath the skin seems to have proved a more appropriate correlate of homeothermy than a coat of hair; perhaps because hair causes drag in an aquatic environment, reducing the efficiency of locomotion, and also wet hair is not as efficient an insulation as fat. Although a variety of aquatic mammals have retained a hair coat (e.g., seals, walruses, otters, mink, muskrats, beaver), these species either live in cold climates or spend a significant amount of time out of the water (or both). Some mammals have replaced part or all of the hair coat with armor of one sort or another (e.g., armadillos, pangolins, anteaters; in some armadillos abundant ventral hair is retained); such forms live only in mild or tropical climates (Walker 1975). Several large, entirely (elephant, rhinoceros) or primarily (hippopotamus) terrestrial mammals have lost a hair coat in favor of a thick, leathery skin. It has been postulated that these tropical forms have a low body surface area in relation to their body mass and therefore have gained by increasing their ability to lose heat through the skin. As predicted from this hypothesis, temperate-zone, montane, and rain-forest-dwelling relatives of these forms (e.g., tapirs) have more hair (Walker 1975). As also predicted, juveniles of these forms have more hair than adults. A few mammal species (suids and some primates) are somewhat intermediate, having lost much of their hair (Lyne and Short 1965). Female mammals that bed down or nest in contact with their young, or carry infants on their venters, frequently have lost much of the hair on their venters and around the mammary glands (e.g., suids, rodents, some primates). In such cases, the youngest juveniles are also either virtually hairless (rodents), relatively so (suids), or only lightly haired on the particular parts of their anatomy that regularly contact the mother (primates).

Only two mammal species besides the above groups have virtually hairless adults and older juveniles: naked mole-rats and humans. Each of these species appears to have evolved nudity independently of any other mammalian forms, since their close relatives are all relatively hairy. The exception to this statement is that most rodent newborns are naked; thus, only the older juveniles and adults of naked mole-rats have diverged in this regard from all other rodents. The hairlessness of non-newborns in naked mole-rats and humans is also similar in that in neither case is there either a dramatically thickened skin or armor (although a relatively thicker skin in naked mole-rats); it differs, of course, in that adult humans have retained abundant hair on the head, in the pelvic region, and in the armpits. The nudity of non-newborns in these two species, representing two of seven or more independent origins of relative hairlessness, seems more reminiscent of the kind of hairlessness of newborn altricial mammals; these two species may also be the only mammals in which nudity in newborns was followed by the evolution of nudity in older juveniles and adults.

Hairlessness in newborns is widespread in mammals, as is the absence or near absence of feathers in newly hatched birds (see discussion of altriciality below). This is probably the reason that hairlessness has been regarded as part of a neotenic trend, which may be a correct view in terms of developmental processes but does not provide an explanation in evolutionary or selective terms. Phenomena such as neoteny and allometry may represent inertial or constraining forces in the sense that natural selection must always operate on ''last-year's model;'' but in the same sense, all genetic, developmental, physiological, and morphological attributes of organisms represent inertial elements for selection. Unless one assumes that natural selection is helpless in the face of such inertias, the search for evolutionary (selective) explanations necessarily continues in approximately the same fashion as in the absence of information about such inertias. The general assumption of such searches is that selection is the *principal* (not the sole) guiding force of evolution.

HAIRLESSNESS AND ECTOTHERMY

Newborn mammals that are both naked and sometimes left by the mother in a nest also tend to be ectothermic, as are altricial vertebrates in general. This implies that there is merit in attempting to relate the evolution of hairlessness to that of altriciality.

An ectothermic organism is one that relies for its body temperature largely or entirely on external sources. Such organisms are often described as having ''poor'' or ''inadequate'' means of thermoregulation. This view is not productive of hypotheses as to the origin and basis of the trait, unless one imagines that selection has somehow been ineffective, and a trait that is disadvantageous has evolved. Such traits do evolve, as in senescence (Williams 1957), but only under conditions such as pleiotropy, with beneficial and deleterious gene effects continuing in concert whenever they derive from the same indivisible chunk of genetic material. Such deleterious traits are saved only because their (currently) inevitable companion traits are sufficiently beneficial to overcome the deleterious effects, and no way of divorcing the two effects has yet appeared. In no organism, apparently, has a reason been generated for regarding ectothermy as a deleterious pleiotropic effect, and, contrary to the situation with senescence, no circumstances seem to exist that make such an explanation likely. Accordingly, it seems parsimonious to assume that ectothermy evolved in naked mole-rats because it is somehow advantageous.

One correlate of naked mole-rat ectothermy is a rather low metabolic rate, and it has sometimes been assumed that this is the source of the advantage, that a lowered metabolic rate simply allows naked mole-rats to subsist on fewer calories and therefore suggests a continuing problem in caloric intake that is greater or of a different nature than that encountered by the usual homeothermic mammal. The more appropriate correlate of problems in locating food in the tropics, however, would seem to be heterothermy, a capability that correlates with seasonal or cyclic food shortages. There are times when it is better to lower one's metabolic rate and enter some kind of resting stage, such as hibernation or estivation, so as to pass through the season of most severe food stress with the least metabolic expense. (Dormancy is probably most frequently regarded as a response to temperature stress, but one may wonder whether the most relevant temperature stress is that to which a species' food supply responds by becoming dormant or otherwise unavailable). Safety from predators must also be ensured (to an appropriate level) before vulnerable resting states can evolve. It seems evident that naked mole-rats must suffer sharply increased food problems during the droughts that are common in the arid tropical regions where they live. Because their predators (e.g., snakes that can move through their burrows) may or may not estivate during droughts, it is difficult to comment on whether naked mole-rats might have evolved ectothermy rather than heterothermy so as not to be entirely vulnerable to predators during seasons of low food availability.

A potentially profitable way to start thinking about the evolution of ectothermy in a previously homeothermic animal is to consider that it has evidently become more efficient, for whatever reason, for that animal to rely upon an external source of warmth. This situation would seem to prevail whenever such external sources are so reliable and effective that the expense of homeothermic machinery is superfluous. To understand when such conditions might exist, one must consider not only the external source of heat itself, but also the nature of threats, such as the inability to obtain food when it is crucial and the inability to escape from predators that are able to maintain a high rate of metabolism and predatory ability when external heat sources are minimal or absent. Naked mole-rats have apparently been largely relieved of homeothermic predators, and perhaps most of the predators that afflicted their less subterranean and larger ancestors; and their tropical burrow systems are relatively stable in both temperature and humidity.

Hairlessness and Altriciality

Altricial juveniles that have come to depend on their parents for food and for virtually all protection from predators (as with naked, ectothermic forms) are in a position to dispense with homeothermy and use their parents as the (primary) external heat source. Such a juvenile might gain by refraining from use of nutrition provided by the parent to maintain a high metabolic rate when the parent is absent, and instead conserve ingested calories for later growth by maintaining a high metabolic rate *only* when external heat (e.g., from a parent or the sun) is available. Parents thus supply calories not only through food, but also through providing heat for metabolism. This set of attributes correlates with both nakedness in the juvenile and nakedness in at least the part of the anatomy of the parent that directly contacts the juvenile during brooding (e.g., brood patches of birds, naked bellies and mammary glands of mammals). It is probably significant that most naked juvenile birds and mammals occur in litters, and single offspring are rarely naked. Nakedness allows rapid absorption of heat from extrinsic sources such as the bodies of other individuals (parents or siblings in the case of altricial littermates, and other colony members and warmed soil on sunny days and into the night in the case of naked mole-rats). Coincidentally, it also causes or allows rapid heat loss to other individuals, which are always close relatives: siblings or offspring in most parental birds or mammals and colony members in naked mole rats.

Naked mole-rats live in the tropics, and they live underground. In laboratory colonies they bask under heat sources, even when the temperature in their tunnels is in the 25°–30°C range (Jarvis, Appendix). This basking often involves large numbers of bodies piled together, and a mole-rat that has been running through the tunnel system, or working, may dash to such a basking group and snuggle against the other bodies (Lacey et al., chap. 8). Similarly, a basking mole-rat may abruptly run off and feed or carry out some housekeeping task. In other words, their ectothermy does not mean that they simply tolerate lower metabolic rates; they obviously behave in ways that take advantage of external sources of heat, even when their surroundings are at a rather high temperature.

One aspect of ectothermy that seems not to have been discussed is that an ectothermic organism not only can survive a decided lowering of its body temperature, but it may also be able to function in a superior way at higher ambient temperatures than homeothermic species. Thus, some insects become so active at high temperatures (e.g., on warm sunny days) that they are virtually impossible to catch. Presumably this happens because their body temperatures are so high that they can move with unusual speed and their reaction times are very short. This effect may also occur in predators in tropical situations, perhaps even in the burrow systems of naked mole-rats. If so, then ectothermy in predators may make ectothermy in prey advantageous as well. Thus, perhaps only a prey animal with a very high temperature, not likely under homeothermy, has the best chance of escaping a predator with a very high temperature. Perhaps some ectothermic juveniles are even able to increase their growth rates by using the sun as an external heat source that causes their body temperatures to rise above that of homeothermic species, enabling them to grow exceedingly fast. Good candidates are ground-nesting birds such as some arctic songbirds that grow and develop exceptionally fast and depart their predator-vulnerable nests within a week or so of hatching.

Why Does Altriciality Evolve?

The words "altricial" and "precocial" actually come from the ornithological literature and are still defined in most dictionaries in terms of their application to newly hatched birds. *Altricial* hatchlings, as with sparrows, starlings, and pigeons, are more or less naked and helpless; food is brought to the nest for them by their parents. In contrast, *precocial* hatchlings, as with chicks, ducklings, pheasants, and quail, are covered with down, agile, and ready to move out alongside their mother and pick up food themselves. There are degrees of intermediacy; for example, goslings (which typically have two or more adults attending them) are somewhat more helpless than ducklings (which are tended only by their mother). In all likelihood, the extreme differences between songbirds and gallinaceous birds and the rarity of intermediacy are the reasons why the terms altricial and precocial were applied so readily to birds and have retained their meaning there.

Other animals, of course, also display the kind of variation found in birds. Newborn rodents, including naked mole-rats, are naked, blind, and helpless, and they are born in a nest where they remain for some time. However, in other mammals such as ungulates, newborn often are able to stand alone within a few minutes and some, such as horses, can gallop alongside their mothers in less than an hour. Newborn ungulates may travel considerable distances with their mothers and may be required to follow a herd in its everyday activities. Again, there are intermediates: canine and feline babies are blind when born, but not naked, and not as helpless as newborn rodents; some ungulate newborns are physically less capable than others. Little attention has been paid to explaining the distribution of these variations in selective terms.

To develop an understanding of the concepts of altricial and precocial, it may be useful to apply them even more widely, for example, to insect juveniles. Maggots and the maggotlike larvae of some insects with complete metamorphosis (e.g., honey bees) can be regarded as altricial. In contrast, the nymphs of insects with incomplete metamorphosis, such as grasshoppers and crickets, are precocial in the same sense as some juvenile mammals and birds. Again, there are intermediates. For example, within the family Gryllidae, including all crickets, most juveniles would be seen as precocial. Their exoskeletons are hard, and they are agile, quick, and seek out their own food right from hatching; there is no parent alive to assist them. But in genera such as *Anurogryllus* and *Gymnogryllus*, in which the female cricket prepares a closed burrow with a food cache before she lays her eggs and then tends her offspring until she dies (feeding them small, apparently unfertilized trophic eggs), the hatchlings are soft and fat, resembling termite juveniles (West and Alexander 1963; Alexander, unpubl. data). Many other examples could be given: thus, caterpillars may be soft and helpless or quick-moving and covered with urticaceous hairs or other defenses. Internal parasites, especially those living in the alimentary tracts of their hosts, tend to have the features of altricial juveniles.

Ricklefs (1974, 1975) began developing a theory to explain altricial and precocial juveniles when he showed that altricial nestlings of birds grow faster than the more precocial nestlings of related species. Faster growth may be the function of altriciality in a wide variety of species, but it is difficult to believe that others, such as human babies, have evolved to be as helpless as they are just so that they can grow faster, even if they do that. Because the special case of altriciality in the human baby is treated elsewhere (Alexander, in press), it is not considered here.

In general it is easy to understand why precocial organisms might gain from having the attributes that cause us to label them precocial. It is obvious why an ungulate would gain from being able to run alongside its mother soon after birth. Precocial birds are most often tended only by their mothers, are hatched in vulnerable nests on the ground, and eat the kind of food that can be captured by moving about on the ground or in the water. Juvenile insects in species with incomplete metamorphosis—that is, the precocial sort—live without parents in dangerous locations, and they are usually able either to run or leap, or else they produce various kinds of poisons or other deterrents to predators. Their abilities to do these things are what causes us to see them as precocial. The same is true of precocial larvae in forms with complete metamorphosis.

The question that remains is, Why, when the selective pressures favoring precociality are removed, do juveniles become soft, helpless, and maggotlike? Is there a general answer, other than the dissatisfying or incomplete one that particular selective pressures are relieved or removed, or that there is some advantage to the parents (e.g., Eisenberg 1981) rather than to the juvenile itself? I think there is a general answer, and I would hypothesize as follows.

Juvenile life, in general, may be said to have two functions: first, to survive to the adult stage, and, second, to become the best possible adult; that is, to be maximally capable of doing whatever an adult has to do in order to reproduce as well or better than anyone else. That one function of the juvenile is to survive to the adult stage implies that it must get past certain dangers or causes of mortality. In general, the things that juveniles do to reduce risks from predators, parasites, food shortages, climate, and weather—to refer to Darwin's hostile forces of nature, or the causes of eventual reproductive failure—cause the juvenile to become the type that we would call precocial. In other words, we usually apply the term precocial to physically capable juveniles that are able to protect themselves in some fashion from sources of mortality.

I propose the following hypothesis as a general explanation of altriciality (for a single exception, marsupials, see below). When relieved of the necessity or any importance of evolving to protect one's self from extrinsic hostile forces of nature, the juvenile organism is freed to devote a greater proportion of its calories to the task of becoming a better adult. This will be true, regardless of the means by which the relief is effected: by direct or continual parental solicitude, or by having been placed in a safe location by a now deceased or departed parent. Protected juveniles are free to evolve, earlier and earlier in their juvenile life, traits and tendencies—and to respond to events (e.g., growing, learning, practicing)—that are devoted solely to causing them to be better competitors as adults. Traits and tendencies that make one a better adult are not necessarily synonymous with traits that enable one to bypass or deal successfully with particular hazards along the pathway to adulthood. What we call precociality represents expensive ways of dealing with hazards that may terminate juvenile life.

One result of altriciality, as with the songbirds studied by Ricklefs, is to allow more calories to be devoted to growth. In songbirds this is possible because the nest is hidden and usually off the ground and because, in general, both parents provide food. Something parallel is true for subterranean rodents and crickets, as well as for species whose larvae are protected, for example, by being injected into wood.

It is possible for even the same activities to conflict with one another in quite different life stages. To choose a worst case—or one least likely to be grasped easily—I expect that even a newborn ungulate's ability to run alongside its mother within an hour after birth actually conflicts to some extent with its ability to run later in the ways and situations that an adult has to run. The reason for expecting this conflict is that expending calories and neurons on running ability as a newborn almost certainly subtracts from the ability of the juvenile to achieve most efficiently the best possible adult size and agility at the right time (e.g., as soon as possible). The fact that the newborn has evolved to run well so quickly compared with other prey species that hide offspring, moreover, implies that its ability is probably not profitable as practice for running well and fast as an adult many months or years later. Most precocial traits would more easily be seen as expenses that detract from the ability to generate optimal adult traits.

As suggested earlier, some special situations must be explained to be understood according to the idea being developed here. One such is the maggots that occur in dung, carrion, fungi, and other short-lived habitats. They are not necessarily protected from sources of mortality. Why, then, do they take on the aspect of being altricial? I hypothesize that because they cannot deter the serious threats in their habitat, their best bet is to get through the dangerous feeding stage and out of the larval habitat as fast as possible. As with internal parasites, which may also be protected, they have evolved to become mere sacks of efficient nutrition-grabbing ability. They load up their "grocery bags" as fast as possible and drop off or crawl out of the dangerous place where they have secured their food, to do nearly all of their development in the so-called pupal, or developmental, stage. They evolve a kind of apparent altriciality that enables them to grow as fast and as safely as possible. In fact, of course, they are highly "precocial" in terms of their ability to ingest their medium rapidly, and presumably many "altricial" juveniles are correspondingly precocial in respects that are not obvious but reflect preparation for assumption of some crucial adult activity or trait. Elsewhere (Alexander, in press), I argue that the human baby has evolved to become physically altricial because it thereby advanced the development (through both ontogeny and learning) of its complex brain, intellect, and social competence.

The single exception suggested to the hypothesis advanced here involves the extremely short gestation periods of marsupials. In some cases, gestation is even shorter than the estrous cycle. Thus, pregnancy does not interfere with the timing of estrus because the fetus is born before estrus recurs. In these cases, characteristic of species living under extremely unpredictable conditions that sometimes involve prolonged droughts, Low (1978) and others seem to have argued successfully that the mother gains by being able to discard a juvenile in the pouch in favor of another embryo in a diapause stage in the uterus, thereby initiating another offspring with minimal delay. Such females can repeatedly initiate embryos at a very high rate and low cost, discarding or saving them according to whether or not rain has been adequate to produce sufficient nutrition to make the effort of rearing an offspring worthwhile.

Why Did Naked Mole-Rats Begin to Live in Groups?

To understand group living in any species, one must eventually address two questions, (1) what selective forces initiated group living, and (2) what selective forces caused it to be maintained or elaborated? The two answers may be the same, but they need not be.

I have argued previously (Alexander 1974, 1977, 1979, 1989; see also Alexander et al., chap. 1) that the number of reasons for the onset of group living is small: (1) protection from predators as a selfish-herd effect (W. D. Hamilton 1971), or as a more efficient alarm system or deterrence; (2) group cooperation in securing some food item that is difficult to locate or capture; or (3) mere clustering on a scarce resource or habitat. I have rejected the notions that group living can evolve because the group serves as an information center (although groups may so serve when other reasons for group living are present) or that group living can be initiated as a cooperative defense of food (e.g., Wrangham and Rubenstein 1986). I assume that such cooperation is unlikely unless groups have already formed for other reasons such as clustering on a scarce but relatively large or clumped food supply.

Here I suggest (see also Alexander, in press) that group living that begins as one or both parents and a brood of offspring (as opposed to those that begin as a cluster of juveniles without parents or a collection of adults without offspring; see Alexander et al., chap. 1) may invariably evolve as a consequence of predator effects. We usually think of parental solicitude as involving primarily the feeding of offspring or perhaps even protection from climate or weather. It seems to me, however, that feeding and protection from the elements may always be secondary, and that what we call parental care may always begin as an effort to reduce the effects of predators on a brood. These efforts cannot take the form of placing offspring in a stationary nest or protected location until the parent has evolved a means of providing food for the juveniles. Once the ability of parents to feed offspring has become elaborate, it is easy to attribute the significance of parental care to feeding and forget that fed offspring typically are placed in hidden nests, and offspring that have the physical apparatus to secure their own food are also mobile. Thus, neither must they remain in a potentially vulnerable location, nor are they completely incapable of reacting to predators in ways that increase their likelihood of escape. It is unlikely that nests or juveniles can best be protected in regions that also have a maximal availability of food, or that areas of high food availability will also be maximally predator-free. In the second situation, parents are required to protect their offspring from predation in more or less direct and obvious fashions; in the first situation, the nature and location of nests (the initial acts of parental care) are determined by predation, even if parental feeding activities (necessitated by keeping the offspring in nests hidden or inaccessible to predators) are more obvious.

Perhaps our tendency to associate parental solicitude with feeding also returns to the significant amount of nutrition incorporated into the fertilized egg. Even here, however, one needs to know if the added nutrition — and sometimes increased time inside the mother's body — does not derive its adaptive significance solely from the decreased vulnerability of the larger (and later) juvenile to certain kinds of predators.

Alexander et al. (chap. 1) argued that parental behavior in termites, naked mole-rats, and some Hymenoptera turned into eusociality because these organisms moved into niches that were (1) food-rich in such ways that they need not be exited (or such that the risks of obtaining food were relatively small), (2) expansible (could accommodate expanding social groups), and (3) relatively predator-safe, yet in which it was also possible for individuals to carry out extremely risky or suicidal antipredator acts that would save enough partial relatives to make such heroism genetically profitable. In this scenario, predator safety would be the driving force, both in naked mole-rats' becoming completely subterranean and in their evolving group living and eusociality (and also in their becoming small, thereby preventing large snakes and mammalian predators from entering the burrow systems).

A second argument regarding reasons for group living by naked mole-rats might be that group living was originally sustained as cooperation to compete against other families of naked mole-rats. This suggestion has been made repeatedly for the maintenance and elaboration of group living in humans (see discussions in Alexander 1979, 1987, 1989, in press), but not for its origin. For naked mole-rats, we must explain why broods of juveniles stayed at home, as they do in many of their relatives that are not eusocial (see Jarvis and Bennett, chap. 3). No one seems to have imagined that stay-at-home juveniles in other rodents are incipient soldiers that defend their parents and siblings against other families of conspecifics. Unless further investigation suggests such a scenario, there would seem to be little support for the hypothesis that naked mole-rats evolved eusociality as a cooperation to compete against conspecifics (note, however, that mole-rats in the laboratory do defend their colonies strongly against intrusions by conspecifics; Lacey and Sherman, chap. 10; Jarvis, chap. 13).

A third scenario (Jarvis and Bennett, chap. 3; Brett, chap. 5) is that naked mole-rats started living in groups because groups are needed to locate food (large subterranean tubers). This hypothesis does not seem to explain why *Heterocephalus glaber* initially began to live in groups. There is every reason to believe that naked mole-rat social groups arose out of parents' tending young juveniles, the juveniles subsequently tending increasingly to stay with the parents as they matured. Initially, these juveniles would not have contributed to parental care but would instead have represented a cost to the parent, not only as tiny nursing juveniles but as larger juveniles eating the same food as the parents. It seems unlikely that difficulty in finding food would cause juveniles to stay at home. Rather, parents would have gained if juveniles had left, allowing the parents sole access to their own food supply, which they would have already located since they were able to produce and raise juveniles. Juveniles may have stayed with parents initially because the parents had a safe burrow with food, and dispersal was risky because it had to take place aboveground. Or, because migrating aboveground was too risky, juveniles may simply have burrowed away from their parents' nest and thereby found food that the parents could also use (even if in this fashion they carried out the first helping behavior). In either case, predation would have been central to the continuing changes leading toward eusociality.

I suggest that food tended to be abundant for naked mole-rats early in the evolution of their social groups, even if it were also expensive (difficult to reach by burrowing) under the relatively predator-safe condition of remaining completely subterranean. I also suggest that the principal benefit realized from group living was the increased possibility of escaping predators in the extensive burrow system. A pair of naked mole-rats with their offspring would surely be much more vulnerable to predation if they tended to stay in one location in a small burrow or chamber, such as near a single large tuber, which might otherwise provide sufficient food for a small family for a long time. Moreover, if the difficulty of locating food and getting to it were the sole explanation for group living, then colonies should be smallest in localities where tubers are most abundant. In such situations, additional colony members would be least valuable and most likely to interfere with reproduction. By far the largest naked mole-rat colony ($n > 295$ animals), however, was located in a garden of yams where food was more abundant than is typical (Brett, chap. 4). In a parallel fashion, I would expect social groups of wolves, hunting dogs, and other group-foraging organisms—groups that apparently do exist because of group-hunting—to be smallest when food is most abundant, whether the variation is geographic or seasonal. This may be the case, since wolf pairs sometimes den alone (presumably they produce young during times of food abundance), rejoining a pack after the young have left the den (Mech 1970).

If food is a problem of the nature or severity that would directly lead to ectothermy because of lowered metabolic rates (see Jarvis and Bennett, chap. 3), then it would seem that group living is even more difficult to explain in naked mole-rats. If naked mole-rats live in groups so as to locate food, for example, then (1) food sources would have to be located more effectively by group efforts (this is likely, since the subterranean life of naked mole-rats means that they must burrow to each new food source), and (2) food sources would have to be large enough to compensate for the losses suffered by having to share them with other group members. In other words, we would have to explain why naked mole-rats did not profit from locating large food sources and exploiting them as small nuclear families, even, perhaps, defending them against other naked mole-rats and forcing juveniles to disperse to different food sources. Some of the tubers on which naked mole-rats feed are indeed quite large and also quite separated from one another (Brett, chap. 5); this represents one set of criteria necessary for group living as a foraging benefit (Alexander 1974, 1977, 1989). As already noted, one reason this scenario was not played out may be that permanently locating near large tubers would cause small groups of naked mole-rats to be vulnerable to digging predators. In any case, most arguments make it seem that predators were either directly or indirectly involved in many aspects of the evolution of naked mole-rat eusociality.

I am led, then, to the following hypothesis: given the relative safety of their burrows, the ectothermy of their principal predators, the uncertainty and expense of maintaining adequate food supplies while also avoiding predators, and the reliability and inexpensiveness of the sun as a source of energy, naked mole-rats have benefited from using the sun and one another (rather than food) as the principal sources of energy for keeping up body temperatures. As a result of this combination of factors, they have shed both the expensive machinery of homeothermy and their hair. Ectothermy and nudity together allow them both to increase and to decrease body temperatures swiftly (the latter, e.g., after digging in extremely warm sites), by movements between warm and cool (e.g., deeper) portions of the burrows and use of one another's body heat. Group living, food-getting, nakedness, ectothermy, and changes in predation thus may all be parts of a set of attributes that are unlikely to be understood unless they are considered *together* and as a response to a general selective situation, rather than individually as if their selective backgrounds consisted of individual and independent selective forces.

I am therefore suggesting that group living in naked mole-rats is essentially a response to predation. This particular response is possible because of the enormous food supply in the region (niche) they entered, and because the soil and habitat are conducive to the mole-rats' becoming almost completely subterranean. Food is essentially everywhere, and finding a big tuber represents nutritional insurance for a while. It is also possible that these tubers, which appear not to be eaten by many other animals, including humans, were not particularly fine food for species not evolved to use them. Moreover, a sacrifice of some sort may have been involved when *H. glaber* started to feed on them, but this sacrifice may have been more than offset by the benefits of predator avoidance in their new subterranean niche. The strategy of naked mole-rats, then, is to keep the burrow system so extensive that they simultaneously obtain access to an abundant food supply and remain prepared to escape whatever predators may still plague them, presumably snakes. This predator-prey relationship may now be rather specific, with certain snakes (from several genera; see Jarvis and Bennett, chap. 3; Brett, chap. 4; Braude, chap. 6) having become specialized at preying on naked mole-rats. The burrow systems may be designed largely to give the mole-rats the greatest relief from snake predators. If these things are all true, the structure of burrow systems should show it, in the form of specializations such as bolt holes (Brett, chap. 5) and rapid plugging capabilities (Brett, chap. 4; Jarvis, Appendix).

To summarize, I see naked mole-rats as "fugitive" eusocialists, essentially defenseless except through flight, and somewhat parallel to tropical wasps whose colonies are continually in jeopardy from army ants, whose queens are not physogastric, and whose queens and workers both flee. The link that is still missing is determining how and when naked mole-rats might carry out heroic acts that save enough relatives to make virtual suicide reproductive (see the appendix to this chapter). We still lack information on whether or not, as I predicted in 1976 at Northern Arizona University (see the Preface), reproductives in this eusocial form tend to senesce at later ages than workers (see Alexander et al., chap. 1; also Jarvis, chap. 13).

Why Do Naked Mole-Rats Remain Aggressive
toward One Another?

The individual naked mole-rats within a colony are apparently extremely similarly genetically (Honeycutt et al., chap. 7; Reeve et al. 1990), yet they show mild aggression almost continually (Reeve and Sherman, chap. 11) and sometimes fight to the death (Pepper et al., chap. 9; Lacey and Sherman, chap. 10; Jarvis, chap. 13; Faulkes et al., chap. 14). How can this be, given W. D. Hamilton's (1964) arguments with regard to maximizing inclusive fitness via close genetic relatives?

Alexander et al. (chap. 1) argued that much of the mild aggression shown in small-colony eusocial forms is actually part of a monitoring process that simultaneously tells the reproductive female whether or not any other females are beginning the morphological and physiological changes that will carry them toward queenship and tells the attacked individuals that their queen is still vigorous and healthy, able to present them with siblings to tend, and unlikely to weaken and die or be replaced as queen by one of their own sisters. This interpretation rests on the assumption that sisters, which are essentially genetically identical, continue to compete rather intensely for reproductive opportunities. As with the more severe aggression that occurs when, for example, two females simultaneously start developing toward reproductive dominance, or when an individual is ostracized and is killed or dies, such mild aggression may be evidence of severe competition between individuals that are essentially identical genetically.

Reeve and Sherman (chap. 11) analyzed "shoving" behavior (assumed to be a form of mild aggression) in captive *Heterocephalus glaber* colonies. They found that the breeding female did most of the shoving and that she shoved less-related larger individuals without regard to their sex. They concluded that, by shoving, a breeding female may incite colony mates to become more active and may also maintain her reproductive dominance. The kind of monitoring behavior postulated in chapter 1 would be expected to be directed preferentially toward larger and less closely related individuals but does not at first seem likely to be directed equally at males and females. However, males approaching breeding condition are a threat to the breeding female if their reproductive condition causes them to mate with other females or to affect the reproductive condition of other females (Jarvis [1981, chap. 13] reported that all males in a colony possess active sperm).

Why does aggression continue in *H. glaber* colonies when genetic differences are minimal or absent? Alexander (1979, p. 130; see also pp. 128–129) argued that "[to understand] why Hamilton (1964) was correct to focus his analysis upon relatedness in genes identical by immediate descent [as opposed to genes identical by nature but not necessarily by immediate descent], . . . one needs only to consider the fates of mutants affecting nepotistic behavior. Such mutants represent the means by which the altruism of nepotism generates, increases, and becomes directed with precision. The successive waves of such mutants will always maximize their own spread by treating relatives as if their own likelihood of occurring in the relative depends upon the proportion of genes identical by immediate descent. This is because each new mutant will at first indeed tend to be present in just those proportions: for this reason better odds will not occur.''

In other words, nepotism spreads and is molded according to the likelihood that any new mutant will be present in any particular relative that is a candidate for nepotistic treatment, not according to the overall similarity of the two genotypes as a result of inbreeding or accidental similarity (see also Dawkins 1979; D. Krebs 1989). Early in the evolutionary trajectory of a mutant, it has approximately a 50% chance of being present in siblings, parents, and offspring of any individual that possesses it. Accordingly, the mutants affecting nepotistic behavior that will spread most rapidly are those that cause their bearer to treat relatives according to their likelihood of possessing it at the outset. Such mutants will tend to accumulate, and with inbreeding there will be a tendency toward genetic identity within populations in regard to all such genes. Contrary to intuition, however, this condition does not lead eventually to tendencies to ignore genetic differences and treat everyone alike. The reason is that no mutant leading to such behavior can invade the system just described. The existing mutations, which tend to cause their bearer to treat certain kinds of relatives as if each of their genes had the likelihood of being present that is given by immediate descent, cannot magically change their messages to the rest of the organism just because different individuals gradually become increasingly alike genetically. Even in populations of such genetically similar individuals, new mutants affecting nepotism will spread according to their tendency to cause treatment appropriate to their own initial likelihoods of presence in other individuals, not their eventual distribution. Accordingly, this argument predicts the condition found in naked mole-rats: regardless of their closeness of relationship as a result of inbreeding, in sexual organisms, relatives tend to treat each other as if they share only the proportion of genes that would be alike as a result of identity through immediate descent.

Why Don't Naked Mole-Rats Allogroom?

It may seem that naked mole-rats do not groom each other because of their hairlessness. Although hairlessness may be partly responsible for the paucity of ectoparasites in *Heterocephalus glaber*, it may not be sufficient to account for the absence of allogrooming. Humans are also hairless over much of the body, yet they groom and massage even the most naked parts of one another a great deal. (Humans not only retain external parasites but also show tendencies toward skin infection, both apparently associated with hair follicles.) In naked mole-rats, the only observations suggesting allogrooming are those involving treatment of juveniles by elders, as with the rapid pushing of pups with the nose and the licking of pups by older colony mates (see Lacey et al., chap. 8; Lacey and Sherman, chap. 10). Naked mole-rats groom or clean themselves, primarily their feet, and they also scratch themselves, probably in response to the presence of subcutaneous mites. These mites may not be accessible through allogrooming, perhaps partly because naked mole-rats live in darkness and are essentially blind.

Naked mole-rats may fail to allogroom, then, because, first, their hairlessness has reduced ectoparasites to subcutaneous mites, and, second, because the locations of irritations by these mites are not readily available to individuals other than the infected one.

Why Have Naked Mole-Rats Not Speciated?

The question "why have naked mole-rats not speciated?" presumes that naked mole-rats all belong to a single species, and not enough is known to establish this as a fact (see Honeycutt et al., chap. 2). Species multiply when different populations diverge sufficiently that interbreeding is irreversibly prevented. Biologists recognize different species when they find distinctive populations living together and maintaining their differences, or when they decide on circumstantial evidence that two allopatric populations would not interbreed if the extrinsic isolation between them were to disappear. Naked mole-rats have a wide geographic range in Kenya, Sudan, and Somalia (Honeycutt et al., chap. 2). They vary considerably in body size and perhaps in other attributes (Brett, chap. 4; Jarvis et al., chap. 12). These differences may or may not be heritable, and they may or may not bear on the question of irreversible divergence. Even if allopatric populations of naked mole-rats were to prove genetically incompatible, the question of why they have not diverged more will remain, as will the question of why apparently only a single species lives in any region. All such questions, however, must remain unanswered until a great deal more information has been gathered.

Summary

The naked mole-rat is an unusually distinctive species, almost as distinctive among rodents as humans are among primates. A large number of the most interesting questions about its existence and how it evolved remain unanswered and will not be resolved until a great deal more information has been gathered about the biology of *Heterocephalus glaber*. In view of the many decades that have been spent studying eusocial insects (in the case of the honey bee, centuries might be more accurate) and the number of significant questions still unanswered — and in view of the much longer lifetimes of naked mole-rats (and therefore cycling times for their colonies) and the greater difficulty and expense of keeping several colonies in the laboratory — we may expect many important questions about naked mole-rats to remain unanswered for a long time.

Acknowledgments

I thank L. K. Alexander, S. H. Braude, R. A. Brett, W. R. Dawson, E. A. Lacey, J. W. Pepper, C. Kagarise Sherman, and P. W. Sherman for discussing the ideas in this chapter at various stages during their development. I am deeply indebted to J.U.M. Jarvis, who provided detailed and helpful comments and expert information on bathyergids; she and I disagree amicably on several points in this chapter, especially the importance of predation in initiating group living and promoting eusociality. S. Finger and S. Isil were unusually fine research associates during the early part of the 10 years that naked mole-rats have been studied at Michigan. I also thank T. A. Vaughan for drawing my attention to naked mole-rats in 1976, and for helping me to establish contact with J.U.M. Jarvis.

Appendix

Lacey and Sherman (chap. 10) described interactions between naked mole-rats and a snake introduced into laboratory colonies, and Brett (chap. 4) and Braude (chap. 6) reported observations of snakes attacking *Heterocephalus glaber* colonies in the field (see also Jarvis and Bennett, chap. 3). Here I report an additional laboratory observation, which seems to me to support the suggestion, central to the above arguments about predation, that snakes within their burrow systems have long been a special jeopardy to naked mole-rats.

In September 1980, my assistant, S. Finger, and I gently introduced a small (ca. 25 cm in length) North American garter snake (*Thamnophis* sp.) into one of the naked mole-rat colonies at Michigan. No mole-rats were disturbed by the introduction, and the snake moved slowly down a straight tunnel about 3 m long (the tunnels at that time were yellow plastic tubes sold as Hamster Habit Trails).

Eventually the snake passed lightly over several basking mole-rats without disturbing them. One mole-rat raised its head slightly and seemed to sniff the air after the snake had passed. Near the corner at the end of this long tunnel, the snake reversed itself and returned along the tunnel. By either touching them or passing near, the snake had by this time caused several basking mole-rats to stir and start to locomote, but none of them seemed oriented toward it. The snake continued back along the tunnel and eventually turned two left corners and entered another section of long tunnel paralleling the first section. After continuing down this tunnel for about 1.5 m, the snake reversed itself again. By this time, we had noticed that a single medium-sized mole-rat was walking slowly down the tunnel in the same direction that the snake had been taking before it reversed itself the second time, although it was about 50 cm behind the snake. This mole-rat walked in a peculiar way, as if it were stalking the snake. Within a few seconds, it became obvious to us that a second mole-rat, on the opposite side of the snake in the tunnel system and 30 cm or so distant from it, was also walking slowly toward the snake in the same fashion. When initially seen, both mole-rats were in regions recently vacated by the snake and presumably had picked up either its odor or the vibrations of its movements or both.

As the two mole-rats approached the snake from either side, it became obvious that the snake was aware of the presence of both mole rats. It moved first one way and then the other, stopping each time when it was near or, in one case, had actually touched one of the stalking mole-rats. As the two mole-rats came nearer to it, the snake increased the speed of its locomotion until it was literally thrashing its way back and forth between the two mole-rats. When the mole-rats were about 30 cm apart, one of them seized the snake's body with its incisors and immediately and very rapidly bit its way down the body for several centimeters (several bites per second). It then released the snake, which writhed as if fatally wounded. The other mole-rat then seized and bit the thrashing snake in a similar fashion, and the snake appeared dead. The biting occupied only a few seconds, and then the two attacking mole-rats moved away from the snake, apparently giving it no further attention. The attack occurred so swiftly that we had no time to interfere.

The dead snake lay for several hours where it had been killed; during intermittent observations, no mole-rats were seen to pay attention to it. At the end of this time, I watched a small colony member pick the snake up with its incisors and carry and drag it to the refuse chamber, about 2 m away. The snake was not examined to count the bites inflicted on it or to discover why they were so quickly fatal.

Although this particular kind of snake was a novelty for the mole-rats, the peculiar rapid bites inflicted by the mole-rats, not observed in any other context, and the almost immediate termination of interest in the snake that had been so attractive to the mole-rats only a few seconds before suggest that *H. glaber* has special responses to potential predators in the burrows. Lacey and Sherman (chap. 10) did not report any fatal attacks on the milk snake they introduced to *H. glaber* tunnel systems. The difference between the outcome that I observed and those reported by Lacey and Sherman may have been caused by the difference in the size or species of the snake (theirs was considerably larger), or the rapid and continual locomotion (or thrashing about) of the snake in this case, which evidently incited the naked mole-rats to seize and bite it.

Appendix: Methods for Capturing, Transporting, and Maintaining Naked Mole-Rats in Captivity

Jennifer U.M. Jarvis

Capture Methods

Naked mole-rats can be captured because of their habit of investigating and then blocking up opened sections of their burrow system. When their burrows are damaged, the animals approach the surface opening and investigate it; this can occur in less than 5 min after the burrow is opened, although it often takes longer. If the mole-rats are not disturbed, they collect soil from within the burrow and kick it to the opening (Jarvis and Sale 1971). After several loads of soil have been transported, the mole-rats use their hind legs to pack the soil and thereby seal the opening. Once the opening is sealed, the animals continue plugging the burrow until 30–100 cm of it has been filled with soil, thereby rendering the system air-tight and predator-proof again. The mole-rats can be captured in the act of approaching the burrow opening and in the first stages of sealing the burrow.

Because *Heterocephalus glaber* seals the side branches leading to a molehill ("volcano") once it has been fully formed (Jarvis and Sale 1971; Brett, chap. 5), it is often difficult to find the deeper-lying "highway" burrows of the system except in the proximity of freshly formed molehills (unweathered ones with an identifiable plug at their bases). Brett (1986, chap. 5) has shown that molehill production and excavation of surface burrows peaks during the rainy seasons in Kenya (March–May and October–December), and consequently these are the best times of the year to capture whole colonies. If surface temperatures are low, the mole-rats will block openings throughout the day, and they can be caught anytime; on hot days, it is possible to capture them only in the early morning and the late afternoon, when it is cooler.

Three capture methods have been used with reasonable success. The first (with various modifications) has been used by all of us (i.e., Alexander, Braude, Brett, Honeycutt, Jarvis, and Sherman). The second and third were developed by Brett (1986) and Braude (pers. comm.), respectively, based on a trap devised by Hickman (1979a). The details of these methods follow.

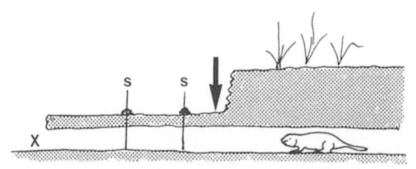

Fig A-1 The standard method of capturing naked mole-rats As the mole-rat appraches the opened portion of the burrow (x) it displaces fine straws (s) inserted into the burrow through its roof When the straw nearest the opening moves, a hoe is driven smartly down at the point of the arrow; see also plate A-1

The first method (described in Hill et al. 1957) involves digging under a group of freshly formed molehills to find the deeper burrows. The openings to each of these burrows is cleared of soil, and openings subsequently visited by mole-rats are identified, either by sitting quietly long enough to see the heads of the animals as they come to investigate the breaks in their tunnel system or by noting small piles of soil being kicked into the opening.

Occupied burrows are then prepared for catching the mole-rats. First, the soil lying above the burrow is scraped away until about 2 cm is left roofing the burrow. Care is taken to ensure that the section of burrow used for capture of the mole-rats is straight and unbranched for at least 60 cm. Then, if the soil is not too hard, two or three very small holes can be made with a sharp instrument (like a strong skewer) along the length of the almost exposed burrow section and fine straws are inserted through the holes into the burrow (fig. A-1). Soil is piled around the base of the straws to seal the holes and prevent air leaks; in extremely hard soil, a fine straw can be inserted directly down the opening to the burrow. The movement of the straws indicates the approach of a mole-rat to the opening. When the straw nearest the opening moves, or the nose of a mole-rat is sighted (plate A-1), a spade, hoe, or knife is driven smartly through the burrow behind the mole-rat, thereby cutting off its retreat into the main burrow system (plate A-1). With the hoe still in place, the captured animal is then carefully dug out.

After removing loose soil from the burrow floor, it is often possible to roof over the damaged section of the burrow with chunks of dirt, topped with loose soil to seal all air leaks. In this way, a prepared burrow can be used several times and, as Brett (1986) pointed out, this also lessens the risk of injury to retreating mole-rats because less force is needed to bisect the burrow at the next capture attempt. Eventually the mole-rats become wary of entering the capture area and block off the burrow behind the trap. When this happens, it becomes necessary to dig back and prepare a fresh section of the burrow.

Plate A-1 *Left,* A naked mole-rat arrives to investigate a tunnel opening created by the investigator. *Right,* A mole-rat catcher waits behind the opened burrow for an animal to appear at the entrance; then the catcher drives the large knife into the soil, cutting off the animal's escape back into its burrow system, and the mole-rat can be dug out by hand Photos: M. Griffin, J.U M Jarvis

If no animals come to a hole for a long time (e.g., 1–2 h), all the mole-rats in that portion of the tunnel system have been captured, or they have blocked the burrow deeper in. In the latter case, more burrow has to be opened up and the blockage cleared. Throughout the capture operations, it is essential to sit, walk, and talk quietly, because naked mole-rats are very sensitive to vibrations and to airborne sounds and will retreat from the opened area if alarmed.

In a day, 30 or more mole-rats can be captured using these techniques, but capture success drops when most of the colony has been trapped. Young pups and the breeding female are usually among the last animals captured (Brett, chap. 4). If entire colonies are to be captured, care should be taken to ensure that all burrow openings encountered are marked and checked daily for signs of occupancy. Furthermore, the capture area should be visited after work has ended for the day because animals that have been separated from their colony in cul-de-sacs can sometimes then be found wandering on the surface.

If some colony members were still not captured by the end of the day, Brett (1986, chap. 4) and Braude (chap. 6) wedged pieces of sweet potato into the ends of the opened burrows and sealed them with loose soil. This generally

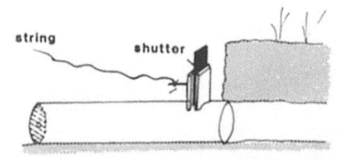

Fig A-2. The shutter trap for capturing naked mole-rats (after Brett 1986).

stopped the mole-rats from extensively blocking these burrows during the night, and tooth marks on the sweet potato the next morning (evidence of feeding) indicated which burrows still contained animals.

In the second method of capturing live *H. glaber*, Brett (1986) constructed shutter traps similar to those described by Hickman (1979a) for capturing *Cryptomys* spp. (fig. A-2). Brett used clear acrylic plastic (plexiglass or perspex) to make his traps. Instead of the mole-rats' springing the trap themselves, they were captured by the manual release of the shutter (by removal of a pin attached to a length of string). By leaving a small section of the tube free of soil, he could observe approaching mole-rats entering the trap and then release the shutter. Brett found that animals would only enter the trap if its floor had been covered with loose soil, and all air leaks between the tube and the burrow had been sealed. During the day the trap was shaded to prevent it from heating up in the sun. Brett (1986, chap. 4) noted that although shutter traps successfully avoided the need for extensive excavation of the burrow system, mole-rats were often wary of entering them, and at best 10–15 animals were captured per day.

The third capture method (developed by Braude) is similar to the second, except that the pin holding the shutter is connected by a thin wire to a piece of bait (a sweet potato) inside the trap. When the mole-rat tugs on the sweet potato, the pin releases the shutter, which falls down behind the animal, thus preventing its return to the burrow. Braude (pers. comm.) is also developing a trap similar to Hickman's, in which the door is electronically closed when the mole-rat crosses a beam of light.

Transporting Naked Mole-Rats

Because naked mole-rats like to huddle together in a small space, they are amenable to being transported (in the field or between research labs) in a fairly

Plate A-2 Entire naked mole-rat colonies can be transported in relatively small containers, constructed of materials (e.g., metal) impervious to the animals' chewing Photo J U.M. Jarvis.

small container. Enough space should be provided to enable the animals to select a preferred temperature and for there to be sufficient wood shavings to absorb all urine voided during the trip. The container should be made of metal or acrylic plastic (plexiglass or perspex; the animals will chew through softer materials) with its top and the upper parts of its sides perforated with air holes (plate A-2). A hole should also be made for inserting a thermometer to allow periodic temperature checks. Adequate nesting material and food (see below) must be provided.

The box housing the animals should be placed in a slightly larger box with air holes and insulating materials (such as a blanket or wood shavings). A hot water bottle (ca. $50°C$) wrapped in toweling, and preferably some other heat source (such as chemical hand warmers or electric blankets), should be placed at one end of the outer box and against the box containing the mole-rats, thereby providing a localized heat source and enabling the animals to select their preferred temperature. This arrangement is sufficient to keep the animals warm for at least 10–12 h. If the mole-rats are to be shipped, however, it is advisable to do a trial run in the lab before sending them. The temperature at the cool end of the container should not exceed $34°C$, and at the termination of the trial run the temperature at the warm end should not have dropped below approximately $24°C$.

Maintenance in Captivity

HOUSING CONDITIONS

Housing conditions for naked mole-rats have been described by Brett (1985); information in Brett's report is summarized and augmented here. Rooms housing naked mole-rats should be soundproof and vibration-proof, shielded especially against sudden low-frequency sounds and vibrations (which invariably produce a panic response). Silence and no vibrations is particularly important for the first 3 wk after the birth of a litter, because the alarmed mole-rats will carry pups out of the nest and often trample them. Speech in the animal rooms should be limited to whispers and all intermittent noises kept to a minimum. The mole-rats habituate to fairly constant noises (such as room ventilation), and the animal rooms at Cape Town have a constant background of low intensity white noise to mask sounds made during routine maintenance. The animals at Cornell are partially isolated from building vibrations by being on floating tables (such as those used by neurobiologists) that do not touch the walls; placing the tables' legs in buckets of sand can also damp vibration. The Michigan animals are housed in a subbasement, where building vibrations and traffic are minimized.

Room temperature should be kept as close as possible to the natural burrow temperatures of 28°–31°C (Brett 1986; Bennett et al. 1988). Additional heat sources (lights) should be placed at points along the burrow system so that the mole-rats can behaviorally thermoregulate by basking. I use adjustable-angle lamps with 40-W bulbs as the supplementary heat source; these can be easily moved about the tunnel system to meet the changing needs of a colony. The Cornell animals are kept in constant darkness, with heat/light provided by 25-W red or yellow light bulbs on gooseneck lamps, adjusted so that they are 10–15 cm above the tunnel surface. Small colonies may require a heated (lighted) nest chamber; however, in large colonies there are enough mole-rats huddling in the nest to maintain warmth, and the animals may refuse to use a heated nest chamber. The requirements of a colony can be determined by trial and error, and minor adjustments must be made as the colony grows in numbers or uses different nest boxes in the tunnel system.

Burrow humidities in the field are high (relative humidities are typically more than 80%; McNab 1966; Withers and Jarvis 1980), and mole-rats housed in an air-conditioned and heated room may experience problems with dry skin. However, if room humidities are kept too high, condensation in the tubes and nest chambers can result in wet nesting material and generally damp, unhealthy conditions. At Cape Town and Cornell the animals are kept at 50%–65% relative humidity. In large colonies, the humidity in the nest chamber is elevated by the animals' respiratory water loss, which is usually sufficient to maintain a good skin condition. In these large colonies, it is often necessary to allow some of the moisture to escape from the chamber by either drilling holes in the top (the number of holes in use can be controlled by covering excess holes with tape) or by having a lid that can be opened slightly. Small colonies should be given a small nest chamber (see below), and if the animals' skins still appear dry, open containers of water with cotton or cloth wicks can be placed close to the nest and at other strategic locations. Nest humidities can also be raised by providing the mole-rats with small quantities of fresh grass, lettuce, or green maize (corn) husks (the latter are used in constructing nests).

HYGIENE

To avoid and contain infections, care should be taken to disinfect one's hands before handling mole-rats. Hands should also be washed between handling mole-rats from different colonies and between cleaning parts of the burrow systems of different colonies. Ideally, shoes should be wiped in antiseptic before entering the room housing the mole-rats, and clothing contaminated by other domestic or wild animals should not be worn in the mole rooms. These simple precautions will save many hours of potential work in treating sick animals. It may be judicious not to allow observers into the rooms if they have a cold or flu and, especially, if they have recently come into contact with sick animals. However, we have no direct evidence that human colds or flu can be transmitted to the mole-rats or that diseases of domestic animals are a threat.

BURROW SYSTEMS

Wild mole-rats live in a sealed burrow system in which they move about in close contact with the earthen walls. In the laboratory it is not possible (or desirable) to maintain an airtight tunnel system, but air currents should be minimized and the inside tunnel diameter should approximate that found in the field (4–7 cm; Brett, chap. 4); however, the dimensions depend on the maximum body size of the colony members. For most colonies, a diameter of 5–6 cm is adequate, but if the breeding female is very large and bears large litters, it may become necessary to increase the tunnel diameter so that she can still move about the system and easily pass other mole-rats when she is in the later stages of pregnancy (see chap. 13); this is especially critical if the colony is alarmed and many animals attempt to pass the breeding female in the tubes.

Burrow lengths in the wild can exceed 3 km (Brett, chap. 5), largely because considerable tunneling is associated with locating the widely dispersed geophytes on which the animals feed. In the laboratory, however, the animals do not have to search for food, and in all cases the total length of the tunnel systems housing our lab colonies is less than 3% of that in nature. None of us has been successful in getting the animals to live in "ant farm" arrangements (e.g., earth sandwiched between sheets of acrylic plastic). The laboratory tun-

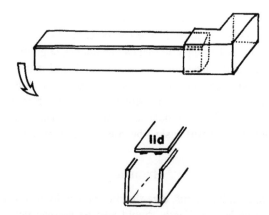

Fig A-3 The runs, made from sheets of acrylic plastic (also known as perspex or plexiglass), used to house captive *Heterocephalus glaber* at Cape Town. These runs have removable lids See also plate 10-1 in Lacey and Sherman (chap 10, p 280)

nel systems must meet three requirements: (1) they must be transparent so that the mole-rats can be observed in all parts of the system; (2) they must be easily cleaned without severely disrupting colony life; and (3) they must be made of a durable material that will withstand persistent scratching and chewing by the mole-rats.

Two basic types of mole-rat tunnels are used in our laboratories. The first type are square runs made from sheets of clear acrylic plastic 3–4 mm thick with inside dimensions of 6 × 6 cm (fig. A-3). The runs have removable lids or tops and can be joined to each other, to corners, T-junctions, and chambers by fitting each into larger-sized sleeves. The removable lids facilitate cleaning and also make it easy to capture an animal without disturbing the rest of the colony. Once a system has been assembled, it is advisable to tape connecting pieces together because the mole-rats tend to push at these places and eventually escape. The major disadvantage of this type of burrow system is that there are numerous corners and edges where the mole-rats can chew. These must be inspected regularly and any holes repaired (i.e., with a quick-setting epoxy putty). In general, the initial hole takes a number of months to make, but once a small opening has been gnawed, the mole-rats soon enlarge it and try to escape.

The second type of housing comprises round runs made from transparent acrylic plastic tubes (4 mm thick, Cornell) or glass tubes (Michigan) with an inside diameter of 4.5 cm. Opaque T-, L- and cross-shaped joints (with clear acrylic plastic tops) are used at Cornell to connect the straight stretches of tunnel (see Lacey and Sherman, chap. 10); at Michigan, glass corners are used. The advantage of acrylic plastic over glass is that it is easier to construct complex, multangular systems and to change their configuration; the disadvantage is that tunnels must be replaced frequently because the animals scratch and chew through them. In contrast to the square tunnels with removable lids (above), long, unjointed stretches of tubing make it harder to capture individual mole-rats without considerable colony disruption.

Several chambers should be linked to both these types of artificial burrows; they should be of several sizes and with more than one entrance. Large chambers (> 30 cm across) are not used by even the largest colonies (> 90 mole-rats), and chamber size can be crucial to the successful rearing of pups. The mole-rats prefer to be well-packed when resting and avoid too large a chamber, especially ones with a lot of headroom. The size of nest chambers must be periodically adjusted if the colony is growing numerically. Mole-rats given a choice of chamber sizes will demonstrate which one best meets their needs. A colony of 40 animals fits into a round nest chamber 22 cm in diameter and 15 cm high, but a newly founded colony (such as a pair and their first litter) prefers lower-roofed chambers with a diameter of 10–14 cm. Some colonies remain faithful to the same nest chamber, but others periodically move; often the move to a new chamber precedes the birth of a litter. Lacey and Sherman (chap. 10) provided each colony with three to four clear acrylic plastic boxes, some with several openings and floored with wood shavings and one having a single opening and containing a layer of sand. This latter chamber served as the toilet. Sherman (pers. comm.) uses movable partitions in the (square) nest chambers. As the colony grows, the partitions can be moved and the inside dimensions of the chamber expanded.

For easy viewing, nest chambers should have glass or clear acrylic plastic tops. A weight or latch on the lid prevents the mole-rats from pushing it off and at the same time enables easy and quiet access to the nest chamber. Whatever lid arrangement is adopted, it is important that it can be lifted off quietly and rapidly so that animals can be captured before they can escape down a tunnel (Sherman also slides an acrylic plastic shutter over nest chamber exits to prevent the animals from escaping capture). By placing the lid so that there is a small opening at one side (or by uncovering holes drilled in the lid), it is possible to regulate the humidity in the nest and also to insert small items of food (for weaning pups) or even to return small pups to the nest without disturbing the resting animals.

Nesting material (fine wood shavings, paper toweling, dried grass, or dried corn husks) should be provided for small colonies but may be unnecessary for larger colonies. Indeed, some nest chambers that I have excavated in the field had little or no nesting material. Cotton (cotton wool) should not be used as nesting material because the pups become entangled in it and strands may also wind around the legs of adults and obstruct their circulation. The birth of pups should be anticipated, and soiled nest chambers cleaned out at least 2 days before parturition, allowing the colony to settle before the pups arrive.

The other chambers in the system are used by the mole-rats as feeding and toilet chambers. Again, loose-fitting or hinged lids are preferable to sliding lids. The mole-rats generally select a chamber with a single entrance or a blind tube as a toilet; this should be made easy to remove from the system so as not to disturb the colony when it is cleaned. The toilet chamber is the only portion of the system that I clean daily, and I clean or rinse it with water but not with soap or disinfectants that would remove the colony's odors. Sherman and Alexander (pers. comm.) clean toilet chambers less frequently (i.e., every 2–3 wk), to avoid disturbing colony chemical communication as much as possible. The frequency of cleaning depends, to some extent, on the number of animals in the system. Unlike the soil of natural toilet areas, those in the laboratory have limited absorbency; in large colonies (> 40 animals), toilet areas become awash with urine in 24 h and require frequent cleaning. Uneaten pieces of spoilable food should be removed daily, although large items, such as half an apple or a sweet potato can be left longer, as long as they do not mold. The rest of the tunnel system should be cleaned only if it becomes very soiled. The mole-rats become unsettled if their normal colony odors are removed and if the system is kept very clean; indeed, they often urinate copiously in newly cleaned tubes and toilets. The animals move through their burrows with their eyes closed and therefore probably use the senses of touch and smell for orientation (Lacey et al., chap. 8); regular removal of landmarks seems to disturb the animals.

Wood shavings or clean sand should be placed in the cleaned toilet areas and in the neighboring tubes. The mole-rats sweep these along the tubes, thereby cleaning them. Some wood shavings will also be carried to the nest area, and others will be eaten. It is therefore essential that the shavings contain no fungicides or other chemicals.

FOOD

In the wild, *H. glaber* feeds on roots, bulbs, and tubers (Jarvis and Sale 1971; Brett, chap. 5). In the lab, the secret to maintaining healthy colonies is to give them a varied diet. In general, naked mole-rats prefer sweet foods (apples, grapes, sweet potatoes, corn on the cob, and bananas), but they will eat a variety of other vegetables—especially of the pumpkin and squash family—jicama, and legumes (peas and beans). Carrot tops, lettuce, freshly cut grass, clover, and dandelions are also highly palatable and especially recommended during lactation. Cabbage, broccoli, and cauliflower should be avoided as they lead to infertility in many mammals (whether or not they have this effect on *H. glaber* is unknown). Occasionally the diet can be supplemented with dried (split) peas, lentils, barley, dried maize, and sunflower seeds; however, these and the fresh legumes may produce bloat (below) and should be discontinued if there is any sign of this. No drinking water is necessary, as naked mole-rats obtain all their water requirements from their food.

When fresh food is in season, it is diced and a mixture given to the mole-rats each day. In addition to this, larger pieces of apple, squash, or sweet potato, which the mole-rats can visit and gnaw, should be continually available. I sterilize all the fresh vegetables by soaking the week's supply (before it has been cut up) for 15 min in water to which bleach has been added. The food is then thoroughly rinsed (ca. 5 min) and stored in a clean refrigerator used solely for mole-rat food. This procedure greatly reduces the incidence of *Escherichia coli* infections.

I have found that for the successful rearing of young, it is often necessary to supplement the diet with high-protein breakfast cereal (e.g., Pronutro), baby food (Fairex) or baby cereal, to which is sometimes added debittered yeast powder and vitamin syrup. Sherman and Alexander also provide their animals with commercial dog treats (Bonz) and occasionally with mealworms. Brett (1986, chap. 5) records instances of wild mole-rats gnawing on a large bone. Although my captive animals ignore bones, they do partly eat dead colony members.

Foreign Odors

The mole-rats become habituated to the odors of people frequenting the animal rooms but often raise their noses and sniff the air in response to visitors and to strong smells (e.g., nicotine, perfumes, and peppermint). Martin (1975) discussed breeding problems associated with endangered mammals, stating that some temperamental species stop breeding if their keeper is changed. He suggested that although there may be no overt response by the animals to conditions outside their cages, many are nevertheless aware of and are responding to these stimuli. Cognizance should therefore be taken of this, and procedures standardized, especially by researchers engaged in behavioral studies on *H. glaber*.

Introducing a Foreign or Escaped Animal to a Burrow System

Mole-rat colonies also have distinctive colony odors, and care must be taken when returning animals that have escaped, especially if they have been handled. Sometimes colonies accept the return of an escapee, and at other times they attack it. To avoid aggression, I usually roll escapees in the damp, soiled wood shavings of the toilet area. Occasionally, the "toilet chamber" treatment is ineffective and the animals are attacked by their own colony mates when they are returned. It is then sometimes necessary to catch the entire colony and put them, together with the escapees, in a small container for about 30 min. During the pell-mell excitement that follows, the returnee may again pick up its colony's odors. Regardless of the mechanism, being treated in this way usually prevents further trouble when all the animals are returned to their burrow system.

Colonies seldom immediately accept a completely foreign mole-rat or an animal that has been away from the colony for several days. The status of the reintroduced mole-rat affects the colony's response: a breeding female is never accepted, but subordinate animals (even from foreign colonies) usually are accepted eventually. For this reason, breeding animals should never be separated from their colonies. Sick or foreign animals that are to be introduced to a colony should have soiled wood shavings from the colony placed with them for 2 days or longer before they are introduced.

On introduction, the colony usually smells the introduced animal all over; it lies passively with its head turned to one side as it is sniffed and nudged. If the introduced animal is attacked, it may be necessary to remove it immediately. Severe damage can be inflicted within a few minutes, and young animals can be killed with one bite. As with escapees, it is sometimes possible to introduce a foreign mole-rat by catching the entire colony and placing it, together with the foreign animal, in a small container. It should be stressed however, that colony response is extremely variable. Some colonies are much more tolerant of the return of escapees and even of foreign animals than are others, and the same colony's responses may differ at different phases of the reproductive cycle of the breeding female. The reasons for these differences are unknown.

CARE OF NEWBORN PUPS

In captive colonies the breeding female gives birth every 76–84 days (Jarvis, chap. 13; Lacey and Sherman, chap. 10). Any disturbing manipulations, such as marking the colony or cleaning the nest, should be completed before the expected birth. Once the young are born, the colony becomes even more sensitive to disturbance and is more excitable than usual. When alarmed, the mole-rats move the pups out of the nest area and may sweep them along the burrows with their hind feet (see Lacey et al., chap. 8). Frequent alarms mean that the pups are rarely in the nest, get cold and trampled on, and fail to get sufficient milk. Some highly excitable colonies even panic at familiar sounds such as those made during the daily feeding and cleaning of the toilet area. It is therefore essential that the colony is disturbed as little as possible and, unless supplementary feeding is being done, the nest area is left strictly alone until the pups are at least 2-wk old.

After a major disturbance of the colony, pups that can be easily retrieved without further alarming the animals should be returned to the nest via the gap in the lid; other pups can be left where they are and the colony will return them to the nest once the alarm has subsided. Newborn pups seem to require at least one good meal a day for survival. If disturbing the nest area is unavoidable, all the pups should be removed and placed in a warm (30°C), moist, chamber and kept there (several hours if necessary) until the colony has settled down. It is inadvisable to put adult mole-rats (including their mother) with the pups because the adults will carry the pups around the chamber, as well as trigger an alarm response when they are reintroduced into the colony. Human odors on the pups may also elicit an alarm response in the colony, and hands should be rubbed in the soiled toilet shavings before the pups are handled.

HAND-REARING PUPS

I have only been successful in rearing pups that are 10 days or older, and I only supplement the feeding done by the pups' mother. I use a standard milk substitute (formulated for human babies), to which a minute quantity of yeast and vitamin syrup is added. I twist a wisp of cotton (wool) around the tip of a toothpick and connect it to a slightly thicker wad of cotton higher up the toothpick shaft. The cotton tip is dipped in the milk solution and gently inserted into the pup's mouth; the pup will usually suck vigorously, and the cotton higher up the toothpick acts as a reservoir to keep the milk flowing. I keep a piece of tissue paper at hand to dry the pups nose and mouth if they get flooded with milk. The milk in the stomach can be seen through the body wall. After the pup is full, I gently rub the anal region to encourage the pup to urinate and defecate. If not immediately returned to the burrow system, the pup should be kept warm and moist. By supplementing feeding, I once successfully raised 22 pups from a litter of 27 (see chap. 12).

Before attempting hand-rearing pups, one must consider the disadvantages of alarming the colony when the pups are removed against the advantages of perhaps rearing some of them. Alarm to the colony can be reduced by removing the pups in the morning and only returning them in the evening. If a small gap is left in the nest-box lid, it is possible to drop the pups in without alerting the colony. The pups should be returned when the colony has just been fed and is busy eating, because the mole-rats do not seem to panic as easily then. The advantage of returning pups each evening is that they retain the colony odor and can suckle from their mother and obtain cecotrophes (when they are old enough).

MAINTAINING AN ACTIVELY RECRUITING COLONY

A problem experienced in all our labs is that after a colony has successfully reared a few litters, the colony fails to rear all or most of its subsequent litters. After multiple failures, the colony sometimes ceases reproducing entirely. Some colonies appear to be better at pup care than others, but even colonies with a good track record tend to stop rearing pups. Reasons for this and solutions to the problem are still largely a matter of conjecture. One possible solution follows.

I suggested (chap. 13) that this problem may be exacerbated in captivity because a lack of predation enables colonies to grow (numerically) too rapidly. This leads to too much interference with the young pups, especially by the subadult animals, and eventually to the demise of the pups. In two of my colonies, numerical growth was dampened by the periodic removal of some pups and subadult animals. This led to a more sustained recruitment to the colonies and the best percentage pup-rearing success of any captive colonies.

IDENTIFICATION OF ANIMALS

Mole-rats can be marked at weaning by clipping unique combinations of toes. Since the toe numbers are not visible enough for behavioral work, the skin of adult (anesthetized) animals may be tattooed with a hand-held machine developed for tattooing laboratory animals. The needles are set to protrude 1–2 mm; black tattoo fluid is used. Young animals cannot be tattooed because the skin is too thin, and the needle may rupture the body wall. Tattoo marks can be made very large or else at several places on the body (sides, rump, head, and abdomen) to facilitate identification even in nest huddles. For short-term identification, marks can be made with a felt-tipped permanent marker. Depending on the colony size (and the amount of rubbing that occurs as animals pass each other), pen marks can last as long as a week.

DETERMINATION OF SEX

The genitalia of both sexes, except those of the breeding female, are very similar, making sexing difficult. It is particularly hard to sex newborn pups accurately. The major difference between adult males and females is the presence of a thin, often darker, red line between the anus and clitoris of the female. This line can be more clearly seen if the area is dampened. The line indicates the position of the vagina, which is sealed (imperforate) in most colony members. In the breeding female, the vagina is frequently open (perforate), and the nipples of the mammary glands are prominent. This latter feature is not always a good indicator though, because in many captive colonies, all the mole-rats (both males and females) develop teats when the breeding female is in the later stages of pregnancy (Jarvis, chap. 13). A few nonbreeeding females in the colony may also become perforate just before the birth of the next litter; however, the breeding female usually has a longer body than other females (Jarvis et al., chap. 12).

ILLNESS, INJURIES, AND OTHER PROBLEMS

Laboratory-maintained naked mole-rats are generally healthy and have long life spans (see chap. 3, 13). Nonetheless, they do sometimes become sick. In all illnesses, one should be aware that once a sick animal is removed from its colony, it will be very difficult to return it. Preferably, the whole colony, not just those showing the symptoms, should be treated if the disease is suspected to be infectious.

The treatments listed below are not intended to be comprehensive but have been found to be effective for commonly observed ailments.

SKIN CARE

Dry skin in an entire colony is a sign that the humidity is too low. If the complaint is confined to one or two animals, it may indicate illness. Dry skin looks scaly, feels tough and thick, loses its plasticity, and cracks. If dry skin is neglected, scabs may form and fungal infections develop.

If the dry skin is widespread in the colony, it can usually be rectified by changing the humidity of the room or the degree of ventilation of the nest chamber (see above). However, if the animals are severely affected, the skin condition can be improved by applying skin creams. This, however should be used only as a last resort because it will affect colony odor; furthermore, the whole colony should be treated (otherwise foreign-smelling, treated colony members may be attacked when reintroduced). Choose a hand cream that is rapidly absorbed and leave the greased animals in a container until all the cream has been absorbed.

Scabs present on the back, especially around the rump and tail, may indicate a fungal infection. Regular (ca. twice weekly) swabbing of the skin with gentian violet solution should cure this problem. Only the affected animals need to be treated since the medication has little odor.

GASTROINTESTINAL AND RESPIRATORY PROBLEMS

These are often interlinked in that a sick animal frequently shows secondary signs of respiratory distress. *Escherichia coli* infections result in acute gastric symptoms with the abdominal region becoming greatly distended on one or both sides. The mole-rats have difficulty in breathing, frequently wipe their noses, and if neglected, die. Oral administration of antibiotics is not recommended because this will also kill the commensal gut fauna needed for cellulose digestion. Preparations containing tetracyclines (such as Engemycin) are well tolerated by naked mole-rats, and I administer them twice daily by subcutaneous injection. Because of the small size of the mole-rats and the difficulty of injecting small amounts with any precision, enough antibiotic for the whole colony can be diluted in saline solution or an electrolyte mixture such as half-strength Darrow's solution, with 5% dextrose (Baxter Travenol Labs., Inc.), and a larger quantity can then be injected. This dilution has an added advantage for sick mole-rats that are not eating and are dehydrated: the electrolyte solution will often keep them alive until the antibiotic can act. As much as 1 ml can be injected twice daily; it forms a large bubble under the skin and is slowly absorbed. *Escherichia coli* infections should not be confused with a pot-bellied stage that most of the weaning pups go through (see chap. 13). This needs no treatment.

Sometimes a change in diet will lead to bloat and the condition can be alleviated by reverting to the original diet and by adding to a cereal supplement veterinary preparations used in treating bloat and faunal imbalances in cattle (such as ones containing dioctol sodium sulphosuccinate).

PHYSICAL INJURIES

Electrolyte/glucose solution on its own can be used to keep injured animals alive until they have recovered enough to begin to eat again. Physical injuries should be treated promptly. The wounds can be bathed with a mixture of gentian violet and Mercurochrome (avoid getting the latter into the animal's mouth), and the mole-rat should be given fluids subcutaneously if it appears dehydrated (e.g., has dry thick skin, and its body cavity is caved in just anterior to the hind limbs). If the wounds become infected, it may be necessary to administer antibiotics; often however, prompt application of the above solutions is all that is necessary.

If the injuries have been sustained through intracolonial fighting, the injured animal often refuses to eat and apparently loses all incentive to live. In these circumstances, it is not unusual to find the animal sitting or lying limply in the toilet, or (rarely) in the nest area, and unless removed from the colony, it dies within a few days. These ostracized animals may have only superficial injuries. When the breeding female encounters these animals, she invariably attacks them (see Lacey et al., chap. 8).

If a sick or injured animal is removed from its colony, it is important to keep it warm and its skin moist. I place the animal in a small aquarium, give it a small tube in which to sit and place a lamp against one side of the aquarium; the mole-rat can then select its preferred temperature. The top of the aquarium is covered with a piece of glass, leaving a small gap for gaseous exchange; a container of water is placed next to this gap to keep the humidity high. As soon as the animal begins to recover, it should be given a companion, preferably of the opposite sex (from the same or another colony). The pair can then be used to form a new colony. If the second animal is not accepted immediately, interchanging the soiled bedding for a couple of days will often familiarize the mole-rats with each other's odors and facilitate acceptance (see above).

ORAL THRUSH

Oral thrush can be a problem in newly weaned pups. The pup has difficulty in swallowing, becomes very thin, and frequently raises its head and makes chewing and yawning movements. If neglected, the pup rapidly becomes dehydrated and dies. Thrush can be treated by washing the mouth out with dilute gentian violet solution. Twice a day I twist a small amount of cotton around the sharp point of a toothpick, dip this into the gentian violet solution and gently insert the tip into the pup's mouth and rub it around the oral cavity. If the pup is very weak, it may also be necessary to inject a small bubble of electrolyte solution subcutaneously until the pup begins to eat again. If the pup is eating cooked high-protein cereals, gentian violet can be mixed in, and this will help contain the infection. In my experience, oral thrush is only a problem among pups from newly captured colonies.

BROKEN INCISORS AND ABNORMAL INCISOR GROWTH

Mole-rats occasionally break their incisors. If both the upper or lower incisors break, the animal cannot bite off pieces of food until the teeth have regrown (ca. 1 wk). Finely grated food, bananas, and baby cereals should be given to the injured mole-rat and, if necessary, these can be supplemented with injections of electrolyte mixture. It may also be necessary to trim the other incisors, which will grow very long in the absence of wear and honing by the broken pair. If untrimmed, these long incisors will also keep the regrowing incisors short. I use a small electric drill with a circular cutting disc (ca. 5 mm in diameter) to trim and shape the incisors.

Undamaged incisors are kept in trim by the mole-rats, who hone their teeth frequently, especially when resting in the nest huddle (see Lacey et al., chap. 8). This honing, together with gnawing on edges and corners of their tunnels (and on commercial dog treats), is usually sufficient to keep the incisors from growing too long.

ANESTHETICS

I use Halothane (Fluothane) to anesthetize naked mole-rats. This is successful the first time as an anesthetic, but if it is used repeatedly, the animals will sometimes hold their breath and die of anoxia rather than succumb to the anesthetic (Buffenstein, pers. comm.). The anesthetic should be administered in a flow-through chamber to ensure that the mole-rats obtain a good supply of oxygen. The anesthetized animals must be kept warm.

Acknowledgments

I thank the Council for Scientific and Industrial Research (South Africa) and the University of Cape Town for funding my research and the National Geographic Society for a grant to collect mole-rats in Kenya. My gratitude also goes to J. Booysen, who has for many years helped me maintain the mole-rats, and to the staff of the Cape Veterinary Hospital, Rondebosch, who have advised me on treating sick animals. I thank P. W. Sherman for commenting on the manuscript and for sharing ideas and information.

Literature Cited

Abbott, D. H. 1984. Behavioral and physiological suppression of fertility in subordinate marmoset monkeys. *Am. J. Primatol.* 6:169–186.

———. 1987. Behaviorally mediated suppression of reproduction in female primates. *J. Zool. (Lond.)* 213:455–470.

Abbott, D. H., and J.P. Hearn. 1978. Physical, hormonal and behavioral aspects of sexual development in the marmoset monkey, *Callithrix jacchus. J. Reprod. Fertil.* 53:155–166.

Abbott, D. H., J. K. Hodges, and L. M. George. 1988. Social status controls LH secretion and ovulation in female marmoset monkeys (*Callithrix jacchus*). *J. Endocrinol.* 117:329–339.

Abbott, D. H., J. Barrett, C. G. Faulkes, and L. M. George. 1989. Social contraception in naked mole-rats and marmoset monkeys. *J. Zool. (Lond.)* 219: 703-710.

Alexander, R. D. 1974. The evolution of social behavior. *Annu. Rev. Ecol. Syst.* 5:325–383.

———. 1977. Natural selection and the analysis of human sociality. In *Changing Scenes in the Natural Sciences: 1776–1976.* C. E. Goulden, ed., pp. 283–337. Bicentennial Symposium Monograph, Spec. Publ. 12. Philadelphia: Philadelphia Academy of Natural Sciences.

———. 1979. *Darwinism and Human Affairs.* Seattle: University of Washington Press.

———. 1987. *The Biology of Moral Systems.* Hawthorne, N.Y.: Aldine De Gruyter.

———. 1989. The evolution of the human psyche. In *The Human Revolution: Behavioral and Biological Perspectives on the Origins of Modern Humans.* P. Mellars and C. Stringer, eds., pp. 455–513. Edinburgh: University of Edinburgh Press.

———. In press. How did humans evolve? Reflections on the uniquely unique species. *Univ. Mich. Mus. Zool. Misc. Publ.*

Allen, G. M. 1912. New African rodents. *Bull. Mus. Comp. Zool.* 54:437–447.

———. 1939. A checklist of African mammals. *Bull. Mus. Comp. Zool.* 83:1–763.

Altmann, J. 1974. Observational study of behavior: Sampling methods. *Behaviour* 49:227–267.

Andersson, M. 1984. The evolution of eusociality. *Annu. Rev. Ecol. Syst.* 15:165–189.

Ansell, W.F.H. 1978. *The Mammals of Zambia.* Chilanga, Zambia: National Parks and Wildlife Service.

Aoki, S. 1977. *Colophina clematis* (Homoptera, Pemphigidae) an aphid species with "soldiers." *Kontyû* 45:276–282.

———. 1979. Further observations on *Astegopteryx styracicola* (Homoptera: Pemphigidae), an aphid species with soldiers biting man. *Kontyû* 47:99–104.

———. 1982. Soldiers and altruistic dispersal in aphids. In *The Biology of Social Insects.* M. D. Breed, C. D. Michener, and H. E. Evans, eds., pp. 154–158. Boulder, Colo.: Westview Press.

Ar, A., R. Arieli, and A. Shkolnik. 1977. Blood-gas properties and function in the fossorial mole-rat under normal and hypoxic-hypercapnic atmospheric conditions. *Respir. Physiol.* 30:201–218.

Arieli, R. 1979. The atmospheric environment of the fossorial mole-rat (*Spalax ehrenbergi*): Effects of season, soil texture, rain, temperature and activity. *Comp. Biochem. Physiol. A, Comp. Physiol.* 63:569–575.

Arieli, R., A. Ar, and A. Shkolnik. 1977. Metabolic responses of a fossorial rodent (*Spalax ehrenbergi*) to simulated burrow conditions. *Physiol. Zool.* 50:61–75.

Aspey, W. P. 1977. Wolf spider sociobiology: I. Agonistic display and dominance-subordinance relations in adult male *Schizocosa crassipes. Behaviour* 62:103–141.

Avise, J. C., C. Giblin-Davidson, J. Laerm, J. C. Patton, and R. A. Lansman. 1979. Mitochondrial DNA clones and matriarchal phylogeny within and among geographic populations of the pocket gopher, *Geomys pinetis. Proc. Natl. Acad. Sci. U.S.A.* 76:6694–6698.

Avise, J. C., J. Arnold, R. M. Ball, E. Bermingham, T. Lamb, J. E. Neigel, C. A. Reeb, and N. C. Saunders. 1987. Intraspecific phylogeography: The mitochondrial DNA bridge between population genetics and systematics. *Annu. Rev. Ecol. Syst.* 18:489–522.

Bakker, E. M. Van Zinderen. 1967. The "Arid Corridor" between southwest Africa and the Horn of Africa. In *Paleoecology of Africa.* Vol. 2. E. M. Van Zinderen Bakker, ed., pp. 76–79. Amsterdam: A. A. Balkema.

Balinsky, B. I. 1962. Patterns of animal distribution on the African continent. *Ann. Cape Prov. Mus.* 2:299–310.

Barnes, R. H., G. Fiala, and E. Kwong. 1963. Decreased growth rate resulting from prevention of coprophagy. *Fed. Proc.* 22:125–133.

Bartz, S. H. 1979. Evolution of eusociality in termites. *Proc. Natl. Acad. Sci. U.S.A.* 76:5764–5768.

Batzli, G. O., L. L. Getz, and S. S. Hurley. 1977. Suppression of growth and reproduction of microtine rodents by social factors. *J. Mammal.* 58:583–591.

Bekoff, M. 1977. Quantitative studies of three areas of classical ethology: Behavioral dominance, behavioral taxonomy, and behavioral variability. In *Quantitative Methods in the Study of Animal Behavior.* B. A. Hazlett, ed., pp. 1–46. New York: Academic Press.

Bennett, N. C. 1988. The trend towards sociality in three species of Southern African mole-rats (Bathyergidae): Causes and consequences. Ph.D. diss., University of Cape Town.

———. 1989. The social structure and reproductive biology of the common mole-rat, *Cryptomys h. hottentotus* and remarks on the trends in reproduction and sociality in the family Bathyergidae. *J. Zool. (Lond.)* 219:45–59.

———. 1990. Behavior and social organization in a colony of the Damaraland mole-rat *Cryptomys damarensis. J. Zool. (Lond.)* 220:225–228.

Bennett, N. C., and J.U.M. Jarvis. 1988a. The reproductive biology of the Cape mole-rat, *Georychus capensis* (Rodentia, Bathyergidae). *J. Zool. (Lond.)* 214:95–106.

———. 1988b. The social structure and reproductive biology of colonies of the mole-rat, *Cryptomys damarensis* (Rodentia, Bathyergidae). *J. Mammal.* 69:293–302.

Bennett, N. C., J.U.M. Jarvis, and K. C. Davies. 1988. Daily and seasonal temperatures in the burrows of African rodent moles. *S. Afr. J. Zool.* 23:189–195.

Bennett, N. C., J.U.M. Jarvis, and D. B. Wallace. 1990. The relative age structure and body masses of complete wild-captured colonies of two social mole-rats, the common mole-rat, *Cryptomys hottentotus hottentotus* and the Damaraland mole-rat, *Cryptomys damarensis. J. Zool. (Lond.)* 220:469–485.

Berreman, G. D. 1962. Pahari polyandry: a comparison. *Am. Anthropol.* 64:60–75.

Bertram, B.C.R. 1975. Social factors influencing reproduction in wild lions. *J. Zool. (Lond.)* 177:463–482.

Betts, B. J. 1976. Behavior in a population of Columbian ground squirrels, *Spermophilus columbianus columbianus. Anim. Behav.* 24:652–680.

Bibb, M. J., R. A. Van Etten, C. T. Wright, M. W. Walberg, and D. A. Clayton. 1981. Sequence and gene organization of mouse mitochondrial DNA. *Cell* 26:167–180.

Bonney, R. C., D. J. Wood, and D. G. Kleiman. 1982. Endocrine correlates of behavioral oestrus in the female giant panda (*Ailuropoda melanoleuca*) and associated hormonal changes in the male. *J. Reprod. Fertil.* 64:209–215.

Borgia, G. 1980. Evolution of haplodiploidy: Models for inbred and outbred systems. *Theor. Popul. Biol.* 17:103–128.

Borror, D. J., and D. M. Delong. 1964. *An Introduction to the Study of Insects.* Rev. ed. New York: Holt, Rinehart, & Winston.

Bottego, V. 1896. Map and account of expedition to Italian Somaliland. *Boll. Soc. Geogr. Ital.* II: 68, and *Memorie* VI (1):149–170.

Bourke, A.F.G. 1988. Worker reproduction in the higher eusocial Hymenoptera. *Q. Rev. Biol.* 63:291–311.

Brandt, J. F. 1855. Beitrage zur nahern Kentnis der Saugethiere Russlands. *Mem. Acad. Imp. St. Petersbourg* 9:1–375.

Braude, S. H. 1983. Digging and associated behaviors of the naked mole-rat, *Heterocephalus glaber.* Undergraduate honors diss., University of Michigan, Ann Arbor.

Breed, M. D. 1975. Sociality and seasonal size variation in halictine bees. *Insectes Soc.* 22:375–380.

———. 1976. The evolution of social behavior in primitively social bees: a multivariate analysis. *Evolution* 30:234–240.

Breed, M. D., and G. J. Gamboa. 1977. Behavioral control of workers by queens in primitively eusocial bees. *Science (Wash., D.C.)* 195:694–696.

Brett, R. A. 1985. Captive breeding and management of naked mole-rats *Heterocephalus glaber. Proc. Symp. Assoc. Br. Wild Animal Keepers* 10:6–11.

———. 1986. The ecology and behavior of the naked mole-rat (*Heterocephalus glaber* Rüppell) (Rodentia: Bathyergidae). Ph.D. diss., University of London.

Broadley, D. G. 1983. *FitzSimons' Snakes of Southern Africa.* Johannesburg: Delta Books.

Brockmann, H. J. 1984. The evolution of social behavior in insects. In *Behavioral Ecology: An Evolutionary Approach,* 2d ed. J. R. Krebs and N. B. Davies, eds., pp. 340–361. Oxford: Blackwell.

Brockmann, H. J., and R. Dawkins. 1979. Joint nesting in a digger wasp as an evolutionarily stable preadaptation to social life. *Behaviour* 71:203–245.

Brown, G. G., and M. V. Simpson. 1981. Intra- and interspecific variation of the mito-chondrial genome in *Rattus norvegicus* and *Rattus rattus*: Restriction enzyme analy-sis of variant mitochondrial DNA molecules and their evolutionary relationships. *Genetics* 97:125–143.

Brown, J. L. 1987. *Helping and Communal Breeding in Birds*. Princeton, N.J.: Prince-ton University Press.

Brown, W. M. 1980. Polymorphism in mitochondrial DNA of humans as revealed by restriction endonuclease analysis. *Proc. Natl. Acad. Sci. U.S.A.* 77:3605–3609.

Buckle, G. R. 1982. Queen-worker behavior and nestmate interactions in young colo-nies of *Lasioglossum zephyrum. Insectes Soc.* 29:125–137.

Buffenstein, R., J.U.M. Jarvis, L. A. Opperman, F. S. Hough, and F. P. Ross. 1989. Vitamin D metabolism and expression in chthonic naked mole-rats. *J. Bone Miner. Res.* 201:S119.

Burda, H. 1989. Reproductive biology (behavior, breeding, and postnatal development) in subterranean mole-rats, *Cryptomys hottentotus* (Bathyergidae). *Z. Säugetierkd.* 54:360–376.

Burnham, L. 1978. Survey of social insects in the fossil record. *Psyche (Camb., Mass.)* 85:85–133.

Capanna, E., and M. S. Merani. 1980. Karyotypes of Somalian rodent populations 1. *Heterocephalus glaber* Rüppell, 1842 (Mammalia Rodentia). *Monit. Zool. Ital.* N. Ser. 3:45–51.

Capranica, R. R., E. Nevo, and A.J.M. Moffat. 1974. Vocal repertoire of a subterranean rodent (*Spalax*). *J. Acoust. Soc. Am.* 55: 383 (abstr.).

Carlin, N. F., and B. Hölldobler. 1986. The kin recognition system of carpenter ants (*Camponotus* spp.). I. Hierarchical cues in small colonies. *Behav. Ecol. Sociobiol.* 19: 123–134.

———. 1987. The kin recognition system of carpenter ants (*Camponotus* spp.). II. Larger colonies. *Behav. Ecol. Sociobiol.* 20:209–217.

Carpenter, F. M. 1930. The fossil ants of North America. *Bull. Mus. Comp. Zool. Harv. Univ.* 70:1–66.

———. 1953. The geological history and evolution of insects. *Am. Sci.* 41:256–270.

Carpenter, J. M. In press. Phylogenetic relationships and the origin of social behavior in Vespidae. In *The Social Biology of Wasps*. K. G. Ross and R. W. Matthews, eds. Ithaca, N.Y.: Cornell University Press.

Carter, C. S., L. L. Getz, and M. Cohen-Parsons. 1986. Relationships between social organisation and behavioral endocrinology in a monogamous mammal. *Adv. Study Behav.* 16:109–145.

Cei, G. 1946. L'occhio di *Heterocephalus glaber* Rüppell. Note anatomo-descrittive e istologiche. *Monit. Zool. Ital.* 55:89–96.

Charnov, E. L. 1978. Evolution of eusocial behavior: offspring choice or parental para-sitism? *J. Theor. Biol.* 75:451–465.

Chase, I. D. 1964. Models of hierarchy formation in animal societies. *Behav. Sci.* 19:374–382.

Cheney, D. L. 1977. The acquisition of rank and the development of reciprocal alliances among free-ranging baboons. *Behav. Ecol. Sociobiol.* 2:303–318.

Clayton, R. N., and K. J. Catt. 1981. Gonadotrophin-releasing hormone receptors: Characterization, physiological regulation, and relationship to reproductive function. *Endocrinol. Rev.* 2:186–209.

Cooke, H.B.S. 1972. The fossil mammal fauna of Africa. In *Evolution, Mammals, and Southern Continents.* A. Keast, F. C. Erk, and B. Glass, eds., pp. 89–139. Albany: State University of New York Press.

Coppola, D. M., and J. G. Vandenbergh. 1987. Induction of a puberty-regulating chemosignal in wild mouse populations. *J. Mammal.* 68:86–91.

Cowan, D. 1978. Behavior, inbreeding, and parental investment in solitary eumenid wasps (Hymenoptera: Vespidae). Ph.D. diss., University of Michigan, Ann Arbor.

Cowley, J. J., and D. R. Wise. 1972. Some effects of mouse urine on neonatal growth and reproduction. *Anim. Behav.* 20:499–506.

Craig, R. 1979. Parental manipulation, kin selection, and the evolution of altruism. *Evolution* 33:319–334.

———. 1980. Sex ratio changes and the evolution of eusociality in the Hymenoptera: Simulation and games theory studies. *J. Theor. Biol.* 87:55–70.

———. 1982. Evolution of male workers in the Hymenoptera. *J. Theor. Biol.* 94:95–105.

Crozier, R. H., and P. Luykx. 1985. The evolution of termite eusociality is unlikely to have been based on a male-haploid analogy. *Am. Nat.* 126:867–869.

Darwin, C. R. 1859 (1967). *On the Origin of Species: A Facsimile of the First Edition with an Introduction by Ernst Mayr.* Cambridge, Mass.: Harvard University Press.

Davies, K. C., and J.U.M. Jarvis. 1986. The burrow systems and burrowing dynamics of the mole-rats *Bathyergus suillus* and *Cryptomys hottentotus* in the fynbos of the south-western Cape, South Africa. *J. Zool. (Lond.)* 209:125–147.

Dawkins, R. 1976. *The Selfish Gene.* Oxford: Oxford University Press.

———. 1979. Twelve misunderstandings of kin selection. *Z. Tierpsychol.* 51:184–200.

De Beaux, 0. 1934. Mammiferi Raccolti dal Prof. G. Scortecci nella Somalia Italiana -1931 (*Heterocephalus glaber scortecci* subsp. nov.) *Atti Soc. Ital. Milano* 73:264–300.

De Feu, C., M. Hounsome, and I. Spence. 1983. A single-session mark/recapture method of population estimation. *Ringing & Migr.* 4:211–226.

De Graaff, G. 1960. A preliminary investigation of the mammalian microfauna in Pleistocene deposits of caves in the Transvaal system. *Palaeontol. Afr.* 7:59–118.

———. 1964a. A systematic revision of the Bathyergidae (Rodentia) of Southern Africa. Ph.D. diss., University of Pretoria.

———. 1964b. On the parasites associated with the Bathyergidae. *Koedoe* 7:113–123.

———. 1971. Family Bathyergidae. In *The Mammals of Africa: An Identification Manual.* J. Meester and H. W. Setzer, eds., pp. 1–5. Washington, D.C.: Smithsonian Institution Press.

———. 1979. Mole-rats (Bathyergidae, Rodentia) in South African national parks: Notes on the taxonomic "isolation" and hystricomorph affinities of the family. *Koedoe* 22:89–107.

———. 1981. *The Rodents of Southern Africa.* Pretoria: Butterworths.

Dempsey, E. W., J.U.M. Jarvis, and M. L. Purkerson. 1974. The location of sulphur in spermatozoa by energy dispersive X-ray analysis and scanning electron microscopy. *Scanning Electron Microsc.* 1974:631–638.

Densmore, L. D., J. W. Wright, and W. M. Brown 1985. Length variation and heteroplasmy are frequent in mitochondrial DNA from parthenogenetic and bisexual lizards (genus *Cnemidophorus*). *Genetics* 110:689–707.

Denys, C. 1985. Laetoli: A pliocene southern savanna fauna in the eastern Rift Valley (Tanzania). Ecological and zoogeographical implications. In *Proceedings of the International Symposium on African Vertebrates.* K.-L. Schuchmann, ed., pp. 35–51. Bonn: Selbstverlag.

Dew, H. E. 1983. Division of labor and queen influence in laboratory colonies of *Polistes metricus* (Hymenoptera; Vespidae). *Z. Tierpsychol.* 61:127–140.

Dittus, W. 1988. An analysis of toque macaque cohesion calls from an ecological perspective. In *Primate Vocal Communication.* D. Todt, P. Goedeking, and D. Symmes, eds., pp. 31–50. Berlin: Springer-Verlag.

Dixon, W. J., ed. 1983. *BMDP Statistical Software.* Berkeley: University of California Press.

Downing, H. A., and R. L. Jeanne. 1985. Communication of status in the social wasp, *Polistes fuscatus* (Hymenoptera: Vespidae). *Z. Tierpsychol.* 67:78–96.

Drake-Brockman, R. E. 1910. *The Mammals of Somaliland.* London: Hurst & Blackett.

Draper, N. R., and H. Smith. 1966. *Applied Regression Analysis.* New York: Wiley.

Dubey, A. K., J. L. Cameron, R. A. Steiner, and T. M. Plant. 1986. Inhibition of gonadotrophin secretion in castrated male rhesus monkeys (*Macaca mulatta*) induced by dietary restriction: Analogy with the prepubertal hiatus of gonadotrophin release. *Endocrinology* 118:518–525.

Dunbar, R.I.M. 1980. Determinants and evolutionary consequences of dominance among female gelada baboons. *Behav. Ecol. Sociobiol.* 7:253–265.

Du Toit, J. T., J.U.M. Jarvis, and R. N. Louw. 1985. Nutrition and burrowing energetics of the Cape mole-rat *Georychus capensis. Oecologia (Berl.)* 66:81–87.

Eberhard, W. G. 1985. *Sexual Selection and Animal Genitalia.* Cambridge, Mass.: Harvard University Press.

Eickwort, G. C. 1981. Presocial insects. In *Social Insects.* Vol. 2. H. R. Hermann, ed., pp. 199–280. New York: Academic Press.

Eisenberg, J. F. 1974. The function and motivational basis of hystricomorph vocalizations. *Symp. Zool. Soc. Lond.* 34:211–244.

———. 1981. *The Mammalian Radiations.* Chicago: University of Chicago Press.

Eisenberg, J. F., and D. G. Kleiman. 1977. Communication in lagomorphs and rodents. In *How Animals Communicate.* T. A. Sebeok, ed., pp. 634–654. Bloomington: Indiana University Press.

Ellerman, J. R. 1940. *The Families and Genera of Living Rodents.* Vol. I. London: Trustees of the British Museum (Natural History).

Ellerman, J. R., T.C.S. Morrison-Scott, and R. W. Hayman. 1953. *Southern African Mammals, 1758–1951: A Reclassification.* London: Trustees of the British Museum of Natural History.

Eloff, G. 1958. The functional and structural degeneration of the eye of the South

African rodent moles, *Cryptomys bigalkei* and *Bathyergus maritimus*. *S. Afr. J. Sci.* 54:293–302.

Emlen, S. T. 1981. Altruism, kinship, and reciprocity in the white-fronted bee-eater. In *Natural Selection and Social Behavior*. R. D. Alexander and D. W. Tinkle, eds., pp. 217–230. New York: Chiron Press.

———. 1982. The evolution of helping. I, II. *Am. Nat.* 119:29–53.

———. 1984. Cooperative breeding in birds and mammals. In *Behavioral Ecology: An Evolutionary Approach*. 2d ed. J. R. Krebs and N. B. Davies, eds., pp. 305–339. Oxford: Blackwell.

Emlen, S. T., and P. H. Wrege. 1986. Forced copulations and intra-specific parasitism: Two costs of social living in the white-fronted bee-eater. *Ethology* 71:2–29.

———. 1988. The role of kinship in helping decisions among white-fronted bee-eaters. *Behav. Ecol. Sociobiol.* 23:305–315.

Estes, R. D., and J. Goodard. 1967. Prey selection and hunting behavior of the African wild dog. *J. Wildl. Manage.* 31:52–70.

Evans, H. E. 1958. The evolution of social life in wasps. *Proc. 10th Int. Congr. Entomol.* 2:449–457.

———. 1977. Extrinsic versus intrinsic factors in the evolution of insect sociality. *BioScience* 27:613–617.

Evans, H. E., and M. J. West-Eberhard. 1970. *The Wasps*. Ann Arbor: University of Michigan Press.

Fain, A. 1968. Un hypope de la famille Hypoderidae Murray 1877 vivant sous la peau d'un rongeur (Hypoderidae: Sarcoptiformes). *Acarologia* 10:111–115.

Faulkes, C. G. 1990. Social suppression of reproduction in the naked mole-rat, *Heterocephalus glaber*. Ph.D. diss., University of London.

Faulkes, C. G., D. H. Abbott, and J.U.M. Jarvis. 1989a. Reproductive suppression in female naked mole-rats, *Heterocephalus glaber*. In *Comparative Reproduction in Mammals and Man*. Proceedings of the N.C.R.R. Conference Nairobi, November, 1987. R. M. Eley, ed., pp. 155–161. Nairobi: Institute for Primate Research/National Museums of Kenya.

Faulkes, C. G., D. H. Abbott, J.U.M. Jarvis, and F. E. Sherriff. 1989b. Social suppression of reproduction in male naked mole-rats, *Heterocephalus glaber*. *J. Reprod. Fertil. Abstr. Ser.* 3:113.

Faulkes, C. G., D. H. Abbott, and J.U.M. Jarvis. 1990a. Social suppression of ovarian cyclicity in captive and wild colonies of naked mole-rats, *Heterocephalus glaber*. *J. Reprod. Fertil.* 88:559–568.

Faulkes, C. G., D. H. Abbott, J.U.M. Jarvis, and F. E. Sherriff. 1990b. LH responses of female naked mole-rats, *Heterocephalus glaber*, to single and multiple doses of exogenous GnRH. *J. Reprod. Fertil.* 89:317–323.

Fawcett, D. W., W. B. Neaves, and M. N. Flores. 1973. Comparative observations on the intertubular lymphatics and the organization of the interstitial tissue of the mammalian testis. *Biol. Reprod.* 9:500–532.

Fisher, R. A. 1930. *The Genetical Theory of Natural Selection*. 2d ed., 1958. New York: Dover.

FitzSimons, V.F.M. 1962. *Snakes of Southern Africa*. Cape Town: Purnell and Sons.

Flinn, M. V. 1981. Uterine vs. agnatic kinship variability and associated cousin marriage preferences: an evolutionary biological analysis. In *Natural Selection and Social Behavior*. R. D. Alexander and D.W. Tinkle, eds., pp. 439–475. New York: Chiron Press.

Foltz, D. W., and J. L. Hoogland. 1981. Analysis of the mating system in the black-tailed prairie dog (*Cynomys ludovicianus*) by likelihood of paternity. *J. Mammal.* 62:706–712.

Frame, L. H., J. R. Malcolm, G. W. Frame, and H. van Lawick. 1979. Social organization of African wild dogs (*Lycaon pictus*) on the Serengeti Plains, Tanzania, 1967–1978. *Z. Tierpsychol.* 50:225–249.

Francaviglia, M. C. 1896. La Pelle dell *Heterocephalus glaber* Rüppell. *Boll. Soc. Zool. Ital.* 5:1–10.

Francis, A. M., J. P. Hailman, and G. E. Woolfenden. 1989. Mobbing by Florida scrub jays: Behavior, sexual asymmetry, role of helpers and ontogeny. *Anim. Behav.* 38:795–816.

French, J. A., D. H. Abbott, and C. S. Snowdon. 1984. The effect of social environment on estrogen excretion, scent marking, and sociosexual behavior in tamarins (*Saguinus oedipus*). *Am. J. Primatol.* 6:155–167.

Gamboa, G. J. 1978. Intraspecific defense: advantage of social cooperation among paper wasp foundresses. *Science (Wash., D.C.)* 199:1463–1465.

Gay, F. J. 1968. Soldier-reproductive intercastes in a species of *Tumulitermes* (Isoptera: Termitidae). *J. Aust. Entomol. Soc.* 7:83–84.

Genelly, R. E. 1965. Ecology of the common mole-rat (*Cryptomys hottentotus*) in Rhodesia. *J. Mammal.* 46:647–665.

George, W. 1979. Conservatism in the karyotypes of two African mole-rats (Rodentia, Bathyergidae). *Z. Säugertierkd.* 44:278–285.

Ghiselin, M. T. 1974. *The Economy of Nature and the Evolution of Sex*. Berkeley: University of California Press.

Gibo, D. L. 1974. A laboratory study on the selective advantage of foundress associations in *Polistes fuscatus* (Hymenoptera: Vespidae). *Can. Entomol.* 106:101–106.

Glasier, A., A. S. McNeilly, and D. T. Baird. 1986. Induction of ovarian activity by pulsitile infusion of LHRH in women with lactational amenorrhea. *Clin. Endocrinol.* 24:243–252.

Godfrey, G. K. 1955. A field study of the activity of the mole (*Talpa europaea*). *Ecology* 36:678–685.

Godfrey, G. K., and P. Crowcroft. 1960. *The Life of the Mole (Talpa europaea Linnaeus)*. London: Museum Press.

Gould, E. 1983. Mechanisms of mammalian auditory communication. In *Advances in the Study of Mammalian Behavior*. J. F. Eisenberg and D. G. Kleiman, eds., pp. 265–342. Am. Soc. Mammal. Spec. Publ. 7. Shippensburg, Pa.: American Society of Mammalogists.

Gouzoules, H., and S. Gouzoules. 1989. Design features and developmental modification of pigtail macaque, *Macaca nemestrina*, agonistic screams. *Anim. Behav.* 37:383–401.

Grafen, A. 1984. Natural selection, kin selection and group selection. In *Behavioral*

Ecology: An Evolutionary Approach. 2d ed. J. R. Krebs and N. B. Davies, eds., pp. 62–84. Oxford: Blackwell.

Greene, P. J. 1978. Promiscuity, paternity, and culture. *Am. Ethnol.* 5:151–159.

Greenway, P. J. 1969. A checklist of plants collected in Tsavo National Park. *J. East Afr. Nat. Hist. Soc. Natl. Mus.* 27:169–209.

Gyllensten, U. B., and H. A. Erlich. 1988. Generation of single-stranded DNA by the polymerase chain reaction and its application to direct sequencing of the HLA-DQA locus. *Proc. Natl. Acad. Sci. U.S.A.* 85:7652–7656.

Hafner, M. S., and L. J. Barkley. 1984. Genetics and natural history of a relictual pocket gopher, *Zygogeomys* (Rodentia: Geomyidae). *J. Mammal.* 65: 474–479.

Hafner, M. S., J. C. Hafner, J. L. Patton, and M. F. Smith. 1987. Macrogeographic patterns of genetic differentiation in the pocket gopher *Thomomys umbrinus. Syst. Zool.* 36:18–34.

Hagen, L. 1985. Metrische untersuchungen an *Heterocephalus glaber* Rüppell 1842. einer sozialen nagetierat. M.Sc. diss., Christian Albrechts University.

Haigh, G. R., B. S. Cushing, and F. H. Bronson. 1988. A novel postcopulatory block of reproduction in white-footed mice. *Biol. Reprod.* 38:623–626.

Haigh, G. R., D. M. Lounsbury, and T. A. Gordon. 1985. Pheromone-induced reproductive inhibition in young female *Peromyscus leucopus. Biol. Reprod.* 33:271–276.

Haim, A., and N. Fairall. 1986. Physiological adaptations to the subterranean environment by the mole-rat *Cryptomys hottentotus. Cimbebasia Ser. A* 8:49–53.

Haim, A., G. Heth, and E. Nevo. 1985. Adaptive thermoregulatory patterns in speciating mole-rats. *Acta Zool. Fenn.* 170:174–184.

Haldane, J.B.S. 1932. *The Causes of Evolution.* London: Longmans, Green. Reprinted 1966. Ithaca, N.Y.: Cornell University Press.

————. 1955. Population genetics. *New Biol.* 18:34–51.

Hamilton, A. C. 1982. *Environmental History of East Africa: A Study of the Quaternary.* London: Academic Press.

Hamilton, W. D. 1964. The genetical evolution of social behavior. I, II. *J. Theor. Biol.* 7:1–52.

————. 1966. The moulding of senescence by natural selection. *J. Theor. Biol.* 12:12–45.

————. 1967. Extraordinary sex ratios. *Science (Wash., D.C.)* 156:477–488.

————. 1971. Geometry for the selfish herd. *J. Theor. Biol.* 31:295–311.

————. 1972. Altruism and related phenomena, mainly in social insects. *Annu. Rev. Ecol. Syst.* 3:193–232.

————. 1976. Haldane and altruism. *New Sci.* 71: 40.

————. 1978. Evolution and diversity under bark. *Symp. R. Entomol. Soc. Lond.* 9: 154–175.

————. 1987. Kinship, recognition, disease, and intelligence: constraints of social evolution. In *Animal Societies: Theories and Facts.* Y. Ito, J. L. Brown, and J. Kikkawa, eds., pp. 81–102. Tokyo: Japan Scientific Society Press.

Hamilton, W. J., Jr. 1928. *Heterocephalus,* the remarkable African burrowing rodent. *Brooklyn Mus. Sci. Bull.* 3:173–184.

Hamilton, W. J., III. 1973. *Life's Color Code.* New York: McGraw-Hill.

Hanken, J., and P. W. Sherman 1981. Multiple paternity in Belding's ground squirrel litters. *Science (Wash., D.C.)* 212:351–353.

Harcourt, A. H. 1988. Alliances in contests and social intelligence. In *Machiavellian Intelligence*. R. Byrne and A. Whiten, eds., pp. 132–152. Oxford: Clarendon Press.

Harlow, C. R., S. Gems, J. K. Hodges, and J. P. Hearn. 1983. The relationship between plasma progesterone and the timing of ovulation and early embryonic development in the marmoset monkey (*Callithrix jacchus*). *J. Zool. (Lond.)* 201:273–282.

Harris, H., and D. A. Hopkinson. 1976. *Handbook of Enzyme Electrophoresis in Human Genetics*. Amsterdam: North-Holland.

Haverty, M. I. 1977. The proportion of soldiers in termite colonies: A list and a bibliography (Isoptera). *Sociobiology* 2:199–216.

Heap, R. B., and A.P.F. Flint. 1984. Pregnancy. In *Reproduction in Mammals: 3. Hormonal Control of Reproduction*. 2d ed. C. R. Austin and R. V. Short, eds., pp. 153–194. Cambridge: Cambridge University Press.

Heatwole, H. 1976. *Reptile Ecology*. St. Lucia, Queensland: University of Queensland Press.

Heinrich, B. 1976. The foraging specializations of individual bumble bees. *Ecol. Monogr.* 46:105–128.

———. 1979. *Bumblebee Economics*. Cambridge, Mass.: Harvard University Press.

Helwig, J. T., and K. A. Council. 1979. *SAS User's Guide*. Cary, N.C.: SAS Institute Inc.

Hermann, H. R. 1979. *Social Insects*. Vols. 1, 2. New York: Academic Press.

Hewitt, P. H., J.J.C. Nel, and S. Conradie. 1969. Preliminary studies on the control of caste formation in the harvester termite *Hodotermes mossambicus* (Hagen). *Insectes Soc.* 16:159–172.

Hickman, G. C. 1978. Reactions of *Cryptomys hottentotus* to water (Rodentia: Bathyergidae). *Zool. Afr.* 13:319–328.

———. 1979a. A live-trap and trapping technique for fossorial mammals. *S. Afr. J. Zool.* 14:9–12.

———. 1979b. Burrow system structure of the bathyergid *Cryptomys hottentotus* in Natal, South Africa. *Z. Säugetierkd.* 44:153–162.

———. 1982. Copulation of *Cryptomys hottentotus* (Bathyergidae), a fossorial rodent. *Mammalia* 46:293–298.

———. 1983. Swimming ability of a naked mole-rat, *Heterocephalus glaber*. *Mammalia* 47:267–269.

Hill, J. E., and T. D. Carter. 1941. The mammals of Angola, Africa. *Bull. Am. Mus. Nat. Hist.* 78:1–211.

Hill, W.C.O. 1953. Notes and exhibitions. *Proc. Zool. Soc. Lond.* 123:711–712.

Hill, W.C.O., A. Porter, R. T. Bloom, J. Seago, and M. D. Southwick. 1957. Field and laboratory studies on the naked mole-rat (*Heterocephalus glaber*). *Proc. Zool. Soc. Lond.* 128:455–513.

Hodges, J. K. 1986. Monitoring changes in reproductive status. *Int. Zoo Yearb.* 24/25:126–130.

Hodges, J. K., S.A.K. Eastman, and N. Jenkins. 1983. Sex steroids and their relationship to binding proteins in the serum of the marmoset monkey (*Callithrix jacchus*). *J. Endocrinol.* 96:443–450.

Hodges, J. K., P. Cottingham, P. M. Summers, and L. Yingnan. 1987. Controlled ovulation in the marmoset monkey (*Callithrix jacchus*) with human chorionic gonadotrophin following prostaglandin-induced luteal regression. *Fertil. Steril.* 48:299–305.

Hölldobler, B. 1987. Communication and competition in ant communities. In *Evolution and Coadaptation in Biotic Communities*. S. Kawano, J. H. Connell, and T. Hidaka, eds., pp. 95–124. Tokyo: University of Tokyo Press.

Hollister, N. 1919. East African mammals in the United States National Museum. *US Natl. Mus. Bull.* 99:159–160.

Honeycutt, R. L., and W. C. Wheeler. 1990. Mitochondrial DNA: Variation in humans and higher primates. In *DNA Systematics: Humans and Higher Primates*, vol. 3. S. K. Dutta and W. Winter, eds., pp. 91–129. Boca Raton, Fla.: CRC Press.

Honeycutt, R. L., and S. L. Williams. 1982. Genic differentiation in pocket gophers of the genus *Pappogeomys*, with comments on intergeneric relationships in the subfamily Geomyinae. *J. Mammal.* 63:208–217.

Honeycutt, R. L., S. V. Edwards, K. Nelson, and E. Nevo. 1987. Mitochondrial DNA variation and the phylogeny of African mole-rats (Rodentia: Bathyergidae). *Syst. Zool.* 36:280–292.

Hoogland, J. L. 1979. The effect of colony size on individual alertness of prairie dogs (Sciuridae: *Cynomys* spp.). *Anim. Behav.* 27:394–407.

Hoogland, J. L., and P. W. Sherman. 1976. Advantages and disadvantages of bank swallow (*Riparia riparia*) coloniality. *Ecol. Monogr.* 46:33–58.

Isil, S. 1983. A study of social behavior in laboratory colonies of the naked mole-rat (*Heterocephalus glaber* Rüppell; Rodentia, Bathyergidae). Master's thesis., University of Michigan, Ann Arbor.

Jarvis, J.U.M. 1969. The breeding season and litter size of African mole-rats. *J. Reprod. Fertil. (Suppl.)* 6:237–248.

———. 1973a. The structure of a population of mole-rats *Tachyoryctes splendens*, (Rodentia: Rhizomyidae). *J. Zool. (Lond.)* 171:1–14.

———. 1973b. Activity patterns in the mole-rats *Tachyoryctes splendens* and *Heliophobius argenteocinereus*. *Zool. Afr.* 8:101–119.

———. 1974. Notes on the golden mole, *Chrysocloris stuhlmanni* Matschie from the Ruwenzori Mountains, Uganda. *East Afr. Wild. J.* 12:163–166.

———. 1978. Energetics of survival in *Heterocephalus glaber* (Rüppell), the naked mole-rat (Rodentia: Bathyergidae). *Bull. Carnegie Mus. Nat. Hist.* 6:81–87.

———. 1981. Eusociality in a mammal: Cooperative breeding in naked mole-rat colonies. *Science (Wash., D.C.)* 212: 571–573.

———. 1982. Chemical control of colony reproductive state in naked mole-rats, *Heterocephalus glaber*. Abstract distributed at 3d International Theriological Congress, Helsinki.

———. 1984. African mole-rats. In *The Encyclopedia of Mammals*. D. Macdonald, ed., pp. 708–711. New York: Facts on File Publications.

———. 1985. Ecological studies on *Heterocephalus glaber*, the naked mole-rat, in Kenya. *Natl. Geogr. Soc. Res. Rep.* 20: 429–437.

Jarvis, J.U.M., and N. C. Bennett. 1990. The evolutionary history, population biology, and social structure of African mole-rats: Family Bathyergidae. In *Evolution of Sub-*

terranean Mammals at the Organismal and Molecular Levels. E. Nevo and O. A. Reig, eds., pp. 97–128. New York: Wiley–Liss.

Jarvis, J.U.M., and J. B. Sale. 1971. Burrowing and burrow patterns of East African mole-rats *Tachyoryctes, Heliophobius,* and *Heterocephalus. J. Zool. (Lond.)* 163: 451–479.

Jeanne, R. L., H. A. Downing, and D. C. Post. 1988. Age polyethism and individual variation in *Polybia occidentalis,* an advanced eusocial wasp. In *Interindividual Behavioral Variability in Social Insects.* R. L. Jeanne, ed., pp. 323–357. Boulder, Colo.: Westview Press.

Johansen, H.G.K., and G.M.O. Maloiy. 1977. Tissue metabolism and enzyme activities in the rodent *Heterocephalus glaber,* a poor temperature regulator. *Comp. Biochem. Physiol. B, Comp. Biochem.* 57:293–296.

Johansen, K., G. Lykkeboe, R. E. Weber, and G.M.O. Maloiy. 1976. Blood respiratory properties in the naked mole-rat *Heterocephalus glaber,* a mammal of low body temperature. *Resp. Physiol.* 28:303–314.

Kappleman, J. 1984. Plio-Pleistocene environments of bed I and lower bed II, Olduvai Gorge, Tanzania. *Palaeogeogr. Paleoclimatol. Palaeoecol.* 48:171–196.

Karsch, F. J., E. L. Bittman, D. L. Foster, R. L. Goodman, S. J. Legan, and J. E. Robinson. 1984. Neuroendocrine basis of seasonal reproduction. *Rec. Prog. Horm. Res.* 40:185–225.

Kaufmann, E. 1986. Vocalizations in the naked mole-rat *(Heterocephalus glaber).* Undergraduate honors thesis, Cornell University, Ithaca, N.Y.

Kayanja, F.I.B., and J.U.M. Jarvis. 1971. Histological observations on the ovary, oviduct and uterus of the naked mole-rat. *Z. Säugetierkd.* 36:114–121.

Kennedy, C. H. 1947. Child labor of the termite society versus adult labor of the ant society. *Sci. Monthly* 65:309–324.

Kennerly, T. E. 1964. Microenvironmental conditions of the pocket gopher burrow. *Texas J. Sci.* 16:395–441.

Keverne, E. B. 1983. Endocrine determinants and constraints on sexual behavior in monkeys. In *Mate Choice.* P.P.G. Bateson, ed., pp. 407–420. Cambridge: Cambridge University Press.

Keverne, E. B., J. A. Eberhart, U. Yodyingyuad, and D. H. Abbott. 1984. Social influences on sex differences in the behavior and endocrine state of talapoin monkeys. *Prog. Brain Res.* 61:331–347.

Kingdon, J. 1974. *East African Mammals, An Atlas of Evolution in Africa.* Vol. II, part B, *Hares and Rodents,* pp. 342–704. New York: Academic Press.

Kirkland, L. E., and E. L. Bradley. 1986. Reproductive inhibition and serum prolactin concentrations in laboratory populations of the prairie deermouse. *Biol. Reprod.* 35:579–586.

Kocher, T. D., W. K. Thomas, A. Meyer, S. V. Edwards, S. Paabo, F. X. Villablanca, and A. C. Wilson. 1989. Dynamics of mitochondrial DNA evolution in animals: Amplification and sequencing with conserved primers. *Proc. Natl. Acad. Sci. USA.* 86:6196–6200.

Koenig, W. D., and F. A. Pitelka. 1981. Ecological factors and kin selection in the evolution of cooperative breeding in birds. In *Natural Selection and Social Behavior.* R. D. Alexander and D. W. Tinkle, eds., pp. 261–280. New York: Chiron Press.

Kolmes, S. A. 1986. Age polyethism in worker honey bees. *Ethology* 71:252–255.

Krebs, D. 1989. Detecting genetic similarity without detecting genetic similarity. *Behav. Brain Sci.* 12:533–534.

Krebs, J. R. 1978. Optimal foraging: decision rules for predators. In *Behavioral Ecology: An Evolutionary Approach.* 1st ed. J. R. Krebs and N. B. Davies, eds., pp. 23–63. Oxford: Blackwell.

Krishna, K., and F. M. Weesner. 1969. *Biology of Termites.* Vols. 1, 2. New York: Academic Press.

Kruczek, M., and A. Marchlewska-Koj. 1986. Puberty delay of bank vole females in a high-density population. *Biol. Reprod.* 35:537–541.

Kukuk, P. F., G. C. Eickwort, M. Ravaret-Richter, B. Alexander, R. Gibson, R. A. Morse, and F. Ratnieks. 1989. Importance of the sting in the evolution of sociality in the Hymenoptera. *Ann. Entomol. Soc. Am.* 82:1–5.

Kurland, J. A. 1979. Paternity, mother's brother, and human sociality. In *Evolutionary Biology and Human Social Behavior: An Anthropological Perspective.* N. A. Chagnon and W. Irons, eds., pp. 145–180. North Scituate, Mass.: Duxbury Press.

Kuyper, M. A. 1979. A biological study of the golden mole (*Amblysomus hottentotus*). Master's thesis, University of Natal, Pietermaritzburg.

———. 1985. The ecology of the golden mole *Amblysomus hottentotus. Mammal. Rev.* 15:3–11

Lacy, R. C. 1980. The evolution of eusociality in termites: A haplodiploid analogy? *Am. Nat.* 116:449–451.

Laerm, J., J. C. Avise, J. C. Patton, and R. A. Lansman. 1982. Genetic determination of the status of an endangered species of pocket gopher in Georgia. *J. Wildl. Manage.* 46:513–518.

La Fage, J. P., and W. L. Nutting. 1978. Nutrient dynamics of termites. In *Production Ecology of Ants and Termites.* M. V. Brian, ed., pp. 165–232. Cambridge: Cambridge University Press.

Lansman, R. A., J. C. Avise, C. F. Aquadro, J. F. Shapira, and S. W. Daniel. 1983. Extensive genetic variation in mitochondrial DNA's among geographic populations of the deer mouse, *Peromyscus maniculatus. Evolution* 37:1–16.

Lasley, B. L. 1985. Methods for evaluating reproductive function in exotic species. *Adv. Vet. Sci. Comp. Med.* 30:209–228.

Lavocat, R. 1978. Rodentia and Lagomorpha. In *Evolution of African Mammals.* V. J. Maglio and H.B.S. Cooke, eds., pp. 69–89. Cambridge, Mass.: Harvard University Press.

Lawick, H. van. 1973. *Solo.* London: Collins.

Lawick, H. van, and J. van Lawick-Goodall. 1970. *Innocent Killers.* London: Collins.

Lawick-Goodall, J. van. 1971. *In the Shadow of Man.* Boston: Houghton-Mifflin.

Lawton, A. D., and J. M. Whitsett. 1979. Inhibition of sexual maturation by a urinary pheromone in male prairie deer mice. *Horm. and Behav.* 13:128–138.

Lee, S. van der, and L. M. Boot. 1955. Spontaneous pseudopregnancy in mice. *Acta Physiol. Pharmacol. Neerl.* 4:442–444.

Legan, S. J., and F. J. Karsch. 1979. Neuroendocrine regulation of the estrous cycle and seasonal breeding in the ewe. *Biol. Reprod.* 20:74–85.

Leger, D. W., S. D. Berney-Key, and P. W. Sherman. 1984. Vocalizations of Belding's ground squirrels (*Spermophilus beldingi*). *Anim. Behav.* 32:753–764.

Lepri, J. J. 1986. Familial chemosignals interfere with reproductive activation in female pine voles, *Microtus pinetorum*. In *Chemical Signals in Vertebrates 4: Ecology, Evolution, and Comparative Biology*. D. Duvall, D. Müller-Schwarze, and R. M. Silverstein, eds., pp. 555–560. London: Plenum Press.

Lepri, J. J., and J. G. Vandenbergh. 1986. Puberty in pine voles, *Microtus pinetorum*, and the influence of chemosignals on female reproduction. *Biol. Reprod.* 34:370–377.

Lin, N., and C. D. Michener. 1972. Evolution of sociality in insects. *Q. Rev. Biol.* 47:131–159.

Lincoln, G. A., and R. V. Short. 1980. Seasonal breeding: Nature's contraceptive. *Rec. Prog. Horm. Res.* 36:1–52.

Livingston, D. A. 1975. Late quaternary climatic change in Africa. *Annu. Rev. Ecol. Syst.* 6:249–280.

Lonnberg, E. 1912. Swedish zoological expedition to East Africa 1911 (*Heterocephalus glaber* progrediens subsp. nov.) Kungl. *Sven. Vetenskapsakademiens Handlingar* 48:102–105.

Lovegrove, B. G. 1986a. Thermoregulation of the subterranean rodent genus *Bathyergus* (Bathyergidae). *S. Afr. J. Zool.* 21:283–288.

–––––––. 1986b. The metabolism of social subterranean rodents: adaptation to aridity. *Oecologia (Berl.)* 69:551–555.

–––––––. 1987a. The energetics of sociality in the mole-rats (Bathyergidae). Ph.D. diss., University of Cape Town.

–––––––. 1987b. Thermoregulation in the subterranean rodent *Georychus capensis* (Rodentia: Bathyergidae). *Physiol. Zool.* 60:174–180.

–––––––. 1989. The cost of burrowing by the social mole-rats (Bathyergidae) *Cryptomys damarensis* and *Heterocephalus glaber*: The role of soil moisture. *Physiol. Zool.* 62:449–469.

Lovegrove, B. G., and J.U.M. Jarvis. 1986. Coevolution between mole-rats (Bathyergidae) and a geophyte, *Micranthus* (Iridaceae). *Cimbebasia Ser. A* 8:79–85.

Lovegrove, B. G., and S. Painting. 1987. Variations in the foraging behavior and burrow structures of the Damara mole-rat *Cryptomys damarensis* in the Kalahari Gemsbok National Park. *Koedoe* 30:149–163.

Lovegrove, B. G., and C. Wissel. 1988. Sociality in mole-rats: Metabolic scaling and the role of risk sensitivity. *Oecologia (Berl.)* 74:600–606.

Low, B. S. 1978. Environmental uncertainty and the parental strategies of marsupials and placentals. *Am. Nat.* 112:197–213.

Luckett, W. P. 1985. Superordinal and intraordinal affinities of rodents: developmental evidence from the dentition and placentation. In *Evolutionary Relationships among Rodents: A Multidisciplinary Analysis*. W. P. Luckett and J.- L. Hartenberger, eds., pp. 227–276. New York: NATO ASI Series, Plenum Press.

Luckett, W. P., and J.- L. Hartenberger. 1985. Evolutionary relationships among rodents: comments and conclusions. In *Evolutionary Relationships among Rodents: A Multidisciplinary Analysis*. W. P. Luckett and J.- L. Hartenberger, eds., pp. 685–712. New York: NATO ASI Series, Plenum Press.

Lüscher, M. 1972. Environmental control of juvenile hormone (JH) secretion and caste differentiation in termites. *Gen. Comp. Endocrinol.*, Suppl. 3:509–514.

–––––––, ed. 1977. *Phase and Caste Determination in Insects: Endocrine Aspects*. 15th Int. Congr. Entomol., Washington, D.C. New York: Pergamon Press.

Lush, J. L. 1947. Family merit and individual merit as bases for selection. *Am. Nat.* 81:241–261.

Lyne, A. G., and B. F. Short. 1965. *Biology of the Skin and Hair Growth*. New York: Elsevier.

Mabelis, A. A. 1979. Wood ant wars. *Neth. J. Zool.* 29:451–620.

McClintock, M. K. 1983a. Synchronizing ovarian and birth cycles by female pheromones. In *Chemical Signals in Vertebrates*. D. Müller-Schwarze and R. M. Silverstein, eds., pp. 159–178. New York: Plenum Press.

———. 1983b. Pheromonal regulation of the ovarian cycle: Enhancement, suppression and synchrony. In *Pheromones and Reproduction in Mammals*. J. G. Vandenbergh, ed., pp. 113–149. New York: Academic Press.

Macdonald, D. W. 1979. "Helpers" in fox society. *Nature (Lond.)* 282:69–71.

———. 1983. The ecology of carnivore social behavior. *Nature (Lond.)* 301:379–384.

Macdonald, D. W., and P. D. Moehlman. 1983. Cooperation, altruism, and restraint in the reproduction of carnivores. In *Perspectives in Ethology*. Vol. 5. P.P.G. Bateson and P. Klopfer, eds., pp. 433–466. New York: Plenum Press.

Mace, G. M., P. H. Harvey, and T. H. Clutton-Brock. 1981. Brain size and ecology in small mammals. *J. Zool. (Lond.)* 193:333–354.

McIntosh, T. K., and L. C. Drickamer. 1977. Excreted urine, bladder urine, and the delay of sexual maturation in female house mice. *Anim. Behav.* 25:999–1004.

McNab, B. K. 1966. The metabolism of fossorial rodents: A study of convergence. *Ecology* 47:712–733.

———. 1968. The influence of fat deposits on the basal rate of metabolism in desert homoiotherms. *Comp. Biochem. Physiol.* 26:337–343.

———. 1979. The influence of body size on the energetics and distribution of fossorial and burrowing mammals. *Ecology* 60:1010–1021.

———. 1980. On estimating thermal conductance in endotherms. *Physiol. Zool.* 53:145–156.

Maier, V., O.A.E. Rasa, and H. Scheich. 1983. Call-system similarity in a ground-living social bird and a mammal in the bush habitat. *Behav. Ecol. Sociobiol.* 12:5–9.

Maier, W., and F. Schrenk. 1987. The hystricomorphy of the Bathyergidae, as determined from ontogenetic evidence. *Z. Säugetierkd.* 52:156–164.

Malcolm, J. R., and K. Marten. 1982. Natural selection and the communal rearing of pups in African wild dogs (*Lycaon pictus*). *Behav. Ecol. Sociobiol.* 10:1–13.

Martin, R. D. 1975. General principles for breeding small mammals in captivity. In *Breeding Endangered Species in Captivity*. R. D. Martin, ed., pp. 143–166. London: Academic Press.

Maruyama, T., and M. Kimura. 1980. Genetic variability and effective population size when local extinction and recolonization of subpopulations are frequent. *Proc. Natl. Acad. Sci. U.S.A.* 77:6710–6714.

Massey, A., and J. G. Vandenbergh. 1980. Puberty delay by a urinary cue from female house mice in feral populations. *Science (Wash., D.C.)* 209:821–822.

Maynard Smith, J. 1964. Group selection and kin selection. *Nature (Lond.)* 201:1145–1147.

———. 1975. Survival through suicide. *New Sci.* 67:496–497.

Maynard Smith, J., and M. G. Ridpath. 1972. Wife sharing in the Tasmanian native hen, *Tribonyx mortierii*: a case of kin selection? *Am. Nat.* 106:447–452.

Mech, L. D. 1970. *The Wolf: The Ecology and Behavior of an Endangered Species.* New York: Natural History Press.

———. 1988. *The Arctic Wolf: Living with the Pack.* Stillwater, Minn.: Voyageur Press.

Medawar, P. B. 1957. *The Uniqueness of the Individual.* London: Methuen.

Metzger, L. H. 1971. Behavioral population regulation in the woodmouse, *Peromyscus leucopus. Am. Midl. Nat.* 86:434–448.

Michener, C. D. 1958. The evolution of social behavior in bees. *Proc. 10th Int. Congr. Entomol.* (Montreal, 1956) 2:441–447.

———. 1969. Comparative social behavior of bees. *Annu. Rev. Entomol.* 14:299–342.

———. 1974. *The Social Behavior of the Bees.* Cambridge, Mass.: Belknap Press of Harvard University Press.

———. 1985. From solitary to eusocial: need there be a series of intervening species? In *Experimental Behavioral Ecology and Sociobiology.* B. Hölldobler and M. Lindauer, eds., pp. 293–305. Sunderland, Mass.: Sinauer.

———. 1988. Reproduction and castes in social halictine bees. In *Social Insects: An Evolutionary Approach to Castes and Reproduction.* W. Engels, ed., pp. 75–115. Berlin: Springer-Verlag.

Michener, C. D., and D. J. Brothers. 1974. Were workers of eusocial Hymenoptera initially altruistic or oppressed? *Proc. Natl. Acad. Sci. U.S.A.* 71:671–674.

Michener, C. D., and D. A. Grimaldi. 1988. The oldest fossil bee: Apoid history, evolutionary stasis, and the antiquity of social behavior. *Proc. Natl. Acad. Sci. U.S.A.* 85:6424–6426.

Miller, E. M. 1969. Caste differentiation in the lower termites. In *Biology of Termites.* Vol. 1. K. Krishna and F. M. Weesner, eds., pp. 283–310. New York: Academic Press.

Moehlman, P. D. 1983. Socioecology of silverbacked and golden jackals (*Canis mesomelas* and *C. aureus*). In *Advances in the Study of Mammalian Behavior.* J. F. Eisenberg and D. G. Kleiman, eds., pp. 423–453. Am. Soc. Mammal. Spec. Publ. 7. Shippensburg, Pa.: American Society of Mammalogists.

———. 1986. Ecology of cooperation in canids. In *Ecological Aspects of Social Evolution.* D. I. Rubenstein and R. W. Wrangham, eds., pp. 64–86. Princeton, N.J.: Princeton University Press.

Moltz, H., and T. M. Lee. 1981. The maternal pheromone of the rat: Identity and functional significance. *Physiol. Behav.* 26:301–306.

Moon, T. W., T. Mustafa, and J. B. Jorgensen. 1981. Metabolism, tissue metabolites and enzyme activities in the fossorial mole-rat *Heterocephalus glaber. Molec. Physiol.* 1:179–194.

Murie, A. 1944. *The Wolves of Mount McKinley.* Washington, D.C.: U. S. Government Printing Office.

Mustafa, T., J. B. Jorgensen, and T. W. Moon. 1981. Isozyme patterns and effect of temperature on pyruvate kinase and lactate dehydrogenase from tissues of the fossorial mole-rat, *Heterocephalus glaber. Molec. Physiol.* 1:195–208.

Nalepa, C. A. 1988. Reproduction in the woodroach *Cryptocercus punctulatus* Scudder

(Dictyoptera: Cryptocercidae): Mating, oviposition, and hatch. *Ann. Entomol. Soc. Am.* 81:637–641.

Nanni, R. F. 1988. The interaction of mole-rats (*Georychus capensis* and *Cryptomys hottentotus*) in the Nottingham Road region of Natal. Master's thesis, University of Natal, Pietermaritzburg.

Naumova, H. I. 1974. Coprophagy of rodents and double-toothed rodents with special reference to the microscopical and histochemical structure of the alimentary canal. *Zool. Zh.* 53:1848–1855.

Needleman, S. B., and C. D. Wunsch. 1970. A general method applicable to the search for similarities in the amino acid sequence of two proteins. *J. Mol. Biol.* 48:443–453.

Nei, M., and D. Graur. 1984. Extent of protein polymorphism and the neutral mutation theory. *Evol. Biol.* 17:73–118.

Nei, M., and W. -H. Li. 1979. Mathematical model for studying genetic variation in terms of restriction endonucleases. *Proc. Natl. Acad. Sci. U.S.A.* 76:5269–5273.

Nei, M., T. Maruyama, and R. Chakraborty. 1975. The bottleneck effect and genetic variability in populations. *Evolution* 29:1–10.

Nevo, E. 1961. Observations on Israeli populations of the mole-rat *Spalax e. ehrenbergi* Nehring 1898. *Mammalia* 25:127–144.

––––––. 1979. Adaptive convergence and divergence of subterranean mammals. *Annu. Rev. Ecol. Syst.* 10:269–308.

––––––. 1982. Speciation in subterranean mammals. In: *Mechanisms of Speciation*. C. Barigozzi, ed., pp. 191–218. New York: Alan Liss.

Nevo, E., and H. Cleve. 1978. Genetic differentiation during speciation. *Nature (Lond.)* 275:125–126.

Nevo, E., and C. R. Shaw. 1972. Genetic variation in a subterranean mammal, *Spalax ehrenbergi. Biochem. Genet.* 7:235–241.

Nevo, E., A. Beiles, and R. Ben-Shlomo. 1984. The evolutionary significance of genetic diversity: ecological, demographic, and life history correlates. *Lecture Notes Biomath.* 53: 13–213.

Nevo, E., R. Ben-Shlomo, A. Beiles, J.U.M. Jarvis, and G. C. Hickman. 1987. Allozyme differentiation and systematics of the endemic subterranean mole-rats of South Africa. *Biochem. Syst. Ecol.* 15:489–502.

Nevo, E., E. Capanna, M. Corti, J.U.M. Jarvis, and G. C. Hickman. 1986. Karyotype differentiation in the endemic subterranean mole-rats of South Africa (Rodentia, Bathyergidae). *Z. Säugetierkd.* 51:36–49.

Nevo, E., Y. J. Kim, C. R. Shaw, and C. S. Thaeler. 1974. Genetic variation, selection, and speciation in *Thomomys talpoides. Evolution* 28:1–23.

Noirot, C. 1969. Formation of castes in the higher termites. In *Biology of Termites*. Vol. 1. K. Krishna and F. M. Weesner, eds., pp. 311–350. New York: Academic Press.

Noirot, C., and J. M. Pasteels. 1987. Ontogenetic development and evolution of the worker caste in termites. *Experientia* 43:851–860.

Noonan, K. M. 1981. Individual strategies of inclusive-fitness-maximizing in *Polistes fuscatus* foundresses. In *Natural Selection and Social Behavior*. R. D. Alexander and D. W. Tinkle, eds., pp. 18–44. New York: Chiron Press.

O'Brien, S. J., D. E. Wildt, D. Goldman, C. R. Merril, and M. Bush. 1983. The cheetah is depauperate in genetic variation. *Science (Wash., D.C.)* 221:459–462.

Oster, G. F., and E. O. Wilson. 1978. *Caste and Ecology in the Social Insects*. Princeton, N.J.: Princeton University Press.

Otte, D. 1974. Effects and functions in the evolution of signaling systems. *Annu. Rev. Ecol. Syst.* 5:385–417.

Owings, D. H., and D. W. Leger. 1980. Chatter vocalizations of California ground squirrels: Predator and social-role specificity. *Z. Tierpsychol.* 54:163–184.

Pamilo, P., and R. H. Crozier. 1982. Measuring genetic relatedness in natural populations: methodology. *Theor. Popul. Biol.* 21:171–193.

Parona, C. 1895. Acari parassiti dell'eterocefalo. *Annali Museo Civico di Storia Naturale Giacomo Doria Ser. 2*, 15:539–547.

Parona, C., and G. Cattaneo. 1893. Note Anatomiche e Zoologiche sull' *Heterocephalus*, Rüppell. *Annali Museo Civico di Storia Naturale Giacomo Doria Ser. 2*, 33:419–445.

Patton, J. L. 1985. Population structure and the genetics of speciation in pocket gophers, genus *Thomomys*. *Acta Zool. Fenn.* 170:109–114.

———. 1990. Divergence patterns in geomyid rodents: History, random processes and selection. In *Evolution of Subterranean Mammals at the Organismal and Molecular Levels*. E. Nevo and O. A. Reig, eds., pp. 49–69. New York: Wiley-Liss.

Patton, J. L., and P. V. Brylski. 1987. Pocket gophers in alfalfa fields: Causes and consequences of habitat-related body size variation. *Am. Nat.* 130:493–506.

Patton, J. L., and J. H. Feder. 1978. Genetic divergence between populations of the pocket gopher, *Thomomys umbrinus* (Richardson). *Z. Säugetierkd.* 43:17–30.

Patton, J. L., and S. Y. Yang. 1977. Genetic variation in *Thomomys bottae* pocket gophers: Macrogeographic patterns. *Evolution* 31:697–720.

Payne, S. F. 1982. Social organization of the naked mole-rat (*Heterocephalus glaber*): Cooperation in colony labor and reproduction. Master's thesis, University of California, Santa Cruz.

Pennisi, E. 1986. Not just another pretty face. *Discover* 7:68–78.

Penrith, M.-L. 1982. A new species of *Paoligena* Pic (Coleoptera: Tenebrionidae: Strongyliini) from Kenya. *Ann. Transvaal Mus.* 33:291–298.

Pilleri, G. 1960. Über das Zentralnervensystem von *Heterocephalus glaber* (Rodentia, Bathyergidae). *Acta Zool.* 41:101–111.

Pitman, C.R.S. 1974. *A Guide to the Snakes of Uganda*. Rev. ed. Codicote, U.K.: Wheldon & Wesley.

Poduschka, W. 1978. Zur Frage der Wahrnehmung von Lichtreizen durch die Mullratte, *Cryptomys hottentotus* (Lesson, 1826). *Säugetierkd. Mitt.* 26:269–274.

Porter, A. 1957. Morphology and affinities of entozoa and endophyta of the naked mole-rat *Heterocephalus glaber*. *Proc. Zool. Soc. Lond.* 128:515–527.

Pyke, G. H. 1978. Are animals efficient harvesters? *Anim. Behav.* 26:241–250.

Quay, W. B. 1981. Pineal atrophy and other neuroendocrine and circumventricular features of the naked mole-rat, *Heterocephalus glaber*, a fossorial, equatorial rodent. *J. Neural Transm.* 52:107–115.

Queller, D. 1989. The evolution of eusociality: Reproductive head starts of workers. *Proc. Natl. Acad. Sci. U.S.A.* 86:3224–3226.

Rasa, O.A.E. 1973. Intra-familial sexual repression in the dwarf mongoose *Helogale parvula*. *Naturwissenschaften* 6:303–304.

————. 1976. Invalid care in the dwarf mongoose (*Helogale undulata rufula*). *Z. Tierpsychol.* 42:337–342.

————. 1977. Differences in group member response to intruding conspecifics and frightening or potentially dangerous stimuli in dwarf mongooses (*Helogale undulata rufula*). *Z. Säugetierkd.* 42:108–112.

————. 1986. Coordinated vigilance in dwarf moongoose family groups: the "watchman's song" hypothesis and the costs of guarding. *Ethology* 71:340–344.

————. 1987. The dwarf mongoose: a study of behavior and social structure in relation to ecology in a small, social carnivore. *Adv. Study Behav.* 17:121–163.

Rasnitzyn, A. P. 1975. Hymenoptera Apocritia of Mesozoic. *Trans. Paleontol. Inst.* 147:1–132.

————. 1977. New Jurassic and Cretaceous hymenopterans of Asia. *Paleontol. J.* 1977:98–108.

Ratnieks, F.L.W. 1988. Reproductive harmony via mutual policing by workers in eusocial Hymenoptera. *Am. Nat.* 132:217–236.

Rees, E. L. 1968. A note on the brain of the Cape dune mole-rat *Bathyergus suillus*. *Acta Anat.* 71:147–153.

Reeve, H. K., and G. J. Gamboa. 1983. Colony activity integration in primitively eusocial wasps: The role of the queen (*Polistes fuscatus*, Hymenoptera: Vespidae). *Behav. Ecol. Sociobiol.* 13:63–74.

————. 1987. Queen regulation of worker foraging in paper wasps: A social feedback control system (*Polistes fuscatus*, Hymenoptera: Vespidae). *Behaviour* 102:147–167.

Reeve, H. K., D. F. Westneat, W. A. Noon, P. W. Sherman, and C. F. Aquadro. 1990. DNA "fingerprinting" reveals high levels of inbreeding in colonies of the eusocial naked mole-rat. *Proc. Natl. Acad. Sci. U.S.A.* 87:2496–2500.

Reichman, O. J., T. G. Whitham, and G. A. Ruffner. 1982. Adaptive geometry of burrow spacing in two pocket gopher populations. *Ecology* 63:687–695.

Reig, O. A. 1970. Ecological notes on the fossorial octodont rodent *Spalacopus cyanus* (Molina). *J. Mammal.* 51:592–601.

Reik, E. F. 1970. Fossil history. In *Insects of Australia*, pp. 168–186. Melbourne: Melbourne University Press.

Ricklefs, R. E. 1974. Energetics of reproduction in birds. In: *Avian Energetics*. R. A. Paynter, Jr., ed., pp. 152–297. Publ. 15. Cambridge, Mass.: Nuttall Ornithological Club.

————. 1975. The evolution of co-operative breeding in birds. *Ibis* 117:531–534.

Rissing, S. W., and G. B. Pollock. 1986. Social interaction among pleometrotic queens of *Veromessor pergandei* (Hymenoptera: Formicidae) during colony foundation. *Anim. Behav.* 34:226–233.

Roberts, A. 1937. The South African antelopes. *S. Afr. J. Sci.* 33:771–787.

————. 1951. *The Mammals of South Africa*. Johannesburg: Trustees of the "Mammals of South Africa" Book Fund.

Rohlf, F. J., and R. R. Sokal. 1969. *Statistical Tables*. San Francisco: W. H. Freeman.

Rood, J. P. 1974. Banded mongoose males guard young. *Nature (Lond.)* 248:176.

————. 1978. Dwarf mongoose helpers at the den. *Z. Tierpsychol.* 48:277–287.

————. 1980. Mating relationships and breeding suppression in the dwarf mongoose. *Anim. Behav.* 28:143–150.

―――. 1983. The social system of the dwarf mongoose. In *Advances in the Study of Mammalian Behavior*. J. F. Eisenberg and D. G. Kleiman, eds., pp. 454–488. Am. Soc. Mammal. Spec. Publ. 7. Shippensburg, Pa.: American Society of Mammalogists.

―――. 1986. Ecology and social evolution in the mongooses. In *Ecological Aspects of Social Evolution*. D. I. Rubenstein and R. W. Wrangham, eds., pp. 131–152. Princeton, N.J.: Princeton University Press.

Rosenthal, C. 1989. The social structure and general behavior of a complete colony of *Cryptomys hottentotus hottentotus*, the common mole-rat from the South Western Cape, South Africa. Honors thesis, University of Cape Town.

Rosevear, D. R. 1969. *The Rodents of West Africa*. London: Trustees of the British Museum (Natural History).

Roth, L. M., and E. R. Willis. 1960. The biotic associations of cockroaches. *Smithson. Misc. Coll.* 141:1–470.

Rubenstein, D. I., and R. W. Wrangham, eds. 1986. *Ecological Aspects of Social Evolution*. Princeton, N.J.: Princeton University Press.

Rüppell, E. 1842. *Heterocephalus* nov. gen. Uber Saugethiere aus de Ordnung de Nager (1834). *Museum Senckenbergianum Abhandlungen* 3:91–116.

Sage, R. D., J. R. Contreras, V. G. Roig, and J. L. Patton. 1986. Genetic variation in the South American burrowing rodents of the genus *Ctenomys* (Rodentia: Ctenomyidae). *Z. Saugetierkd.* 51:158–172.

Saiki, R. K., D. H. Gelfand, S. Stoffel, S. J. Scharf, R. Higuchi, G. T. Horn, K. B. Mullis, and H. A. Erlich. 1988. Primer-directed enzymatic amplification of DNA with a thermostable DNA polymerase. *Science (Wash., D.C.)* 239:487–491.

Sands, W. A. 1972. The soldierless termites of Africa (Isoptera: Termitidae). *Bull. Br. Mus. Nat. Hist. Suppl.* 18:1–244.

Sarich, V. M. 1985. Rodent macromolecular systematics. In *Evolutionary Relationships among Rodents: A Multidisciplinary Analysis*. W. P. Luckett and J. -L. Hartenberger, eds., pp. 423–452. New York: NATO ASI Series, Plenum Press.

Saunders, C. D. 1988. Ecological, social, and evolutionary aspects of baboon (*Papio cynocephalus*) grooming behavior. Ph.D. diss., Cornell University, Ithaca, N.Y.

Schaller, G. 1972. *The Serengeti Lion*. Chicago: University of Chicago Press.

Schilling, A., and M. Perret. 1987. Chemical signals and reproductive capacity in a male prosimian primate (*Microcebus murinus*). *Chem. Sens.* 12:143–158.

Schnell, G. D., and R. K. Selander. 1981. Environmental and morphological correlates of genetic variation in mammals. In *Mammalian Population Genetics*. M. H. Smith and J. Joule, eds., pp. 60–99. Athens: University of Georgia Press.

Seeley, T. D. 1979. Queen substance dispersal by messenger workers in honey bee colonies. *Behav. Ecol. Sociobiol.* 5:391–415.

―――. 1982. Adaptive significance of the age polyethism schedule in honey bee colonies. *Behav. Ecol. Sociobiol.* 11:287–293.

―――. 1985. *Honeybee Ecology*. Princeton, N.J.: Princeton University Press.

―――. 1986. Division of labor among worker honey bees. *Ethology* 71:249–251.

Selander, R. K., and T. S. Whittam. 1983. Protein polymorphism and the genetic structure of populations. In *Evolution of Genes and Proteins*. M. Nei and R. K. Koehn, eds., pp. 89–114. Sunderland, Mass.: Sinauer.

Selander, R. K., D. W. Kaufman, R. J. Baker, and S. L. Williams. 1974. Genic and chromosomal differentiation in pocket gophers of the *Geomys bursarius* group. *Evolution* 28:557–564.

Selander, R. K., M. H. Smith, S. Y. Yang, W. E. Johnson, and J. B. Gentry. 1971. Biochemical polymorphism and systematics in the genus *Peromyscus*. I. Variation in the old-field mouse (*Peromyscus polionotis*). *Univ. Texas Publ. 7103. Stud. Genet.* 6:49–90.

Senna, A. 1915. Sull' *Heterocephalus glaber* Rüppell. *Monit. Zool. Ital. Firenze.* 26:1–7.

Seyfarth, R. M. 1977. A model of social grooming among adult female monkeys. *J. Theor. Biol.* 65:671–698.

———. 1983. Grooming and social competition in primates. In *Primate Social Relationships*. R. A. Hinde, ed., pp. 182–190. Sunderland, Mass.: Sinauer.

Shaw, C. R., and R. Prasad. 1970. Starch gel electrophoresis of enzymes – a compilation of recipes. *Biochem. Genet.* 4:297–320.

Sherman, P. W. 1989. Mate guarding as paternity insurance in Idaho ground squirrels. *Nature (Lond.)* 338:418–420.

Sherman, P. W., M. L. Morton, L. M. Hoopes, J. Bochantin, and J. M. Watt. 1985. The use of tail collagen strength to estimate age in Belding's ground squirrels. *J. Wildl. Manage.* 49:874–879.

Short, R. V. 1984. Oestrus and menstrual cycles. In *Reproduction in Mammals: 3-Hormone Control of Reproduction*. 2d ed. C. R. Austin and R. V. Short, eds., pp. 115–152. Cambridge: Cambridge University Press.

Shortridge, G. C. 1934. *The Mammals of South West Africa*. London: Heinemann.

Silk, J. B. 1982. Altruism among female *Macaca radiata*: Explanations and analysis of patterns of grooming and coalition formation. *Behaviour* 79:162–188.

Sillén-Tullberg, B. 1988. Evolution of gregariousness in aposematic butterfly larvae: a phylogenetic analysis. *Evolution* 42:293–305.

Simpson, G. G. 1945. The principles of classification and a classification of mammals. *Bull. Am. Mus. Nat. Hist.* 85:1–350.

Sivinski, J. 1980. Sexual selection and insect sperm. *Fla. Entomol.* 63:99–111.

Smith, J.N.M. 1974. The food searching behavior of two European thrushes. 2: The adaptiveness of the search patterns. *Behaviour* 49:1–61.

Smithers, R.H.N. 1971. *The Mammals of Botswana*. Mem. Natl. Mus. Rhodesia 4.

———. 1983. *The Mammals of the Southern African Subregion*. Pretoria: University of Pretoria Press.

Smithers, R.H.N., and J. Wilson. 1979. *Check List and Atlas of the Mammals of Zimbabwe Rhodesia*. Mem. Natl. Mus. Rhodesia 9.

Snodgrass, R. E. 1935. *Principles of Insect Morphology*. New York: McGraw-Hill.

Sokal, R. R., and F. J. Rohlf. 1981. *Biometry*. San Francisco: W.H. Freeman.

Southwood, T.R.E. 1966. *Ecological Methods, with Particular Reference to the Study of Insect Populations*. London: Methuen.

Sparks, J. 1967. Allogrooming in primates: A review. In *Primate Ethology*. D. Morris, ed., pp. 148–175. Chicago: Aldine Press.

Spradbery, J. P. 1973. *Wasps*. Seattle: University of Washington Press.

Stark, D. 1957. Beobachtungen an *Heterocephalus glaber* Rüppell 1842 (Rodentia, Bathyergidae in der Provinz Harar). *Z. Säugetierkd.* 22:50–56.

Starr, C. K. 1985. Enabling mechanism in the origin of sociality in the Hymenoptera—the sting's the thing. *Ann. Entomol. Soc. Am.* 78:836–840.

———. 1989. In reply, is the sting the thing? *Ann. Entomol. Soc. Am.* 82:6–8.

Stevenson, M. F. 1978. The behavior and ecology of the common marmoset (*Callithrix jacchus jacchus*) in its natural environment. In *Biology and Behavior of Marmosets.* H. Rothe, H. J. Wolters, and J. P. Hearn, eds., p. 278. Cuttingen, W. Germany: Eigenverlag Harmut Rothe.

Stone, R. D., and M. L. Gorman. 1985. Social organization of the European mole (*Talpa europaea*) and the Pyrenean desman (*Galemys pyrenaicus*). *Mammal. Rev.* 15:35–42.

Strassmann, J. E., and D. C. Meyer. 1983. Gerontocracy in the social wasp, *Polistes exclamans. Anim. Behav.* 31:431–438.

Tam, W. H. 1974. The synthesis of progesterone in some hystricomorph rodents. *Symp. Zool. Soc. Lond.* 34:363–384.

Taylor, P. J., J.U.M. Jarvis, T. M. Crowe, and K. C. Davies. 1985. Age determination in the Cape mole-rat, *Georychus capensis. S. Afr. J. Zool.* 20:261–267.

Taylor, V. A. 1978. A winged elite in a subcortical beetle as a model for a prototermite. *Nature (Lond.)* 276:73–75.

Terman, C. R. 1973. Recovery of reproductive function by prairie deermice (*Peromyscus maniculatus bairdii*) from asymptotic populations. *Anim. Behav.* 21:443–448.

———. 1984. Sexual maturation of male and female white-footed mice (*Peromyscus leucopus noveboracensis*): Influence of physical or urine contact with adults. *J. Mammal.* 65:97–102.

———. 1987. Intrinsic behavioral and physiological differences among laboratory populations of prairie deermice. *Am. Zool.* 27:853–866.

Thacker, E. J., and C. S. Brandt. 1955. Coprophagy in the rabbit. *J. Nutr.* 55:375–385.

Thigpen, L. W. 1940. Histology of the skin of a normally hairless rodent. *J. Mammal.* 21:449–456.

Thomas, O. 1885a. Exhibition and remarks on a burrowing rodent allied to *Heterocephalus glaber* (*Heterocephalus phillipsi* sp. nov.). *Proc. Zool. Soc. Lond.* 611–612.

———. 1885b. Notes on the rodent genus *Heterocephalus. Proc. Zool. Soc. Lond.* 845–849.

———. 1895. Examination of specimens from the first Expedition of Capt. Bottego in Museo Civico, Genoa. *Annali Museo Civico di Storia Naturale Giacomo Doria* 35(2): 3–6.

———. 1897. Mammals from the second Bottego Expedition, collected at Lugh and Brava, Italian Somaliland. *Annali Museo Civico di Storia Naturale Giacomo Doria* 37:106.

———. On specimens of naked rodents from East Africa (*Heterocephalus ansorgei* sp. nov. and *Fornarina* nov. gen.). *Proc. Zool. Soc. Lond.* 2:336–337.

———. 1904. On mammals of Somaliland (*Fornarina phillipsi* nov. gen.). *Ann. Mag. Nat. Hist.* 7:104.

———. 1909. New African mammals in the British Museum collection (*Heterocephalus dunni* sp. nov.). *Ann. Mag. Nat. Hist.* (Ser. 8) 4:109–112.

_____. 1910. List of mammals from Mount Kilimanjaro, obtained by Mr. Robin Kemp, and presented to the British Museum by Mr. C. D. Rudd. *Ann. Mag. Nat. Hist.* 8:303–316.

_____. 1917. Notes on *Georychus* and its allies. *Ann. Mag. Nat. Hist.* 20:441–444.

Trivers, R. L., and H. Hare. 1976. Haplodiploidy and the evolution of the social insects. *Science (Wash., D.C.* 191:249–263.

Trudeau, A. M., G. R. Haigh, and S. H. Vessey. 1980. The use of nest boxes to study the behavioral ecology of *Peromyscus leucopus* (abstr.). *Ohio J. Sci.* 80:91.

Tucker, R. 1981. Digging behavior and skin differentiations in *Heterocephalus glaber*. *J. Morphol.* 168:51–71.

Tullberg, T. 1899. Uber das System der Nagethiere: eine phylogenetische studie. *Nova Acta Regiae Soc. Sci. Uppsala* 18:1–514.

Van Couvering, J.A.H. 1980. Community evolution in East Africa during the late Cenozoic. In *Fossils in the Making: Vertebrate Taphonomy and Paleocecology*. A. K. Behrensmeyer and A. P. Hill, eds., pp. 272–298. Chicago: University of Chicago Press.

Van Couvering, J.A.H., and J. A. Van Couvering. 1976. Early Miocene mammal fossils from East Africa: aspects of geology, faunistics and paleontology. In *Human Origins: Louis Leakey and the East African Evidence*. G. L. Isaac and E. R. McCown, eds., pp. 155–207. Menlo Park, Calif.: W. A. Benjamin.

Van Damme, M.-P., D. M. Robertson, and E. Diczfalusy. 1974. An improved in vitro bioassay method for measuring luteinizing hormone (LH) activity using mouse Leydig cell preparations. *Acta Endocrinol.* 77:655–671.

Vandenbergh, J. G. 1986. The suppression of ovarian function by chemosignals. In *Chemical Signals in Vertebrates 4: Ecology, Evolution, and Comparative Biology*. D. Duvall, D. Müller-Schwarze, and R. M. Silverstein, eds., pp. 423–432. London: Plenum Press.

Vandenbergh, J. G., L. C. Drickamer, and D. R. Colby. 1972. Social and dietary factors in the sexual maturation of female mice. *J. Reprod. Fertil.* 28:397–405.

van der Horst, G. 1970. Seasonal effects on the anatomy and histology of the reproductive tract of the male rodent mole *Bathyergus suillus suillus* (Schreber). Master's thesis, University of Stellenbosch.

Varvio, S.-L., R. Chakraborty, and M. Nei. 1986. Genetic variation in subdivided populations and conservation genetics. *Heredity* 57:189–198.

Vehrencamp, S. L. 1983. Optimal degree of skew in cooperative societies. *Am. Zool.* 23: 327–335.

Verdcourt, B. 1969. The arid corridor between the northeast and southwest areas of Africa. In *Paleoecology of Africa*. Vol. 4. E. M. Van Zinderen Bakker, ed., pp. 140–144. Amsterdam: A. A. Balkema.

Vleck, D. 1979. The energy cost of burrowing by the pocket gopher *Thomomys bottae*. *Physiol. Zool.* 52:122–135.

_____. 1981. Burrow structure and foraging costs in the fossorial rodent *Thomomys bottae*. *Oecologia (Berl.)* 49:391–396.

Walker, E. P. 1975. *Mammals of the World*. Vols. 1, 2. 3d ed. Baltimore, Md.: Johns Hopkins University Press.

Waller, D. A., and J. P. La Fage. 1987. Nutritional ecology of termites. In *Nutritional*

Ecology of Insects, Mites, and Spiders. F. Slansky, Jr. and J. G. Rodriguez, eds., pp. 487–532. New York: Wiley & Sons.

Wanyonyi, K. 1974. The influence of the juvenile hormone analogue ZR 512 (Zoecon) on caste development in *Zootermopsis nevadensis* (Hagen) (Isoptera). *Insectes Soc.* 21:35–44.

Wasser, S. K., and D. P. Barash. 1983. Reproductive suppression among female mammals: Implications for biomedicine and sexual selection theory. *Q. Rev. Biol.* 58: 513–538.

Watson, J.A.L., and H. M. Abbey. 1977. The development of reproductives in *Nasutitermes exitiosus* (Hill) (Isoptera: Termitidae). *J. Aust. Entomol. Soc.* 16:161–164.

Watson, J.A.L., E. C. Metcalf, and J. J. Sewell. 1977. A re-examination of the development of castes in *Mastotermes darwiniensis* Froggatt (Isoptera). *Aust. J. Zool.* 25: 25–42.

Weir, B. J. 1974. Reproductive characteristics of hystricomorph rodents. *Symp. Zool. Soc. Lond.* 34:265–299.

Weir, B. J., and I. W. Rowlands. 1974. Functional anatomy of the hystricomorph ovary. *Symp. Zool. Soc. Lond.* 34:303–332.

Wells, L. H. 1967. N.E.-S.W. 'arid corridor' in Africa. *S. Afr. J. Sci.* 63: 480–481.

Wesselman, H. B. 1984. The Omo micromammals. In *Contributions to Vertebrate Evolution.* Vol. 7. M. K. Hecht and F. S. Szalay, eds., pp. 1–219. Basel, Switzerland: S. Karger.

West, M. J., and R. D. Alexander. 1963. Subsocial behavior in a burrowing cricket *Anurogryllus muticus* (De Geer) (Orthoptera: Gryllidae). *Ohio J. Sci.* 63:19–24.

West-Eberhard, M. J. 1969. The social biology of polistine wasps. *Univ. Mich. Mus. Zool. Misc. Publ.* 140:1–101.

———. 1975. The evolution of social behavior by kin selection. *Q. Rev. Biol.* 50:1–33.

———. 1978a. Polygyny and the evolution of social behavior in wasps. *J. Kans. Entomol. Soc.* 51:832–856.

———. 1978b. Temporary queens in *Metapolybia* wasps: nonreproductive helpers without altruism? *Science (Wash., D.C.)* 200:441–443.

———. 1979. Sexual selection, social competition, and evolution. *Proc. Am. Philos. Soc.* 123:222–234.

———. 1981. Intragroup selection and the evolution of insect societies. In *Natural Selection and Social Behavior.* R. D. Alexander and D. W. Tinkle, eds., pp. 3–17. New York: Chiron Press.

Wheeler, W. M. 1910. *Ants: Their Structure, Development, and Behavior.* New York: Columbia University Press.

———. 1923. *Social Life among the Insects.* New York: Harcourt, Brace.

———. 1928. *The Social Insects, Their Origin and Evolution.* New York: Harcourt, Brace.

Whitten, W. K. 1959. Occurrence of anoestrus in mice caged in groups. *J. Endocrinol.* 18:102–107.

Williams, G. C. 1957. Pleiotropy, natural selection, and the evolution of senescence. *Evolution* 11:398–411.

———. 1966. *Adaptation and Natural Selection.* Princeton, N.J.: Princeton University Press.

Williams, G. C., and D. C. Williams. 1957. Natural selection of individually harmful social adaptations among sibs with special reference to social insects. *Evolution* 11: 32–39.

Williams, S. L., D. A. Schlitter, and L. W. Robbins. 1983. Morphological variation in a natural population of *Cryptomys* (Rodentia: Bathyergidae) from Cameroon. *Ann. Kon. Mus. Mid. Afr., Zool. Wetensch.* 237:159–172.

Wilson, D. S. 1980. *The Natural Selection of Populations and Communities.* Menlo Park, Calif.: Benjamin/Cummings.

Wilson, E. O. 1971. *The Insect Societies.* Cambridge, Mass.: Belknap Press of Harvard University Press.

———. 1975. *Sociobiology: The New Synthesis.* Cambridge, Mass.: Belknap Press of Harvard University Press.

———. 1976. Behavioral discretization and the number of castes in an ant species. *Behav. Ecol. Sociobiol.* 1:141–154.

———. 1985. The sociogenesis of insect colonies. *Science (Wash., D.C.)* 228:1489–1495.

Wilson, E. O., and B. Hölldobler. 1980. Sex difference in cooperative silk-spinning by weaver ant larvae. *Proc. Natl. Acad. Sci. U.S.A.* 77:2343–2347.

Winston, M. L. 1987. *The Biology of the Honey Bee.* Cambridge, Mass.: Belknap Press of Harvard University Press.

Winterbottom, J. M. 1967. Climatological implications of avifaunal resemblances between southwestern Africa and Somaliland. In *Paleoecology of Africa.* Vol. 2. E. M. Van Zinderen Bakker, ed., pp. 77–79. Amsterdam: A. A. Balkema.

Withers, P. C., and J.U.M. Jarvis. 1980. The effect of huddling on thermoregulation and oxygen consumption for the naked mole-rat. *Comp. Biochem. Physiol. A, Comp. Physiol.* 66:215–219.

Wolpoff, M. H. 1980. *Paleoanthropology.* New York: Knopf.

Wood, A. E. 1985. The relationships, origin, and dispersal of the hystricognathus rodents. In *Evolutionary Relationships among Rodents: A Multidisciplinary Analysis.* W. P. Luckett and J. -L. Hartenberger, eds., pp. 475–513. New York: NATO ASI Series, Plenum Press.

Woods, C. A. 1975. The hyoid, laryngeal, and pharyngeal regions of bathyergid and other selected rodents. *J. Morphol.* 147:229–250.

Woolfenden, G. E. 1975. Florida scrub jay helpers at the nest. *Auk* 92:1–15.

Woolfenden, G. E., and J. W. Fitzpatrick. 1978. The inheritance of territory in group-breeding birds. *BioScience* 28:104–108.

———. 1984. *The Florida Scrub Jay.* Princeton, N.J.: Princeton University Press.

Wrangham, R. W., and D. I. Rubenstein. 1986. Social evolution in birds and mammals. In *Ecological Aspects of Social Evolution.* D. I. Rubenstein and R. W. Wrangham, eds., pp. 452–470. Princeton, N.J.: Princeton University Press.

Wright, S. 1940. Breeding structure of populations in relation to speciation. *Am. Nat.* 74:232–248.

Yahav, S., R. Buffenstein, J.U.M. Jarvis, and D. Mitchell. 1989. Thermoregulation and evaporative water loss in the naked mole-rat, *Heterocephalus glaber. S. Afr. J. Sci.* 85:340.

Zar, J. H. 1984. *Biostatistical Analysis.* 2d ed. Englewood Cliffs, N.J.: Prentice-Hall.

Index